PRENTICE-HALL ELECTRICAL ENGINEERING SERIES
William L. Everitt, editor

INFORMATION THEORY SERIES
Thomas Kailath, editor

BARTON
Radar System Analysis

DI FRANCO AND RUBIN
Radar Detection

DOWNING
Modulation Systems and Noise

DUBES
The Theory of Applied Probability

GOLOMB, BAUMERT, EASTERLING, STIFFLER, AND VITERBI
Digital Communications with Space Applications

RAEMER
Statistical Communication Theory and Applications

PRENTICE-HALL INTERNATIONAL, INC., *London*
PRENTICE-HALL OF AUSTRALIA, PTY. LTD., *Sydney*
PRENTICE-HALL OF CANADA, LTD., *Toronto*
PRENTICE-HALL OF INDIA PRIVATE LTD., *New Delhi*
PRENTICE-HALL OF JAPAN, INC., *Tokyo*

Harold R. Raemer

Professor of Electrical Engineering
Chairman, Department of Electrical Engineering
Northeastern University

STATISTICAL COMMUNICATION THEORY AND APPLICATIONS

Prentice-Hall, Inc., Englewood Cliffs, New Jersey

Englewood Cliffs, New Jersey

Current printing (last digit):
10 9 8 7 6 5 4 3 2 1

Library of Congress Catalog Card Number:
69-12130

Printed in the United States of America

To Paulyne,
Danny, Liane, and Diane

PREFACE

Enough books on statistical communication theory have appeared in recent years to warrant a justification for still another. A little personal history will be invoked to justify this one.

Inspiration for this book derived from several years as an engineering analyst in industrial research and development laboratories. The problems that concerned me largely involved detection, location, and communication systems employing radio frequency electromagnetic waves, e.g., radar seekers, RF underground object detectors, and digital radio communication systems. Having some background in statistical communication theory, I was often called upon to apply it to such system problems, and eventually it became evident that knowledge of this discipline is an important part of the systems engineer's tool kit. Many treatments are so permeated with abstraction, however, that they appeal only to those interested in communication theory as a field of research activity. There was a need for a book that would integrate elementary aspects of communication theory with key ideas of systems engineering for the benefit of applications-oriented readers.

This book was undertaken with the idea of helping to fill that need. In the intervening period, there have been books on radar systems, on digital communication systems, and on elementary aspects of statistical communication theory with illustrative engineering applications. This volume draws together some of the elementary concepts of the theory and discussions of applications of these concepts within the contexts of both radar systems and radio communication systems. In order to accomplish this, radar and radio communication systems engineering ideas were introduced in greater detail than is usual in a communication theory text. While communication theory is the central issue, peripheral sections on such topics as radio propagation, antennas, etc. appear both as background for treatment of system applications (primarily in Chapter 5) and integrated into the communication theory discussions (primarily in Chapters 9 through 12). The inclusion of some of these topics points up the difference between the perspective of the professional communication theorist and that of the systems engineer. The former usually focuses on the mathematics of communication theory as an end in itself, while the latter sees the theory as a tool in analyzing real-world engineering problems. If he must delve into propagation and antennas in order to treat a system as an integrated whole, he does so without worrying about crossing boundary lines between different research fraternities or IEEE professional groups.

This book can be used as a college textbook and as a self-study or reference book for practicing engineers. Chapters 2, 3, 4, 6, 7, and 8 and selected portions of other chapters form the basis for an undergraduate E.E. course in introductory statistical communication theory. Such a course would prepare the student for a graduate course in which the subject is presented on a more abstract level. A student not headed for graduate school would find this course useful in laying the groundwork for an elementary understanding of the use of the theory in the engineering areas that he will encounter on the job. The mathematics used in the book is very elementary. The mathematical arguments presented are generally not rigorous and can be followed by anyone with a knowledge of basic calculus and a few standard mathematical tools now familiar to undergraduate E.E. students, such as Fourier series and Fourier transforms. Topics requiring extensive mathematical background have been excluded, usually in favor of a reference to the literature or a somewhat heuristic or qualitative discussion to introduce the reader to the topic.

The choice of topics here will not necessarily be agreeable to all readers. First, the applications are all in the areas of radio frequency systems. In order to discuss applications in some depth, it was necessary to focus on a specific area. I naturally chose radio systems because it is the most closely related to my experience. With minor modifications, much of this material would carry over into areas such as coherent light and underwater sound. Obviously, what is referred to here as a "radio signal" is really a narrowband signal, and could as well represent a sonar or coherent light signal.

Another point that should be mentioned is the fact that most of the theory in the book falls into the category now called "detection theory." The information-theoretic viewpoint is virtually omitted except for a rather cursory treatment of a few topics in Section 3.6. Along with this omission goes omission of the topic of coding, which is usually treated from the information theory viewpoint. Still another set of topics treated superficially is that which relates to multiplicative noise, including fading, inter-symbol interference in digital communication systems, and diversity reception. There are undoubtedly many other topics that some readers will regard as having been covered much too lightly and still others that they will regard as superfluous. For example, much of Chapter 1 and Chapter 5 could be omitted by many readers without disturbing continuity; however, inclusion of this material is designed to render the book as self-contained as possible and to reduce the amount of material that must be looked up in outside references in order to facilitate understanding of certain discussions of systems applications of communication theory.

To summarize points about choice and depth of coverage of various topics, I can only say that it was necessary to keep the book's length within reasonable bounds and therefore necessary to omit or to cover only lightly a good many topics of great interest and importance to communication theorists. Because of the book's special orientation, it was also necessary to cover topics that are not a part of communication theory. I make no apologies for these choices; I only hope that the book stimulates further study of topics that it does not cover extensively.

I wish to acknowledge the help I received from Mr. David Potter and Mr. Yash Pal Verma, both of whom performed computation and curve-plotting tasks for some of the figures, to Mrs. Ann Kellner and the Misses Ward, Pasek, and Steinberg of the Electrical Engineering office staff at Northeastern for typing and for other tasks associated with the terminal stages of manuscript preparation and proof editing. Finally, appreciation is expressed to my wife, without whose patience and encouragement this book would not have been completed.

Harold R. Raemer
Boston, Massachusetts

CONTENTS

1

INTRODUCTORY BACKGROUND TOPICS

This book is concerned with applications of the discipline known as *statistical communication theory*† to various aspects of radio and radar systems technology. The program begins with the presentation of basic ideas of the theory itself and culminates in discussions of ways in which these ideas can be used to provide guidelines for engineering design decisions in radar detection and location of targets and in radio communication.

The "systems approach" will be stressed throughout the book. In discussions of applications of the theory to system problems, the statistical or "random" aspects of these problems will usually be integrated with the nonstatistical or "deterministic" aspects. Thus, while the emphasis will be on applications of the theory of random processes, it should be remembered that the theory is being applied to physical situations in which performance is as often limited by deterministic considerations as it is by random or "noisy" effects.

To illustrate the point, it is possible to analyze a pulse radar detection problem from the viewpoint of statistical communication theory without detailed understanding of radio wave propagation, antenna theory, and receiver circuitry. However, without at least some elementary acquaintance with these topics, it is difficult to apply the result of the analysis meaning-

† Broadly defined here as the application of the theory of random processes to the study of intelligence-bearing signals in the presence of unwanted random disturbances and the ways in which the effects of these disturbances can be processed out with a minimum loss of signal intelligence.

fully to an actual radar design problem. When the statistical aspects of the problem are not properly integrated with propagation and circuitry considerations, the conclusions of the analysis may well be naive or incomplete.

In the first four chapters of the book, an attempt will be made to lay the groundwork for an understanding of the technological applications by presenting the elements of noise and statistical communication theory (Chapters 2 and 3), and some important nonstatistical background topics, such as passive linear system analysis, resolution and wave propagation (Chapter 1). The statistical theory of optimum detection and measurement processes is presented in Chapter 4. Chapter 5 is a brief and cursory summary of radio system analysis. Chapters 6, 7, and 8 treat applications of statistical communication theory to processing of radio signals from the signal-to-noise ratio, detection, and information extraction viewpoints, respectively. Chapters 9 and 10 treat some applications to monostatic, bistatic, and passive radar systems and Chapters 11 and 12 concern certain applications to radio communication systems.

To define the scope of this book, we might say that it concerns applications of statistical communication theory to "information systems" using radio waves. An information system is not necessarily a "communication system," the latter being a special subclass of the former.† An *information system* is defined here as any physical device or combination of devices whose purpose is the conveyance of some form of information to human beings. Assuming for the moment that the term "information" has its qualitative or popular meaning,‡ it should be noted that, in the above definition, it is conveyed to a human and does not necessarily originate with another human. It may originate in inanimate nature, but its recipient is always man. It is man who constructs the device, motivated by his desire to extract information either from nature or from other humans.

The concept of "information" will be discussed below. Once the reader has this concept in mind, he will see that many kinds of devices fall under our definition of an "information system." The scope of this definition is, in fact, so large that we must confine ourselves to only a small fraction of the possible classes of such systems. By introducing general principles common to many classes of systems, we will attempt to include a reasonable number of special applications within the covers of this book.

---

† A *communication system*, as defined in Chapters 11 and 12, is specifically a system in which a deliberate attempt is made to transmit a message from one party to another. This excludes target detection and location systems from the class of "communication systems," as defined in this book.

‡ That is, a change in someone's state of knowledge about something or other.

1.1 INFORMATION

The word "information" must be used in a carefully qualified sense to avoid ambiguity in meaning. In a qualitative sense, an increase or decrease of information means, respectively, a reduction or enhancement of the state of uncertainty regarding a specific quantity. The question "Who is how uncertain about what at time t_1, and how much less uncertain is he at a later time t_2?" must be answered before one can discuss quantitatively a change in someone's store of information during the time interval from t_1 to t_2. For example, consider two stock market investors A and B. Suppose A has just learned that Acme Tools, Inc. went up 20 points today. He informs B of this fact. Suppose B had never previously heard of Acme. Having received A's message, he now knows that (1) Acme exists and (2) its stock went up 20 points today. In transmitting this message, A has neither added to nor subtracted from his own store of knowledge. He has, however, increased that of B. In the popular sense of the word, "information" has been conveyed to B. Phrasing this somewhat differently, B's uncertainty has been reduced or, equivalently, his information has been increased.

We now know *who* (B) was uncertain at time t_1 (before receiving A's message), and that he is less uncertain at time t_2 (after receiving the message), but we don't know *how much* less uncertain, or exactly about *what.* In attempting to measure the change in his uncertainty, we must enter B's consciousness before time t_1, and ask a specific question. Suppose the question is: "What is the state of the stock market?" or "Which tool manufacturing stocks are doing well these days?" or "What was last year's birth rate?" This message provides absolutely no information relative to the third question, and only fragmentary information about the first two. However, measurement of the change in information is possible if B asks the question, "Is there a company called Acme Tools, Inc.?" or "Assuming that Acme exists, has the value of its stock decreased or increased within the past day?" or "Assuming an increase, is it more or less than 15 points?"

The message has provided answers to the latter three questions and reduced B's uncertainty about these specific questions from some finite state to zero. His initial uncertainty, of course, depends on his education on stock market affairs and possibly on many other factors. But until a very specific question can be asked and the initial uncertainty about this question can be measured in some way and compared with B's final uncertainty, nothing quantitative can be said about the amount of information conveyed by A to B.

The vast number of possible questions that B could have asked before time t_1, and the enormous possible variation in his initial uncertainty, would appear to make the problem of specifying information quantitatively a hopeless one, even in such a simple case as that of our example. However, in the late 1940's, Claude Shannon formulated a theory by which some progress was made toward a satisfactory method of numerical specification of the quantity called "information."† This "information theory" has been useful to communications engineers in a conceptual sense, in that it has provided tools of thought by which a designer can optimize the performance of his communication systems. The utility of the theory in precise design calculations is sometimes questioned. However, in cases where it is of no help in analyzing a system, it is only because the problem contains too many variables to be analyzed in a precise sense, not because the concepts are unsound. The fact remains that the ability to define "gain or loss of information" in Shannon's sense provides us with a quantitative basis for comparing system quality without recourse to subjective concepts of information.

Very little reference to information theory as such will be made in this book, aside from the extremely cursory glance at the subject provided by the rudimentary discussion in Section 3.6. However, a general philosophical notion about the quantitative concept of information appears either explicitly or implicitly in a number of places throughout the book. This is the notion that one pays a price to acquire information. If a process of detection, measurement, or communication begins with very little information available, and ends with a great deal more information, it will usually be found that a high price has been paid in time, equipment complexity, or some other commodity. The more the information acquired, the higher the price. Thus in many discussions in the book, although no explicit mention is made of information theory, if one approached the process from an information-theoretic viewpoint, he would draw qualitative conclusions consistent with those obtained through other lines of reasoning.

1.2 THE PRINCIPLES OF PASSIVE LINEAR SYSTEM ANALYSIS

To analyze problems of information transmission and processing, we must know how to analyze the physical systems used to accomplish the processing. To a large extent these systems can be considered for practical purposes as *linear*, i.e., (1) the output in response to a sum of inputs is the sum of the output responses to the individual inputs, and (2) multiplying the input by a constant scale factor does not change the functional form of the output, except for introduction of the scale factor. The methods of analyzing systems

† See Shannon, 1948, or Shannon and Weaver, 1949.

for which these two conditions hold are extremely well-developed. This set of analysis techniques is based on the mathematical theory of linear integro-differential equations, because these systems can be represented by such equations. Electrical engineers know this discipline as linear network analysis and mechanical engineers know it as the theory of linear dynamical systems. It is generically known as *linear system analysis*, and will be so designated here. The theory has been thoroughly exploited in connection with various technologies, and a spate of literature exists on it. Some of its highlights will be briefly reviewed here for the benefit of readers not thoroughly familiar with it. References for the reader interested in pursuing it further are cited at the end of this chapter.†

Nonlinear systems are defined generally as those systems not obeying the above two conditions. No completely general theory of such systems exists, and each narrow class of nonlinear system requires an essentially new analysis technique. We will confine ourselves in any discussions of nonlinear systems in the book to those directly related to very specific problems of interest.

1.2.1 Lumped Constant, Passive Linear Systems

A lumped parameter linear system can be characterized by a linear integro-differential equation involving its output time function $x_o(t)$ and its input time function $x_i(t)$, i.e., in general,

$$\begin{aligned}[a_{-n}(t)x_o(t)]^{(-n)} &+ \cdots + [a_{-1}(t)x_o(t)]^{(-1)} + a_o(t)x_o(t) \\ &+ a_1(t)x_o^{(1)}(t) + \cdots + a_n(t)x_o^{(n)}(t) = [b_{-n}(t)x_i(t)]^{(-n)} + \cdots \\ &\qquad + [b_{-1}(t)x_i(t)]^{(-1)} + b_o(t)x_o(t) \\ &\qquad + b_1(t)x_i^{(1)}(t) + \cdots \\ &\qquad + b_n(t)x_i^{(n)}(t)\end{aligned} \tag{1.1}$$

where $f^{(k)}$ and $f^{(-k)}$, respectively, denote the kth time derivative and kth time integral of a function $f(t)$ and it is indicated that the coefficients $a_k(t)$ and $b_k(t)$ may be functions of time.

The types of systems for which the theory has been most thoroughly developed are those with constant parameters a_k and b_k ("lumped constant" systems) and those which are passive (i.e., there are no energy sources within the system), not because such systems are necessarily common in nature, but because their analysis is feasible. Many common system elements, such

† It will be assumed that nearly all readers will have had some exposure to linear system analysis. The coverage here is inadequate to teach the subject for the first time. It is intended only as a brief review and for the reader's convenience in later discussions.

as resistors, capacitors, and inductors in electrical networks, are sufficiently well approximated as constants to justify this simplification under many conditions. The same remarks apply to the assumption that the coefficients are independent of the input, the output, and their derivatives and integrals, the key assumption that renders a system "linear." Many practical systems have this property over a wide range of conditions.

Systems represented by linear integro-differential equations with constant coefficients can be analyzed either in the time domain or the frequency domain. Time domain analysis consists of the actual solution of the equation for the output $x_o(t)$ in terms of a known input $x_i(t)$ by any available method. Frequency domain analysis is the process of finding the response of the system to a sinusoidal function of arbitrary frequency. Modern engineers who must deal with electrical or dynamical systems should be conversant with both methods.

1.2.1.1 Frequency Domain Analysis–The Steady State

To analyze a linear system in the frequency domain, we specify an input $X_i(\omega) \exp(j\omega t)$ and postulate that the output will be of the form $X_o(\omega) \exp(j\omega t)$, where ω is the angular frequency, equal to $2\pi f$, f being the actual frequency in cycles per second or hertz.†

The solution of (1.1), under these assumptions, is

$$\begin{aligned} X_o(\omega)[(j\omega)^{-n}a_{-n} + \cdots + (j\omega)^{-1}a_{-1} + a_o + j\omega a_1 + \cdots + (j\omega)^n a_a] \\ = X_i(\omega)[(j\omega)^{-n}b_{-n} + \cdots + (j\omega)^{-1}b_{-1} + b_o + j\omega b_1 + \cdots + (j\omega)^n b_n] \end{aligned} \tag{1.2}$$

The ratio $X_o(\omega)/X_i(\omega)$ is known as the *frequency response function* of the system and will be designated as $F(\omega)$.

Its absolute square $|F(\omega)|^2$ is the *spectrum* of the system, acting as a measure of the relative power per unit of frequency passed by the system as a function of frequency. The arctangent of the ratio of the imaginary to the real part of $F(\omega)$ at a given frequency ω is a measure of the phase shift between output and input introduced by the system at that frequency.

This type of analysis is directly useful in predicting the response of a system to any periodic function, since such a function can always be expanded in a complex Fourier series, i.e., a series of waves of discrete frequencies ω_n, each term of which is of the form $X_i(\omega_n)\, e^{(j\omega_n t)}$. The frequency response

† The exponential representations of $\cos \omega t$ and $\sin \omega t$, based on the relation $e^{j\omega t} = \cos \omega t + j \sin \omega t$, will be used here for mathematical convenience. Through this relation, any function $(a \cos \omega t + b \sin \omega t)$ can be represented as $(c\, e^{j\omega t} + d\, e^{-j\omega t})$, where c and d are complex in general. Complex outputs $x_o(t)$ must be added to their conjugates $x_o^*(t)$ after solution to return to the physical world, where all quantities must be real.

function $F(\omega)$ can be used to designate the response to the entire input function through superposition of the responses to each of the Fourier terms.

For a function of time $x(t)$, whether periodic or not, the integral

$$F\{x(t)\} = X(\omega) = \int_{-\infty}^{\infty} x(t)\, e^{\mp j\omega t}\, dt \tag{1.3}$$

is defined as the *Fourier transform* of $x(t)$, and it follows from an extension of the above argument regarding Fourier series that

$$F\{x_o(t)\} = X_o(\omega) = F(\omega)X_i(\omega) = F\{f(t)\}F\{x_i(t)\} \tag{1.4}$$

where†

$$F\{\underset{i}{x_o}(t)\} = \underset{i}{X_o}(\omega) = \int_{-\infty}^{\infty} x(t)\, e^{\mp j\omega t}\, dt$$

the Fourier transform of the output. In other words, knowledge of the frequency response function of the system, with the aid of (1.4), will establish the relationship between input and output waveforms and thereby completely characterize the system's response to an arbitrary input.

A periodic input signal is by definition presumed to "go on forever," that is, from $-\infty$ to $+\infty$. Thus the analysis of a system's response to such a signal is a "steady state" analysis, applicable to a situation where the transient disturbances due to switching of elements in or out of the system have long ago decayed to negligible magnitude. This is the kind of analysis we are performing when we find $F(\omega)$, Fourier analyze the input signal, and apply (1.4) to obtain the output Fourier components.

The Fourier transform analysis described above, on the other hand, is a transient analysis, in that it concentrates on the system's response to an input with a finite lifetime. More will be said of transient analysis in Section 1.2.1.3.

1.2.1.2 *Time Domain Analysis*

Inversion of the Fourier transform $X_o(\omega)$ specified in (1.4) results in

$$x_o(t) = \int_{-\infty}^{\infty} dt'\, f(t-t')x_i(t') = \int_{-\infty}^{\infty} dt'\, f(t')x_i(t-t') \tag{1.5}$$

where

$$f(t) = \frac{1}{2\pi}\int_{-\infty}^{\infty} d\omega\, e^{\pm j\omega t}F(\omega)$$

the inverse Fourier transform of $F(\omega)$.

---

† Note the two possible signs on the exponentials in the Fourier transform integrals given in (1.3). Either sign may be used in the time integral in (1.3) provided the opposite sign is used in the frequency integral in (1.5).

Equation (1.5) is the well-known *convolution theorem* for Fourier transforms. To understand the physical significance of this theorem, consider the input $x_i(t) = \delta(t)$, i.e., the unit impulse or *delta function*, whose basic properties are:

$$\delta(t) = \infty \quad \text{for} \quad t = 0, \tag{1.6 a}$$

$$\delta(t) = 0 \quad \text{for} \quad t \neq 0, \tag{1.6 b}$$

$$\int_{-\infty}^{\infty} dt' \, \delta(t') = 1, \tag{1.6 c}$$

$$\int_{-\infty}^{\infty} dt' \, \delta(t' - t_o) f(t') = f(t_o) \tag{1.6 d}$$

$$\int_{-\infty}^{\infty} dt' \, \delta(t') \, e^{\mp j\omega t} = 1, \tag{1.6 e}$$

$$\frac{1}{2\pi} \int_{-\infty}^{\infty} d\omega \, e^{\pm j\omega t} = \delta(t) \tag{1.6 f}$$

The output in response to this input is $f(t)$, i.e., the function $f(t)$ is the response of the system to a unit impulse, or *impulse response* of the system. It is also known as the *time weighting function* of the system, for the following reason: The input can be thought of as a sequence of impulse functions at all time instants previous to time t. The output as given by (1.5) is a superposition of responses to all of these impulses, each impulse at time t' being weighted with its response $f(t - t')$. In general, those impulses occurring the most recently are given the heaviest weighting and those far back in the past have a negligibly small weighting. In this sense, then, the impulse response can be thought of as a representation of the system's "memory," i.e., the system "remembers" only the inputs it received in the recent past and "forgets" long past inputs.

Our intuition tells us that no physical system can respond to future inputs, that is, an effect cannot precede its cause. An impulse response $f(t - t')$, to be "causal," must vanish for $t < t'$. This condition becomes a basic hypothesis of linear system theory, and is known as the *causality condition*.

It is noted that knowledge of either $F(\omega)$ or $f(t)$ implies knowledge of the other, since these two functions are reciprocal Fourier transforms. Thus, in principle, we can analyze a system in either the frequency or time domain by solving for $f(t)$ or $F(\omega)$, the choice depending on which is simplest to obtain, and can get the other by Fourier transformation. Time and frequency domain analysis thus produce equivalent information about the system. However, sometimes one or the other point of view is the most convenient in a particular analysis. These two analytical approaches can be regarded as different languages. Every word in each language has an exact translation into the other.

1.2.1.3 Transient Analysis

Analysis of transient disturbances in linear systems is best accomplished with the aid of the single-sided Laplace transform, the advantage of this approach being that the initial conditions appear naturally in the solutions. The principles of Laplace transform analysis are covered in many texts, and the reader is referred to references cited at the end of this chapter for an elaboration of them. For our purposes, it suffices to say that any linear integro-differential equation with constant coefficients can be solved by Laplace transforming both sides and inserting values for the boundary conditions that naturally appear in the transform solution. Inverting the transform of $x_o(t)$ will result in the time domain solution of the equation, that is, $x_o(t)$ for a given $x_i(t)$.

The single-sided Laplace transform of a function $f(t)$ is

$$\mathscr{L}\{f(t)\} \equiv F(-js) = \int_0^\infty dt\, e^{-st} f(t) \tag{1.7}$$

The transforms of derivatives and integrals of $f(t)$ are given by

$$\mathscr{L}\{f^{(n)}(t)\} = s^n F(-js) - s^{n-1} f(0_+) - s^{n-2} f^{(1)}(0_+) - \cdots - s f^{(n-2)}(0_+) - f^{(n-1)}(0_+) \tag{1.8}$$

$$\mathscr{L}\{f^{(-n)}(t)\} = s^{-n} F(-js) + s^{-n} f^{(-1)}(0_+) + s^{-n+1} f^{(-2)}(0_+) + \cdots + s^{-2} f^{(-n+1)}(0_+) + s^{-1} f^{(-n)}(0_+) \tag{1.9}$$

where the initial conditions are specified at $t = 0_+$, that is, at an instant immediately following $t = 0$.

Solution of an integro-differential equation of the form (1.1) with constant coefficients, using (1.8) and (1.9), leads to the results

$$\begin{aligned}
&[a_{-n}s^{-n} + \cdots + a_{-1}s^{-1} + a_0 + a_1 s + \cdots + a_n s^n] X_o(-js) \\
&\quad = [b_{-n}s^{-n} + \cdots + b_{-1}s^{-1} + b_0 + b_1 s + \cdots + b_n s^n] X_i(-js) \\
&\quad + \sum_{k=1}^{n} \sum_{j=0}^{k-1} \{[a_k s^{k-1-j} x_o^{(j)}(0_+) - a_{-k} s^{-k+j} x_o^{-(j+1)}(0_+)] \\
&\quad - [b_k s^{k-1-j} x_i^{(j)}(0_+) - b_{-k} s^{-k+j} x_i^{-(j+1)}(0_+)]\}
\end{aligned} \tag{1.10}$$

where $X_o(-js)$ and $X_i(-js)$ are the Laplace transforms of the output $x_o(t)$ and the input $x_i(t)$, respectively.

Solution of the original equation (1.1) for an arbitrary input $x_i(t)$ is accomplished by solving (1.10) for the ratio $X_o(-js)/X_i(-js)$ and inverting.†

† Inversion is in many cases a difficult process. However, commonly occurring transform pairs are extensively tabulated. Thus in practical use of Laplace transforms, the inversion operation is seldom performed by the user. The Laplace transform method of transient analysis contains a great deal of "table-look up" exercise.

This ratio can be expanded in partial fractions if the denominator polynomial roots $s_1, \ldots, s_n$ can be found, leading to an expression of the form (if the denominator polynomial is of higher order than the numerator polynomial; otherwise additional terms involving positive powers of s exist)

$$\frac{X_o(-js)}{X_i(-js)} = \frac{A_1}{(s - s_1)} + \cdots + \frac{A_{k-1}}{(s - s_{k-1})} + \sum_{n=k}^{k+J-1} \left[\frac{A_{n,1}}{(s - s_n)} + \cdots + \frac{A_{n,r_n}}{(s - s_n)^{r_n}} \right] \tag{1.11}$$

where the A_n's are constants and the possibility of multiple roots is accounted for by the summation terms in brackets. The first through the $(k - 1)$st roots are *simple* or nonrepeated. Each term group in the summation corresponds to a root repeated r_n times. There are J such groups.

Assuming the input to be an impulse function $\delta(t)$ and the system to be initially at rest (i.e., initial values of $x_i(t)$ and $x_o(t)$ and all their derivatives and integrals vanish), we can invert the right-hand side of (1.11), resulting in

$$x_o(t) = A_1 e^{s_1 t} + \cdots + A_{k-1} e^{s_{k-1} t} + \sum_{n=k}^{k+J-1} e^{s_n t} \left[A_{n,1} + \cdots + \frac{A_{n,r_n}}{(r_n - 1)!} t^{r_n - 1} \right] \tag{1.12}$$

If the system is passive, we can conclude from (1.12) that the real part of each of the denominator polynomial roots s_n must be negative. If this is not the case, there will be either a constant amplitude oscillation (if $\text{Re}(s_n) = 0$ and $\text{Im}(s_n) \neq 0$) or an unlimited growth of the output as time progresses (if $\text{Re}(s_n) > 0$), indicating that the system is either operating as a "perpetual motion machine of the first kind," i.e., violating the law of conservation of energy, or receiving energy for its growth from a mysterious phantom source. A negative value of $\text{Re}(s_n)$ for all s_n indicates that each term eventually decays to a negligible value after the excitation has ceased, which is what intuition tells us must happen in a passive system.

Noting the above remarks, we observe that the properties of the roots of the denominator polynomial are indicative of the characteristic behavior of the system in response to arbitrary inputs, that is, by studying these roots, we can completely analyze the system. To do this, we think in terms of the system's *transfer function*, $T(-js)$, defined as the ratio of the Laplace transform of the output to that of the input when the input and output and all their derivatives and integrals are initially zero. By referring back to Section 1.2.1.1 and comparing Eq. (1.2) with Eq. (1.10), we observe that if the complex variable s had no real part,† the transfer function would degenerate into the frequency response function $F(\omega)$. Thus the study of $T(-js)$ is a generalization of the frequency domain analysis technique outlined in

---

† The variable $-s$ is known as *complex frequency*. The imaginary part is the actual angular frequency.

Section 1.2.1.1. Moreover, the convolution theorem for Fourier transforms, Eq. (1.5), has a counterpart for Laplace transforms in the form

$$x_o(t) = \int_0^t dt' \hat{f}(t - t')x_i(t') \tag{1.13}$$

where $\hat{f}(t)$ is the inverse of the transfer function $T(-js)$, and $\hat{f}(t)$ again is the impulse response of the system, i.e.,

$$\hat{f}(t) = f(t) \tag{1.14}$$

The only difference between (1.5) and (1.13) is that of the limits of integration. The lower limit of zero appears because single-sided Laplace transform theory automatically incorporates into the analysis the fact that the system began operating at $t = 0$. The upper limit of t automatically assures causality of the system.

Equation (1.14) tells us that $F(\omega)$ and $T(-js)$ are Fourier and Laplace transforms, respectively, of the same function $f(t)$. This suggests a relationship between the two types of transforms. Indeed, if we write

$$s = \text{Re}\,(s) + j\,\text{Im}\,(s) \tag{1.15}$$

and allow Re (s) to vanish, and if, moreover, we stipulate that the function $f(t)$ must be zero for $t < 0$, the two transforms become identical. The difference between them is that the real part of s, which must be negative in a passive system, introduces an automatic convergence factor into the transform integral. A physical function will nearly always have a finite Laplace transform. In the case of most such functions, even those which approach infinity as $t \rightarrow \infty$, the product of the function and $e^{\{[\text{Re}\,(s)]t\}}$ approaches zero as $t \rightarrow \infty$ with sufficient rapidity to ensure the convergence of the Laplace transform integral to a finite value. Many common functions, on the other hand, do not have a Fourier transform that is everywhere finite. For example, the Fourier transform of $\cos \omega_o t$ is $\frac{1}{2}\{\delta(\omega - \omega_o) + \delta(\omega + \omega_o)\}$, which is infinite at $\omega = \pm\omega_o$, while its Laplace transform is $s/(s^2 + \omega_o^2)$, which is everywhere finite.

The general statement that can be made about the way in which the transforms are related is

$$F\{u(t)f(t)\} = \lim_{\text{Re}(s)\rightarrow 0} \mathscr{L}\{f(t)\} \tag{1.16}$$

where $u(t)$ is the *unit step function*, which is unity for $t \geq 0$, zero for $t < 0$.

Because of (1.16), we can do frequency domain analysis of a system by calculating its transfer function and replacing s by $j\omega$. We can also do analysis of transient disturbances in the time domain by inverting the transfer function to determine the impulse response $f(t)$, carrying out the integration indicated in (1.13). However, it is not usually necessary to do this, because the location of the denominator and numerator polynomial roots, that is, the poles and zeros of the transfer function in the complex plane, will often tell the engineer enough about the system to serve as an adequate basis for his design work.

1.2.1.4 Analysis of Linear Electrical Filters

Discussions in later chapters will often involve electrical circuit concepts. The circuits that enter the discussions will usually be treated as *black boxes*, i.e., we will not be interested in the implementation of the circuit but only in the function it performs as reflected in its impulse response or its frequency response. However, it is useful to bear in mind the principles of circuit analysis even when treating circuits as black boxes.

Let us apply the ideas of Sections 1.2.1.1, 1.2.1.2, and 1.2.1.3 to the simplest types of filters used in signal processing systems. This will be done under the assumption that the reader is familiar with basic rules of circuit analysis, for example, Kirchhoff's current and voltage laws and the concepts of impedance and admittance.

We will briefly sketch the elementary rules of circuit analysis by which the response of a filter can be determined. First a circuit can be analyzed by equating the sum of emf's to the sum of voltage drops in each independent loop of the circuit. It can be alternatively analyzed by equating the currents entering each junction point to the currents leaving the junction point. In using the first method, it is necessary to know that (1) the voltage drop across each inductance L is $L(di/dt)$, (2) the drop across each capacitance C is $1/C \int i(t)\,dt$ and (3) the drop across each resistance is $Ri(t)$, where $i(t)$ is in each case the current through the element. The currents $i(t)$ through these elements in terms of the voltages $v(t)$ across the elements, needed in the second method of analysis, are $(1/L) \int v(t)\,dt$, $C(dv/dt)$, and $v(t)/R$, respectively. In the loop analysis, the result of applying the rules is a set of simultaneous integro-differential equations in the currents $i_n(t)$ with the known applied voltages $v_k(t)$. In the current junction type of analysis, the set of differential equations is in the voltages $v_n(t)$ at each junction point, with respect to a specified zero reference level of voltage, and the inputs are given as current sources. In either case, the first order integro-differential equations resulting from application of the circuit laws can be solved for sinusoidal inputs by the methods discussed in Section 1.2.1.1 to find the frequency response function or they can be solved by Laplace transform methods, in accordance with the theory outlined in Section 1.2.1.3 to find the response to arbitrary transient inputs or to find the transfer function $T(-js)$. In the loop analysis, $T(-js)$ is $[I_o(-js)/V_i(-js)]$, where $I_o(-js)$ and $V_i(-js)$ represent transforms of output current and input voltage, respectively. This type of transfer function is a *transfer admittance*, denoted by $Y(-js)$. In the junction point analysis, the transfer function is a ratio of output voltage to input current, or a *transfer impedance*, denoted by $Z(-js)$. Transfer functions may also be ratios of voltages or ratios of currents.

In transient analysis, initial conditions are required. These are supplied by the requirements that (1) the voltage across a capacitor cannot be changed instantaneously and (2) the current through an inductor cannot be changed instantaneously. Thus at the instant the input is switched in ($t = 0_+$), each inductor carries the current it had immediately before switching and each capacitor retains its preswitching voltage.

1.2.1.4.1. The Low Pass Filter

A simple type of low pass filter is an *RC* circuit shown in Figure 1-1(a). An alternative type is an *RL* circuit (Figure 1-1(b)). The impulse response and spectral response of either type of low pass filter are sketched respectively in (c) and (d) of Figure 1-1. The derivations can be found in standard textbooks on network analysis.† In the figure, V_i and V_o represent input and output voltages, respectively.

The *RC* low pass filter has an impulse response‡

$$f(t) = e^{-t/RC}u(t) \tag{1.17}$$

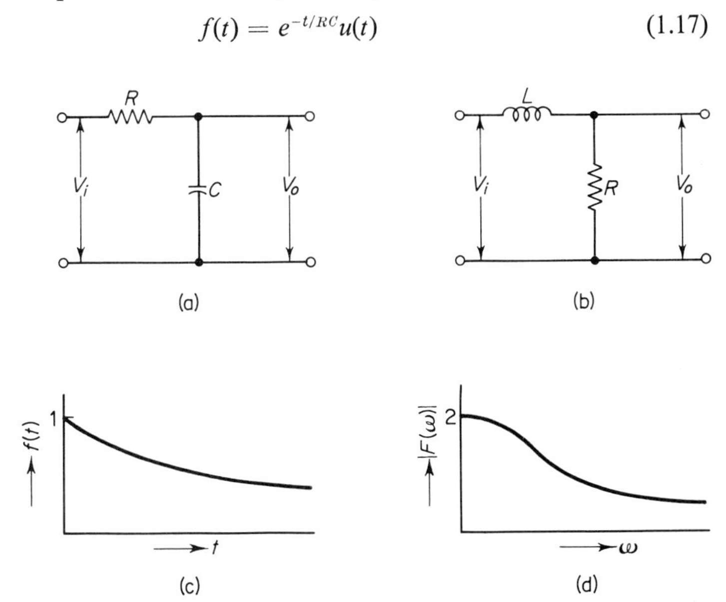

Figure 1-1 Low pass filters.

† See Scott, 1960, Chapter 11.

‡ Scott, 1960, Section 11.5, pp. 358–362. When an *RC* circuit is used as a low pass filter, the output voltage is that across the capacitor, as shown in Figure 1-1(a).

(where $u(t)$ is the unit step function) and a spectral response (obtained by Fourier transformation of (1.17) and the observation made in Section 1.2.1.2 that the spectral response $F(\omega)$ is the Fourier transform of the impulse response $f(t)$).

$$|F(\omega)|^2 = \left|\frac{V_o(\omega)}{V_i(\omega)}\right|^2 \propto \left|\frac{1}{1 + j\omega RC}\right|^2 = \frac{1}{1 + \omega^2(RC)^2} \tag{1.18}$$

while the RL low pass filter has impulse response†

$$f(t) = e^{-t/(L/R)}u(t) \tag{1.19}$$

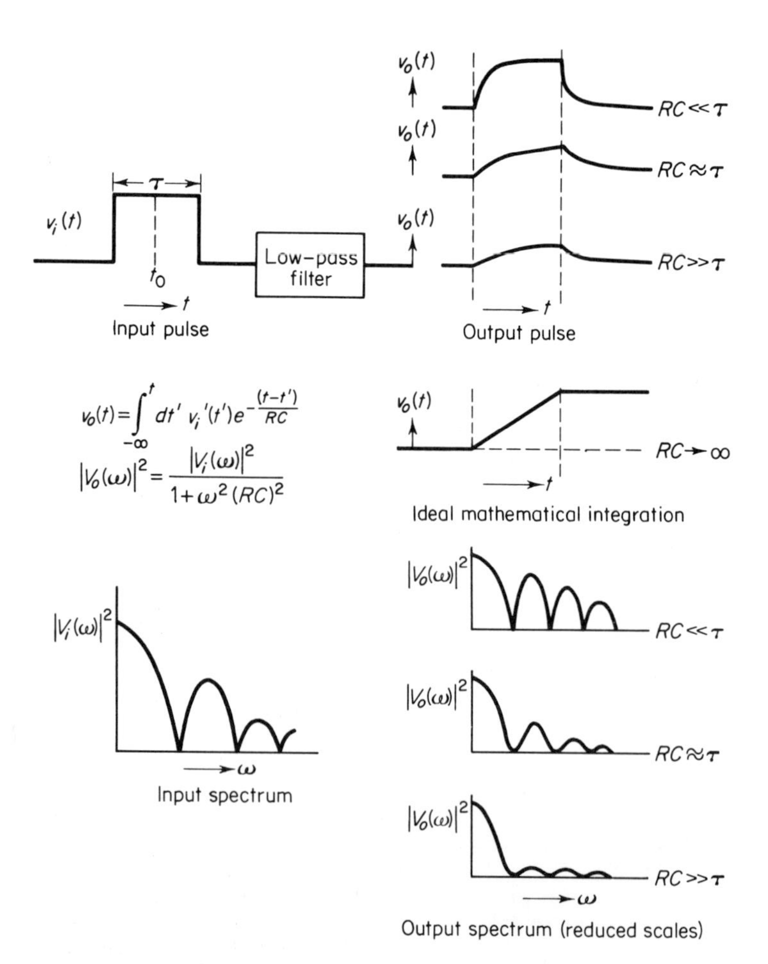

Figure 1-2 *RC* low pass filtering of square pulse. Note that these curves are qualitative and are not drawn to scale.

† See Scott, 1960, Section 11.2, pp. 350–353. When the *RL* circuit is used as a low pass filter, the output voltage is across the resistor as shown in Figure 1-1(b).

and spectral response

$$|F(\omega)|^2 = \left|\frac{V_o(\omega)}{V_i(\omega)}\right|^2 \propto \left|\frac{1}{1 + j\omega(L/R)}\right|^2 = \frac{1}{1 + \omega^2(L/R)^2} \tag{1.20}$$

The time constants in the RC and RL cases are, respectively (RC) and (L/R). If the time constant is large compared to the duration of the input voltage waveform, then the low pass filter simulates a mathematical integration, and could in this case be called an *integrator.*

The action of an RC low pass filter on a square pulse is shown graphically in Section 1.4.1 (Figure 1-2) in connection with discussion of time resolution of waveforms. The interpretation of the curves of Figure 1-2 is discussed in Section 1.4.1. The comparison of the filter outputs with those of an idealized mathematical integration is also shown in Figure 1-2. It is evident from the curves that the filter approaches a true integrator as the ratio of time constant to pulse duration becomes extremely large.

The reciprocal of the time constant in RC and RL low pass filters is the half-power point on the spectral curve and is usually designated as the *bandwidth* (or equivalently *cutoff frequency*) of the filter. This is an arbitrary designation, however, and in some applications it might be desirable to define the cutoff frequency or bandwidth as a fraction or a multiple of the reciprocal time constant.

1.2.1.4.2. The High Pass Filter

High pass filtering can be accomplished easily with RL or RC arrangements as shown in Figure 1-3(a) and (b), respectively. The impulse response and spectrum of either type of high pass filter are illustrated in Figure 1-3(c) and (d), respectively.

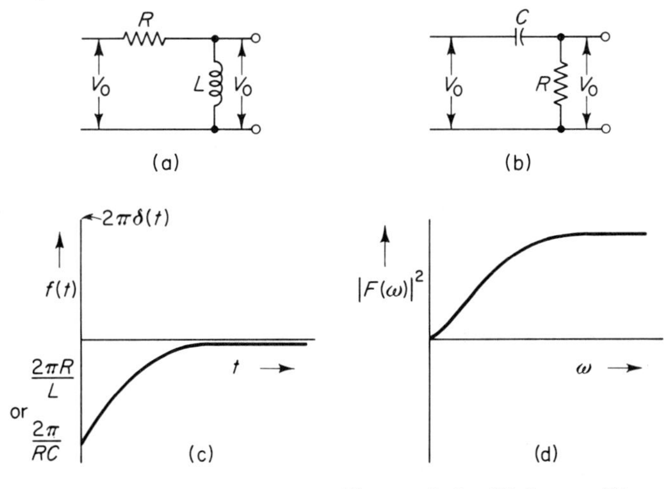

Figure 1-3 High pass filters.

Impulse responses and spectra of these filters are:

1. *RL* high pass filter.†

$$f(t) = \left\{\delta(t) - \left(\frac{R}{L}\right) e^{-Rt/L} u(t)\right\} \tag{1.21}$$

$$|F(\omega)|^2 = \left|\frac{V_o(\omega)}{V_i(\omega)}\right|^2 \propto \left|\frac{j\omega L}{R + j\omega L}\right|^2 = \left(\frac{\omega L}{R}\right)^2 \frac{1}{1 + \omega^2(L/R)} \tag{1.22}$$

2. *RC* high pass filter.‡

$$f(t) = \left\{\delta(t) - \left(\frac{1}{RC}\right) e^{-t/RC} u(t)\right\} \tag{1.23}$$

$$|F(\omega)|^2 = \left|\frac{V_o(\omega)}{V_i(\omega)}\right|^2 \propto \left|\frac{j\omega RC}{1 + j\omega RC}\right|^2 = (RC)^2 \frac{1}{1 + \omega^2(RC)^2} \tag{1.24}$$

Note that a high pass filter with a small time constant simulates mathematical differentiation and can therefore be used as a differentiator in processing circuits. Again, in analogy with the low pass filter case, the low cutoff frequency can be defined as the reciprocal of the time constant. The action of a high pass filter on a square pulse is shown in Figure 1-4. The

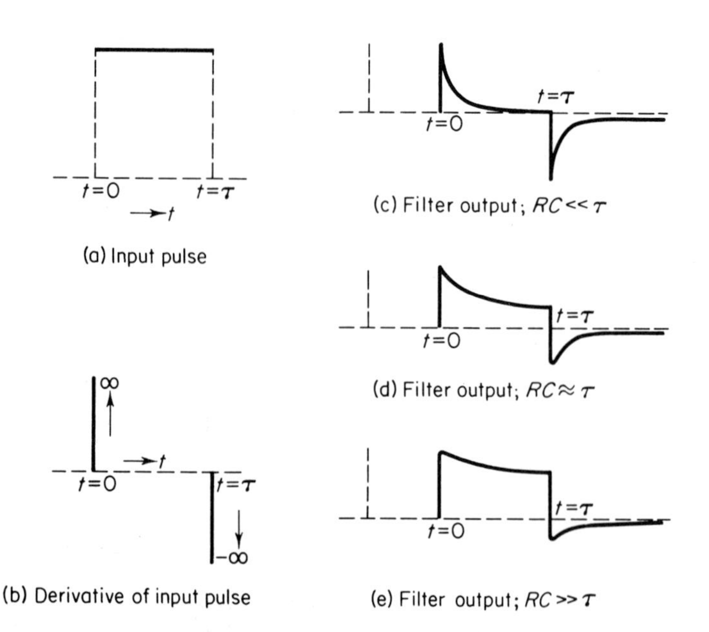

Figure 1-4 High pass filtering and mathematical differentiation.

† See Scott, 1960, Section 11.2. The output voltage in the high pass case is across the inductance, as shown in Figure 1-3(a).

‡ Scott, 1960, Section 11.5. The output voltage is across the resistance, as shown in Figure 1-3(b).

input pulse is shown in (a), ideal mathematical differentiation in (b), and the outputs of an RC differentiator in (c), (d), and (e), where cases (c) through (e) correspond to time constants that are (c) short compared with pulse duration τ, (d) comparable to τ, and (e) long compared with τ. It is clear from these sketches, as would be expected, that the differentiation is approached as the ratio of time constant to pulse duration becomes negligibly small, or equivalently, as bandwidth approaches infinity.

1.2.1.4.3. Bandpass Filters

Series RLC and parallel RLC bandpass filters are shown in Figure 1-5(a) and (b), respectively. Sketches of the impulse response and spectrum of both types of bandpass filter in the case where the passband is narrow compared with the central frequency are shown in (c) and (d) of Figure 1-5. Impulse responses and spectra of these filters are:

1. Series RLC bandpass filter.†

$$f(t) = \frac{e^{-Rt/2L}}{2L}\left\{\left[1 - j\frac{RC}{\sqrt{4LC - R^2C^2}}\right] e^{-j\sqrt{\frac{1}{LC}-\left(\frac{R}{2L}\right)^2}t} + \left[1 + j\frac{RC}{\sqrt{4LC - R^2C^2}}\right] e^{j\sqrt{\frac{1}{LC}-\left(\frac{R}{2L}\right)^2}t}\right\}u(t) \tag{1.25}$$

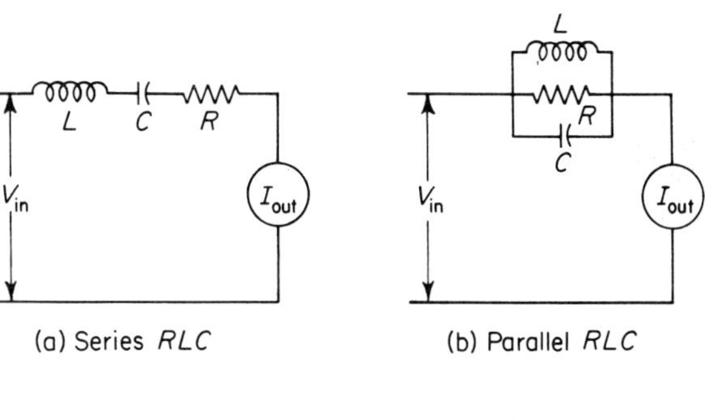

(a) Series RLC (b) Parallel RLC

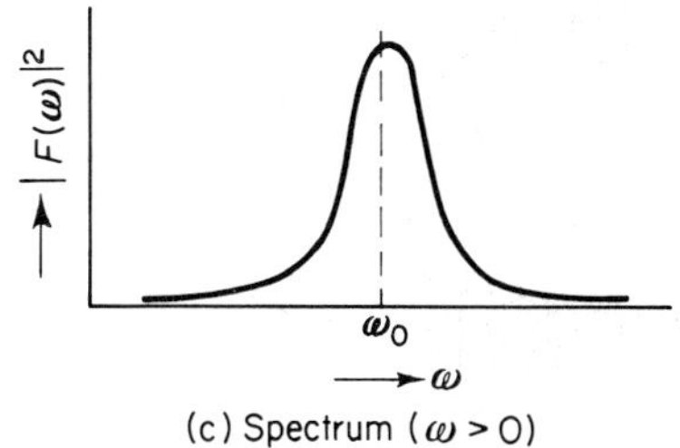

(c) Spectrum ($\omega > 0$)

Figure 1-5 Bandpass filters.

† For derivations, see Scott, 1960, Chapter 12, especially Section 12.4, pp. 392–398.

$$|F(\omega)|^2 = \left|\frac{I_o(\omega)}{V_i(\omega)}\right|^2 = |Y(\omega)|^2 \propto \left|\frac{j\omega C}{1 + j\omega RC + (j\omega)^2 LC}\right|^2 \tag{1.26}$$

2. Parallel RLC bandpass filter.†

$$f(t) = \frac{e^{-t/2RC}}{2C}\left\{\left[1 - j\frac{LC}{4R^2C^2 - LC}\right]e^{-j\sqrt{\frac{1}{LC}-\left(\frac{1}{2RC}\right)^2}\,t} + \left[1 + j\frac{LC}{(4R^2C^2 - LC)}\right]e^{j\sqrt{\frac{1}{LC}-\left(\frac{1}{2RC}\right)^2}\,t}\right\}u(t) \tag{1.27}$$

$$|F(\omega)|^2 = \left|\frac{V_o(\omega)}{I_i(\omega)}\right|^2 = |Z(j\omega)|^2\left|\frac{j\omega L}{1 + j\omega\frac{L}{R} + (j\omega)^2 LC}\right|^2 \tag{1.28}$$

Further discussion of bandpass filters will be found in Section 6.7, in the specific context of signal-to-noise ratio enhancement.

1.3 WAVES

Many of the processes with which we will deal in this book involve the conveyance of information between two points in space by means of waves. A wave, for our purposes, is any disturbance which propagates through space from one point to another.

The source may be an acceleration of electric charge at a point which generates electromagnetic waves, or a local agitation of air molecules, giving rise to sound waves. In either case, the local disturbance generates another disturbance in the adjacent region, which in turn propagates to its neighboring region, etc., until the effect literally "travels" out into space.

To illustrate this, consider the case of electromagnetic waves. The Maxwell equations provide the theoretical description of why such waves propagate by showing how an electric field varying with time at a point in space must give rise to a magnetic field at the same point, and how the latter "curls" around the vector in the direction of the time rate of change of the electric field and is of magnitude proportional to this rate of change. The same effect exists respectively between a time-varying magnetic field and an electric field "curling" around the time rate of change vector. A moving charge, by Ampere's circuital law (the integral form of one of the Maxwell equations), can be shown to be (roughly)‡ surrounded by a magnetic field whose lines form concentric circles around the charge in the plane

---

† See Scott, 1960, Section 12.5, pp. 398–401.

‡ This statement is only a very rough approximation to the truth, but useful in visualizing the generation of a wave by an accelerating charge. Strictly speaking, the concentric circles only exist with an infinite axially symmetric current-carrying structure and then only in the static regime.

normal to its motion. If the velocity of the charge varies periodically with time, the magnitude of its surrounding magnetic field will vary at the same frequency. This time variation gives rise to a corresponding time variation of the "curling" magnetic field. It is not difficult to see that the time variations of electric and magnetic fields continue to feed back on each other as long as the charge continues to oscillate, and that the disturbance will propagate out into the space around the accelerating charge. It is also apparent that far from the charge, the electric and magnetic fields are mutually perpendicular, the direction of propagation being perpendicular to both. These effects can be shown to be present by a detailed mathematical analysis, based on formal solution of Maxwell's equations.

A type of wave motion that is physically different but represented mathematically in a similar manner is the propagation of sound. Sound, as is well known to freshman physics students, requires a medium in which to propagate, whereas electromagnetic waves can propagate in a vacuum. Moreover, sound waves are longitudinal, i.e., the agitation of the medium giving rise to compressions and rarefactions is in the direction of propagation. Electromagnetic waves propagating in an unbounded medium are transverse, i.e., the oscillations of electric and magnetic fields are in a plane normal to the direction of propagation.

1.3.1 The Wave Equation†

The mathematical evidence that a form of wave motion is present is provided by the *wave equation*, obeyed by electric and magnetic fields, pressure in a gas, or other parameters that propagate in the form of waves.

This equation is a direct consequence of Maxwell's equations in the case of electromagnetic waves or of the laws of hydrodynamics in the case of sound waves. Its form is

$$\nabla^2 \boldsymbol{\psi}(\mathbf{r}, t) - \frac{1}{v^2}\frac{\partial^2 \boldsymbol{\psi}(\mathbf{r}, t)}{\partial t^2} = \mathbf{S}(\mathbf{r}) \tag{1.29}$$

where $\boldsymbol{\psi}(\mathbf{r}, t)$ is a vector whose components are in general complex (its magnitude will be called *complex amplitude* in what follows) that may represent an electric field, a magnetic field, a sound pressure or density, etc., $\mathbf{S}(\mathbf{r})$ is a source function, for example, an electric current, a tuning fork vibration, etc., v is the phase velocity of the wave in the medium, $\mathbf{r}$ and t are, respectively, the position vector of a point and the time, and ∇^2 is the Laplacian operator, given in rectangular coordinates ($\mathbf{r} = (x, y, z)$) by

$$\nabla^2 = \frac{\partial^2}{\partial x^2} + \frac{\partial^2}{\partial y^2} + \frac{\partial^2}{\partial z^2} \tag{1.30}$$

† While no direct use is made of the wave equation in later discussions in this book, its solutions form the mathematical basis on which the physics of wave motion can be understood; hence the inclusion of this brief discussion. See Moore, 1964, Chapters 2 and 3.

To illustrate how wave motion follows from (1.29), consider the simplest case, that for which $\boldsymbol{\psi}(\mathbf{r}, t)$ can be resolved into rectangular components, each of which is a function of only a single coordinate, for example, x in the one-dimensional case. If $\psi(x, t)$ is a component of $\boldsymbol{\psi}(x, t)$ and we are in a source-free region, Eq. (1.29) becomes

$$\frac{\partial^2 \psi}{\partial x^2} - \frac{1}{v^2}\frac{\partial^2 \psi}{\partial t^2} = 0 \tag{1.31}$$

Equation (1.31) is satisfied by general solutions of the form

$$\psi(x, t) = g(x \pm vt) \tag{1.32}$$

An example of such a solution is

$$\psi(x, t) = \psi(x)\, e^{\pm j\omega t} \quad \text{or} \quad \psi^*(x, t) = \psi^*(x)\, e^{j\omega t} \tag{1.33}$$

where $\psi(x) = e^{\pm jkx}$. Here k is the *propagation constant* of the wave, equal to ω/v, and the $+$ and $-$ signs, in a conventional rectangular coordinate system, represent waves traveling to the right and to the left, respectively. The function $\psi(x)$ is *an harmonic solution* of (1.31), that is, the time dependence is understood to be of the form $e^{\pm j\omega t}$. To get a complete harmonic solution for a particular value of k, one introduces factors e^{jkx} and e^{-jkx} with appropriate coefficients obtained from the spatial boundary conditions. In a given problem, one also has temporal boundary conditions. These conditions determine a complex coefficient to be multiplied by $e^{\pm j\omega t}$ in constructing a complete solution of (1.31). The real part of $\psi(x, t)$ is the actual physical variable undergoing wave motion, such as electric field or pressure. Note that in general

$$\begin{aligned}\operatorname{Re}\{\psi(x, t)\} = C_1 \cos(\omega t - kx) + C_2 \sin(\omega t - kx) \\ + C_3 \cos(\omega t + kx) + C_4 \sin(\omega t + kx)\end{aligned} \tag{1.34}$$

where the constants C_1 through C_4 are determined from both spatial and temporal boundary conditions.

1.3.2 Huyghens' Principle and the Concept of Aperture

Huyghens' principle is a well-known physical law that has important implications in information transmission. Briefly, the principle states that "Given a source of waves, every point in space can be considered as a source of secondary spherical wavelets." (See Figure 1-6.) Without going into the analytical details (except to mention that it follows from the wave equation (1.29)),† we can state that Huyghens' principle allows us to take

† See Silver, 1949, Section 5.12, pp. 160–162, also pp. 172, 198, and 199. For a much more elementary treatment of Huyghens' principle, see Sears, Vol. III, 1948, Chapter 9, especially pp. 221–223.

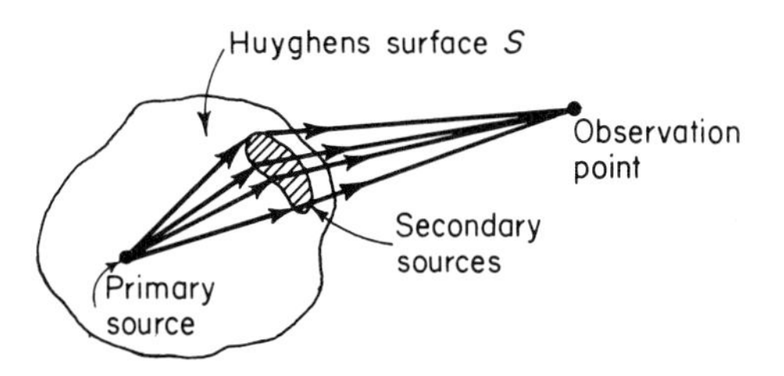

Figure 1-6 Huyghens' principle.

a closed surface S surrounding the source of waves and consider the complex amplitude ψ of the wave at an observation point as a superposition of contributions from each point on S. Each point on S is thus considered as if it were a source. If the contributions from all such sources are added, weighted with their proper phase factors, the correct complex amplitude at the observation point results. This reduces the problem of calculating the complex amplitude distribution along the Huyghens' surface S and then performing the superposition integration.

If the surface S is broken, as shown in Figure 1-7, so that it is opaque to the radiation emitted by the source except over the area S', the latter is referred to as an *aperture*. The complex amplitude in this case is nonvanishing only over the aperture, and we can determine ψ at 0 by integrating only over S'.

Huyghens' principle can be derived from the wave equation. For electromagnetic waves it is quite complicated in general, involving the specification of the field at an observation point as a convolution of a "dyadic Green's function" of the surface relative to the observer, and a distribution of electric and magnetic currents on the surface S. Under conditions where the aperture area is large in terms of square wavelengths and the observer is in the "far zone" of the aperture, that is, many wavelengths away, and under similar conditions for acoustic waves, it is quite simple, reducing to

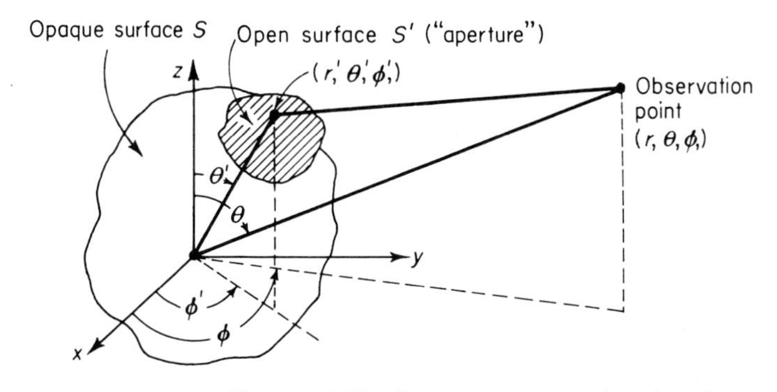

Figure 1-7 An aperture on a closed surface.

a familiar form well-known in physical optics, microwave antenna theory, and elementary acoustics. We will limit ourselves to this simple form in the discussions to follow.

Briefly (see Figure 1-7), the harmonic complex amplitude ψ_o at a point characterized by spherical coordinates r, θ, ϕ is given by

$$\psi_o(r, \theta, \phi) = \int_{S'} dS' \hat{\psi}_i(x', y', z') \frac{e^{jk|\mathbf{r}-\mathbf{r}'|}}{|\mathbf{r} - \mathbf{r}'|} \tag{1.35}$$

where ψ_o and $\hat{\psi}_i$ are respectively the harmonic complex amplitudes† at 0 and the *illumination* at an arbitrary point whose rectangular position coordinates are x', y', z' on the aperture S', k is the propagation constant of the medium, and $|\mathbf{r} - \mathbf{r}'|$ is the separation distance between the points r, θ, ϕ and x', y', z'.

For a complicated aperture geometry and distribution $\hat{\psi}_i$, the integration indicated in (1.35) may be prohibitively difficult. There are two cases, however, for which it is quite simple, and we will therefore work these out and use the results as a basis for later discussions. These are (1) the rectangular aperture with uniform excitation and (2) the circular aperture with uniform excitation. Before specializing to these cases, we will assume that the entire aperture is sufficiently far from the observer so that it subtends an extremely small angle at the observer's position. As a consequence of this assumption, we can consider the amplitude factor $1/|\mathbf{r} - \mathbf{r}'|$ as being effectively constant and equal to $1/r$ over the aperture and can assume with negligible error that the waves from the aperture all propagate toward the

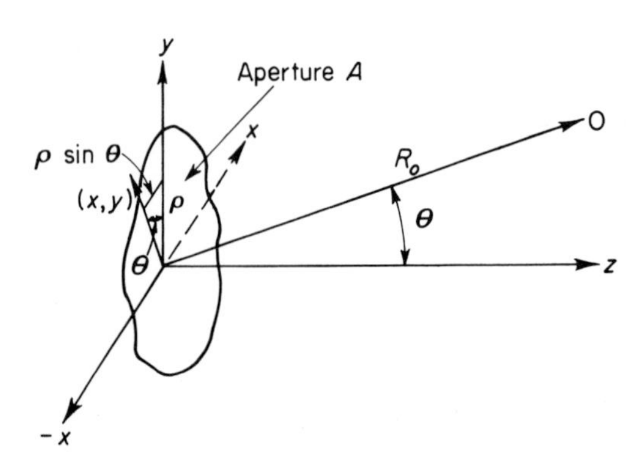

Figure 1-8 Geometry of plane aperture–Fraunhofer diffraction.

† The circumflex on $\hat{\psi}_i$ is used to indicate that the complex amplitude ψ_o and the excitation $\hat{\psi}_i$ may be associated with entirely different quantities, e.g., $\hat{\psi}_i$ may be an electric current and ψ_o an electric field.

observer at the same angle (Fraunhofer diffraction).† We will further assume that the aperture lies in a plane, which can be designated as the x, y-plane if we adopt a rectangular coordinate system with z-axis normal to the aperture plane (see Figure 1-8). Under these assumptions, if the phase factor due to path length differences is unity for the aperture point at the origin of coordinates, that associated with an aperture point ρ' units from the center is $e^{-jk\rho'\sin\theta}$. The complex amplitude at r, θ, ϕ, then, is approximately‡

$$\psi_{r_o}(r, \theta, \phi) \simeq \frac{e^{jkR_0}}{R_0} \int_{\text{aperture}} dx'\, dy'\, e^{-jk\rho'\sin\theta} \hat{\psi}_{r_i}(x', y', 0) \tag{1.36}$$

where R_0 is the distance between (r, θ, ϕ) and the origin of the x', y'-coordinate system on the aperture plane and θ is the angle between the normal to the aperture and the line connecting the center of the aperture with the point of observation.

The form (1.36) has the advantage of simplicity over (1.35) and is still sufficiently general for most "rough" engineering analysis purposes.

1.3.2.1 The Rectangular Aperture

We consider the case of a rectangular aperture of dimensions A, B, as shown in Figure 1-9, where the observer is in the Fraunhofer diffraction region and at a small angle off of the central axis. This is the z-axis in Figure

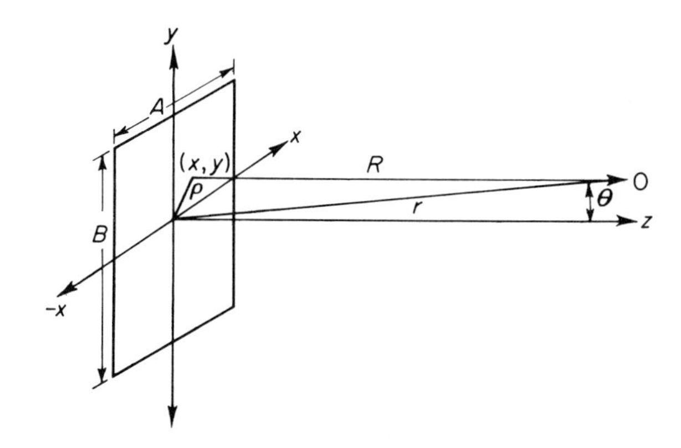

Figure 1-9 The rectangular aperture.

† See Sears, Vol. III, 1948, p. 225 or Jenkins and White, 1937, Chapter 5.

‡ The function $|\mathbf{r} - \mathbf{r}'|$ is expanded in a Taylor series in (x', y', z'), the first two terms of which are $r - [(x' \cos\phi + y' \sin\phi) \sin\theta - z' \cos\theta]$. The stated assumption justifies neglect of terms beyond these. Both of the terms must be retained in the phase factor $e^{jk|\mathbf{r}-\mathbf{r}'|}$ in (1.35). Only the first term $1/r$ is retained in the amplitude factor $1/|\mathbf{r} - \mathbf{r}'|$, because this factor is far less sensitive to small changes in (x', y', z') than is the phase.

1-9, x, y, z being the rectangular coordinates of the observer's position. The coordinates of an arbitrary point on the aperture are designated as x' and y'. "Small angle" here implies that $|x - x'| \ll |z|$, $|y - y'| \ll |z|$, and $(x'^2 + y'^2) \ll r^2$. Under these conditions, where the observation point is given in terms of spherical coordinates r, θ, ϕ, the distance between an aperture point $x', y', 0$ and the observation point is

$$|\mathbf{r} - \mathbf{r}'| = \sqrt{(r \sin\theta \cos\phi - x')^2 + (r \sin\theta \sin\phi - y')^2 + z^2} \tag{1.37}$$

Expansion of $|\mathbf{r} - \mathbf{r}'|$ in powers of $x' \sin\theta \cos\phi/r$ and $y' \sin\theta \sin\phi/r$ and retention only of linear terms results in

$$|\mathbf{r} - \mathbf{r}'| \simeq r - (x' \sin\theta \cos\phi + y' \sin\theta \sin\phi) \tag{1.38}$$

Substituting (1.38) into (1.35) yields

$$\psi_o(r, \theta, \phi) \simeq \frac{e^{jkr}}{r} \int_{-B/2}^{B/2} \int_{-A/2}^{A/2} dx'\, dy'\, \hat{\psi}_i(x', y', 0)\, e^{-jk \sin\theta (x' \cos\phi + y' \sin\phi)} \tag{1.39}$$

If the illumination function $\hat{\psi}_i(x_1, y_1, 0)$ is uniform along the aperture and made equal to $1/AB$ for notational convenience,† integration of (1.39) is easily accomplished and leads to

$$\psi_o(r, \theta, \phi) \simeq \frac{e^{jkr}}{r} \psi(\theta, \phi) \tag{1.40}$$

where

$$\psi(\theta, \phi) = \frac{\sin[(kA/2) \sin\theta \cos\phi]}{(kA/2) \sin\theta \cos\phi} \frac{\sin[(kB/2) \sin\theta \sin\phi]}{(kB/2) \sin\theta \sin\phi}$$

Thus we see that $\psi(\theta, \phi)$, which may be called the *radiation pattern* of the rectangular aperture, is a product of two radiation patterns, one along the vertical or *elevation* angle θ, the other along the horizontal or *azimuth* angle ϕ. Each can be considered separately, and most often studied are the x, z-plane pattern ($\phi = 0$) and the y, z-plane pattern ($\phi = \pi/2$). The former pattern is that which would arise from a slit of infinite width in the y-direction and width B in the x-direction and the latter would result from a slit of infinite width in the x-direction and width A in the y-direction.

A normalized plot of $|\psi(\theta, 0)|^2$ against $(kA/2) \sin\theta$, or, equivalently, of $|\psi(\theta, \pi/2)|^2$ against $(kB/2) \sin\theta$ is presented in Figure 1-10.

The function plotted in Figure 1-10 is

$$\psi'(\theta) = \left| \frac{\sin\alpha(\theta)}{\alpha(\theta)} \right|^2 \tag{1.41}$$

† For this discussion we are not interested in absolute values of $\psi_o(r, \theta, \phi)$, only its functional form. Therefore the absolute magnitude of $\hat{\psi}$ may be chosen arbitrarily.

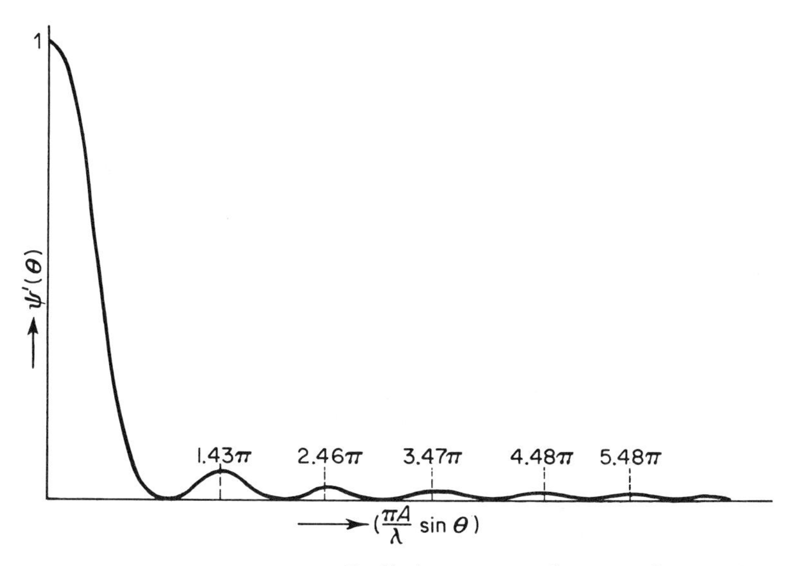

Figure 1-10 Radiation pattern of rectangular aperture.

where

$$\alpha(\theta) \equiv \frac{\pi A}{\lambda} \sin\theta \quad \text{or} \quad \frac{\pi B}{\lambda} \sin\theta$$

$$\lambda = \text{wavelength, equal to } \frac{2\pi}{k}$$

We observe that this function has a maximum value of unity, reached at $\theta = 0$, has its first zero at $\alpha(\theta) = \pi$ and a lower maximum at $\alpha(\theta) = 1.43\pi$. Beyond that point it reaches a zero at all values of θ for which $\alpha(\theta) = n\pi$ and a maximum at $\alpha(\theta)$ equal to $\tan\alpha(\theta)$, with the envelope of the maxima decaying as $1/|\alpha(\theta)|^2$. The first lobe of the pattern is designated as the *main lobe*, the others are called *side lobes*.

1.3.2.2 The Circular Aperture with Uniform Illumination

The circular aperture with uniform illumination is a natural model for radar "dish" antenna reflectors, optical lenses, and other circular configurations used in various technologies. From an appropriate modification of (1.39) and Figure 1-8, we have for this case

$$\psi_o(r, \theta, \phi) \simeq \frac{e^{jkr}}{r} \hat{\psi}_i \int_0^{2\pi} d\phi' \int_0^{\bar{R}} dr' r' \, e^{-jk[\sin\theta\cos(\phi'-\phi)]r'} \tag{1.42}$$

where we have made the assumption that the observation point is close to the z-axis and have converted aperture plane coordinates from rectangular x' and y' to polar r' and ϕ' and where $\bar{R}$ is the radius of the aperture.

The integration in (1.42) can be carried out by a well-known technique involving the Fourier expansion

$$e^{jkx\cos\theta} = \sum_{n=0}^{\infty} J_n(kx)\cos n\theta \tag{1.43}$$

where $J_n(kx)$ is the nth order Bessel function of the first kind. If $\hat{\psi}_i$ is set equal to unity, this results in

$$\psi_o(r, \theta, \phi) \simeq \frac{e^{jkr}}{r}\psi(\theta) \tag{1.44}$$

where

$$\psi(\theta) = 2\pi \int_0^{\bar{R}} dr'\, r' J_0(kr' \sin\theta)$$

The integral of (1.44) is a standard Bessel function integral and its evaluation leads to†

$$\psi(\theta) = \frac{J_1[(2\pi\bar{R}/\lambda)\sin\theta]}{[(2\pi\bar{R}/\lambda)\sin\theta]} \tag{1.45}$$

The first order Bessel function in (1.45) is a tabulated function. The pattern is shown plotted in Figure 1-11.

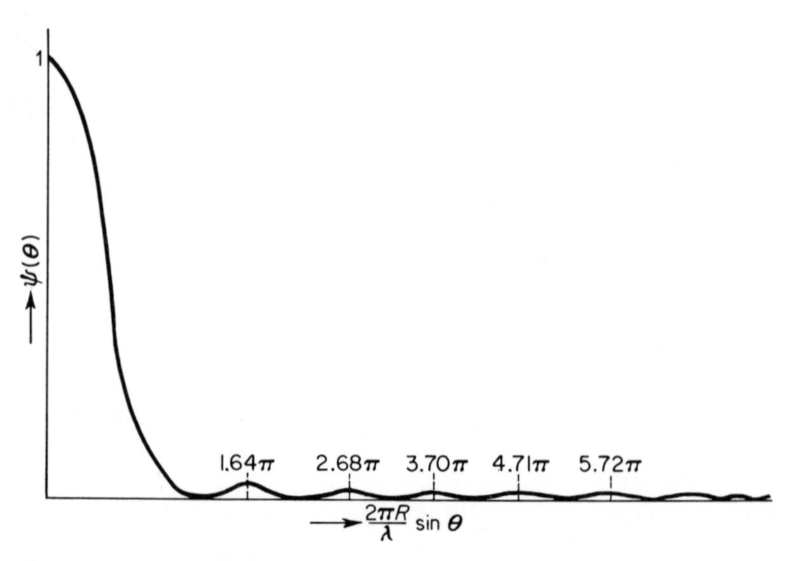

Figure 1-11 Radiation pattern of circular aperture.

† See Silver, 1949, pp. 193–194.

1.4 RESOLUTION AND BANDWIDTH

Resolution, for our purposes, will be defined as the ability to distinguish the separate parts of a composite entity. The entity may be, for example, a pair of overlapping pulses on an oscilloscope screen. In this case, the ability to separate out the two pulses is a measure of the time resolution capability of the human eye. The ability of the ear to pick out two separate tones closely spaced in the spectrum is a measure of the ear's frequency resolution properties. A telescope that can pick out two closely spaced stars which the ordinary telescope would see as a single "blob" of light is said to have high angular resolution. A radar that does not confuse a return from two slightly separated targets with a return from a single target at the same angle is a high angular resolution radar.

The resolution capabilities of any device are related to its *bandwidth*, whether it be bandwidth in the usual time-frequency sense or a spatial or angular analog of bandwidth. The present section will be devoted to discussion of various kinds of resolution and the relationship of each to a corresponding analog of bandwidth. Noise will be neglected in the discussion, although it should be remarked that noise is the ultimate limiting factor in all kinds of resolution.†

1.4.1 Time Resolution of Waveforms

In general, large bandwidth is a requirement for good resolution of waveforms in time. For example, it is an important factor in resolving pairs of closely spaced pulses. To begin the discussion, consider a square pulse of duration τ, amplitude $1/\tau$, and centered at t_0 (see Figure 1-12)

$$v(t) = \frac{1}{\tau} u\left(t - \left[t_0 - \frac{\tau}{2}\right]\right) - u\left(t - \left[t_0 + \frac{\tau}{2}\right]\right) \tag{1.46}$$

where $u(t)$ is the unit step function.

The power spectrum of such a pulse (see Section 1.2.1.1) is proportional to

$$|V(\omega)|^2 = \text{sinc}\left(\frac{\omega\tau}{2}\right)^2 \tag{1.47}$$

where $V(\omega)$ is the Fourier transform of $v_i(t)$ and, in general, sinc $(x) = \sin x/x$.

It may be desired for certain applications to pinpoint the location of the leading or trailing edge of this pulse. This is no problem with a perfectly square pulse, since the edges are sharply defined. It would appear from (1.46) that one could obtain perfect definition of the pulse edges if

† See Helstrom, 1960, Chapter X, Sections 1, 2, 3, pp. 267–288.

no bandlimiting were applied to the pulse. However, an infinite bandwidth is not a practical possibility, so it is important to study the effect of bandlimiting on the sharpness of the pulse edges.

Consider the pulse represented by (1.46) driven through a low pass filter (see Figure 1-12). A typical low pass filter (or integrator) is an RC circuit, whose transfer function (see Section 1.2.1.4.1, Eqs. (1.17) and (1.18)) is

$$T(-jp) = \frac{1}{1 + p(RC)} \tag{1.48}$$

and whose impulse response is

$$f(t) = e^{-t/RC}u(t) \tag{1.49}$$

If the input $v_i(t)$ is the pulse represented by (1.46), the filter output, from (1.5) and (1.49), is

$$v_o(t) = \frac{RC}{\tau}\left[1 - e^{\frac{1}{RC}\left(t-\left[t_0-\frac{\tau}{2}\right]\right)}\right] u\left(t - \left[t_0 - \frac{\tau}{2}\right]\right) - \left[1 - e^{-\frac{1}{RC}\left(t-\left[t_0+\frac{\tau}{2}\right]\right)}\right] u\left(t - \left[t_0 + \frac{\tau}{2}\right]\right) \tag{1.50}$$

while the output spectrum (see Eqs. (1.18) or (1.47)) is proportional to

$$|V_o(\omega)|^2 = \frac{|V_i(\omega)|^2}{1 + \omega^2(RC)^2} \tag{1.51}$$

where

$$|V_i(\omega)|^2 = \left[\text{sinc}\left(\frac{\omega\tau}{2}\right)\right]^2$$

The effect of the filter on the pulse shape is shown in Figure 1-12, where the filter time constant RC is a variable parameter. If RC is much less than the pulse duration τ, then the pulse emerges almost undistorted. But if RC is comparable to or greater than τ, it is apparent that the distortion of the pulse is substantial.

Consider the bandlimiting operation provided by the RC low pass filter. The slope of the pulse edges is a decreasing function of bandwidth, or equivalently, an increasing function of filter time constant. The time required for the pulse to reach 63 percent of its maximum value is given by RC, the filter time constant, and is defined as the *rise time* of the pulse. This is the same time as is required for the pulse to decay from its maximum value to 37 percent of maximum, and in this context it is known as the *decay time*. In either case, it is a measure of the degree to which it is possible to define the edges of the pulse, and is seen to be the reciprocal of the filter bandwidth.

These considerations apply when we are attempting to resolve two narrow square pulses of equal duration and amplitude that are close together on the time scale, i.e., to perceive them as two separate pulses rather than a single extended pulse. Clearly, with no bandlimiting the two pulses will always be easily resolved provided their centers are more than a pulse width apart.

If the pulses, their centers separated by T_s, are passed through the *RC* low pass filter, the output obtained by superposition of the two outputs given individually by (1.50) will be as shown in Figure 1-12 (for filter bandwidths of infinity, $100/T_s$, $10/T_s$, $1/T_s$, $0.1/T_s$, and $0.01/T_s$, given in (a), (b), (c), (d), (e), and (f) of the figure, respectively). The exact limit of resolution (i.e., the minimum separation distance required for resolution of the two pulses) is arbitrary, but a satisfactory criterion would be to regard the pulses as "just resolved" in the case (d), where the separation distance is the reciprocal of the filter bandwidth. Thus we see that the minimum separation required for adequate time resolution of two adjacent pulses passed through a low pass filter is inversely proportional to the bandwidth of the filter.

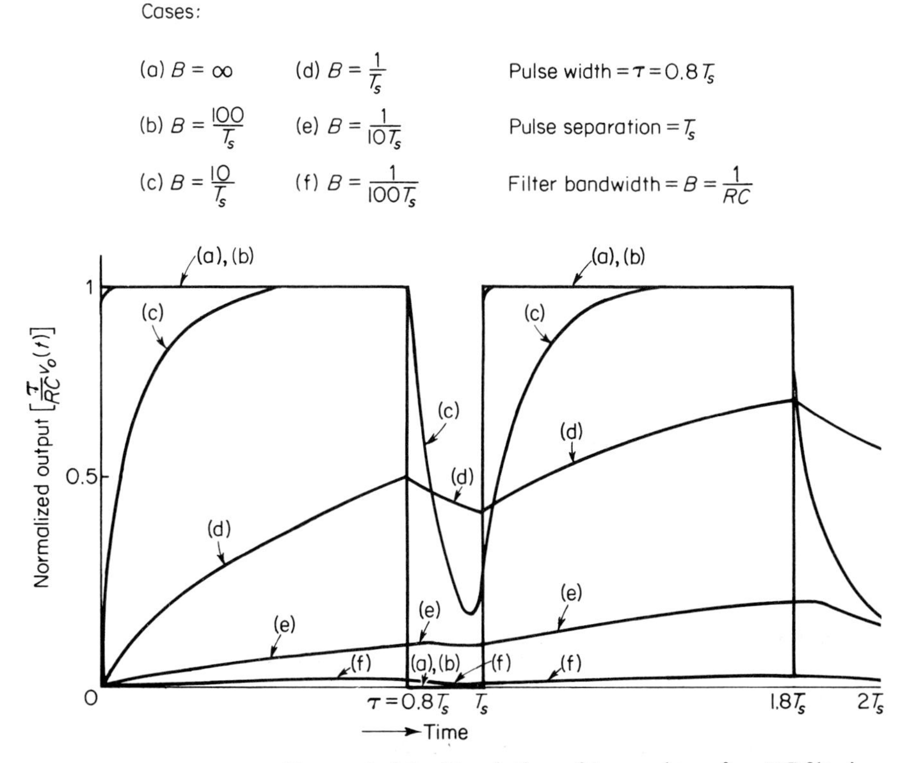

Figure 1-12 Resolution of two pulses after *RC* filtering.

Many more examples like those above could be cited, but they would all result in the same general proposition that the ability of a linear system to time-resolve a waveform is directly proportional to its bandwidth. Of course, it should be noted that, in the absence of noise, there is no real theoretical limit to perfect time resolution,† but in practice there is at least an increase in the difficulty required to resolve time waveforms as the bandwidth is reduced.

1.4.2 Frequency Resolution of Waveforms

If two sine waves of equal amplitude and slightly separated frequencies ω_1 and ω_2 are allowed to run for an indefinite period of time, their superposition will have a double line spectrum as shown in (a) of Figure 1-13, and the two sine waves can be frequency-resolved no matter how small their separation $(\omega_2 - \omega_1)$. If the running time is limited to a period T, the spectrum will have the form shown in (b) of the figure. A shorter period, for example, $T/2$, will show a spectrum like that of (c). As the running time decreases, the spectrum of the superposition of the two sine waves will

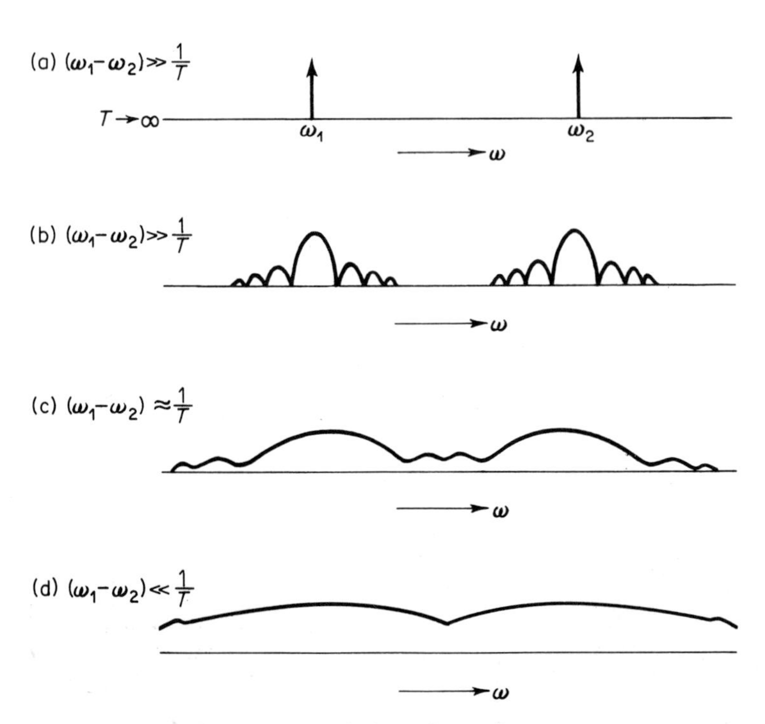

Figure 1-13 Frequency resolution of two sine waves.

† See Helstrom, 1960, Chapter X, Sections 1, 2, 3, pp. 267–288.

begin to look more like the spectrum of a single sine wave switched on for a finite time. The smallest frequency difference between the two sine waves that will allow the spectrum of their superposition to be discerned as that of two separate sine waves is the *limit of frequency resolution.* As in the case of time resolution, this limit is arbitrary, but a reasonable choice is that $(\omega_1 - \omega_2)$ be roughly equal to $1/T$. It will be seen in Section 1.4.4 that this criterion of resolvability is analogous to the Rayleigh limit of resolution in optics.

1.4.3 The Duration Bandwidth Uncertainty Principle

There is an interesting concept in communication theory known as the *duration bandwidth uncertainty principle.*† It is so-called in analogy to the uncertainty principle of quantum mechanics,‡ and can be derived in an analogous way.

Let us postulate a time waveform $v(t)$ with Fourier transform $V(\omega)$. The principle states that, for this waveform,

$$\sqrt{\overline{(\Delta t)^2}\,\overline{(\Delta\omega)^2}} \geq 1 \tag{1.52}$$

where $\overline{(\Delta t)^2}$ and $\overline{(\Delta\omega)^2}$ are defined by:

$$\overline{(\Delta t)^2} = 2\int_{-\infty}^{\infty} (t' - \bar{t})^2 p(t')\, dt' \tag{1.53}$$

$$\overline{(\Delta\omega)^2} = 2\int_{-\infty}^{\infty} (\omega' - \bar{\omega})^2 p(\omega')\, d\omega' \tag{1.54}$$

and where

$$p(t) \equiv \frac{|v(t)|^2}{\int_{-\infty}^{\infty} dt'\, |v(t')|^2}, \qquad \int_{-\infty}^{\infty} dt'\, p(t') = 1, \qquad p(t) \to 0 \quad \text{as} \quad t \to \pm\infty$$

$$p(\omega) = \frac{|V(\omega' - \bar{\omega})|^2}{\int_{-\infty}^{\infty} d\omega'\, |V(\omega')|^2}, \qquad \int_{-\infty}^{\infty} d\omega'\, p(\omega') = 1,$$

$$p(\omega) \to 0 \quad \text{as} \quad \omega \to \pm\infty$$

$$\bar{t} \equiv \int_{-\infty}^{\infty} dt'\, t'\, p(t')$$

$$\bar{\omega} \equiv \int_{-\infty}^{\infty} d\omega'\, \omega'\, p(\omega')$$

After probability density functions have been defined in Section 2.3, it will become apparent to the reader that $p(t)$ and $p(\omega)$ have the properties of probability density functions, although strictly speaking there is nothing necessarily statistical about this discussion, since $v(t)$ may be an entirely

---

† See Brillouin, 1956, pp. 89–93.

‡ See Schiff, 1949, pp. 6–15 or Harris and Loeb, 1963, Chapter 3.

deterministic waveform. However, $\bar{t}$, $\bar{\omega}$, $\overline{(\Delta t)^2}$ and $\overline{(\Delta\omega)^2}$ may be thought of as moments, analogous to those discussed in Section 2.4.2; for while $p(t)$ and $p(\omega)$ are not exactly probability densities, they are measures of weighting of the waveform in time and angular frequency space, respectively.

The Schwartz inequality, Eq. (3.23), will now be applied in the form

$$\int_{-\infty}^{\infty} dt' \, |g(t')|^2 \int_{-\infty}^{\infty} dt'' \, |h(t'')|^2 \geq \left| \int_{-\infty}^{\infty} dt' \, g(t')h(t') \right|^2 \qquad (1.55)$$

where

$$g(t) = (t - \bar{t})v(t) \quad \text{and} \quad h(t) = \frac{dv(t)}{dt}$$

The result of carrying out the indicated integrations on both sides of (1.55), noting that both $g(t)$ and $h(t)$ are real, and using integration by parts and the definitions given in (1.53) and (1.54), is the time bandwidth uncertainty relation (1.52).

The meaning of (1.52) is that a waveform cannot simultaneously have an arbitrarily small bandwidth (where $\sqrt{\overline{(\Delta\omega)^2}}$ plays the role of bandwidth) and an arbitrarily small duration. This is not to say that bandwidth and duration cannot both be arbitrarily large. It has been shown in Section 1.4.1 that accuracy in measurement of the time of occurrence of some chosen part of a waveform (e.g., the leading edge of a pulse) requires a large bandwidth. Precise measurement of a chosen frequency in a waveform (e.g., the central frequency of a pulsed sine wave), on the other hand, requires a long duration (see Section 1.4.2). The principle (1.52) does not imply the incompatibility of these two features and therefore imposes no limitations on simultaneous measurement of both time and frequency. It does, however, place a limitation on the simultaneous confinement of a waveform to a narrow range of frequencies and a small interval of time.

1.4.4 Limit of Angular or Spatial Resolution

One of the most important limiting factors in information systems based on wave motion is that of spatial (or, equivalently, angular) resolution. By *spatial resolution* we mean the degree to which a system can recognize two point sources of radiation in close proximity as two separate sources in contrast to erroneously deciding that the pair is a single source. *Angular resolution* is the same concept, where the distance between the objects is replaced by the angle subtended at the observer's position.

We ordinarily associate the limit of angular resolution with optics, as it is a standard topic in elementary treatments of that subject. However, the concept is basic and exists as well with acoustic or radio waves, in fact, with all types of waves.

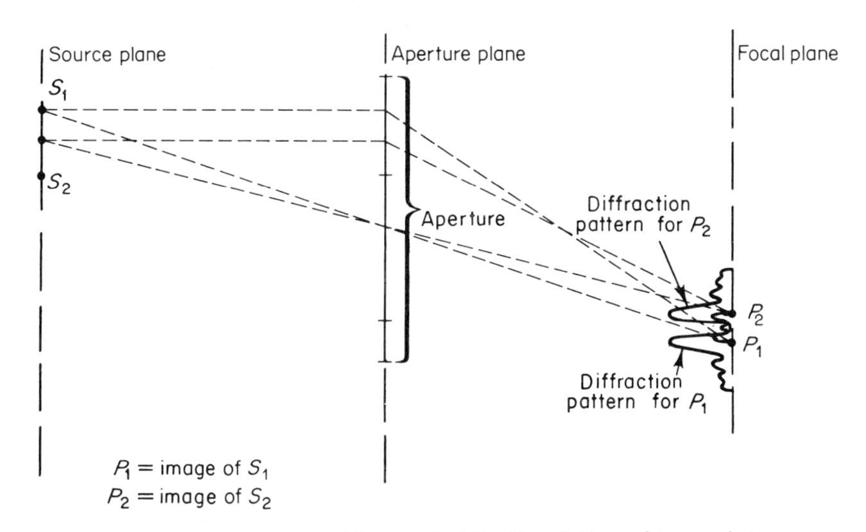

Figure 1-14 Resolution of two point sources.

To approach the idea of angular resolution, we can use either the theory of the circular aperture, discussed in Section 1.3.2.2, or that of the rectangular "slit" aperture, as discussed in Section 1.3.2.1. In either case, a point source in the "far zone" relative to the direction normal to the aperture plane (see Figure 1-14) when focussed by a lens at a point P_1, will produce a diffraction pattern on the focal plane as shown in the figure. The source S_2 will produce the pattern shown at P_2. The sharpness of both patterns will increase with aperture size. The sharper the pattern, the more readily the two point sources can be resolved.

Conceptually there is nothing restrictive about the use of an optical lens here. If, for example, we are working with microwaves, a parabolic reflector fulfills the same function, and the feed acts as the focal point. In acoustics, a reflector and receiving element have the corresponding functions. Regardless of the nature of the propagating waves (acoustic, microwave, etc.) the important point is that the radiation from each source be intercepted by a finite aperture and focussed at a point.

The two points are easily resolved if the centers of the diffraction patterns corresponding to the two sources are sufficiently separated so that no overlap occurs between the main lobes of the two patterns, as illustrated in Figure 1-15.

A standard criterion used to define the limit of resolution in optics is the Rayleigh limit criterion,† wherein two equally bright point sources

† See Sears, Vol. III, 1948, Section 10.1, pp. 257–260, or Jenkins and White, 1937, pp. 118–120.

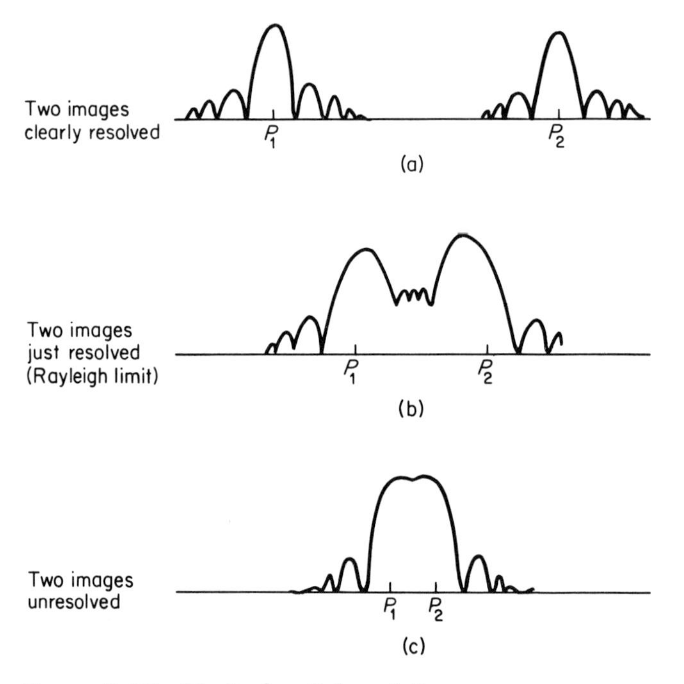

Figure 1-15 Limit of spatial resolution.

whose light is intercepted by a circular aperture are said to be just resolved if the central maximum of the image diffraction pattern of one source coincides with the first minimum of that of the other. From the plot of Eq. (1.45) in Figure 1-11 it is evident that the minimum resolvable angle $\theta_{\min}$ is

$$\theta_{\min} = \sin^{-1}\left[\frac{0.61\lambda}{\bar{R}}\right] \tag{1.56}$$

where $\bar{R}$ is the aperture radius and λ is the wavelength. Equivalently, the minimum resolvable separation distance $d_{\min}$ at a distance D from the aperture, for small angles (i.e., where $\sin\theta \simeq \theta \simeq \tan\theta$), is approximated by

$$d_{\min} \simeq \frac{0.61\lambda D}{\bar{R}} \tag{1.57}$$

1.4.5 The Ambiguity Function

An analytical device to formalize the ideas of resolution limitations of systems is the *ambiguity function*, introduced by Woodward† for application to radar. To illustrate its application, consider a time function $v(t)$. Two values of this function, at instants t and $t + \tau$, are just resolved if there is

† Woodward, 1953, Chapter 7, pp. 115–125.

a perceptible difference between $v(t+\tau)$ and $v(t)$. In a practical sense, the ability to time-resolve $v(t)$ is an increasing function of the absolute square difference $|v(t+\tau)-v(t)|^2$. A waveform for which this squared difference is large for all τ except over a very small range near $\tau = 0$ can be said to have good time resolution.

With this criterion, a convenient measure of time resolution is

$$\Delta(\tau) = \int_{-\infty}^{\infty} dt'\, |v(t'+\tau) - v(t')|^2 = 2\int_{-\infty}^{\infty} dt'\, |v(t')|^2\{1 - \mathrm{Re}\,[\rho(\tau)]\} \tag{1.58}$$

where

$$\rho(\tau) = \frac{\int_{-\infty}^{\infty} dt'\, v(t')v^*(t'+\tau)}{\int_{-\infty}^{\infty} dt'\, |v(t')|^2}$$

The function $\rho(\tau)$ is the NACF (normalized autocorrelation function, see Section 2.4.2.3) of $v(t)$ where $v(t)$ has its complex representation. It takes on its maximum value of unity at $\tau = 0$ and may take on negative values between 0 and -1 throughout the range of τ. If its absolute square or *modulus* $|\rho(\tau)|^2$ is very small or vanishing except over a very small region near $\tau = 0$, then $v(t)$ is said to have good time resolution. If it is uniformly large over a wide range of values of τ, then $v(t)$ is considered to have poor resolution.

Ambiguity in frequency is defined analogously to ambiguity in time, as follows:

$$\Psi(f) = \frac{\int_{-\infty}^{\infty} df'\, V(2\pi f')V^*(2\pi(f'+f))}{\int_{-\infty}^{\infty} df'\, |V(2\pi f')|^2} \tag{1.59}$$

where $V(2\pi f)$ is the Fourier transform of $v(t)$ as defined by Eq. (1.3). This is the frequency domain analog of the normalized autocorrelation function. Good frequency resolution is indicated by large values of $\Psi(f)$ over a very small range of f and vanishingly small values of $\Psi(f)$ elsewhere.

The most important ambiguity function is that which involves the coupling between time and frequency resolution, as follows:

$$\begin{aligned}\Lambda(\tau, f) &= \frac{\int_{-\infty}^{\infty} dt'\, v(t')v^*(t'+\tau)e^{-2\pi j f t'}}{\int_{-\infty}^{\infty} |v(t')|^2\, dt'} \\ &= \frac{\int_{-\infty}^{\infty} df'\, V^*(2\pi f')V(2\pi[f'+f])e^{-2\pi j f'\tau}}{\int_{-\infty}^{\infty} df'\, |V(2\pi f')|^2}\end{aligned} \tag{1.60}$$

The integrals

$$T = \int_{-\infty}^{\infty} |\rho(\tau)|^2\, d\tau = \frac{1}{S_\omega} \tag{1.61}$$

$$B = \int_{-\infty}^{\infty} |\Psi(f)|^2 \, df = \frac{1}{S_T} \tag{1.62}$$

$$R = \int_{-\infty}^{\infty} |\Lambda(\tau, f)|^2 \, d\tau \, df \tag{1.63}$$

are called by Woodward the *time resolution constant*, the *frequency resolution constant*, and the *area of ambiguity*, respectively. The reciprocals of (1.61) and (1.62) are called, respectively, the *frequency span* and *time span*. The time span S_T is roughly the total amount of time occupied by the signal and the frequency span S_f is the total frequency space occupied by the signal, both measures being independent of whether time or frequency intervals are continuous or separated.

For example, a pulse train with pulse duration of 0.1 ms and a repetition frequency of 1000/s running for 100,000 s has a time span of 10,000 s, not 100,000 s. The same considerations apply to frequency span. That is, a waveform with 10 flat frequency bands each 1 MHz wide, the center of each band separated from that of the next by 10 MHz, has the same frequency span as a waveform occupying a continuous 10 MHz band, although by the usual definition of bandwidth, the former has 10 times the bandwidth of the latter.

Because of the normalization of (1.60), the area of ambiguity has a value of unity, as can be shown by carrying out the integration indicated in (1.63).

1.5 THE SAMPLING THEOREM

The sampling theorem tells us that a time function of limited duration T and highest significant frequency B can be completely characterized in terms of $2BT$ equally spaced samples.† To show this, consider a time function $f(t)$ extending from $t = -T/2$ to $t = T/2$ and vanishing outside the interval $-T/2$ to $T/2$. Inside the interval such a function can be treated as if it were periodic and expanded in a complex Fourier series with fundamental period T. In particular, if all Fourier components beyond the Nth harmonic are negligible, then

$$\begin{aligned} f(t) &\simeq \sum_{n=-N}^{N} c_n \, e^{2\pi n j t/T}, & -\frac{T}{2} \leq t \leq \frac{T}{2} \\ &= 0, & |t| > \frac{T}{2} \end{aligned} \tag{1.64}$$

† The obvious practical implication of the sampling theorem is that a signal waveform known to have no significant frequencies beyond B can be sampled at a rate of $2B$ per second and the complete waveform can then be recovered from the samples. The rate $2B$ per second is known as the *Nyquist sampling rate* (Nyquist, 1928).

where

$$c_n = \frac{1}{T}\int_{-T/2}^{T/2} f(t')\, e^{-2\pi njt'/T}\, dt'$$

According to (1.64), if it is a priori known that $f(t)$ vanishes outside the interval $-T/2$ to $T/2$ and is limited to frequencies below N/T, then determination of the $(2N+1)$ Fourier coefficients c_n will provide us with complete knowledge of the shape of the function.

The Fourier transform of $f(t)$ is

$$F(\omega) = \int_{-T/2}^{T/2} dt'\, e^{-j\omega t'} f(t') \tag{1.65}$$

which for $\omega = 2\pi m/T$; $m = 0, 1, 2, \ldots, 2N$, becomes, by virtue of (1.64),

$$F\left(\frac{2\pi m}{T}\right) = \int_{-T/2}^{T/2} dt'\, e^{-2\pi njt'/T} f(t') = Tc_m \tag{1.66}$$

Since $f(t)$ has been assumed confined to the frequency region below N/T, then in analogy to the steps leading from (1.64) to (1.66) we can regard $F(\omega)$ as if it were a periodic function of ω with fundamental period $2\pi N/T$ and expand it in a Fourier series of the form

$$F(\omega) = \sum_{n=-N}^{N} \tilde{C}_n\, e^{-nj\omega T/N} \tag{1.67}$$

where

$$\tilde{C}_n = \frac{T}{2\pi N}\int_{-\pi N/T}^{\pi N/T} d\omega'\, F(\omega')\, e^{nj\omega' T/N}$$

It follows from (1.65) and (1.67) that†

$$f\left(\frac{mT}{N}\right) = \frac{N}{T}\tilde{C}_m \tag{1.68}$$

Sampling theorems in the frequency and time domains can be developed from Eqs. (1.64) through (1.68). To show this, we first invoke (1.64), (1.65), and (1.66) to derive the result‡

$$\begin{aligned} F(\omega) &= \int_{-T/2}^{T/2} dt'\, e^{-j\omega t'}\left[\frac{2\pi}{T}\sum_{n=-N}^{N} F\left(\frac{2\pi n}{T}\right) e^{2\pi njt'/T}\right] \\ &= \sum_{n=-N}^{N} F\left(\frac{2\pi n}{T}\right) \text{sinc}\left[\frac{\omega T}{2} - n\pi\right] \end{aligned} \tag{1.69}$$

† Use the reciprocal Fourier transform relationship

$$f\left(\frac{mT}{2N}\right) = \frac{1}{2\pi}\int_{-\pi N/T}^{\pi N/T} d\omega' F(\omega')\, e^{j\omega' mT/2N}$$

and the definition of $\tilde{C}_n$ given in (1.67).

‡ Substitute $f(t')$ as given by (1.64) into the integrand of (1.65), defining c_n through (1.66).

where $\operatorname{sinc} x = \sin x/x$. We then make use of the inversion of (1.65) and Eqs. (1.67) and (1.68), from which it follows that

$$\begin{aligned} f(t) &= \frac{1}{2\pi}\int_{-\pi N/T}^{\pi N/T} d\omega' \, e^{j\omega t}\left[\frac{T}{N}\sum_{n=-N}^{N} f\left(\frac{nT}{N}\right)e^{-nj\omega T/N}\right] \\ &= \sum_{n=-N}^{N} f\left(\frac{nT}{N}\right)\operatorname{sinc}\left[\frac{\pi N}{T}\left(t-\frac{nT}{N}\right)\right] \end{aligned} \tag{1.70}$$

The frequency domain sampling theorem (1.69) tells us that the spectrum of a bandlimited time function of finite duration T with highest frequency N/T can be represented as a linear combination of its sample values at frequencies $\omega/2\pi = 0, \pm 2/T, \ldots, \pm N/T$. Analogously, the time domain sampling theorem (1.70) tells us that such a time function can be represented as a linear combination of its sample values at times $0, \pm T/N, \pm 2T/N, \pm 3T/N, \ldots, \pm(N-1)T/N, \pm T$. In either case, it requires $(2N + 1)$ real numbers to completely specify the function, exactly the number of independent real numbers in the coefficients of its Fourier expansion.

It is guaranteed that the samples $f(nT/N)$ in (1.70) will be real numbers if $f(t)$ is defined as a physical function (e.g., an actual voltage or a current). Thus in (1.70), the $(2N + 1)$ real numbers referred to above are the samples $f(nT/N)$ themselves. In the case of (1.69) it should be noted that the samples $F(2\pi n/T)$ are complex in general, and hence a total of $(4N + 1)$ numbers are involved in all the samples of (1.69), i.e., real and imaginary parts of each sample. However, it follows from (1.68) and the assumption that $f(t)$ is real that $F(\omega) = F^*(-\omega)$. Therefore, $F(2\pi n/T)$ and $F(-2\pi n/T)$ are mutually dependent and only $(2N + 1)$ independent real numbers are actually needed to specify the samples.

To remove negative frequencies from (1.69), we write it in the form

$$\begin{aligned} F(\omega) &= F(0)\operatorname{sinc}\left(\frac{\omega T}{2}\right) \\ &\quad + \sum_{n=1}^{N}\left[F\left(\frac{2\pi n}{T}\right)\operatorname{sinc}\left(\frac{\omega T}{2} - n\pi\right) + F^*\left(\frac{2\pi n}{T}\right)\operatorname{sinc}\left(\frac{\omega T}{2}\right) + n\pi\right] \end{aligned} \tag{1.71}$$

which, after a little manipulation, becomes

$$F(\omega) = \sum_{n=0}^{N}\epsilon_n \frac{\cos n\pi \sin(\omega T/2)}{[(\omega T/2)^2 - (n\pi)^2]}(\omega T)\operatorname{Re}\left[F\left(\frac{2\pi n}{T}\right)\right] + 2\pi nj\operatorname{Im}\left[F\left(\frac{2\pi n}{T}\right)\right] \tag{1.72}$$

where $\epsilon_n = \frac{1}{2}$ for $n = 0$, $\epsilon_n = 1$ for $n > 0$.

The form (1.72) can be used to represent $F(\omega)$ in terms of the real and imaginary parts of its samples $F(2\pi n/T)$ at *positive* frequencies. This eliminates the necessity for dealing with the fictitious mathematical device of negative frequency in implementing the sampling theorem.

REFERENCES†

Passive Linear System Analysis, Electrical Networks, and Fourier and Laplace Transforms

Aseltine, 1958

Bohn, 1963

Churchill, 1944

Craig, 1964

Guillemin, 1953

Guillemin, 1963

Lathi, 1965; Chapters 1 through 10

Lynch and Truxal, 1961

Schwarz and Friedland, 1965

Scott, 1960

Sneddon, 1958; Chapter 1

Van Valkenburg, 1955

Widder, 1946; Chapter VI

Waves (Electromagnetic, Acoustic, etc.)

Beranek, 1954; Chapter 2

Coulson, 1941

Jordan, 1950; especially Chapter 5

Moore, 1964; Chapters 1, 2, 3

Ramo and Whinnery, 1953; especially Chapters 1 through 4 and Chapters 7 and 12

Sears, Vol. III, 1948; Chapter 9, especially pp. 221–223 for elementary discussion of Huyghens' principle

Silver, 1949; Chapters 4 and 5, especially Section 5.12, pp. 160–162, p. 172, pp. 198–199 (Huyghens' principle, etc.)

Skilling, 1942 (good elementary physically oriented treatment of electromagnetic waves)

Resolution, Uncertainty Principles, Ambiguity Functions

Brillouin, 1956; pp. 89–93 (time bandwidth uncertainty principle)

Harris and Loeb, 1963; Chapter 3 (quantum mechanical uncertainty principle)

Sears, Vol. III, 1948; Chapter 10

Schiff, 1949; pp. 6–15 (quantum mechanical uncertainty principle)

Woodward, 1953; especially Chapter 7 (ambiguity function)

---

† See the general list at the end of the book for bibliographical details on references cited.

Sampling Theorem

Black, 1953; Chapter 4

Davenport and Root, 1958; Chapter 5

Lathi, 1965; Section 11.2, pp. 435–443

Schwartz, 1959; Chapter 4, especially Section 4.5

2

INTRODUCTORY NOISE AND STATISTICAL COMMUNICATION THEORY–I

In this chapter some of the elementary aspects of noise and statistical communication theory will be covered. The treatment is somewhat descriptive and no attempt is made to be mathematically rigorous. The reader is referred to texts and articles listed at the end of this chapter for elaboration on points requiring more extensive mathematical analysis for their explanation. Emphasis is on topics that will come up repeatedly in later chapters.

2.1 INTRODUCTORY PROBABILITY THEORY

The foundation of the study of statistical communication theory is the *theory of probability*. This discipline provides the mathematical tools by which problems involving noise and uncertainty can be formulated.

The mathematically correct formulation of probability theory begins with the theory of sets. It defines probability as a measure on a sample space. A set of axioms delineating the postulated properties of the sample space and the probability measure places the theory on a firm logical foundation and allows rigorously correct results to be obtained by the straightforward methods of mathematics.

An intuitive basis for probability theory rests on the *frequency* definition. Given a single experiment, one imagines the experiment to be repeated a large number of times under the same conditions and without mutual interdependence between repetitions. Suppose the experiment can have any one of K possible mutually exclusive outcomes. These outcomes, or *events*, are labeled $A_1, A_2, \ldots, A_K$. Let $n(A_m)$ be the number of repetitions of the experiment in which the outcome is A_m. The probability of occurrence of the event A_m is then defined as

$$P(A_m) = \lim_{N\to\infty} \frac{n(A_m)}{N}, \qquad m = 1\ldots, K \tag{2.1}$$

where N is the total number of independent repetitions of the experiment.

This definition is unsatisfactory from the point of view of the mathematician who does research on fundamental concepts in probability theory. However, as an aid to intuitive thought on the subject, and as a tool for engineering analysis of practical systems, the concept of the probability of an event as a ratio of the number of times the event occurs to the number of *trials* or repetitions of an experiment is not only a good basis but is usually essential to interpretation of results. Moreover, one can neglect the precise limiting process and think only of a large number of trials, assuming that the probability is independent of the number of trials if the latter is sufficiently large. For those who like to think in terms of concrete physical processes rather than abstract entities, this provides a tool of thought, which, while not mathematically precise, is nearly always sufficient for intuitive reasoning about real-world statistical processes. One word of caution: Remember that the nondependence of the probability on the number of independent trials is just an assumption, which is only fulfilled if (a) the number of trials is very large and (b) the conditions are the same during each trial. The validity of (b) in a particular application depends on the rule by which probabilities are formulated. In physical problems, the rule referred to is a physical law that can be specified quantitatively. In human processes (sometimes involved in engineering problems), the rule is sometimes based on psychological or physiological data that are not well understood quantitatively, e.g., visual or hearing mechanisms in signal detection or communication. Failure to recognize the limitations of the assumption that the ratio in (2.1) is not dependent on the number of trials is sometimes responsible for misapplication of probability theory and misinterpretation of its results.

2.1.1 Discrete Probabilities

Consider the case where only a finite number of specific events can occur. To illustrate the ideas, we will use the familiar dice game examples. The basic experiment or trial is a throw of two dice. The events are the various numbers that can turn up, i.e., 2 through 12. There are $6 \times 6 = 36$ ways in which the dice can turn up. It is assumed that the six possible results for a single die are equally probable. (Note that this assumption is based on the laws of mechanics, the symmetry of the die, and the supposition that the die is not "loaded.") On this basis, if a die is thrown many times, we expect that each possible number will turn up on $\frac{1}{6}$th of the throws. Thus the probability of each outcome is $\frac{1}{6}$. Again using concepts of mechanics, we can reasonably assume the results of different throws to be mutually independent.

The probabilities for the outcomes of a throw of two dice are as follows:†

$$P(n) \equiv \text{Prob}\,\{n\} = \begin{cases} 0, & n = 1 \\ \dfrac{n-1}{36}, & n = 2, \ldots, 7 \\ \dfrac{13-n}{36}, & n = 7, \ldots, 12 \end{cases} \tag{2.2}$$

where $P(n)$ and Prob $\{n\}$ are equivalent shorthand notations for "probability that the number n turns up."‡ Once such a result as (2.2) has been established, the task of calculating a probability for a given event becomes routine.

Another popular means of illustrating probability concepts is the use of coin tossing examples. If a coin is not "weighted," the laws of mechanics should give an equal probability of heads or tails. It is reasonable to assume that in a succession of coin tosses, the result of a single toss is not influenced by the results of previous tosses. Then, for example, the probability that on a series of 10 tosses, the first 3 will be heads and the remaining 7 will be tails is $(\frac{1}{2})^3(\frac{1}{2})^7 = (\frac{1}{2})^{10} = \frac{1}{1024}$. Since in this case the outcome "3 heads and 7 tails" has the same probability of occurrence regardless of the sequence of heads and tails, the probability of 3 heads and 7 tails is $\frac{1}{1024}$ multiplied by the total number of ways in which this event can occur; thus,§

$$\text{Prob}\,\{3 \text{ heads and } 7 \text{ tails}\} = \frac{1}{1024}\frac{10!}{7!\,3!} = \frac{15}{128}$$

Probabilities for all combinations of heads and tails could be constructed for any given number of tosses.

---

† The probability that one die turns up a given number, say 5, and the other turns up a given number, say 3, is the product of these two probabilities, which is $\frac{1}{6} \times \frac{1}{6} = \frac{1}{36}$. This is because the numbers turned up on the two dice are mutually independent; hence the probability of a 3 and a 5, or any other combination of two numbers, is the product of the individual probabilities of given numbers turning up, always equal to $(\frac{1}{6})^2$ or $\frac{1}{36}$. The probability of the two numbers summing to n is the product of $\frac{1}{36}$ and the number of ways in which they can sum to n. The reader can easily determine that this line of reasoning leads to (2.2). For example, the outcome $n = 4$ can be reached through (3, 1), (1, 3) or (2, 2), a total of three ways.

‡ The notations Prob $\{A\}$ and $P(A)$ will be used interchangeably throughout this book to denote "probability of occurrence of the event A."

§ Given a set of N boxes and n objects, where $n \leq N$, there are $N!/(N-n)!n!$ possible arrangements of the n objects in the N boxes. In different terms, this is the number of "combinations of N things taken n at a time." See standard textbooks on elementary algebra under "Permutations and Combinations."

2.1.2 Probability Concepts and Definitions

Below are brief statements and illustrations of some key concepts used in probability theory.

1. Joint probability.

The *joint probability* of K events $A_1, A_2, \ldots, A_K$ is the probability that all of these events occur jointly. We will denote this probability by $P(A_1, \ldots, A_K)$. To illustrate, let A_1 be the event "a 7 turns up on the first throw of two dice," A_2 the same event on the second throw, etc. The probability that a 7 turns up on every one of the first five throws is $P(A_1, \ldots, A_5)$.

2. Statistical independence of two events.

Two events A_1 and A_2 are *statistically independent* if the joint probability $P(A_1, A_2)$ is the product of the probabilities $P(A_1)$ and $P(A_2)$. In the dice throwing illustration of Section 2.1.1, statistical independence between throws is assumed. Suppose the events "7 on the first throw" and "7 on the second throw" are denoted by A_1 and A_2 ,respectively. The joint probability $P(A_1, A_2)$ then is $P(A_1)P(A_2)$, equal to $[P(A_1)]^2$ because $P(A_1)$ and $P(A_2)$ are equal. The probability $P(A_1)$ being $\frac{1}{6}$, the joint probability turns out to be $\frac{1}{36}$, a much smaller number. This corresponds to the intuitive thought that two 7's in a row is a much less likely event than a 7 on only a single throw.

3. Conditional probability.

The probability of event B given A is denoted by $P(B/A)$. This is called the *conditional probability* that B occurs, where the "condition" referred to is the occurrence of A. Consider the probability that two heads and one tail will be the outcome of a sequence of three coin tosses. This can happen in three ways, namely *HHT*, *THH*, or *HTH*. Before the coin tossing begins, the probability of occurrence of the event "2 heads and 1 tail" is $(\frac{1}{2})^3 3!/2!1! = \frac{3}{8}$. Suppose, however, that the first coin has been tossed and heads has turned up. The probability of the outcome in question is now $(\frac{1}{2})^2$ multiplied by 2, the number of ways in which 1 head and 1 tail can occur on the last two tosses. The probability of 2 heads and a tail on three tosses, conditional on the hypothesis that a head occurs on the first toss, is $\frac{1}{2}$. By similar reasoning, this conditional probability is seen to be $\frac{1}{4}$ if a tail has turned up on the first toss. The probability of 2 heads and 1 tail on the condition that the first two tosses are heads is $\frac{1}{2}$. If the first two tosses are tails, the conditional probability is again $\frac{1}{2}$. All of the conditional probabilities in this particular example are different from the "unconditional" probability $\frac{3}{8}$. The conditional probabilities $P(B/A)$ and $P(A/B)$ and the joint probability $P(A, B)$ are related by the statement

$$P(A, B) = P(B)P(A/B) = P(A)P(B/A) \tag{2.3}$$

known formally as *Bayes' theorem* and actually a straightforward logical consequence of the intuitive definition of probability. Referring to the above coin tossing example, and letting A represent the event "2 heads and 1 tail" and B the event "a head on the first toss," note that $P(B)$ is $\frac{1}{2}$ and $P(A/B)$ is $\frac{1}{2}$. According to Bayes' theorem, the joint probability $P(A, B)$ is $\frac{1}{4}$. To check this, note that the joint event "a head on the first toss and the outcome of 2 heads and 1 tail on the three tosses" can occur in one of two ways, namely HTH and HHT. With eight possible outcomes of the three tosses, this gives a joint probability of $\frac{2}{8}$, or $\frac{1}{4}$. The probabilities $P(A)$ and $P(B/A)$ are, respectively, $\frac{3}{8}$ and $\frac{2}{3}$, whose product is equal to $\frac{1}{4}$.† This illustrates the validity of the relationship between joint and conditional probability for this particular set of events. Its validity can easily be shown for other sets of events in the coin tossing example.

4. Statistical independence of many events.

A set of N events $A_1, \ldots, A_N$ is said to be statistically independent if (1) the joint probability of occurrence of the N events is the product of the probabilities of the individual events, and (2) if the same statement can be made for any subset of the events chosen at random, e.g., A_1 and A_2, or A_1, A_5, and A_7.

5. Mutually exclusive events.

If the N events $A_1, \ldots, A_N$ are mutually exclusive, then the probability that one of them occurs is the sum of individual probabilities of occurrence. Dice throwing provides a good example of this result. Two dice can only turn up to one of 11 possible numbers. It is obvious that all outcomes are mutually exclusive, e.g., if a 4 turns up, a 3 cannot possibly turn up. The ways in which either a 3 or a 4 can turn up in a single throw of two dice are: 1 and 2, 2 and 1, 1 and 3, 3 and 1, 2 and 2. The probability that either a 3 or a 4 will turn up is therefore $\frac{5}{36}$. But this is the sum of $P(3)$ and $P(4)$ or $\frac{1}{18} + \frac{1}{12}$. Note that the probability of a 3 *or* a 4 on two throws is a very different probability from that of a 3 *and* a 4, which is $2(\frac{1}{18})(\frac{1}{12}) = \frac{1}{108}$.

Note that the probability that one of N mutually exclusive events $A_1, \ldots, A_N$ occurs is unity in the case where it is initially assumed that at least one of the events must occur. This is true by definition, since a probability of unity constitutes certainty, which was initially assumed. We can call this the *normalization rule* for probabilities. A probability is normalized by setting it equal to unity for an outcome which is certain. For this case,

$$P(A_1) + \ldots + P(A_N) = 1 \tag{2.4}$$

If the events are the possible outcomes of a throw of a single die, then we can write

$$P(1) + P(2) + \ldots + P(6) = 1 \tag{2.5}$$

---

† Event A can occur in three ways, namely HHT, HTH, and THH. Two of these ways involve event B. In a large number of tosses on which event A occurred, event B would occur approximately two-thirds of the time, i.e., $P(B/A) = \frac{2}{3}$.

Since $P(n) = \frac{1}{6}$ for $n = 1, 2, \ldots, 6$, it is clear that the normalization rule holds in this case. Also, note that for two mutually exclusive events, $P(A, B) = 0$. By definition of mutual exclusiveness, these two events A and B cannot occur jointly.

2.1.3 Random Processes and Random Variables

Statistical communication theory is a special branch of the *theory of random processes*. A random process is a physical phenomenon which cannot be specified by a precise mathematical law but only by probabilities. The inability to so specify it may be due to human ignorance or to peculiar properties of nature, but the basic reasons are of philosophical interest and our attention will not be focussed on them. The main feature of these processes from our point of view is that we cannot predict their behavior exactly but can assign probabilities to various modes of behavior. Even if we know a precise physical law describing the process, one or more of the parameters in the mathematical expression of the law can be described only by probabilities.

An obvious example is a simple circuit containing a battery, a resistance R, and an ammeter in series. Ohm's law can ordinarily provide all desired information about this circuit. For example, it tells us that the ammeter will read 1 A if the battery voltage is 6 V and the resistance of R is 6 Ω. But suppose that the battery voltage is unknown and for some reason cannot be determined. Even though we know a law relating current and voltage, we cannot predict the circuit current because the voltage cannot be specified. If, however, the battery voltage has some statistical regularity so that probabilities for various voltage values can be assigned (i.e., a probability law for the battery voltage exists and is known), we can at least specify a probability law for the current. For example, if we know that the battery voltage is equally likely to have any value between 2 and 4 V but cannot have a value less than 2 or greater than 4, we know that the current must be between $\frac{1}{3}$ and $\frac{2}{3}$ A and is equally likely to have any value within this range. To know this is less satisfactory than knowing exactly what the current will be, but it is better than no knowledge at all. In case where random processes are involved it may represent all the knowledge we can possibly attain.

A *random variable* is a quantity (such as voltage or current in the above example) which cannot be specified exactly but for which a probability law is known. In studying a function of time, such as a current waveform at the output of an electronic device, one considers the value of the function at a specific instant of time as a random variable. The time function itself is known as a *sample function of a random process*. The philosophical idea behind this terminology is as follows. One imagines the particular time function as a member of a very large (infinite in the mathematical sense) collection of hypothetical functions which makes up the random process. Each of these functions is imagined to be obtained from an experiment performed under conditions identical to those prevailing in the actual experiment and independent of other experiments in the collection. The actual function, then, can be thought of as a *sample* picked at random from the collection of functions known as the *ensemble*.

2.2 SCOPE OF STATISTICAL COMMUNICATION THEORY

The class of random processes of concern to statistical communication theory is that occurring in connection with the transmission of intelligence. The intelligence may be carried on waves propagated through a medium, as in radio, radar or signaling with visible light, infrared, or acoustic waves. It may be traveling through a confined set of electronic boxes, as in a computer or a servo system. The only requirement for inclusion in statistical communication theory as it is defined in this book is that the intelligence be corrupted by noise.

The word "noise," as used here, refers to an electrical signal which: (1) must be considered as a sample function of a random process at the point where intelligence is to be extracted, and (2) does not contain the desired intelligence. This excludes the intelligence bearing signal itself, which in some cases satisfies (1) but by definition never satisfies (2). It includes jamming, which satisfies (2) and which also satisfies (1) even though it may be a completely deterministic signal from the point of view of the sender.

2.3 CONTINUOUS PROBABILITIES

In analyzing noise problems, it is necessary to consider a continuum of possible events. For example, the probability that a current has a value I cannot be computed from the results applicable to discrete probabilities. An extension of these results for the continuous case is required if noise is to be treated by probability methods.

The extension can be made by first dividing the possible range of values of a random variable (e.g., the value of noise current at some given instant of time) into small regions and allowing these regions to become infinitesimally small. To illustrate, consider a noise current which can vary between -10 A and $+10$ A. The scale between -10 and $+10$ can be divided into 20 equal regions each of 1 A, -10 to -9, -9 to $-8, \ldots, 9$ to 10. We can speak of 20 possible "events" in the sense of discrete probability theory. The event A_{-10}^{-9} then is "the current falls between -10 and -9." The event A_{-9}^{-8} is "current falls between -9 and -8," etc., up to A_9^{10}, which is "current falls between 9 and 10." We can then define probabilities $p(A_{-10}^{-9}), \ldots, p(A_9^{10})$ with statements such as "$p(A_0^1)$ is the probability that the current falls between 0 and 1." In this framework, all the results from discrete probability theory become applicable. Moreover, we can define a *probability density* as the ratio of the probability that the current falls within a given range of values, e.g., between 7 and 8 A, to the size of this range of values. The probability density $p(A_7^8)$, for example, is defined as $p(A_7^8)/(8 - 7) = p(A_7^8)$, given in units of reciprocal amperes.

Let the current scale be divided into a larger number of regions of smaller width, i.e., the width of a division ΔI is allowed to approach zero while the number of divisions N becomes infinite, such that the product $N\,\Delta I$ remains constant and equal to I_0, the total range of possible current values. The discrete probability results still hold for the individual probabilities $P(A_k^{k+1})$, but there are an infinite number of such probabilities, and consequently these results must be modified in order to render the continuous case tractable. The solution to this problem is in defining two new quantities, the "probability distribution function" and the "probability density function." The first of these is a true probability. The second is the derivative of the distribution function and hence is not a true probability and does not have the properties required of a probability. For example, its value can exceed unity. This is an important point about probability densities, and failure to recognize it can lead to confusion among beginners treating continuum statistics problems.

Using the above example of the statistics of the current I, and the statements made in (5) in Section 2.1.2 applicable to the occurrence of at least one of a set of mutually exclusive events, we can define a function

$$P\{A_{-K}^{-K+1} \text{ or } A_{-K+1}^{-K+2} \text{ or } \ldots \text{ or } A_{K-1}^{K}\} = \sum_{\ell=-K}^{K} P(A_\ell^{\ell+1}) \tag{2.6}$$

where $K \leq N/2$.

Recalling that $A_\ell^{\ell+1}$ is the event "the current falls within an interval $\ell\,\Delta I$ and $(\ell + 1)\,\Delta I$," we can write

$$P\{A_{-K}^{-K+1} \text{ or } A_{-K+1}^{-K+2} \text{ or } \ldots \text{ or } A_{K}^{K+1}\} = \sum_{\ell=1}^{K} p(I_\ell)\,\Delta I \tag{2.7}$$

where $p\,(I_\ell)$ is the ratio $P(A_\ell^{\ell+1})/\Delta I$, which we define as the *probability density function* of the current within the interval $\ell\,\Delta I$ to $(\ell + 1)\,\Delta I$.

When we let $K\,\Delta I = I$, and invoke the limiting process mentioned above, the sum in (2.7) becomes

$$P(I) - P(-I) = \int_{-I}^{I} p(I')\,dI' = \int_{-\infty}^{I} p(I')\,dI' - \int_{-\infty}^{-I} p(I')\,dI' \tag{2.8}$$

where the function $P(I)$,

$$\int_{-\infty}^{I} p(I')\, dI'$$

which we call the *probability distribution function*, is the probability that the current falls below the value I.

The integrand $p(I)$ is called the probability density function of I. The probability that the current has a value between I and $I + dI$ is given by $p(I)\, dI$. Note that $p(I)\, dI$ has the properties of a probability, but $p(I)$ itself does not.

Since obviously the current must have some value between $-\infty$ and $+\infty$, the normalization rule for $p(I)$ is

$$\int_{-\infty}^{\infty} p(I)\, dI = 1 \tag{2.9}$$

or equivalently in terms of the distribution function

$$P(\infty) = 1 \tag{2.10}$$

2.4 STATISTICAL ANALYSIS OF NOISE WAVEFORMS

Analysis of noise waveforms must be handled by the methods of continuum statistics. Consider a noise current $x(t)$ as shown in Figure 2-1. We denote the values of $x(t)$ at arbitrary instants of time, $t_1, t_2, t_3, \ldots$, by $x_1, x_2, x_3, \ldots$. We consider these quantities as random variables and define probability density functions for them, denoted by $p(x_1), p(x_2), p(x_3), \ldots$. We also define joint and conditional probability densities $p(x_1, x_2)$, $p(x_2/x_1)$, etc.

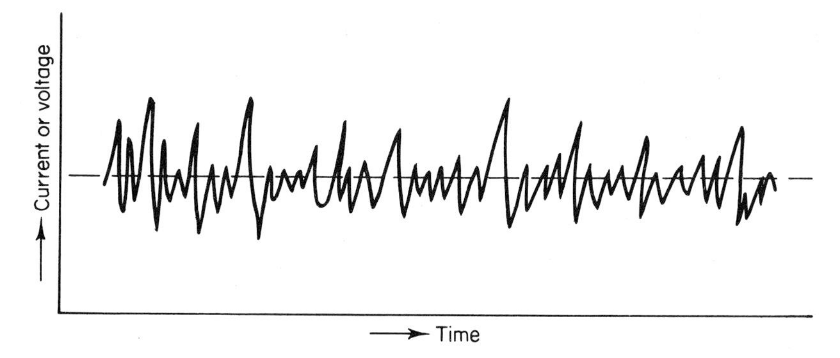

Figure 2-1 Noise voltage or current waveform.

2.4.1 Orders of Statistical Information

The following discussion of orders of statistical information makes use of the above noise waveform for illustrative purposes but can be applied to any function of time which is a sample function of a random process.

1. First order statistics.

The first order statistics of $x(t_k)$ are contained in $p(x_k)\,dx_k$, the probability that $x(t_k)$ has a value between x_k and $x_k + dx_k$ at time t_k. This is the probability from which some of the standard statistical quantities such as mean and variance can be calculated. Note, however, that no information about frequency is contained in the first order statistics.

2. Second order statistics.

A function denoted by $p(x_j, x_k)\,dx_j\,dx_k$ is defined as the joint probability that $x(t)$ has a value between x_j and $x_j + dx_j$ at time t_j and a value between x_k and $x_k + dx_k$ at time t_k. A knowledge of this function constitutes the second order statistics of $x(t)$. Frequency information is contained therein. The autocovariance function (Section 2.4.2) and power spectrum (Section 2.4.4) are calculable from a knowledge of $p(x_j, x_k)$. The conditional probability that $x(t)$ is between x_k and $x_k + dx_k$ at time t_k if it is equal to x_j at time t_j is denoted by $p(x_k/x_j)\,dx_k$. Using (2.3), the relationship between joint and conditional probabilities previously mentioned for discrete statistics (Bayes' theorem), the conditional probability density function $p(x_k/x_j)$ is given by

$$p(x_k/x_j) = \frac{p(x_j, x_k)}{p(x_j)} \tag{2.11}$$

A noteworthy relationship, following from the discussion in (5) of Section 2.1.2,† is

$$p(x_k) = \int_{-\infty}^{\infty} p(x_j, x_k)\,dx_j \tag{2.12}$$

Thus, the first order statistics of $x(t)$ can be inferred from a knowledge of its second order statistics by integrating the right hand side of (2.12).

3. Higher order statistics.

To completely specify the statistics of the current waveform $x(t)$, it would be necessary to know the probability that the entire waveform takes on a particular functional form. This is tantamount to a knowledge of an Nth order joint probability density function for $x(t)$ at times $t_1, \ldots, t_N$, where

† Referring to the first paragraph of (5) in Section 2.1.2, we can think of one of the events $A_1, \ldots, A_N$ as the joint appearance of $x(t_j)$ in a region between x_j and $x_j + dx_j$ and of $x(t_k)$ in a region between x_k and $x_k + dx_k$. If we divide the (x_j, x_k) plane into such regions (events) then the pair $x(t_j)$, $x(t_k)$ cannot be in more than one of them; hence the events are mutually exclusive. The probability density function $p(x_r)$ is then obtained by summing over all the possible x_j events wherein $x(t_r)$ is at x_k. This is the meaning of (2.12).

enough instants of time t_k are used to completely characterize the function over its entire duration. This function can be denoted by $p(x_1, \ldots, x_N)$. The number of time instants which must be included depends on the highest frequency components in the waveform. This was discussed in Section 1.5 in connection with the sampling theorem. In many practical noise problems the highest order statistics of concern are the second order statistics. Third and higher order probabilities for noise not having *Gaussian* statistics† are rarely dealt with in practice, largely because they are too difficult to handle mathematically and the additional statistical information provided by calculating them is not usually considered worth the trouble. In the case of Gaussian noise, a knowledge of second order statistics is sufficient to completely characterize the process.‡ It should be noted that a knowledge of the Nth order statistics of $x(t)$ inplies knowledge of the $(N-1)$st order statistics. This is by virtue of the relation

$$p(x_1, \ldots, x_{j-1}, x_{j+1}, \ldots, x_N) = \int_{-\infty}^{\infty} dx_j\, p(x_1, \ldots, x_N) \tag{2.13}$$

which is a generalization of (2.12). It is true by virtue of statements made in (5) Section 2.1.2, whereby the probability of the event

$$\{x_1 \leq x(t_1) \leq x_1 + dx_1, \ldots, x_{j-1} \leq x(t_{j-1}) \leq x_{j-1} + dx_{j-1}, x_{j+1} \leq x(t_{j+1}) \leq x_{j+1} + dx_{j+1}, \ldots, x_N \leq x(t_N) \leq x_N + dx_N\}$$

is the sum of the probabilities of this event occurring jointly with each possible value of $x(t_j)$. If the scale of values of $x(t_j)$ is divided up into any number of small regions, $R_1, R_2, \ldots$, then $x(t_j)$ can only fall in one of those regions, i.e., the event "$x(t_j)$ falls in R_k" is mutually exclusive with each of the events "$x(t_j)$ falls in R_ℓ, $\ell \neq k$." Consequently the total probability of the distribution of values between $x_1, \ldots, x_{j-1}, x_N$ and $(x_1 + dx_1), \ldots, (x_{j-1} + dx_{j-1}), (x_{j+1} + dx_{j+1}), \ldots, (x_N + dx_N)$ is the sum of the probabilities for x_1–$(x_1 + dx_1), \ldots, x_N$–$(x_N + dx_N)$ over all the mutually exclusive regions R_k into which $x(t_j)$ can fall.

2.4.2 Averages and Moments

The statistical quantities of greatest interest in practical problems are the averages of random variables. Certain special types of averages are called *moments*. The averages of concern here are those derived from first

† Noise with Gaussian statistics is discussed in Section 3.1. The first and second order probability density functions are defined in (3.1) and (3.4), respectively.

‡ This is inherent in the definition of a Gaussian random process. See Davenport and Root, 1958, Section 8.4, especially Eqs. (8.58), (8.59), and (8.60). The elements of the covariance matrix λ_{mn} defined in Davenport and Root's Eq. (8.60) are second order statistical quantities. The parameters defined in their Eq. (8.59) are first order statistical quantities. Hence only first and second order quantities are needed to completely characterize the general Gaussian random process.

and second order probabilities, which will be called first and second order averages, respectively. Higher order averages will not be considered.

2.4.2.1 *First Order Ensemble Averages*

Let $f(x_k)$ be any function of the random variable $x_k = x(t_k)$. The first order ensemble average of $f(x_k)$, i.e., the average over the ensemble of all functions which make up the random process, is denoted by $\langle f(x_k)\rangle$ and is given by

$$\langle f(x_k)\rangle = \int_{-\infty}^{\infty} f(x_k)p(x_k)\,dx_k \tag{2.14}$$

This is the general first order average.

Special first order averages of great practical interest are the moments about zero, $\langle x_k^n\rangle$, and the moments about the mean, $\langle [x_k - \langle x_k\rangle]^n\rangle$. These are given by (2.14), where $f(x_k)$ is set equal to x_k^n and $[x_k - \langle x_k\rangle]^n$, respectively. Some of the more important first order moments and quantities derived from them are given in Table 2-1. It is understood that $x(t)$ is a noise current or voltage.

In the case where the noise is *wide sense stationary* (to be discussed in Section 2.4.3) the quantities in the table are independent of the time t_k and therefore of the index k, and may be designated as

$$\langle x\rangle, \qquad \langle x^2\rangle, \qquad \langle (x - \langle x\rangle)^2\rangle$$

2.4.2.2 *First Order Time Averages*

Another type of averaging is possible for noise waveforms, namely, time averaging. It is nonstatistical in character and can be performed on any type of waveform whether or not it is a sample function of a random process.

If $x(t)$ runs from $-T$ to $+T$, the time average of a function of $x(t)$, which function can be denoted by $f[x(t)]$, is defined as

$$\overline{f[x(t)]} = \frac{1}{2T}\int_{-T}^{T} dt'\,\{f[x(t')]\}$$

or, if $x(t)$ is of unlimited duration,

$$\lim_{T\to\infty} \frac{1}{2T}\int_{-T}^{T} dt'\,f[x(t')] \tag{2.15}$$

As an example of a time average, consider the average power in a noise voltage or current waveform $x(t)$, of duration $2T$, given by a quantity proportional to

$$\overline{x^2(t)} = \frac{1}{2T}\int_{-T}^{T} x^2(t)\,dt \tag{2.16}$$

All of the ensemble average derived quantities in Table 2-1 have analogous time average definitions. It must be remembered, however, that the time average of a quantity for any one sample function of a random process

Table 2-1 First Order Moments and Associated Quantities

Quantity	Usual Designation Symbol	Usual Designation Name	Physical Interpretation
$\langle x_k \rangle$	m_k	Mean or average noise at time t_k	dc value of noise voltage or current
$\langle x_k^2 \rangle$	—	Mean square noise at time t_k	Proportional to average ac noise, or noise fluctuation
$\sqrt{\langle x_k^2 \rangle}$	—	rms noise at time t_k	Same as above, in voltage or current units
$\langle [x_k - \langle x_k \rangle]^2 \rangle$	σ_k^2	Variance of noise at time t_k	Measure of average "spread" of noise voltage from its dc value, in power units
$\sqrt{\langle [x_k - \langle x_k \rangle]^2 \rangle}$	σ_k	Standard deviation of noise at time t_k	Same as above, in voltage or current units
$\langle [x_k - \langle x_k \rangle]^3 \rangle$	—	Skewness of noise	Measure of deviation of the noise probability density function from symmetry about the mean
$\dfrac{\langle [x_k^2 - \langle x_k^2 \rangle]^2 \rangle}{\langle x_k^2 \rangle^2}$	—	Power fluctuation	Measure of average spread of the noise power from its mean value

is not necessarily equal to the ensemble average of that same quantity at any given instant of time. It is not even necessarily true that the time averages for different sample functions of the same random process are equal. As an example, consider 1000 noise generators putting out noise voltages $x_1(t), \ldots, x_{1000}(t)$. If Nos. 3 and 5 are at different operating temperatures, for example, the time average noise power output of No. 3, proportional to $\overline{x_3^2(t)}$, and that of No. 5, proportional to $\overline{x_5^2(t)}$, are different. Also, in general the ensemble average $\langle x^2 \rangle$ over all 1000 generators at a time 5 s after the generators are switched on will be different from both $\overline{x_3^2(t)}$ and $\overline{x_5^2(t)}$. This is a case of a nonergodic random process, which is not easily analyzed by the routine methods of random noise theory. However, in spite of the difficulty of analysis, it is important to recognize that such processes are not uncommon. Failure to obtain agreement between theory and experiment could result from the unconscious assumption that time and ensemble averages must be equivalent. Such failure would occur, for example, if a calculated ensemble average and a measured time average of a nonergodic noise were compared.

2.4.2.3 Second Order Ensemble Averages

A second order ensemble average is an average derived from the second probability density function $p_2(x_1, x_2)$. If $f(x_1, x_2)$ is any function of the two random variables, its ensemble average can be written

$$\langle f(x_1, t_1; x_2, t_2)\rangle = \iint_{-\infty}^{\infty} dx_1\, dx_2 f(x_1, x_2) p_2(x_1, t_1; x_2, t_2) \tag{2.17}$$

where t_1 and t_2 are inserted in f and p_2 to account for the dependence of these functions on t_1 and t_2.

A most important second order average is the *autocorrelation function* or *autocovariance function*† of the noise current or voltage x_1 and x_2. Of some importance is the same function of x_1^2 and x_2^2, proportional to the noise power at time t_1 and t_2, respectively. These functions can be defined with respect to zero or with respect to the mean, as follows:

$$R_x(t_1, t_2) = \iint_{-\infty}^{\infty} dx_1\, dx_2\, x_1 x_2\, p_2(x_1 t_1; x_2 t_2) \tag{2.18}$$

$$\begin{aligned} R_{x0}(t_1, t_2) &= \iint_{-\infty}^{\infty} dx_1\, dx_2\, (x_1 - \langle x_1\rangle)(x_2 - \langle x_2\rangle) p_2(x_1, t_1; x_2, t_2) \\ &= R_x(t_1, t_2) - \langle x_1\rangle\langle x_2\rangle \end{aligned} \tag{2.19}$$

$$R_x^{(2)}(t_1, t_2) = \iint_{-\infty}^{\infty} dx_1\, dx_2\, x_1^2 x_2^2\, p_2(x_1, t_1; x_2, t_2) \tag{2.20}$$

$$R_{x0}^{(2)}(t_1, t_2) = \iint_{-\infty}^{\infty} dx_1\, dx_2\, p_2(x_1, t_1; x_2, t_2)[x_1 - \langle x_1\rangle]^2 [x_2 - \langle x_2\rangle]^2 \tag{2.21}$$

It is often convenient to normalize the autocovariance or autocorrelation function, i.e., to define it in such a way that its value is unity for $t_1 = t_2$. The normalized forms corresponding to (2.18) through (2.21) are:

$$\rho_x(t_1, t_2) = \frac{R_x(t_1, t_2)}{\sqrt{R_x(t_1, t_1) R_x(t_2, t_2)}} \tag{2.22}$$

$$\rho_{x0}(t_1, t_2) = \frac{R_{x0}(t_1, t_2)}{\sqrt{R_{x0}(t_1, t_1) R_{x0}(t_2, t_2)}} \tag{2.23}$$

$$\rho_x^{(2)}(t_1, t_2) = \frac{R_x^{(2)}(t_1, t_2)}{\sqrt{R_x^{(2)}(t_1, t_1) R_x^{(2)}(t_2, t_2)}} \tag{2.24}$$

$$\rho_{x0}^{(2)}(t_1, t_2) = \frac{R_{x0}^{(2)}(t_1, t_2)}{\sqrt{R_{x0}^{(2)}(t_1, t_1) R_{x0}^{(2)}(t_2, t_2)}} \tag{2.25}$$

For a wide sense stationary noise (to be defined and discussed in Section 2.4.3), the second probability density function $p_2(x_1, t_1; x_2, t_2)$ and therefore the autocorrelation function is in all cases independent of the exact values of t_1 and t_2, but depends only on the time difference $t_2 - t_1$ which will be denoted by τ. In this case, averages like $\langle x\rangle$ and $\langle x^2\rangle$ are independent of time. We will denote the autocorrelation functions by $R_x(\tau)$, $R_x^{(2)}(\tau)$, $\rho_x(\tau)$,

† The terminology "autocovariance function" is usually reserved for the case where the mean is subtracted out, whereas "autocorrelation function" is the general term used for the quantity defined here. The abbreviation ACF will be used frequently.

etc. We can write, for the wide sense stationary case

$$R_{x0}(\tau) = R_x(\tau) - \langle x \rangle^2 \tag{2.26}$$

$$R_{x0}^{(2)}(\tau) = R_x^{(2)}(\tau) - 4\langle x \rangle \langle x_1 \, x_2^2 \rangle + 2\langle x^2 \rangle \langle x \rangle^2 + 4\langle x \rangle^2 R_x(\tau) - 3\langle x \rangle^4 \tag{2.27}$$

In this case, all normalized autocorrelation functions† are written in the form

$$\rho(\tau) = \frac{R(\tau)}{R(0)} \tag{2.28}$$

2.4.2.4 Second Order Time Averages

If $x(t)$ is a sample function of a random process, defined in the interval $-T$ to $+T$, a function $f[x(t_1, x(t_2)]$ can be averaged over time as follows:

$$f[x(t_1), x(t_2)] = \frac{1}{2T}\int_{-T}^{T} dt' \, f[x(t'), x(t' + (t_2 - t_1))] \tag{2.29}$$

The relationship between this type of time average and the second order ensemble average for certain special functions, i.e., $f[x(t_1), x(t_2)] = x(t_1)x(t_2)$ and $f[x(t_1), x(t_2)] = x^2(t_1)x^2(t_2)$, will be discussed in Section 2.4.3 for the ergodic case.

2.4.3 Stationary and Ergodic Noise

To render noise problems tractable, it is almost required that the random process of which the noise is a sample function have a property called *stationarity*. A stationary random process can be defined roughly as one whose statistical properties are independent of time. A noise current waveform is stationary in the strict sense if all of its probability functions $p(x_1)$, $p_2(x_1, x_2)$, $p_3(x_1, x_2, x_3)$, etc., are independent of the first time instant t_1 but depend only on the time difference $t_2 - t_1$, $t_3 - t_2$, etc. Stated another way, the time origin can be shifted by any amount without changing the form of a probability density function of any order. A more restricted type of stationarity, known as *wide sense*, requires that moments up to a specified order, e.g., first and second order, are independent of the time origin. It can be shown that strict sense stationarity implies wide sense stationarity, but the converse is not necessarily true. It is true in the special case of the Gaussian random process,‡ to be discussed in Section 2.5.

In most standard problems, only stationary noise is considered. This means that all the action takes place during an interval in which the noise

---

† The normalized autocorrelation function will often be called NACF.

‡ See Davenport and Root, 1958, Section 8.4, p. 154.

statistics do not change. A very important property of some stationary processes from our point of view is that of *ergodicity*.† Stationary processes that are also ergodic have the property that time averages are equivalent to ensemble averages. Given a quantity dependent on values of the sample function at a given instant of time, its average over the entire ensemble of functions making up the random process is equal to its time average over the duration of the sample function. The latter average is the same for all sample functions of the process, and the former average is the same regardless of the instant of time at which the ensemble averaging is done.

For ergodic processes, by virtue of this equivalence of time and ensemble averaging, the autocovariance function defined in Section 2.4.2 can be equivalently written as a time average. For example, if $x(t)$ is a sample function of unlimited duration

$$R_x(\tau) = \lim_{T\to\infty} \frac{1}{2T} \int_{-T}^{T} x(t)x(t+\tau)\,dt \tag{2.30}$$

and

$$R_x^{(2)}(\tau) = \lim_{T\to\infty} \frac{1}{2T} \int_{-T}^{T} x^2(t)x^2(t+\tau)\,dt \tag{2.31}$$

In dealing with ergodic processes, then, we will use the time average and ensemble average interchangeably.

A point worthy of mention is that any time function, random or nonrandom, can be time averaged. The finite time average of a quantity derived from a sample function of a random process is itself a sample from a random process. When it is averaged over the entire ensemble making up the process, the result may be different from that obtained by ensemble averaging followed by time averaging. For an ergodic random process, however, the operation of ensemble averaging followed by time averaging should give the same result as that of time averaging followed by ensemble averaging. In symbols,

$$\langle \bar{f} \rangle = \overline{\langle f \rangle} \tag{2.32}$$

for ergodic processes.

As a special case of (2.32), note that, for an ergodic process

$$R_x(\tau) = \overline{x(t)x(t+\tau)} = \overline{\langle x(t)x(t+\tau)\rangle} = \langle \overline{x(t)x(t+\tau)} \rangle = \langle x(t)x(t+\tau)\rangle \tag{2.33}$$

Stated in words, the autocorrelation function obtained by time averaging the product $x(t)x(t+\tau)$ with one sample function is numerically equal to that obtained by time averaging with any other sample function of the same random process. Therefore, ensemble averaging after time averaging

---

† See Davenport and Root, 1958, p. 67. Basic mathematical treatises on ergodicity are referred to in the footnotes on p. 67 of Davenport and Root.

cannot change the result. Also, the autocovariance function obtained by ensemble averaging the product $x(t)x(t + \tau)$ will be independent of time. Therefore, following this operation by averaging over the time t cannot change the result. Since it was postulated that time and ensemble averaging yield the same result for an ergodic process, the statement (2.33) follows.

2.4.4 The Power Spectrum of Noise

An important property of a noise voltage or current waveform is the average power contained in a unit interval of frequency. This is called the spectral density or power spectrum. It is defined for both random and deterministic waveforms, but should be given special treatment in the former case because of the added feature of ensemble averaging.

2.4.4.1 The Wiener-Khintchine Theorem

A theorem relating the autocorrelation function and the power spectrum of a voltage or current waveform is known as the Wiener-Khintchine theorem. It states that the power spectrum and autocorrelation function are reciprocal Fourier transforms.

To prove the theorem, consider a voltage or current $x(t)$ which is nonzero only during the interval $-T$ to $+T$ (where T will become infinite in the final step of the analysis). Regardless of its functional form, it can always be represented within this interval by an exponential Fourier series with fundamental frequency

$$f_0 = \frac{1}{2T} = \frac{\omega_0}{2\pi}$$

$$\begin{aligned} x(t) &= \sum_{n=-\infty}^{\infty} c_n e^{jn\omega_0 t}, && -T \leq t \leq T \\ &= 0, && t < -T, t > T \end{aligned} \tag{2.34}$$

Making use of (2.34), and the fact that $x(t)$ must be a real function, we obtain

$$\frac{1}{2T}\int_{-T}^{T} dt' x(t')x(t' + \tau) = \sum_{m,n=-\infty}^{\infty} c_m c_n^* e^{-\pi jn\tau/T} \left(\frac{1}{2T}\int_{-T}^{T} dt' \, e^{\pi j(m-n)t'/T}\right) \tag{2.35}$$

Evaluating the integral in (2.35) and invoking the definition (2.30), we obtain

$$R_x(\tau) = \lim_{T\to\infty} \sum_{m,n=-\infty}^{\infty} c_m c_n^* e^{-\pi jn\tau/T} \left\{\frac{\sin \pi(m - n)}{\pi(m - n)}\right\} \tag{2.36}$$

The bracketed quantity is unity if $m = n$, zero if $m \neq n$. Thus

$$R_x(\tau) = \lim_{T\to\infty} \sum_{n=-\infty}^{\infty} |c_n|^2 e^{-\pi jn\tau/T} = \lim_{\omega_0\to 0} \sum_{n=-\infty}^{\infty} |c_n|^2 e^{-j(n\omega_0)\tau} \tag{2.37}$$

Now ω_0 is allowed to approach zero in such a manner that $(2\pi|c_n|^2/\omega_0)$ approaches a finite limit. Denoting this limit by $G_x(\omega)$, where $n\omega_0 = \omega$, we have

$$R_x(\tau) = \frac{1}{2\pi}\int_{-\infty}^{\infty} d\omega\, e^{-j\omega\tau}\, G_x(\omega) \tag{2.38}$$

If we multiply both sides of (2.38) by $e^{j\omega'\tau}$ and integrate over τ, reversing the order of integration on the right-hand side, we obtain

$$\int_{-\infty}^{\infty} d\tau\, R_x(\tau)\, e^{j\omega'\tau} = \frac{1}{2\pi}\int_{-\infty}^{\infty} d\omega\, G_x(\omega) \int_{-\infty}^{\infty} d\tau\, e^{j(\omega'-\omega)\tau} \tag{2.39}$$

Recognizing that the unit impulse function $\delta(x)$ is the Fourier transform of unity [see Eqs. (1.6)], that is,

$$\frac{1}{2\pi}\int_{-\infty}^{\infty} d\tau\, e^{\pm jx\tau} = \delta(x) \tag{2.40}$$

we have

$$G_x(\omega) = \int_{-\infty}^{\infty} d\tau\, R_x(\tau)\, e^{j\omega\tau} \tag{2.41}$$

Equation (2.41) states that the power spectrum is the Fourier transform of the autocorrelation function. Equation (2.38) states that the autocorrelation function is the inverse Fourier transform of the power spectrum. Thus (2.38) and (2.41) together constitute a statement of the Wiener-Khintchine theorem.

Note that the theorem applies to a general voltage or current waveform, whether or not it is a sample function of a random process. If not, the above statement of the theorem is adequate. If so, however, this is not the whole story. Both sides of (2.38) or (2.41) must be averaged over all sample functions of the random process. If the process is ergodic, then by virtue of (2.32), this ensemble averaging does not affect $R_x(\tau)$. Therefore, (2.41) tells us that it also does not affect $G_x(\omega)$.

In the case of an ergodic random process, then, (2.38) and (2.41) constitute an adequate statement of the theorem.

The Wiener-Khintchine theorem can be stated in terms of Fourier cosine transforms. To show this for an ergodic random process observe that $R_x(\tau)$ must be real (by definition) and note that $\langle\overline{x(t)x(t+\tau)}\rangle$ and $\langle\overline{x(t-\tau)x(t)}\rangle$ are equivalent, since one is obtained from the other by a simple time translation which, by assumption of an ergodic process, cannot affect the value. Thus $R_x(\tau)$ is an even function of τ. Using this fact, Eq. (2.41) and Euler's formula ($e^{jx} = \cos x + j \sin x$), we have

$$G_x(\omega) = \int_{-\infty}^{\infty} R_x(\tau)\cos\omega\tau\, d\tau + j\int_{-\infty}^{\infty} R_x(\tau)\sin\omega\tau\, d\tau = 2\int_{0}^{\infty} R_x(\tau)\cos\omega\tau\, d\tau \tag{2.42}$$

The second integral vanishes because its integrand is the product of an even and an odd function of τ and the integration is from $-\infty$ to $+\infty$.

The power spectrum $G_x(\omega)$ is real, as shown by (2.42). Therefore, (2.38) becomes

$$R_x(\tau) = \frac{1}{2\pi}\int_{-\infty}^{\infty} G_x(\omega)\cos\omega\tau\,d\omega - \frac{j}{2\pi}\int_{-\infty}^{\infty} G_x(\omega)\sin\omega\tau\,d\omega = \frac{1}{\pi}\int_{0}^{\infty} G_x(\omega)\cos\omega\tau\,d\omega = R_x(-\tau) \tag{2.43}$$

The second integral must vanish because $R_x(\tau)$ is also real.

Using (2.42) and (2.43) together as a statement of the Wiener-Khintchine theorem for an ergodic random process (or equivalently (2.38) and (2.41)), we note that

$$G_x(\omega) = G_x(-\omega) \tag{2.44}$$

$$G_x(0) = 2\int_{0}^{\infty} R_x(\tau)\,d\tau \tag{2.45}$$

$$R_x(0) = \frac{1}{\pi}\int_{0}^{\infty} G_x(\omega)\,d\omega \tag{2.46}$$

It is convenient to define a *normalized spectrum* $S_x(\omega)$ with the property that $\frac{1}{2\pi}\int_{-\infty}^{\infty} S_x(\omega)\,d\omega = 1$. It is given by

$$S_x(\omega) = \frac{G_x(\omega)}{\frac{1}{2\pi}\int_{-\infty}^{\infty} G_x(\omega)\,d\omega} \tag{2.47}$$

Thus, we obtain from (2.38)

$$\frac{R_x(\tau)}{R_x(0)} = \frac{1}{2\pi}\int_{-\infty}^{\infty} d\omega\, S_x(\omega)\, e^{-j\omega\tau} = \frac{1}{\pi}\int_{0}^{\infty} d\omega\, S_x(\omega)\cos\omega\tau = \rho_x(\tau) \tag{2.48}$$

The normalized spectrum and normalized autocorrelation function, then, are reciprocal Fourier transforms.

The Wiener-Khintchine theorem is very useful in noise theory. One of its most important applications is in providing a tool for computation of the spectrum of an experimentally derived waveform by first calculating its autocorrelation function and then taking the Fourier cosine transform. This technique is a simple and accurate way to get the spectrum of such a waveform.

The relationship also provides a conceptual link between the *correlation time*, τ_c (i.e., the value of τ beyond which $R_x(\tau)$ can be considered negligibly small) and the bandwidth of a noise waveform. Generally speaking, the latter is small when the former is large and vice versa (see Section 2.4.4.2).

2.4.4.2 White Noise

Many common types of noise, such as thermal and shot noise in radio receivers, have an approximately *white* spectrum, i.e., the average power is distributed uniformly throughout most of the radio spectrum.† In actual practice, since all receivers have a finite passband, it is more nearly correct to speak of *bandlimited white noise.* The normalized spectrum of such noise (in complex representation, so that the negative frequency domain must be included) is given by

$$S_x(\omega) = \frac{\pi}{2W}, \qquad \begin{Bmatrix} -(\omega_0 + W) \leq \omega \leq -(\omega_0 - W) \\ (\omega_0 - W) \leq \omega \leq (\omega_0 + W) \end{Bmatrix} \tag{2.49}$$

$$= 0, \qquad \text{otherwise}$$

where ω_0 is the center angular frequency of the passband and W is the bandwidth.‡

By (2.48) and (2.49) (dropping the subscripts x), the normalized autocorrelation function of bandlimited white noise is

$$\rho(\tau) = \frac{1}{2\pi}\int_{-\infty}^{\infty} S(\omega)e^{-j\omega\tau}\,d\omega = \frac{1}{4W}\int_{-(\omega_0+W)}^{-(\omega_0-W)} d\omega\, e^{-j\omega\tau} + \frac{1}{4W}\int_{(\omega_0-W)}^{(\omega_0+W)} d\omega\, e^{-j\omega\tau} = \cos\omega_0\tau\,\operatorname{sinc}(W\tau) \tag{2.50}$$

The function is sketched in Figure 2-2, together with a sketch of the spectrum. As W becomes extremely large, the envelope of the function becomes a very narrow function of τ, approaching a unit impulse function in the limit as $W \longrightarrow \infty$. Physically, this means that as the bandwidth of the noise increases to very large values, the correlation time becomes extremely short, and the noise can be considered uncorrelated between two instants of time which are even slightly separated. One can say that the correlation time of bandlimited white noise is roughly equal to the reciprocal of the cycle per second bandwidth. With a 1 Mc/s band of white noise, for example, a sample at time t_1 can be considered as roughly uncorrelated with a sample at time $t_1 + \tau$ if τ is greater than 1 μs. This is a very

† See Section 3.4.

‡ Throughout the book, bandwidths in angular frequency ω will carry the symbol W, while bandwidths in cycle frequency f will be denoted by B. The word "frequency" will be used interchangeably to mean both cycle and angular frequency. Note also that "bandwidth" as defined here means one half of the actual spread from the lowest to the highest frequency in the passband. This definition will be used throughout the book, although in many references "bandwidth" means twice the number used here. In many cases we will refer to the bandwidth as defined here as *lowpass* bandwidth and to twice its value as *bandpass* bandwidth.

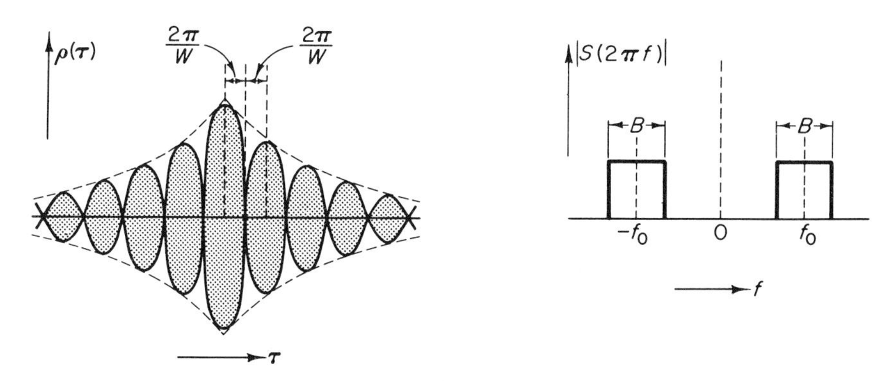

Figure 2-2 Sketch of autocorrelation function and spectrum of bandlimited white noise.

important consideration, for example, in radar pulse integration. If two pulses in bandlimited white noise are not separated by an interval at least equal to the noise correlation time, the noise samples coincident with the two pulses cannot be considered as approximately uncorrelated and the effectiveness of pulse integration will be degraded.†

It is noteworthy that the zeroes of the function sinc $(W\tau)$ occur at the points $\tau = \pi n/W$, where n is any integer. This implies that noise samples separated by *exactly* $\pi n/W$ are uncorrelated, not approximately, but exactly. The argument in the paragraph above is based on the behavior of the *envelope* of the sinc function and concerns *approximately* uncorrelated noise samples. Arguments about *exactly* uncorrelated noise samples based on the occurrence of the zeroes of the sinc are used in optimal detection and estimation theory, and will appear later in connection with those topics.‡

2.4.4.3 Spectrum after a Linear Filter

Referring to Section 1.2.1.1, we note that the absolute square of the Fourier transform of the output of a linear filter is linearly related to that of its input, i.e., taking the absolute square of (1.4),

$$|X_o(\omega)|^2 = |X_i(\omega)|^2\,|F(\omega)|^2 \tag{2.51}$$

Equation (2.51) can be used to relate input and output spectra of deterministic signals. In the case of a stationary random noise, we can determine the relationship between input and output spectra through (1.3), (1.4),

† See Section 10.1.3.

‡ See Sections 4.1.3.1 and 4.1.3.2.

(1.6.f), (2.33), 2.38) and the fact that the impulse response of a filter must be real.†

The result is:

$$G_{x_o}(\omega) = |F(\omega)|^2 G_{x_i}(\omega) \tag{2.52}$$

or equivalently

$$S_{x_o}(\omega) = |F(\omega)|^2 S_{x_i}(\omega)\left(\frac{P_i}{P_o}\right) \tag{2.53}$$

where P_i and P_o are total input power and total output power, respectively.

2.4.4.4 Typical Noise Spectra

If stationary white noise is passed through a simple RC low pass filter (see Section 1.4.1, Eqs. (1.17) and (1.18)), the spectral density at the output of the filter is

$$G_o(\omega) = \left|\frac{1}{1 + j\omega T_c}\right|^2 G_i(\omega) = \frac{G_i}{1 + \omega^2 T_c^2} \tag{2.54}$$

where T_c is the filter time constant RC and G_i is the input spectral density which by assumption is constant over the bandpass of the filter. Thus a white noise passed through an RC integrator or low pass filter has the normalized spectrum

$$S(\omega) = \frac{2T_c}{1 + \omega^2 T_c^2} \tag{2.55}$$

The NACF corresponding to this spectrum can be calculated from (2.48). It is

$$\rho(\tau) = e^{-|\tau|/T_c} \tag{2.56}$$

The correlation time τ_c can be defined in this case as the time required for the autocorrelation function to decay to $1/e$ of its value at $\tau = 0$. With

† The calculation proceeds as follows: From (1.4) and (2.33), we can write:

$$R_{x_o}(\tau) = \langle x_o(t')x_o(t' + \tau)\rangle = \int_{-\infty}^{\infty}\int_{-\infty}^{\infty} dt'\,dt''\,f(t - t')f(t + \tau - t'')R_{x_i}(t'' - t')$$

which takes the form (when all time functions are expressed in terms of their Fourier transforms and (2.38) is used)

$$R_{x_o}(\tau) = \frac{1}{2\pi}\int_{-\infty}^{\infty}\int\int d\omega_1\,d\omega_2\,d\omega_3\,G_{x_i}(\omega_1)F(\omega_2)F(\omega_3)\,e^{j(\omega_2+\omega_3)t + j\omega_3\tau}$$
$$\cdot\left(\frac{1}{2\pi}\int_{-\infty}^{\infty} dt'\,e^{-j(\omega_1+\omega_2)t'}\right)\left(\frac{1}{2\pi}\int_{-\infty}^{\infty} dt''\,e^{j(\omega_1-\omega_3)t''}\right)$$

Now invoking (1.6.f), we have

$$R_{x_o}(\tau) = \frac{1}{2\pi}\int_{-\infty}^{\infty} d\omega'\,G_{x_i}(\omega')F(-\omega)F(\omega')e^{j\omega'\tau}$$

Finally, because $f(t)$ is real, its Fourier transform must have the property $F(-\omega) = F^*(\omega)$; hence through (2.38), Eq. (2.52) follows.

this definition the correlation time is equal to the filter time constant. It is also equal to the reciprocal bandwidth, if bandwidth is defined as the spacing between zero frequency and the half-power point on the spectral curve.

Another typical noise spectrum is that arising when white noise is passed through an *RLC* filter. In this case the normalized spectrum of the output is

$$S_o(\omega) = \frac{1}{W}\left\{\frac{1}{\left(1 + \left[\dfrac{\omega - \omega_0}{W}\right]^2\right)} + \frac{1}{\left(1 + \left[\dfrac{\omega + \omega_0}{W}\right]^2\right)}\right\} \tag{2.57}$$

where W, equal to half the spread between half-power points in the regions around ω_0 or $-\omega_0$ is defined as the filter bandwidth.†

The NACF, as given by (2.48) and (2.57), is

$$\rho(\tau) = \cos \omega_0\tau \, e^{-W|\tau|} \tag{2.58}$$

From (2.58), the NACF for this case is a cosine function of the same frequency as the center of the filter passband with an exponentially decaying envelope. One can define the correlation time as $\tau_c = 1/W$ in this case.

A noise spectrum not necessarily characteristic of a particular circuit, but sometimes used in analysis because of its mathematical simplicity, is the Gaussian bandpass spectrum

$$S(\omega) = \frac{\sqrt{\pi}}{W}\{e^{-(\omega-\omega_0)^2/W^2} + e^{-(\omega+\omega_0)^2/W^2}\} \tag{2.59}$$

where $2W$ is the frequency spread between $1/e$ points in the regions around ω_0 and $-\omega_0$.

The NACF corresponding to this normalized spectrum, as calculated from (2.48), is

$$\rho(\tau) = \cos \omega_0\tau \, e^{-W^2\tau^2} \tag{2.60}$$

Thus $\rho(\tau)$ is a cosine function of the same frequency as that of the center of the passband, whose envelope has a Gaussian shape and which decays more rapidly with τ as the bandwidth increases. The Gaussian low pass spectrum

$$S(\omega) = \frac{2\sqrt{\pi}}{W} e^{-\omega^2/W^2} \tag{2.61}$$

with its corresponding NACF

$$\rho(\tau) = e^{-W^2\tau^2} \tag{2.62}$$

is a special case of the spectrum (2.60) for $\omega_0 = 0$. With either of these

† The frequency response function of series or parallel *RLC* filters, whose absolute squares are given by (1.26) and (1.28), respectively, has the approximate form (2.57) under the narrowband conditions $|\omega - \omega_0| \ll \omega_0$, and $|\omega + \omega_0| \ll \omega_0$, which are assumed to hold in this discussion. White noise passed through such a filter, by virtue of (2.52), will have the same output spectral shape as the filter itself.

spectra, the correlation time can be defined as the time for which $\rho(\tau)$ decays to $1/e$ of its peak value, i.e.,

$$\tau_c = \frac{1}{W} \tag{2.63}$$

A point which sometimes confuses beginners is the relationship between

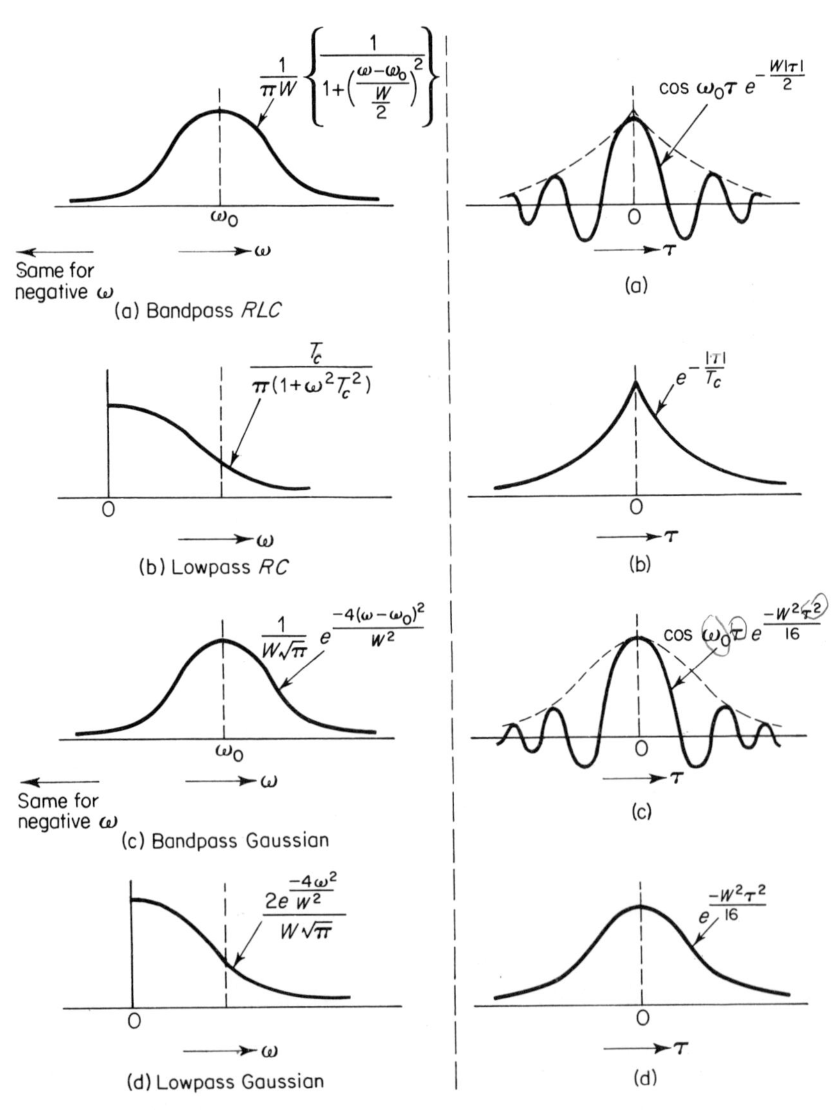

Figure 2-3 Some common noise spectra and their corresponding autocorrelation functions.

a noise with a Gaussian spectral shape and a noise with a first order Gaussian probability density function. The truth is that there is no such relationship. The Gaussian spectral shape is merely a convenient approximation to certain realistic spectra and has nothing to do with Gaussian statistics. Speaking more generally, there is no information in a noise spectrum which tells us the first order noise probability density. Moreover, knowledge of the first order probability function gives no information on the form of the spectrum. The spectrum and the first order probability density function are two different entities, and the fact that a spectrum is Gaussian, or white, or of any other specified shape, tells nothing about its first probability density function.

Figure 2-3 shows a few of the common noise spectra and their corresponding autocorrelation functions.

2.4.4.5 The Crosscorrelation Function

The crosscorrelation function (CCF) of two waveforms $x(t)$ and $y(t + \tau)$ is defined as the time average of the product of $x(t)$ and $y(t + \tau)$. It will be denoted here by $R_{xy}(\tau)$. Like the autocorrelation function, it is defined for both random and deterministic time functions. In symbols, for ergodic random processes

$$R_{xy}(\tau) = \lim_{T \to \infty} \frac{1}{2T} \int_{-T}^{T} dt'\, x(t')y(t' + \tau) \tag{2.64}$$

The crosscorrelation function is a measure of the degree to which two functions tend toward similarity in shape. In general, it will be high if the functions have some key feature in common (e.g., frequency composition) and if the value of τ is chosen to bring the two waveforms into approximate time coincidence. It will be low even for very similar functions $x(t)$ and $y(t)$ if the value of τ does not bring about coincidence. For functions which have no important similarities in their key properties, it will be low for all values of τ.

Two waveforms $x(t)$ and $y(t + \tau)$ are said to be highly correlated at certain values of τ if $R_{xy}(\tau)$ is high for those values of τ, and are said to be uncorrelated if $R_{xy}(\tau)$ is negligibly small for all values of τ.

To get a feel for the magnitudes of $R_{xy}(\tau)$ associated with correlated and uncorrelated waveforms, one can define a *normalized crosscorrelation function* (NCCF) of two sample functions of stationary random processes as

$$\rho_{xy}(\tau) = \frac{R_{xy}(\tau)}{\sqrt{R_{xx}(0)R_{yy}(0)}} \tag{2.65}$$

If $x(t)$ and $y(t)$ are related linearly, i.e.,

$$y(t) = kx(t + \tau_0) \tag{2.66}$$

where k is a constant, then

$$R_{xy}(\tau) = kR_{xx}(\tau + \tau_0) \tag{2.67}$$

$$R_{yy}(t) = k^2 R_{xx}(\tau) \tag{2.68}$$

and

$$\rho_{xy}(\tau - \tau_0) = \rho_{xx}(\tau) \tag{2.69}$$

Values of the magnitude of the NCCF near unity, e.g., 0.8 or 0.9, can be taken to indicate a high degree of correlation between $x(t)$ and $y(t)$, while values below about 0.3 can be taken to indicate a very poor correlation. To study the correlation of x and y, compute $\rho_{xy}(\tau)$ for a range of values of τ and try to observe a significant variation with τ. Poorly correlated functions will have a uniformly low value of $|\rho_{xy}(\tau)|$ over all τ, while highly correlated functions should show a substantial sensitivity to the value of τ, until a value τ_0 is reached where the NCCF attains its largest magnitude. A magnitude above about 0.8 would be regarded by most people as high correlation; a value below about 0.4 would usually be considered as low correlation.

REFERENCES

Statistical Communication Theory and Noise

Bendat, 1958

Bennett, 1956

Bennett, 1960

Davenport and Root, 1958

Fano, 1961

Freeman, 1958

Hancock, 1963

Harman, 1963

Kotelnikov, 1960

Laning and Battin, 1956

Lawson and Uhlenbeck, 1950

Lee, 1960

Middleton, 1960

Rice, 1945, in Wax, 1954

Schwartz, Bennett, and Stein, 1966; Part I (by M. Schwartz), especially pp. 3–85

Wainstein and Zubakov, 1962

Wozencraft and Jacobs, 1965; Chapters 2, 3

Yaglom, 1962

Probability Theory and its Use in Communication Theory

Bendat, 1958; Chapter 3

Cramer, 1951

Cramer, 1955

Davenport and Root, 1958; Chapter 3

Feller, 1957

Freeman, 1958; Chapters 2, 3

Fry, 1928

Lee, 1960; Chapter 4

Laning and Battin, 1956; Chapter 2

Mood, 1950

Neyman, 1950

Parzen, 1960

Woodward, 1953; Chapter 1

Wozencraft and Jacobs, 1956; Chapter 2

Statistical Analysis of Noise Waveforms, Averages, Autocorrelation, Power Spectra, Ergodicity

Bennett, 1956

Blackman and Tukey, 1958

Davenport and Root, 1958; Chapters 3, 4, 5, 6

Lee, 1960; Chapters 2, 5, 7

Middleton, 1960; Chapters 1, 2, 3, 4

See also the reference list above on statistical communication theory and noise. Material on averages, spectra, etc., is contained in most of the references cited.

3

INTRODUCTORY NOISE AND STATISTICAL COMMUNICATION THEORY-II

3.1 THE GAUSSIAN NOISE

A most frequently occurring type of noise is that whose probability density functions are *Gaussian*. For example, such statistics characterize shot and thermal noise in radio receivers prior to rectification.†

It is fortunate that Gaussian noise occurs frequently because it is easy to analyze and is amenable to the development of general theory. It is unfortunate, however, that in some noise theory work, Gaussian noise is assumed because it is easy to analyze and not because it is typical of the physical noise under investigation. This is understandable after one encounters the difficulties in attempting calculations with noise distributions departing radically from Gaussian. A result obtained with the assumption of Gaussian noise, even if not completely correct, is sometimes better than no result at all, particularly if one is looking for a general functional trend rather than precise numerical agreement with experiment. In many (but by no means all) cases, even though the true noise is not strictly Gaussian, the results of an analysis with assumed Gaussian noise will show the same or similar rough functional dependence on key system parameters.‡

For a noise waveform to be designated as Gaussian, the joint probability density function of an arbitrary number of its sample values should be

---

† See Section 3.4.

‡ This notion is particularly *inapplicable* to impulse noise, where analysis based on the assumption of Gaussian noise will surely provide misleading results.

multivariate Gaussian. For example, the PDF of a value of the waveform at a single instant of time t is the familiar bell-shaped or *normal curve* to be defined in (3.1); the joint PDF of sample values at t_1 and t_2 is the *bivariate Gaussian* function defined in (3.4); and the joint PDF of N sample values at $(t_1, t_2, \ldots, t_N)$, that is, the multivariate Gaussian PDF, is a generalization of (3.4) involving an expression of the form $[x_1^2 + x_2^2 + \cdots + x_N^2 - 2(\rho_{12}x_1x_2 + \rho_{13}x_1x_3 + \cdots + \rho_{N-1,N}x_{N-1}x_N)]$ in the exponential. The latter in its general form will be found in several references listed at the back of this chapter.

3.1.1 The First Order Gaussian Function

Given a noise waveform $x(t)$ that is a sample function of a Gaussian random process, the first order probability density function for $x_1 \equiv x(t_1)$ is given by

$$p(x) = \frac{1}{\sqrt{2\pi\sigma_n^2}} e^{-(x-m)^2/2\sigma_n^2} \tag{3.1}$$

where

$$m = \langle x \rangle$$
$$\sigma_n^2 = \langle (x - m)^2 \rangle$$

The function $p(x)$ is sketched in Figure 3-1. It is observed that it has certain mathematically convenient properties. First, it is a completely smooth function, i.e., it has no sharp discontinuities. Second, it is symmetrical about the mean value m. Third, its width, or spread, can be easily defined as the size of the interval between the points where the function decays to $1/\sqrt{e}$ of its peak value. Twice the value of σ is the measure of spread corresponding to this definition.

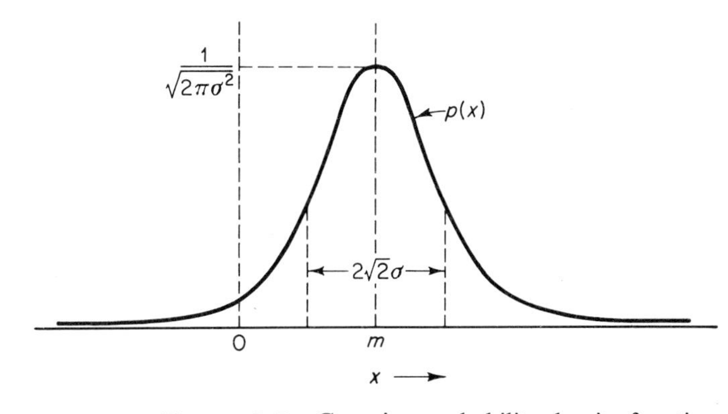

Figure 3-1 Gaussian probability density function.

It is often useful to deal with the probability distribution function

$$P(x) = \frac{1}{\sqrt{2\pi\sigma_n^2}} \int_{-\infty}^{x} \exp\left[-\frac{(x' - m)^2}{2\sigma_n^2}\right] dx' = \frac{1}{2}\left\{1 + \text{erf}\left[\frac{x - m}{\sqrt{2\sigma_n^2}}\right]\right\} \tag{3.2}$$

where the function

$$\text{erf}\,(Y) = \frac{2}{\sqrt{\pi}} \int_0^Y e^{-y^2}\, dy \tag{3.3}$$

known as the *error function* or *probability integral* is extensively tabulated.† The probability distribution function, which was defined in Section 2.3‡ as the probability that the noise is below a given level X, is plotted in Figure 3-2 for the Gaussian case.

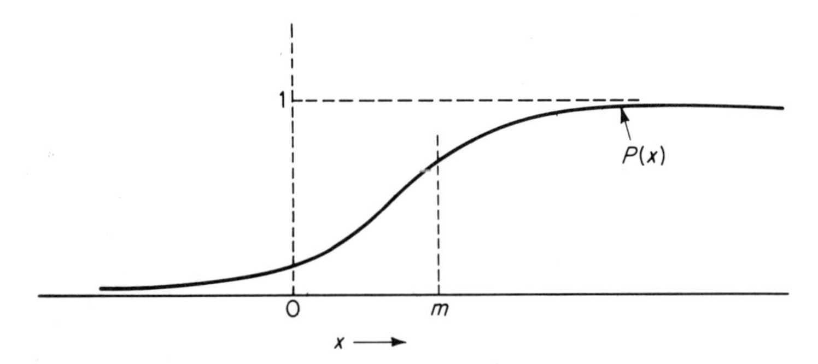

Figure 3-2 Gaussian distribution function.

3.1.2 The Bivariate Gaussian Function

The probability that a noise waveform $x(t)$ with Gaussian statistics and zero mean§ has a value between x_1 and $x_1 + dx_1$ at time t and a value between x_2 and $x_2 + dx_2$ at time $t + \tau$ is given by

$$p_2(x_1, x_2; \tau)\, dx_1\, dx_2 = \frac{dx_1 dx_2}{2\pi\sigma_n^2[1 - \rho^2(\tau)]^{1/2}} e^{-\left[\frac{x_1^2 + x_2^2 - 2\rho(\tau)x_1x_2}{2\sigma_n^2[1-\rho^2(\tau)]}\right]} \tag{3.4}$$

where σ_n^2 is the mean square and $\rho(\tau)$ is the NACF of $x(t)$.

A waveform which can be designated as a Gaussian noise has this second order probability function, known as the *bivariate Gaussian function.*

Without carrying out involved analysis on (3.4), certain of its important properties can be discussed. First note that, if $\rho(\tau) = 0$, the density function becomes a product of the form

† See Pierce, 1929, pp. 116–120 or Abramowitz and Stegun, 1965, pp. 310–311.

‡ See Eq. (2.8).

§ No loss in generality is incurred by assuming a zero mean since the noise current or voltage can always be referred to its mean value.

$$p_2(x_1, x_2; \tau) = \frac{1}{\sqrt{2\pi\sigma_n^2}} e^{\left[-\frac{x_1^2}{2\sigma_n^2}\right]} \frac{1}{\sqrt{2\pi\sigma_n^2}} e^{\left[-\frac{x_2^2}{2\sigma_n^2}\right]} \tag{3.5}$$

that is, the vanishing of the autocorrelation function guarantees that the joint probability of a value between x_1 and $x_1 + dx_1$ at time t and a value between x_2 and $x_2 + dx_2$ at time $t + \tau$ is a product of the first order probabilities for $x(t)$ and $x(t + \tau)$. Stated in another way, if the noise is Gaussian and its value at a given time t is uncorrelated with its value at a time τ later (i.e., $\rho(\tau) = 0$), then the noise at time t and that at time $t + \tau$ are statistically *independent*. This may seem like a trivial or obvious statement, but it definitely is not. In fact, it is not generally true for non-Gaussian noise.

However, for Gaussian and other types of stationary noise, statistical independence between $x(t)$ and $x(t + \tau)$ always implies that $x(t)$ and $x(t + \tau)$ are uncorrelated. To prove this, note that $\rho(\tau)$ for a stationary noise can be obtained by ensemble averaging (see Section 2.4.2 and 2.4.3), that is,

$$\rho(\tau) = \frac{1}{\sigma^2} \int_{-\infty}^{\infty} dx_1\, dx_2\, p_2(x_1, x_2; \tau) x_1 x_2 \tag{3.6}$$

By definition of statistical independence,

$$p_2(x_1, x_2; \tau) = p(x_1) \cdot p(x_2) \tag{3.7}$$

Substituting (3.7) into (3.6),

$$\rho(\tau) = \frac{\langle x_1 \rangle \langle x_2 \rangle}{\sigma_n^2} \tag{3.8}$$

which vanishes by the assumption of zero mean.

To prove the converse, i.e., that lack of correlation implies statistical independence, it would be necessary to show from (3.6) that the condition $\rho(\tau) = 0$ implies (3.7). This clearly cannot be shown, since as stated above it is not necessarily true. For a Gaussian function, however, it *is* true, and therein lies one of the special properties of a Gaussian noise which simplifies its analysis.

The other extreme case, that where $\rho(\tau)$† is very close to unity, i.e., where

† We express $\rho(\tau)$ in the form $\rho(\tau) = 1 - \epsilon$, where $0 < \epsilon \ll 1$. Then, neglecting terms $0(\epsilon^2)$ in (3.4), we have

$$p_2(x_1, x_2; \tau) = \frac{1}{2\pi\sigma_n^2 [2\epsilon - \epsilon^2]^{1/2}} e^{-(x_1 - x_2)^2/2\sigma_n^2(2\epsilon - \epsilon^2)} \cdot e^{-2\epsilon x_1 x_2/2\sigma_n^2(2\epsilon - \epsilon^2)}$$

$$\simeq \frac{1}{\sqrt{2\pi(2\sigma_n^2 \epsilon)}} e^{-(x_1 - x_2)^2/2(2\sigma_n^2\epsilon)} \frac{1}{\sqrt{2\pi\sigma_n^2}} e^{-x_1 x_2/2\sigma_n^2}$$

which approaches the right-hand side of (3.9) as ϵ approaches zero. The last step can be justified by the reader if he satisfies himself that the bracketed function has the properties indicated in Eqs. (1.6.a) through (1.6.d).

$x(t)$ and $x(t + \tau)$ are completely correlated, can be shown to lead to the expression

$$p_2(x_1, x_2; \tau) = \frac{1}{\sqrt{2\pi\sigma_n^2}} e^{-x_1^2/2\sigma_n^2}\, \delta(x_2 - x_1) \tag{3.9}$$

where $\delta(x_2 - x_1)$ is the unit impulse function. This result implies that the condition $\rho(\tau) = 1$ is equivalent to the condition that x_1 and x_2 have precisely the same time variations. If the normalized autocorrelation function of a Gaussian noise is nearly "flat on top," i.e., retains a value close to unity for a range of values of τ other than $\tau = 0$, then for this range of τ, $x(t)$ and $x(t + \tau)$ will follow each other's time variations closely.

3.1.3 Why Gaussian Noise

It seems at first glance to be a fortunate circumstance for analysts that so many common types of noise, such as shot and thermal noise in radio receivers, are Gaussian. The reason for this lies in the *central limit theorem* of probability theory.† It can be shown with the aid of this theorem that, under a wide range of conditions, the statistics of a noise waveform which is a linear superposition of a large number of statistically independent noise waveforms approaches Gaussian as the number becomes infinite. This result is essentially independent of the statistics of the individual noise waveforms. As a practical matter, the requirement that the number be extremely large (to simulate the theoretical approach to infinity) is often satisfied by numbers as small as 10. For this reason, many noise waveforms originating in nature and in man-made devices are approximately Gaussian.

Shot noise arises because of the statistical nature of electron emission in an electronic device. Each electron emitted generates an anode current pulse. The total anode current is a linear superposition of a large number of these pulses, each of which can be considered statistically independent of those due to other electrons. The anode current is, therefore, approximately Gaussian.‡

A thermal noise current is also a linear superposition of the currents arising from fluctuating motions of a large number of electrons in a material. Since the motions of the individual electrons can be considered statistically independent, thermal noise is also approximately Gaussian.‡

For these same reasons, noise in radar receivers arising from scattering by assemblies of independent randomly moving particles (e.g., rain clutter or chaff) for practical purposes can also be considered Gaussian.§

---

† See Lawson and Uhlenbeck, 1950, Sections 3.5, 3.6, pp. 46–53, or Middleton, 1960, Section 7.7.3, pp. 362–367.

‡ See Section 3.4.

§ See Lawson and Uhlenbeck, 1950, Chapter 6, especially pp. 124–132 and Kerr, 1951, Sections 6.19, 6.21, pp. 553–587, Chapter 7, pp. 588–640.

Another important consequence of the central limit theorem is that a train of pulses passed through a linear filter with a very long time constant may be approximately Gaussian at the filter output.† This is because the output in this case is a linear superposition of a large number of past inputs which will be approximately statistically independent if the spacing between pulses is sufficiently wide. If the time constant is long enough (equivalently, if the bandwidth is sufficiently small), the output noise is roughly Gaussian.

It is worth noting also that any linear superposition of statistically independent *Gaussian* noises is Gaussian, regardless of whether the number is large or small. This is a consequence of straightforward analysis and does not require the central limit theorem.

To show this for the sum of two noises, let x and y be two independent Gaussian random variables. The sum $z = x + y$ then has probability density‡

$$
\begin{aligned}
p(z) &= \int_{-\infty}^{\infty} p(x)p(z-x)\,dx = \frac{1}{\sqrt{2\pi\sigma_x^2}}\frac{1}{\sqrt{\pi\sigma_y^2}}\int_{-\infty}^{\infty} e^{\left[-\frac{x^2}{2\sigma_x{}^2}\right]} e^{\left[-\frac{(z-x)^2}{2\sigma_y{}^2}\right]}\,dx \\
&= e\frac{e^{-\frac{z^2}{2(\sigma_x{}^2+\sigma_y{}^2)}}}{\sqrt{2\pi(\sigma_x^2+\sigma_y^2)}}
\end{aligned}
\tag{3.10}
$$

Thus, the sum z is Gaussian and moreover the mean square of z is the sum of the mean squares of x and y, which is a natural consequence of statistical independence.

3.1.4 Passage of Gaussian Noise through Linear Filters

An important property of a stationary Gaussian noise is that it remains Gaussian after passage through *linear* filtering operations. This includes any number of stages of differentiation or integration. Thus, time derivatives and integrals of a stationary Gaussian noise are also stationary Gaussian, as is the output of a linear bandpass, low pass, or high pass filtering operation on a Gaussian noise.

To demonstrate this we can refer to the classical papers on noise theory by S. O. Rice,§ in which a noise waveform $n_i(t)$ that is a sample function of a Gaussian random process and has zero mean is represented within an interval from $t = -T/2$ to $t = T/2$ by a Fourier series of the form

$$
n_i(t) = \sum_{k=1}^{\infty}\left[a_k^{(i)}\cos\left(\frac{2\pi kt}{T}\right) + b_k^{(i)}\sin\left(\frac{2\pi kt}{T}\right)\right], \qquad -\frac{T}{2} \leq t \leq \frac{T}{2} \tag{3.11}
$$

---

† See Bello, 1961.

‡ The integral is equivalent to the statement, "The probability density at z is the probability that x and y sum to the value between z and $z + dz$ (given by the product of the probability density for x and the probability density for $z - x$) integrated over all possible values of x."

§ Rice, 1945, Section 1.7, pp. 25–29 (pp. 157–161 of Wax, 1954).

where a_k and b_k are Gaussian random variables which are "approximately" statistically independent and approach independence more closely as T becomes extremely large,† and whose PDF's are

$$p(a_k^{(i)}) = \frac{1}{\sqrt{2\pi[\sigma_k^{(i)}]^2}} e^{-[a_k^{(i)}]^2/2[\sigma_k^{(i)}]^2} \tag{3.12.a}$$

$$p(b_k^{(i)}) = \frac{1}{\sqrt{2\pi[\sigma_k^{(i)}]^2}} e^{-[b_k^{(i)}]^2/2[\sigma_k^{(i)}]^2} \tag{3.12.b}$$

where

$$\langle a_k^{(i)} \rangle = 0, \qquad \text{all } k$$
$$[\sigma_k^{(i)}]^2 \equiv \langle [a_k^{(i)}]^2 \rangle = \langle [b_k^{(i)}]^2 \rangle, \qquad \text{all } k$$
$$\langle a_j^{(i)} a_k^{(i)} \rangle = \langle b_j^{(i)} b_k^{(i)} \rangle = 0, \qquad j \neq k$$
$$\langle a_j^{(i)} b_k^{(i)} \rangle = 0, \qquad \text{all } j, k$$

The period T is allowed to become infinite at the termination of the analysis.

Using (1.5) and the representation (3.11) for the noise at the input to a linear filter with impulse response $f(t)$, the output waveform is

$$n_o(t) = \sum_{k=1}^{\infty} \left[a_k^{(o)} \cos\left(\frac{2\pi kt}{T}\right) + b_k^{(o)} \sin\left(\frac{2\pi kt}{T}\right) \right] \tag{3.13}$$

where

$$a_k^{(o)} = a_k^{(i)} v_{ck} - b_k^{(i)} v_{sk}$$
$$b_k^{(o)} = a_k^{(i)} v_{sk} + b_k^{(i)} v_{ck}$$
$$v_{ck} \equiv \int_{-\infty}^{\infty} dt' f(t') \cos\left(\frac{2\pi kt'}{T}\right)$$
$$v_{sk} \equiv \int_{-\infty}^{\infty} dt' f(t') \sin\left(\frac{2\pi kt'}{T}\right)$$

The output Fourier coefficients $a_k^{(o)}$ and $b_k^{(o)}$ are linear combinations of the input coefficients $a_k^{(i)}$ and $b_k^{(i)}$. The latter are Gaussian random variables and are statistically independent. Hence from (3.10) we conclude that the output coefficients are also Gaussian random variables. One of the properties of the output coefficients, obtained from those of the input coefficients as given below Eqs. (3.12.a) and (3.12.b) and the relationship between input and output coefficients as given below Eq. (3.13), is that of zero mean, i.e.

$$\langle a_k^{(o)} \rangle = \langle b_k^{(o)} \rangle = 0 \tag{3.14}$$

Other properties of the output coefficients, based on those of the input

---

† See Papoulis, 1965, pp. 454–457. It is shown that exact statistical independence holds only if $n_i(t)$ is periodic. Since the $n_i(t)$ we are discussing here is not periodic, the independence of the coefficients can only be regarded as a convenient approximation which becomes nearly exact for sufficiently large T.

coefficients, the input-output relationships, and the zero-mean property (3.14) are:

$$\langle a_j^{(o)} a_k^{(o)} \rangle = \langle b_j^{(o)} b_k^{(o)} \rangle = [\sigma_k^{(i)}]^2 (v_{ck}^2 + v_{sk}^2) \delta_{jk} \tag{3.15.a}$$

$$\langle a_j^{(o)} b_k^{(o)} \rangle = 0 \tag{3.15.b}$$

From Eqs. (3.14) through (3.15.b), we see that each of the output Fourier coefficients has properties exactly equivalent to those of its corresponding input coefficient with the exception indicated by (3.15.a).

Thus, except for the factor $(v_{ck}^2 + v_{sk}^2)$ on the mean-square value of each Fourier coefficient, the Fourier series representing the output has the same statistical properties as that representing the input. The output is Gaussian but differs from the input in its mean-square value and its power spectrum.

We could carry this analysis further by calculating the autocorrelation functions of $n_i(t)$ and $n_o(t)$, allowing T to become infinite, and thereby demonstrating that the quantities $[\sigma_k^{(i)}]^2$ and $\{[\sigma_k^{(i)}]^2(v_{ck}^2 + v_{sk}^2)\}$ approach spectral densities of input and output waveforms, respectively. It is noted that this has already been done in Section 2.4.4.1 for a general waveform $x(t)$ using an exponential Fourier series. The argument based on the real Fourier series representation (3.11) proceeds along precisely the same lines and its presentation here would be redundant.

3.2 SIGNAL-TO-NOISE RATIO–THE MATCHED FILTER

The most familiar criterion of merit of an information system is *signal-to-noise ratio*, to be denoted throughout this book as ρ or as SNR. Signal-to-noise ratio is not necessarily an adequate measure of system performance, in the sense that an x dB signal-to-noise ratio does not necessarily correspond numerically to a given value of some measure of performance that is meaningful to operators of equipment (such as the average number of errors per second in a digital communication system, or the probability of target detection in a radar). Also, the definition of SNR for nonlinear systems is ambiguous. In the outputs of such systems, there are cross terms between input signal and input noise. The question of whether these terms belong to the signal or noise part of the output depends on how "signal" and "noise" are defined. However, in certain cases there is a one-to-one correspondence between SNR and certain probabilistic measures of performance. For example, in a digital radio communication system, where additive Gaussian noise is the interfering agent, error probability is a monotonically decreasing function of the signal-to-noise ratio at the receiver input.

There are other reasons why SNR is often a useful measure of performance, even if it has no simple relation to operational performance criteria.

A very practical reason is that it is usually the simplest measure of performance to evaluate analytically, and in many cases is the only such measure that is analytically tractable. It is usually true that an increase in SNR improves a system and consequently it is nearly always worthwhile to know that the system A provides an x dB higher signal-to-noise ratio than system B. Conservatively, it can be said that this is usually better than no knowledge at all about comparative system performance.

3.2.1 Signal-to-Noise Ratio Optimization

In many classes of systems it may be desired to process a signal with a linear filter in such a manner that the SNR at the output is maximized for a given input SNR. The method of determining the characteristics of such an *optimum linear filter* will be presented below.†

The input to a linear filter with impulse response and frequency response function $f(t)$ and $F(\omega)$, respectively, is a superposition of a deterministic signal $s_i(t)$ and a stationary random noise $n_i(t)$. The signal and noise terms at the output of the filter are (see Eq. (1.5))

$$s_o(t) = \int_{-\infty}^{\infty} dt' f(t - t') s_i(t') \tag{3.16}$$

$$n_o(t) = \int_{-\infty}^{\infty} dt' f(t - t') n_i(t') \tag{3.17}$$

The ensemble average of the output power is proportional to

$$P_o = \langle [s_o(t) + n_o(t)]^2 \rangle = [s_o(t)]^2 + \langle [n_o(t)]^2 \rangle + 2\langle s_o(t) n_o(t) \rangle \tag{3.18}$$

The noise and signal will be assumed uncorrelated, so that the cross term $\langle s_o(t) n_o(t) \rangle$ takes the form $s(t)\langle n_o(t) \rangle$. It will also be assumed that the mean value of noise is zero. Therefore,

$$\langle s_o(t) n_o(t) \rangle = 0 \tag{3.19}$$

and consequently

$$P_o = \int_{-\infty}^{\infty}\int_{-\infty}^{\infty} dt'\, dt''\, f(t - t') f(t - t'') s_i(t')\, s_i(t'') + \int_{-\infty}^{\infty}\int_{-\infty}^{\infty} dt'\, dt''\, f(t - t') n_i(t') n_i(t'') f(t - t'') \tag{3.20}$$

since the ensemble average of the integral is equivalent to the integral of the ensemble average.

---

† More detailed discussions of this topic will be found in many places. See, e.g., Davenport and Root, 1958, Section 11.7, pp. 244–247; or Schwartz, Bennett, and Stein, 1966, Section 2.4, pp. 63–68; or Middleton, 1960, Section 16.3, pp. 714–721.

The output signal-to-noise ratio ρ_o at time t_1 is (from (3.18) and (2.33))†

$$\rho_o = \rho_o(t_1) = \frac{\left|\int_{-\infty}^{\infty} dt' f(t_1 - t')s_1(t')\right|^2}{\int_{-\infty}^{\infty} dt' \int_{-\infty}^{\infty} d\tau f(t' - \tau)R_n(\tau)f(t')} \tag{3.21}$$

where $R_n(\tau)$ is the ACF of the input noise.

Expressing $s_i(t)$ in terms of its Fourier transform $S_i(\omega)$ and using the Wiener-Khintchine theorem (2.38), we can derive an equivalent form of (3.21) as follows:‡

$$\rho_o = \rho_o(t_1) = \frac{1}{2\pi} \frac{\left|\int_{-\infty}^{\infty} d\omega' \, e^{j\omega' t_1} F(\omega')S_i(\omega')\right|^2}{\int_{-\infty}^{\infty} d\omega' \, |F(\omega')|^2 \, G_n(\omega')} \tag{3.22}$$

There are methods available for finding the impulse response of the optimum linear filter through (3.21) or the frequency response function through (3.22).§ Since either function will provide all necessary information about the filter, we will use the frequency approach, which is quite simple to handle mathematically.

One particularly simple method of finding the frequency response function of the optimum filter (to be denoted by $F^{(o)}(\omega)$) is to invoke the Schwartz inequality||

$$\int_{-\infty}^{\infty} d\omega' \, |g(\omega')|^2 \int_{-\infty}^{\infty} d\omega'' \, |h(\omega'')|^2 \geq \left|\int_{-\infty}^{\infty} d\omega' g(\omega')h(\omega')\right|^2 \tag{3.23}$$

where

$$g(\omega) \equiv F(\omega)\sqrt{G_n(\omega)}$$

$$h(\omega) \equiv \frac{S_i(\omega)e^{j\omega t_1}}{\sqrt{G_n(\omega)}}$$

---

† Note that the ensemble mean of the second term of (3.20) is equivalent to $\iint_{-\infty}^{\infty} dt' \, dt'' \, f(t')\langle n_i(t-t')n_i(t-t'')\rangle f(t'')$ which, since $\langle n_i(t-t')n_i(t-t'')\rangle = R_n(t'-t'')$ for stationary noise, is equivalent to the denominator of (3.21).

‡ The equivalence of (3.21) and (3.22) is based on the fact that the limits on the integrals in (3.21) have been set at $\pm\infty$. This presupposes an infinite signal observation time. If the filter $f(t)$ is causal the true upper limit of the integrals is t_1. However, we do not consider this here. In using (3.22) to develop our arguments further, we will be seeking the linear filter that maximizes SNR, *not* the linear *causal* filter that maximizes SNR.

§ In some treatments, e.g., Davenport and Root, 1958, Section 11.7, pp. 244–246, an integral equation is developed from a form equivalent to (3.21). Solution of this integral equation yields the impulse response of the optimum linear filter.

|| See Lathi, 1965, pp. 158–159 or Bendat, 1958, p. 55 for simple proofs of the Schwartz inequality.

The numerator of (3.22) is $\left|\int_{-\infty}^{\infty} d\omega'\, g(\omega')h(\omega')\right|^2$. According to (3.23), then,

$$\left|\int_{-\infty}^{\infty} d\omega'\, g(\omega')h(\omega')\right|^2 \leq \int_{-\infty}^{\infty} d\omega'\, |g(\omega')|^2 \int_{-\infty}^{\infty} d\omega''\, |h(\omega'')|^2$$
$$= \int_{-\infty}^{\infty} d\omega'\, |F(\omega)|^2\, G_n(\omega') \int_{-\infty}^{\infty} d\omega''\, \frac{|S_i(\omega'')|^2}{G_n(\omega'')} \tag{3.24}$$

from which it follows, with the aid of (3.22), that the maximum value of ρ_o is

$$(\rho_o)_{\max} = \frac{1}{2\pi} \int_{-\infty}^{\infty} d\omega'\, \frac{|S_i(\omega')|^2}{G_n(\omega')} \tag{3.25}$$

According to (3.25),

$$F(\omega) = F^{(0)}(\omega) = \frac{KS_i^*(\omega)\, e^{-j\omega t_1}}{G_n(\omega)} \tag{3.26}$$

where K is a constant.

The physical interpretation of (3.26) is rather simple. The *optimum* or *matched filter* (as it is usually called) has a frequency response function shaped like the complex conjugate of the Fourier transform of the signal if the noise is white. If the noise is colored, the frequencies for which noise is most prominent are depressed in the filter output by weighting the filter's frequency response with the reciprocal of the noise components. Thus the filter preferentially passes the frequencies with the highest signal-to-noise ratios and depresses those with the lowest. It is "matched" to the signal; hence the designation "matched filter."

The signal-to-noise ratio at the output of a matched filter is calculated by substituting the right-hand side of (3.26) into (3.22), resulting in

$$\rho_o^{(0)} = \frac{1}{2\pi} \int_{-\infty}^{\infty} d\omega\, \frac{|S_i(\omega)|^2}{G_n(\omega)} \tag{3.27}$$

Let us assume that the noise is bandlimited white with mean power σ_n^2, central frequency f_0, and a bandpass bandwidth (in c/s) of $2B_n$, that is,

$$G_n(2\pi f) = \frac{\sigma_n^2}{2B_n}, \qquad f_0 - B_n \leq f \leq f_0 + B_n$$
$$= 0, \qquad f < f_0 - B_n,\ \ f > f_0 + B_n \tag{3.28}$$

If the signal's bandpass bandwidth $2B_s$ is smaller than the bandpass noise bandwidth $2B_n$, then from (3.27) and (3.28), we obtain

$$\rho_o^{(0)} = \frac{E_s}{\left[\dfrac{\sigma_n^2}{2B_n}\right]} \tag{3.29}$$

where

$$E_s = \frac{1}{2\pi} \int_{-\infty}^{\infty} d\omega\, |S_i(\omega)|^2 = \int_{-\infty}^{\infty} dt'\, s_i^2(t') = \text{total signal energy}$$

or alternatively

$$\frac{\rho_o^{(0)}}{\rho_i} \simeq 2T_s B_n \tag{3.30}$$

where ρ_i is the input signal-to-noise ratio, equal to $(\overline{[s_i(t)]^2}/\sigma_n^2)$, and where $\overline{[s_i(t)]^2} \equiv 1/T_s \int_{-\infty}^{\infty} dt' \, s_i^2(t')$, T_s being the signal duration, which is assumed to be finite for practical purposes.†

The conclusion to be drawn from (3.29) is that the signal-to-noise ratio at the output of a matched filter (where the input noise is bandlimited white) depends on the total signal energy, and is independent of the way in which this energy is distributed in the signal waveform. The form (3.30) shows that matched filtering can increase the signal-to-noise ratio by an amount no greater than the product of noise bandwidth and signal duration.

Up to this point we have discussed the matched filter only in the frequency domain. It is important to consider its time domain behavior as well, i.e., its impulse response $f^{(0)}(t)$. Inversion of $F^{(0)}(\omega)$ gives the impulse response, which can become quite involved in the case of colored noise, the complexity arising because of the requirement of causality.

We will not concern ourselves with colored noise at the moment, but will consider only white noise, which, in practice, is noise that is flat over a band very large compared to the signal band.

For noise of flat spectral density G_n the matched filter impulse response, easily obtained from (3.26), is (noting that $s_i(t)$ is real)

$$f^{(0)}(t) = \frac{K}{G_n} s_i^*(t_1 - t) = \frac{K}{G_n} s_i(t_1 - t) \tag{3.31}$$

Since the absolute magnitude of $f^{(0)}(t)$ is not relevant to the optimization, we can conveniently set K/G_n equal to unity. The output of the filter for an input $v_i(t)$ is

$$v_o(t) = \int dt' \, s_i(t' - [t - t_1]) v_i(t') \tag{3.32}$$

The filter represented by (3.32) is sometimes called a *crosscorrelator*. The reason for the name is that the filter computes an approximation to the crosscorrelation function (see Section 2.4.4.5) between the input waveform $v(t')$ and the conjugate reference waveform $s_i^*(t' - [t - t_1])$. Since signal and noise are generally uncorrelated, the signal-noise contribution to the crosscorrelator output will be small while the signal-signal contribution will be large.

More will be said of the crosscorrelator in later chapters. Because of its great utility in all types of signal detection and processing, it will come up

† That is, signal energy is negligibly small outside of a given finite time interval.

in discussions on specific applications. In particular, in discussing optimal detection theory in Chapter 4, it will be shown how the crosscorrelator is an optimum filter according to decision theory criteria under certain conditions (specifically for Gaussian noise).

3.3 PASSAGE OF NOISE THROUGH RECTIFIERS

When noise passes through a nonlinear device, whether or not it is accompanied by a deterministic signal, it may undergo operations that render its exact analysis very difficult. For example, signals entering a radio receiver must be rectified before they can be sensed either aurally or visually. Rectification is a nonlinear operation regardless of how it is accomplished (linear, square law, half wave, full wave, synchronous detection, etc.); therefore its analysis requires consideration of the passage of noisy signals through a restricted class of nonlinear devices. Since the signal is sensed *after* rectification, the *post rectification* statistics of the noisy signal must be determined given a knowledge of its *prerectification* statistics. It is characteristic of nonlinear devices such as rectifiers that these two types of statistics are not the same. Three examples of how noisy waveforms are affected by typical idealized rectifier responses will be given. These are the full wave linear envelope rectifier, the full wave square law rectifier, and the half wave linear envelope rectifier. The input voltage is, in all cases, assumed to be composed of a sinusoidal signal of angular frequency ω_c and amplitude a_s, narrowband Gaussian noise with zero mean and a power spectrum that is symmetrical about ω_c. This is a typical model of the prerectification waveform in a radio receiver.†

This voltage waveform can be represented by‡

$$v(t) = a_s \cos \omega_c t + x_n(t) \cos \omega_c t + y_n(t) \sin \omega_c t \tag{3.33}$$

where $x_n(t)$ and $y_n(t)$ are assumed to be independent Gaussian random functions,§, || slowly varying compared to $\cos \omega_c t$ and $\sin \omega_c t$, whose mean

† See Section 6.1.

‡ No loss of generality is incurred by assuming the signal to have no sine component, since this merely involves a choice of time reference.

§ $x_n(t)$ is in phase with the signal, hence it is called the *in phase* noise component. $y_n(t)$ is 90° out of phase with the signal, or *in quadrature* with it, hence it is called the *quadrature* component. The two components $x_n(t)$ and $y_n(t)$ together are often called the *quadrature components* of $n(t)$.

|| The assumption of independence between $x_n(t)$ and $y_n(t)$ implies that the spectrum of the noise process is symmetric about ω_c. To demonstrate this as well as other properties of $x_n(t)$ and $y_n(t)$, we can use the Fourier series representation (3.11) for the noise, which leads to

$$x_n(t) = \sum_{k=1}^{\infty} \left[a_k \cos \left(\left(\frac{2\pi k}{T} - \omega_c\right)t\right) + b_k \sin \left(\left(\frac{2\pi k}{T} - \omega_c\right)t\right)\right]$$

$$y_n(t) = \sum_{k=1}^{\infty} \left[-a_k \sin \left(\left(\frac{2\pi k}{T} - \omega_c\right)t\right) + b_k \cos \left(\left(\frac{2\pi k}{T} - \omega_c\right)t\right)\right]$$

values are zero and whose mean square values $\langle x_n^2 \rangle$ and $\langle y_n^2 \rangle$ are equal.

Consider first the zero-signal case, i.e., $a_s = 0$. The PDF's of x_n and y_n, by virtue of the assumption of Gaussian noise with zero mean (see Section 3.1.1, Eq. (3.1)), are given by

$$p(x_n) = \frac{1}{\sqrt{2\pi\sigma_n^2}} e^{\left[-\frac{x_n^2}{2\sigma_n^2}\right]} \tag{3.34}$$

$$p(y_n) = \frac{1}{\sqrt{2\pi\sigma_n^2}} e^{\left[-\frac{y_n^2}{2\sigma_n^2}\right]} \tag{3.35}$$

where $\sigma_n^2 = \langle x_n^2 \rangle = \langle y_n^2 \rangle$.

Because $x_n(t)$ and $y_n(t)$ are independent, the joint PDF $p_2(x_n, y_n)$ is

$$p_2(x_n, y_n) = \frac{1}{2\pi\sigma_n^2} e^{\left[-\frac{(x_n^2+y_n^2)}{2\sigma_n^2}\right]} \tag{3.36}$$

The probabilities that the amplitude and phase of $v_n(t)$ lie within certain specified incremental limits are obtained by first writing

$$x_n(t)\cos\omega_c t + y_n(t)\sin\omega_c t = |v_n(t)|\cos(\omega_c t + \phi_n(t)) \tag{3.37}$$

where the amplitude

$$|v_n(t)| = \sqrt{[x_n(t)]^2 + [y_n(t)]^2} \tag{3.38}$$

and the phase

$$\phi_n(t) = \tan^{-1}\left[-\frac{y_n(t)}{x_n(t)}\right] \tag{3.39}$$

are both random functions which vary slowly compared with the carrier.

If one thinks of x_n and y_n as rectangular coordinates in a plane, as shown in Figure 3-3, the amplitude and phase can be thought of as the corresponding polar coordinates r_n and ϕ_n. The probability of a point x_n, y_n falling within a narrow ring whose inner radius is r_n and whose outer radius is $r_n + dr_n$ is related to the area of that ring, $2\pi r_n\,dr_n$. The probability that the amplitude is between r_n and $r_n + dr_n$, then, is equivalent to the probability per unit area for $(x_n^2 + y_n^2)$ to lie within the ring, multiplied by the area of the ring. From (3.36) it follows that this product is $(1/2\pi\sigma_n^2)e^{-r_n^2/2\sigma_n^2}2r_n\,dr_n$, i.e.,

$$p(r_n) = \frac{r_n}{\sigma_n^2} e^{\left[-\frac{r_n^2}{2\sigma_n^2}\right]} \tag{3.40}$$

The function given in (3.40) is sketched in Figure 3-4. It is called the *Rayleigh probability density function.* It limits the amplitude to positive values,

---

From Papoulis, 1965, pp. 454–457 (see the footnote on page 74), the Fourier coefficients become approximately statistically independent as T becomes extremely large. These coefficients, in fact, take on the properties indicated on page 74 for sufficiently large T. It follows immediately that $\langle x_n^2 \rangle = \langle y_n^2 \rangle = \sigma_n^2$ and $\langle x_n(t)y_n(t) \rangle = 0$. Basic to these results is the condition $\langle a_j b_k \rangle = 0$ for $j \neq k$. This presupposes spectral symmetry about $\omega = 0$ for the noise, which in turn implies spectral symmetry about ω_c for each of the quadrature components $x_n(t)$ and $y_n(t)$.

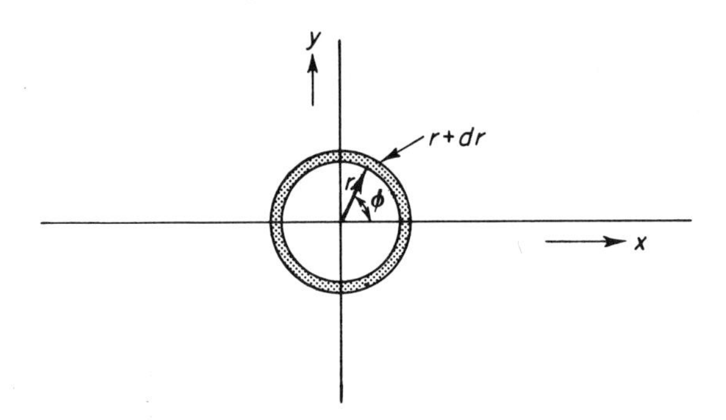

Figure 3-3 Amplitude and phase as polar coordinates.

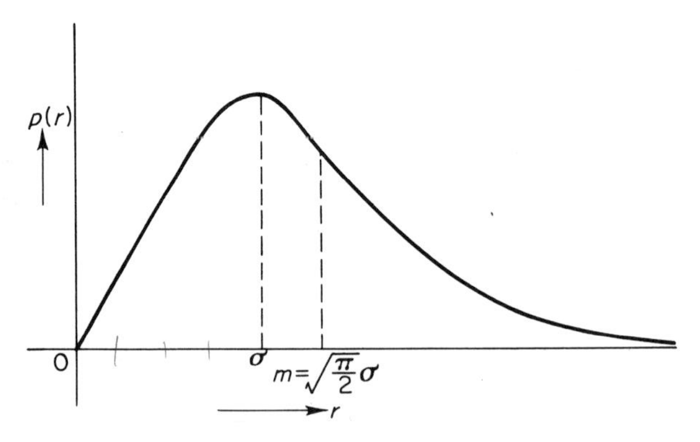

Figure 3-4 Rayleigh PDF for noise amplitude.

as would be intuitively obvious, and has a peak† at $r_n = \sigma_n$, the original rms noise, and a mean value‡ at $r_n = \sqrt{(\pi/2)}\,\sigma_n$, which is slightly greater than the peak or most probable value. The rms value§ is $\sqrt{2\sigma_n^2}$, and the standard deviation (defined in Table 2-1) is therefore equal to $\sigma_n\sqrt{2-(\pi/2)} \simeq 0.66\sigma_n$, somewhat smaller than the standard deviation of the original noise.

One might be interested in the PDF for the *short-time averaged* noise power,|| which is proportional to the square of the amplitude. If the power $\mathscr{P}_n$ is given by

$$\mathscr{P}_n = Cr_n^2 \tag{3.41}$$

† Set $dp/dr_n = 0$ to show this.

‡ Integration of $\int_0^\infty dr\,(r^2/\sigma_n^2)e(-r^2/2\sigma_n^2)$ yields $\sqrt{\pi/2}\sigma_n$. Differentiate $\int_0^\infty du\, e(-\beta u^2)$ with respect to β and set β equal to unity.

§ Integration of $\int_0^\infty dr\,(r^3/\sigma_n^2)e(-r^2/2\sigma_n^2)$ yields $2\sigma_n^2$, whose square root is the rms noise.

|| The time averaging referred to here eliminates contributions at the sine wave frequency.

where C is a constant, then the probability that the power is between $\mathscr{P}_n$ and $\mathscr{P}_n + d\mathscr{P}_n$ is equal to the probability that r_n is between $\sqrt{\mathscr{P}_n/C}$ and $\sqrt{(\mathscr{P}_n + d\mathscr{P}_n)/C}$. We can write

$$p(r_n)\,dr_n = p(\mathscr{P}_n)\,d\mathscr{P}_n \tag{3.42}$$

where $p(\mathscr{P}_n)$ is the PDF for power.

Substituting (3.40) into (3.42),

$$p(\mathscr{P}_n) = \frac{1}{2C\sigma_n^2}\,e^{\left[-\frac{\mathscr{P}_n}{2C\sigma_n^2}\right]} \tag{3.43}$$

This PDF is sketched in Figure 3-5.

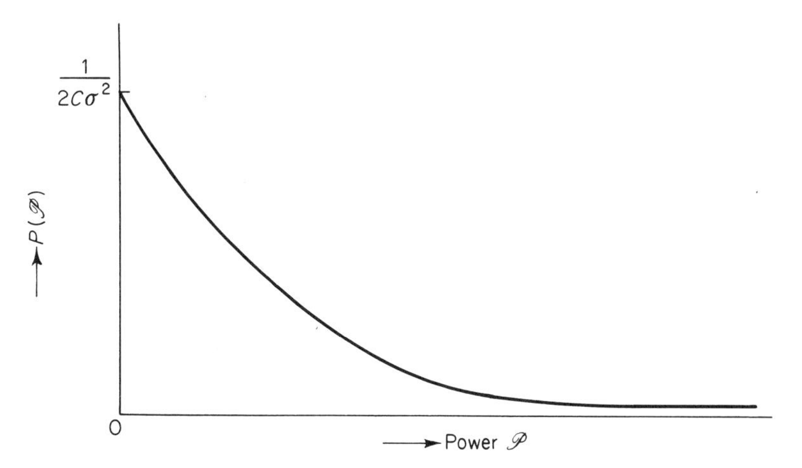

Figure 3-5 PDF for noise power with Rayleigh amplitude distribution.

Another PDF of interest is that of the amplitude when a signal is present, i.e., when $a_s \neq 0$ in (3.33). In this case, the inphase and quadrature components of the signal-plus-noise are denoted by $x(t) = x_n(t) + a_s$ and $y(t) = y_n(t)$, respectively. The counterpart of (3.36) is

$$p_2(x, y) = \frac{1}{2\pi\sigma_n^2}\,e^{\left[\frac{1}{2\sigma_n^2}[(x-a_s)^2+y^2]\right]} \tag{3.44}$$

and that of (3.40), as obtained by the same arguments, is

$$p(r) = \frac{1}{2\pi}\,e^{\left[-\frac{a_s}{2\sigma_n^2}\right]}\int_0^{2\pi} d\phi\,\frac{r}{\sigma_n^2}\,e^{\left[-\frac{1}{2\sigma_n^2}(r^2-2ra_2\cos\phi)\right]} \tag{3.45}$$

where $r = \sqrt{x^2 + y^2}$ and $\phi = \tan^{-1}(-y/x)$.

The interpretation of (3.45) is as follows. The probability of an amplitude between r and $r + dr$ is the probability of this amplitude with a given phase between ϕ and $\phi + d\phi$, integrated over all phases from 0 to 2π.

Integration of (3.45) is accomplished by recognizing that an exponential integral of the form $1/2\pi \int_0^{2\pi} e^{a \cos \phi}\, d\phi$ (where a is a constant) is equal to $I_0(a)$, the zero order modified Bessel function with argument a, which is extensively tabulated.†

The result of the integration is

$$p(r) = \frac{r}{\sigma_n^2} e^{\left[-\frac{(r^2 + a_s^2)}{2\sigma_n^2}\right]} I_0 \left[\frac{a_s^r}{\sigma_n^2}\right] \tag{3.46}$$

The variation of this function with increasing signal-to-noise ratio $(a_s/\sqrt{2}\sigma_n)$ is sketched in Figure 3-6. As would be expected, the effect of the presence of the signal is (1) to increase the mean value by adding contributions from cross-beating of signal and noise and a pure signal contribution, and (2) decrease the standard deviation which is a measure of randomness or uncertainty in signal amplitude.

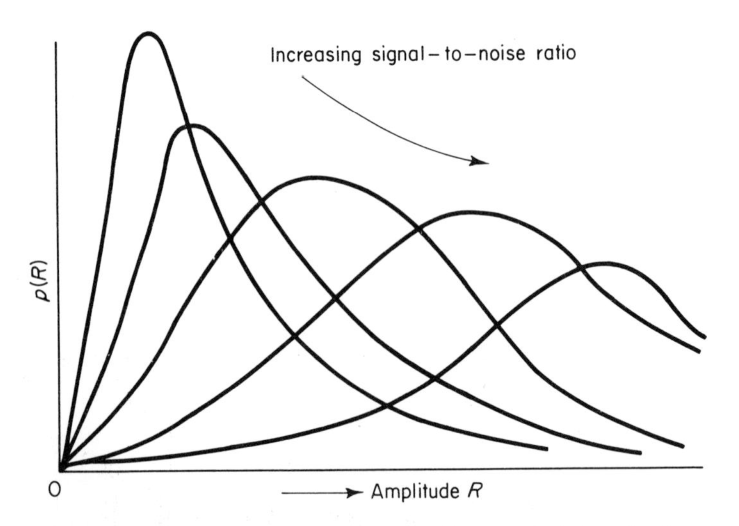

Figure 3-6 Amplitude PDF of signal-plus-noise (Rice distribution).

In the limit of infinite a_s, the amplitude would approach a_s with certainty, or a probability of unity, i.e., the curve would become a unit impulse function at $r = a_s$.

In the limit of zero signal-to-noise ratio, the modified Bessel function $I_0(0)$ attains the value of unity,‡ and the PDF given by (3.46) degenerates into that of (3.40), applicable to the "noise alone" case.

† See Abramowitz and Stegun, 1965, pp. 416–429 and No. 9.6.16, p. 376.

‡ See Abramowitz and Stegun, 1965, pp. 416–429.

3.3.1 Ideal Full Wave Linear Envelope Rectifier

The ideal full wave linear envelope rectifier takes the absolute value of the input and filters out frequencies in the neighborhood of the carrier. This results in a low frequency output of the form

$$v_o(t) = \sqrt{[x_n(t) + a_s]^2 + [y_n(t)]^2} \tag{3.47}$$

which is the pure signal-plus-noise amplitude r discussed above. Therefore the first order probability density function for the output of this type of rectifier is given by (3.46), where $v_o = r$, i.e.,

$$p(v_o) = \frac{v_o}{\sigma_n^2} e^{\left[-\frac{(v_o^2 + a_s^2)}{2\sigma_n^2}\right]} I_0\left(\left[\frac{a_s v_o}{\sigma_n^2}\right]\right) \tag{3.48}$$

or, with no signal present, (3.40) applies, i.e.,

$$p(v_o) = \frac{v_o}{\sigma_n^2} e^{\left[-\frac{v_o^2}{2\sigma_n^2}\right]} \tag{3.49}$$

Thus, since the output of the ideal full wave linear envelope rectifier is the amplitude of the input waveform, the output statistics are those of the input amplitude.

3.3.2 Ideal Full Wave Square Law Envelope Rectifier

The ideal full wave square law envelope rectifier squares the signal envelope. Therefore, its statistics, except for constant normalization factors, are the same as those of the short-time averaged power, as given for the noise alone case by (3.43). If a signal is present, the same arguments as were used to obtain (3.43) from (3.40) will produce the output PDF

$$p(v_o) = \frac{1}{2k\sigma_n^2} e^{-\left[\left(\frac{v_o + ka_s^2}{2k\sigma_n^2}\right)\right]} I_0\left(\frac{a_s}{\sigma_n^2}\sqrt{\frac{v_o}{k}}\right) \tag{3.50}$$

where v_o, the rectifier output, is

$$v_o(t) = k\{[x_n(t) + a_s]^2 + [y_n(t)]^2\} \tag{3.51}$$

k being a constant.

As in the linear envelope rectifier case, the presence of a signal decreases the standard deviation, or, equivalently, the uncertainty or randomness in the output, and adds additional terms to the mean value of the output.

3.3.3 Ideal Half Wave Linear Envelope Rectifier

The half wave linear envelope rectifier, which accepts only positive half-cycles of its input, has essentially the same effect on the first order statistics of a sinusoidal signal superposed on narrowband noise as its full wave counterpart. Since the noise components $x_n(t)$ and $y_n(t)$ and the signal amplitude a_s do not change significantly over a cycle, each negative half cycle that the rectifier rejects has nearly the same noise statistics as the previous positive half cycle that it passed. The decrease in signal power due to rejection of negative half cycles is accompanied by a corresponding decrease in noise power. Hence the absence of the negative half cycles does not seriously affect the relationship between signal and noise (except for multiplicative constants in the probability density functions) or the statistical properties of the output signal-plus-noise. The output statistics for this type of rectifier can therefore be inferred from a knowledge of those of the full wave linear envelope rectifier and the two are very nearly equivalent. Although the equivalence is not exact, it is usually close enough for rough engineering analysis purposes.

3.3.4 The Effect of Audio or Video Filtering

The mathematical calculation of the first and second order output probability functions for a system consisting of a linear filter (IF amplifier in a radio receiver), rectifier, and another linear filter (audio or video), has become a somewhat classical problem in noise theory.† The result of much of the work done on it consists of integral equations whose solution can be obtained easily only for degenerate cases.

The reason that audio or video filtering increases the difficulty of the problem so enormously is that the filter acts on all past inputs to produce its output. Therefore, it is necessary to know the complete statistics of the rectifier output, i.e., first, second, . . . , up to Nth order statistics (where N is the number of samples required to completely specify the waveform) in order to calculate even the simplest (e.g., first order) statistics of the output of the audio or video filter that follows the rectifier.

Some quantitative theoretical results exist for the square law rectifier with Gaussian-shaped IF and audio (or video) filters.‡ Otherwise, one can only make a few qualitative statements about the statistics of the output of the final filter. First, if the ratio of IF to audio (or video) bandwidth is very small, then the postrectification low pass filtering does not significantly reject any of the input noise. The output statistics for this case therefore do not differ appreciably from those of the rectifier without low pass filtering. If the ratio of IF to audio (or video) bandwidth is very high, only a small amount of the original input noise is passed. The statistics of the output tend toward Gaussian as this ratio becomes very high, because a lowering of the low pass filter's bandwidth is equivalent to an increase in its effective integration time. The output begins to simulate a sum of a large number

† See Kac and Siegert, 1947, Siegert, 1954, and Emerson, 1953.

‡ See Emerson, 1953.

of noise sources, which because of the central limit theorem (see Section 3.5) becomes nearer to Gaussian as the integration time (and therefore the effective number of noise sources) is increased.

3.4 NOISE IN ELECTRICAL AND ELECTRONIC DEVICES

Many of the problems to be discussed in later chapters will involve the kinds of noise that arise naturally in electrical and electronic devices. To deal with such noise, it is important to know something about its physics. It is particularly important that the reader understand that the statistical properties of noise usually assumed in analysis (e.g., Gaussian) have some resemblance to the actual noise encountered in devices, and that there are sound physical reasons why the noise has these properties.

3.4.1 Thermal Noise†

In any substance with high conductivity there are many electrons free to move in response to the application of an electric field. If these electrons never collided with each other or with positive ions and neutral molecules present in the material, and if the applied field were constant, they would accelerate uniformly in a direction opposite to that of the field. Because of collisions the electrons acquire a random component of motion which is independent of the uniformly accelerated linear motion due to the applied field. Between collisions, their motion is uniformly accelerated and linear.

Let the applied field **E** be constant and in the x-direction, within a rectangular volume of length (x-directed) L and cross-sectional area A containing N_e electrons. From the electrical equation of continuity (or, equivalently, the law of conservation of electric charge)‡ we can write

$$i(t) = \sum_{\ell=1}^{N_e} \frac{qv_{x\ell}(t)A}{(AL)} \tag{3.52}$$

where

$$\begin{aligned} i(t) &= \text{electron current} \\ v_{x\ell}(t) &= x\text{-component of velocity of } \ell\text{th electron} \\ q &= \text{electron charge} \end{aligned}$$

To derive the spectrum of the noise due to electron collisions,§ known

---

† Often called *Johnson noise*, after J. B. Johnson, who measured it in a classic experiment in 1928. (See Freeman, 1958, Section 4.7, pp. 120–122.)

‡ See any text on electromagnetic theory, e.g. Corson and Lorrain, 1962, p. 191.

§ This "kinetic" derivation of the thermal noise spectrum follows closely the development in Freeman, 1958, pp. 108–113. This derivation is classical and can be found in many places.

as *thermal noise*, we will calculate the ACF of $i(t)$ and take its Fourier transform. The key points in the derivation are the assumptions that (1) the electrons are mutually statistically independent and (2) the state of motion of any single electron remains completely correlated with its previous state of motion until a collision occurs, after which its motion becomes completely uncorrelated with its motion prior to the collision. In symbols,

$$\langle \overline{v_{x\ell}(t)v_{xm}(t+\tau)} \rangle = 0, \qquad \text{for all } \ell \neq m \text{ and for all } \tau\text{; also for } \ell = m \text{ if a collision occurs between } t \text{ and } t+\tau \tag{3.53}$$

$$\langle \overline{v_{x\ell}(t)v_{x\ell}(t+\tau)} \rangle = \langle v_{x\ell}^2 \rangle = \langle v_x^2 \rangle, \qquad \text{(assumed to be the same for all electrons) if no collision occurs in the interval } t,\ t+\tau \tag{3.54}$$

Application of (3.52) through (3.54) to the calculation of the ACF of the current (denoted by $R_i(\tau)$) results in

$$R_i(\tau) = \langle \overline{i(t)i(t+\tau)} \rangle = \frac{q^2}{L^2} \langle v_x^2 \rangle N_e P_{c0}(\tau) \tag{3.55}$$

where $P_{c0}(\tau)$ is the *no collision probability* for the interval $t - (t+\tau)$, i.e., the ratio of the average number of electrons not undergoing collisions during the interval $t - (t+\tau)$ to the total number of electrons N_e.

According to the equipartition theorem of statistical mechanics† each degree of freedom of an electron has kinetic energy $kT/2$, where k is Boltzmann's constant and T is absolute temperature. This presupposes a Maxwellian distribution of electron velocities,‡ i.e., the probability that the x-velocity component is between v_x and $(v_x + dv_x)$ is

$$p(v_x)\,dv_x = \sqrt{\frac{m}{2\pi kT}}\,e^{-mv_x^2/2kT}\,dv_x \tag{3.56}$$

The choice of distribution (3.56) is a physically well founded assumption.§ Multiplying (3.56) by v_x^2 and integrating from $v_x = -\infty$ to $v_x = \infty$, we obtain the result $v_x^2 = kT/m$. Substituting this result into (3.55), we have

$$R_i(\tau) = \left[\frac{q^2 N_e}{mL^2}\right] kTP_{c0}(\tau) \tag{3.57}$$

Between collisions, each electron undergoes uniformly accelerated motion, under the influence of the constant x-directed electric field $\mathbf{E}$. From arguments based on Newton's second law of motion, Ohm's law, and (3.57)

† See Lindsay, 1941, Chapter IV, especially Section 5, pp. 67–69.
‡ Lindsay, 1941, Chapter V, Section 5, pp. 83–86.
§ Lindsay, 1941, Chapters IV and V.

it can be deduced† that (for a unit volume of material)

$$R_i(\tau) = \frac{kT}{R_m}\frac{P_{c0}(\tau)}{\langle t_f \rangle} \tag{3.58}$$

where $\langle t_f \rangle$ is the *mean free time*, or average time interval between collisions, and R_m is the resistance of the piece of material to which the field is applied.

To determine $P_{c0}(\tau)$, we divide the interval τ into M equal subintervals and assume that (1) the probability of a collision in any subinterval is independent of collisions in other subintervals, (2) the subintervals are short enough to preclude the possibility of more than a single collision in any one of them, and (3) the probability of a collision within a subinterval is proportional to the length of the subinterval, or, in symbols, $P_c(\tau/M) = \alpha\tau/M$, where $P_c(\tau/M)$ is the collision probability of exactly m collisions within τ.‡ Note that the bracketed quantity approaches $1/m!$ and $(1 - \alpha\tau/M)^{M-m}$ approaches $(1 - \alpha\tau/M)^M$ as M becomes very large compared to m. Note also that $(1 - \alpha\tau/M)$ becomes an increasingly better approximation to $e^{(-\alpha\tau/M)}$ as M increases. If the number of subintervals M increases to an extremely large number for a fixed value of m, we approach the condition

$$P_c(m, \tau) = \frac{(\alpha\tau)^m e^{-\alpha\tau}}{m!} \tag{3.59}$$

known as the *Poisson distribution*.§ It can easily be shown that α is the average number of collisions per second|| and is the reciprocal of $\langle t_f \rangle$.††

---

† Using a "billiard ball" collision rule, Newton's law of motion applied to electron motion in one dimension gives us the differential equation $m(dv/dt) = qE - mv/\langle t_f \rangle$, where v is election speed. Noting that $E = V/L$, where V is applied voltage, and also noting that $V = iR_m$ and $i = qN_eAv$, where i is current, the result (3.58) follows (for a unit volume of material) from the steady state ($t \to \infty$) solution of the differential equation.

‡ $P_c(m, \tau)$ is the product of (1) the probability of exactly m collisions in a subinterval (i.e., $(\alpha\tau/M)^m$), (2) the probability of exactly $(M - m)$ collisionless subintervals (i.e., $(1 - \alpha\tau/M)^{M-m}$), and (3) the number of ways in which this set of events can occur (i.e., $M!/m!(M - m)!$).

§ See Davenport and Root, 1958, pp. 115–117, Lee, 1960, pp. 177–186, Rice, 1944–45, Section 1.1, pp. 14–15 (pp. 146–147 of Wax, 1954).

|| The average number of collisions per unit time is, by definition,

$$\frac{1}{\tau}\sum_{m=0}^{\infty} mP_c(m, \tau)$$

which according to (3.59) is

$$\frac{(\alpha\tau)}{\tau}e^{-\alpha\tau}\sum_{m=1}^{\infty}\frac{(\alpha\tau)^{m-1}}{(m-1)!} = \alpha \cdot e^{-\alpha\tau} \cdot e^{\alpha\tau} = \alpha$$

†† Assume that a collision took place at $t = 0$. The probability that the time interval between collisions is in the region $(t_f - (t_f + dt_f)$ is the product of (1) the probability of a collision within that interval and (2) the probability of no collision between 0 and t_f. According to (3.59), the probability (2) given by $P_c(0, t_f)$, is equal to $e^{(-\alpha t_f)}$. By the basic assumption that collision probability during an infinitesimal interval is proportional to the interval, the probability (1) is αdt_f. The product is $\alpha e^{(-\alpha t_f)}\, dt_f$. Thus

$$\langle t_f \rangle = \int_0^{\infty} dt_f t_f \alpha\, e^{-\alpha t_f} = \frac{1}{\alpha}$$

If the collision distribution is of the form (3.58) then the probability $P_c(0, \tau)$ is $e^{-\alpha\tau}$, from which it follows (since $\alpha = 1/\langle t_f \rangle$) that (3.58) takes the form

$$R_i(\tau) = \frac{kT}{R_m} \frac{e^{-|\tau|/\langle t_f \rangle}}{\langle t_f \rangle} \tag{3.60}$$

The spectral density of current, denoted by $G_i(2\pi f)$ (in terms of cycle frequency), is the Fourier transform of (3.60) (see Eq. (2.41)), i.e.,

$$G_i(2\pi f) = \frac{2kT}{R_m} \frac{1}{\{1 + [2\pi f \langle t_f \rangle]^2\}} \tag{3.61}$$

if it is assumed (based on experimental evidence) that mean free times are of the order of 10^{-12} seconds at most. For frequencies below tens of gigahertz, then, the spectral density $G_i(2\pi f)$ is, for all practical purposes, constant and equal to $4kT/R_m$. Note that this is the spectral density for current; that for voltage is obtained from (3.61) through multiplication by R_m^2, and the noise power P_n delivered to a matched load with bandpass bandwidth† $2B_n = B$, obtained through multiplication of the voltage spectrum by $1/4R_m$ (see Appendix I, Eq. I.5) and integration over frequency space, is approximately

$$P_n = kTB \tag{3.62}$$

It can be concluded from (3.62) and is experimentally verified that any resistive element in a circuit will act as a generator of white noise of spectral density kT. This thermal noise is a basic property of all electrical networks and is always increased by increasing bandwidth and temperature. Since some minimum bandwidth is always a requirement for adequate information transmission, and the environment is generally at a high absolute temperature, typically about 300°K, this type of noise is a basic limiting factor in the design of information systems.

The probability density function of thermal noise current can be calculated from (3.52), (3.56), (3.62), and the assumption of statistical independence between electrons, as follows:‡

---

† We introduce $2B_n$ here merely to be notationally consistent with the previous use of the symbol B_n in this book. $2B_n$ has previously referred to bandpass bandwidth. In the present discussion, we substitute the symbol B for $2B_n$ in designating the bandpass bandwidth of the noise.

‡ If the electron velocities are statistically independent, the joint probability that the N electrons have x-velocity components within the ranges $v_{x1} - v_{x1} + dv_{x1}$, $v_{x2} + dv_{x2}$, $\ldots, v_{xN} - v_{xN} + dv_{xN}$ is the product of the individual probabilities as given by (3.56). The PDF for a current i (current being a weighted sum of these x-velocity components) is a weighted sum of these joint PDF's (all combinations of x-velocity components being mutually exclusive of all others, see (e) of Section 2.1.2), where the only combinations that contribute to this sum are those for which (3.52) holds. The impulse function within the integrand of (3.63) excludes from the integral all sets (v_{x1}, v_{xN}) that do not obey (3.52) and gives all sets that *do* obey (3.52) a weighting of unity.

$$p(i)\,di = di \int_{-\infty}^{\infty} \frac{dv_{x1} \ldots dv_{xN_e}}{\left[\frac{2\pi kT}{m}\right]^{\frac{N_e}{2}}} e^{-mv^2_{x1}/2kT} \ldots e^{-mv^2_{xN_e}/2kT}\, \delta\left(i - \frac{q}{L} \sum_{\ell=1}^{N_e} v_{x\ell}\right)$$

$$= \frac{e^{-\frac{i^2}{2\left[\frac{kT}{R_m\langle t_f\rangle}\right]}}}{\sqrt{\frac{2\pi kT}{R_m\langle t_f\rangle}}} \tag{3.63}$$

As is well confirmed experimentally, thermal noise turns out to be Gaussian.† It can be shown by a somewhat more tedious and detailed argument that the second probability density function $p(i_1, i_2)$ is a bivariate Gaussian function.‡

3.4.2 Shot Noise

Shot noise arises in electron devices. In tubes, it is defined as the noise due to statistical fluctuations in the emission times of electrons. Its analog also exists in solid state electron devices, or in any device whose action is based on travel of electrons or ions between points at different potentials.

To keep the discussion on a concrete level, consider a temperature limited diode, i.e., one operating at a high enough voltage level so that its anode current is limited by emitter temperature rather than space charge. The electrons leave the emitter at different times and different velocities, the velocity spread being an increasing function of temperature. They will arrive at the anode at different times. The time of arrival of an electron at the anode must be regarded as a random function, the randomness being due to both the spread in emission times and that in initial velocities. The noise due to the latter mechanism is a type of thermal noise and its statistics are determined from the Maxwellian distribution of velocities. This thermal component of shot noise will be neglected in what follows and attention will be devoted exclusively to noise arising from fluctuations in emission times.

The anode current induced during the flight of an electron emitted at time t_k from cathode to anode takes the form of a pulse $i(t - t_k)$. All such pulses, being derived from the same physical law, are identical except for a time shift, provided that interactions between electrons in flight are assumed negligible. If N_e is again the number of electrons emitted in a specified time interval, the total shot current is

$$i_s(t) = \sum_{n=1}^{N_e} \hat{i}_s(t - t_n) \tag{3.64}$$

† See Lawson and Uhlenbeck, 1950, Sections 4.1–4.5, pp. 64–79, especially Section 4.2, pp. 66–68.

‡ Lawson and Uhlenbeck, 1950, Section 4.2, pp. 66–68. Two classic references on Brownian motion (G. E. Uhlenbeck and L. S. Ornstein, "On the Theory of Brownian Motion I," in Wax, 1954, pp. 93–111 and M. C. Wang and G. E. Uhlenbeck, "On the Theory of Brownian Motion II," in Wax, 1954, pp. 113–132) are cited by Lawson and Uhlenbeck to provide the basis for the Gaussian character of thermal noise.

If the emission times t_k are assumed to be statistically independent and uniformly distributed throughout an interval $t - T \leq t_k \leq t$, where t is the instant of observation, then the average shot current at time t is

$$\langle i_s(t) \rangle = \sum_{n=1}^{N_e} \frac{1}{T} \int_{t-T}^{t} i_s(t - t_n)\, dt_n = N_e \frac{1}{T} \int_0^T dt'\, i_s(t') = \langle N_e \rangle \langle i_s \rangle \tag{3.65}$$

The power spectrum $G_{i_s}(2\pi f)$ can be calculated from (2.41), (3.64), and (3.65) as follows:

$$\begin{aligned} G_{i_s}(2\pi f) = \int_{-\infty}^{\infty} d\tau\, e^{2\pi j f \tau} R_{i_s}(\tau) &= \sum_{\substack{m,n=1 \\ m \neq n}}^{N_e} \frac{1}{T^2} \int_{-\infty}^{\infty} d\tau\, e^{2\pi j f \tau} \\ &\quad \cdot \int_{t-T}^{t} dt_m \int_{t-T}^{t} dt_n\, i(t - t_m) i(t - t_n + \tau) \\ &\quad + \sum_{n=1}^{N_e} \frac{1}{T} \int_{-\infty}^{\infty} d\tau\, e^{2\pi j f \tau} \\ &\quad \cdot \int_{t-T}^{t} dt_n\, i(t - t_n)\, i(t - t_n + \tau) \\ &= \langle N_e(N_e - 1) \rangle \langle \hat{i}_s \rangle^2 \delta(f) + \langle N_e \rangle G_{\hat{i}_s}(2\pi f) \end{aligned} \tag{3.66}$$

where $G_{\hat{i}_s}(2\pi f)$ is the spectrum of an individual electron current pulse.

The term proportional to $\langle N_e(N_e - 1) \rangle$ is a dc spike, arising from cross-terms between different electrons in the ACF. This may be called the *coherent term*. The *incoherent term*, i.e., that proportional to $\langle N_e \rangle$, is obtained by direct superposition of the spectra associated with each of the electron current pulses.

The mean square of $i_s(t)$ is

$$\begin{aligned} \langle i_s^2(t) \rangle = &\sum_{\substack{m,n=1 \\ m \neq n}}^{N_e} \frac{1}{T^2} \int_{t-T}^{t} dt_m\, \hat{i}_s(t - t_m) \int_{t-T}^{t} dt_n \hat{i}_s(t - t_n) \\ &+ \sum_{n=1}^{N_e} \frac{1}{T} \int_{t-T}^{t} dt_n [\hat{i}(t - t_n)]^2 = \langle N_e(N_e - 1) \rangle \langle \hat{i}_s \rangle^2 + \langle N_e \rangle \langle \hat{i}_s^2 \rangle \end{aligned} \tag{3.67}$$

Again coherent and incoherent terms are present. For large N_e, we can substitute N_e^2 for $N_e(N_e - 1)$, and from (3.65) and (3.67) the variance of the shot noise current becomes

$$\begin{aligned} \sigma_s^2 = \langle [i_s - \langle i_s \rangle]^2 \rangle = \langle i_s^2 \rangle - \langle i_s \rangle^2 \\ = \langle N_e \rangle \langle \hat{i}_s^2 \rangle + [\langle N_e^2 \rangle - \langle N_e \rangle^2] \langle \hat{i}_s \rangle^2 \end{aligned} \tag{3.68}$$

Note that the time interval T drops out in all of these expressions. It is merely a convenience in the derivation to avoid discussion of infinite periods of time, and all shot noise statistics are independent of it, except insofar as it indirectly affects the statistics of N_e.

To derive the probability density functions for shot noise, an assumption must be made about the statistics of N_e. Under the same assumptions as were made in Section 3.4.1 for thermal noise, N_e turns out to be Poisson distributed.† The PDF's can be determined for arbitrary N_e but the calculation is rather cumbersome. In practice it can always be safely assumed that $\langle N_e \rangle$ is an extremely large number. The shot noise current then becomes a superposition of a large number of statistically independent electron currents, and according to the central limit theorem,‡ such a noise is approximately Gaussian. This is found both theoretically and experimentally to be the case for shot noise.§

3.4.3 Noise Temperature

The phrase *noise temperature* is common in the terminology of electronics. It is defined as the ratio of P_a (the power available to a load from a source) to kB. It is based on the thermal noise power given by (3.62). Noise sources whose output spectrum is colored have noise temperatures defined in terms of a half-power bandwidth or some other bandwidth criterion agreed upon in advance. In symbols

$$T_n = \frac{P_a}{kB} \tag{3.69}$$

where T_n is noise temperature, $B = 2B_n$ as defined above Eq. (3.62) and P_a is the power dissipated across the load input impedance when the source is impedance matched to the load. According to Eq. (I.5) in Appendix I, P_a is given by $V_n^2/4R_m$, where R_m is the input resistance and V_n^2 is the mean square noise voltage impressed across the input.

3.5 THE CENTRAL LIMIT THEOREM

A special case of the central limit theorem of probability theory|| is the proposition that the sum of a large number of statistically independent random variables with arbitrary individual statistics is approximately Gaussian. Because of the great importance of this proposition in noise theory, some discussion of it is in order.

† See Davenport and Root, 1958, pp. 115–117.

‡ See Section 3.5.

§ For more extensive discussions on this point and other aspects of shot noise see Rice, 1945, in Wax, 1954, Section 1.6, pp. 13–25, especially pp. 24–25 (i.e., pp. 145–157 of Wax, especially pp. 156–157). Also see Lawson and Uhlenbeck, 1950, Sections 4.6–4.8, pp. 79–90, or Davenport and Root, 1958, pp. 119–138, especially pp. 124–128.

|| Basic references on the central limit theorem from the mathematician's viewpoint are: Doob, 1953, pp. 137–147, or Loéve, 1955, especially Chapter 6. Material on the central limit theorem and random walk can be found in virtually any comprehensive text on probability theory or stochastic processes.

Derivations of the one- and two-dimensional forms of the theorem, in terms of the theory of the random walk, were given by Lawson and Uhlenbeck.† In the one-dimensional case, it is shown that, if X is a sum of N independent random variables $x_1/\sqrt{N}, \ldots, x_N/\sqrt{N}$, each having zero mean and variance σ^2/N and the same PDF, i.e.,

$$X = \frac{x_1 + \cdots + x_N}{\sqrt{N}} \tag{3.70}$$

then the PDF of X is

$$p(X) = \frac{1}{\sqrt{2\pi\sigma^2}} e^{-X_2/2\sigma^2} \left\{1 + 0\left(\frac{1}{\sqrt{N}}\right)\right\} \tag{3.71}$$

where $0(1/\sqrt{N})$ means terms involving powers of $(1/\sqrt{N})$. The moments $\langle x_k^n \rangle$ are the same for all k, because of the assumption that all x_k have the same statistics.

The two-dimensional form of the theorem, whose proof is also given by Lawson and Uhlenbeck,‡ involves the joint PDF of the two sums

$$\begin{aligned} X &= \frac{x_1 + \cdots + x_N}{\sqrt{N}} \\ Y &= \frac{y_1 + \cdots + y_N}{\sqrt{N}} \end{aligned} \tag{3.72}$$

where $x_1, \ldots, x_N$ and $y_1, \ldots, y_N$ are statistically independent random variables with (qualifiedly) arbitrary individual statistics. The joint PDF of X and Y is shown to be

$$p_2(X, Y) = \frac{1}{\pi \langle r_k^2 \rangle} e^{-(X^2+Y^2)/\langle r_k^2 \rangle} \left\{1 + 0\left(\frac{1}{N}\right)\right\} \tag{3.73}$$

in the case where the joint PDF's of the individual x_k and y_k pairs are identical and isotropic (random phase), i.e.,

$$p_2(x_k, y_k)\, dx_k\, dy_k = \frac{1}{2\pi} p(r_k)\, dr_k\, d\phi_k, \qquad k = 1, 2, \ldots, N \tag{3.74}$$

where $p(r_k)$ is the PDF of $r_k = x_k^2 + y_k^2$, and the moments of r_k in (3.73), the same for all k, are defined by

$$\langle r_k^n \rangle = \int_{-\infty}^{\infty} dr_k' r_k'^n p(r_k') \tag{3.75}$$

The important conclusion to be drawn from (3.71) is that $p(X)$ approaches the Gaussian PDF (see Eq. (3.1)) as N becomes very large, all moments of X remaining constant, i.e.,

---

† Lawson and Uhlenbeck, 1950, Sections 3.5, 3.6, pp. 46–53.

‡ Lawson and Uhlenbeck, 1950, pp. 52–53.

$$\lim_{N\to\infty} p(X) = \frac{1}{\sqrt{2\pi\sigma^2}} e^{-X^2/2\sigma^2} \tag{3.76}$$

where $\langle x_k^n \rangle$ is constant for all n.

From (3.73), it follows that as N becomes large, $p_2(X, Y)$ approaches the two-dimensional Gaussian PDF for independent X and Y (see Eq. (3.5)), i.e.,

$$\lim_{N\to\infty} p_2(X, Y) = \frac{1}{\pi\langle r_k^2 \rangle} e^{-(X^2+Y^2)/\langle r_k^2 \rangle} \tag{3.77}$$

where r_k^n is constant for all n.

It follows in turn from (3.77) that the amplitude $R = \sqrt{X^2 + Y^2}$ has the Rayleigh distribution (see Eq. (3.40)).

The remarkable features of the forms (3.76) or (3.77) are that:

1. They hold for essentially arbitrary statistics of the individual constituents of the sum. It is not necessarily required that the individual PDF's be identical or isotropic in order that (3.77) shall hold.†

2. The value of N need not be very large in order to assure the approximate validity of either (3.76) or (3.77); $N \geq 7$ is sufficient in many cases of practical interest.

Because of these features, (3.76) and (3.77) constitute the justification for the assertion that a particular noise is Gaussian in cases where the physics of the problem indicates that the noise arises from a superposition of a large number of independent random sources. We will have occasion to use this assertion in many discussions throughout this book.

3.6 REMARKS ON INFORMATION THEORY

The *information theory* advanced by Shannon in 1949,‡ to which Hartley,§ Nyquist,|| Gabor†† and others made earlier contributions, was mentioned in Section 1.1. We will briefly discuss here only some of its very simplest aspects.

As indicated in Section 1.1, very little explicit use will be made of infor-

---

† Referring back to Lawson and Uhlenbeck, p. 53, we note that the isotropic assumption, made after Eq. (49), simplifies Eq. (50) from which the final result (51) follows. If this assumption were not made, the effect on Eq. (50) would be to change $e^{-(x^2+y^2)/I_0}$ to an expression of the general form $e^{-(ax^2+bxy+cy^2)}$ which can easily be transformed into $e^{-(x'^2+y'^2)/I_0}$, where x' and y' are linear combinations of x and y. The important point in going from (50) to (51) is the fact that terms beyond the first are in powers of $1/N$; thus when N increases without limit, the lead term only survives, and this term is of a generalized Gaussian form with or without the isotropicity assumption.

‡ Shannon, 1948, Shannon and Weaver, 1949.

§ Hartley, 1928.

|| Nyquist, 1928.

†† Gabor, 1946.

mation theory as such in this book. However, it was felt that the discussion below, as it is, would serve to provide the reader with a cursory glance at the subject. For the reader interested in pursuing information-theoretic approaches to communication problems, several excellent references are cited at the end of this chapter.

3.6.1 The Definition of Information

The basis of the theory is the definition of average information I as a *negative average uncertainty*, where the definition of average U is analogous to the statistical mechanics definition of physical entropy,† as follows:

$$U = \sum_{n=1}^{n} P_n \log_2 P_n = -I \tag{3.78}$$

where P_n is the probability of a particular event, designated as event n, out of a possible N events that could occur. Suppose a man (designated as A) is about to perform an experiment. He knows that his experiment will have one of N possible outcomes, and his state of knowledge is such that he can assign a set of probabilities $(P_1, \ldots, P_N)$ to this set of outcomes. After the experiment has been performed, he knows which of these has occurred, and if it is, for example, the kth outcome, his set of probabilities is now $(\underbrace{0, 0, \ldots, 0}_{(k-1)}, 1, \underbrace{0, \ldots, 0}_{(N-k)})$. Thus as a result of performing the experiment, he has reduced his uncertainty by an amount

$$\Delta U = \sum_{n=1}^{N} P_n^{(1)} \log_2 P_n^{(1)} - \sum_{n=1}^{N} P_n^{(0)} \log_2 P_n^{(0)} \tag{3.79}$$

where $P_n^{(0)}$ is the *before* (a more sophisticated term is *a priori*) probability of occurrence of the nth possible outcome of the experiment and $P_n^{(1)}$ is the corresponding *after* (or *a posteriori*) probability. The a priori probabilities are $(P_1^{(0)}, \ldots, P_k^{(0)}, \ldots, P_N^{(0)})$ and the a posteriori probabilities are $(\underbrace{0, 0, \ldots}_{(k-1)}, 1, \underbrace{0, \ldots, 0}_{(N-k)})$. Therefore the a posteriori uncertainty is $(1) \log_2 (1) = 0$, and the uncertainty has been reduced by an amount

$$\Delta U = -\sum_{n=1}^{N} P_n^{(0)} \log_2 P_n^{(0)} \tag{3.80}$$

Suppose A performs his experiment in stages, such that after the first stage he does not know which outcome will occur, but has more data from which to predict the final result. After this stage, he has eliminated some of the possible outcomes, reducing the total to M, where $M < N$. Among the remaining M possible outcomes, he may have new probabilities $P_1^{(1)}$,

† See Brillouin, 1956, Chapter 9, pp. 114–127, Chapter 17, pp. 245–258, especially pp. 254–258.

$\ldots, P_M^{(1)}$. In this case, the reduction in uncertainty resulting from stage 1 of the experiment is

$$\Delta U^{(1)} = \sum_{n=1}^{M} P_n^{(1)} \log_2 P_n^{(1)} - \sum_{n=1}^{N} P_n^{(0)} \log_2 P_n^{(0)} \tag{3.81}$$

If stage 2 results in an exact specification of one of the original N possible results, then the reduction in uncertainty due to stage 2 is

$$\Delta U^{(2)} = -\sum_{n=1}^{M} P_n^{(1)} \log_2 P_n^{(1)} \tag{3.82}$$

The total uncertainty reduction in the two-stage experiment is

$$\Delta U = \Delta U^{(1)} + \Delta U^{(2)} = -\sum_{n=1}^{N} P_n^{(0)} \log_2 P_n^{(0)} \tag{3.83}$$

which is consistent with (3.80).

To put these ideas on a quantitative basis, we must first define a unit of uncertainty. To this end, consider a *binary* or two-alternative experiment, where the two possible outcomes are equally probable, i.e., $P_1^{(0)} = P_2^{(0)} = \frac{1}{2}$. In this case, according to (3.79) the uncertainty reduction in performing the experiment is

$$\Delta U = -\sum_{n=1}^{2} \left(\frac{1}{2}\right) \log_2 \left(\frac{1}{2}\right) = 1 \tag{3.84}$$

Thus, by definition (3.78), we have lost one unit of uncertainty when we have determined which of two alternative equally likely events has occurred. This *unit uncertainty* is known as a *binit* or more popularly, as a *bit*. Since an *uncertainty loss* is defined as an *information gain*, we say that the experiment has resulted in the acquisition of one bit of information.

We now consider an N-alternative experiment, assuming that all N outcomes are equally probable, i.e., $P_1^{(0)} = P_2^{(0)} = \ldots = P_N^{(0)} = 1/N$. Equation (3.79) now reduces to the form

$$\Delta U = \log_2 N \tag{3.85}$$

In the two-stage experiment described above, where initially there are N equally likely alternatives, and the first stage eliminates $(N - M)$ of them, Eqs. (3.81) through (3.83) reduce to

$$\Delta U^{(1)} = \log_2 N - \frac{M}{N} \log_2 N \tag{3.86}$$

$$\Delta U^{(2)} = \frac{M}{N} \log_2 N \tag{3.87}$$

$$\Delta U = (\Delta U^{(1)} + \Delta U^{(2)}) = \log_2 N \tag{3.88}$$

For example, if $N = 100$ and $M = 50$, then 3.32 bits of information are acquired in the first stage of the experiment, and 3.32 bits during the second, totaling 6.64 bits.

It can be shown that the maximum uncertainty exists when the intial alternatives are equally probable. This is intuitively evident when one reflects on the fact that we know least about a situation when we find it impossible to decide between a set of alternatives. Suppose, for example, that a student taking a multiple choice test with five possible answers to each question encounters a question on which he is totally ignorant. He might then assign a probability of $\frac{1}{5}$ to each of the five answer choices, all of which appear (to him) equally likely to be correct, and make a "random guess" on the correct answer. If he is partially informed on the question, he might be able to assign heavier probability weightings to some of the answers, or possibly to eliminate some (say two) answers. If he knows the answer, then obviously he can choose the correct alternative out of the five that were initially possible. In the first case, the student would require $\log_2 5 = 2.33$ bits of information to answer the question correctly. In the second case, if he eliminates two answers, he needs only $\log_2 3 = 1.57$ bits, while in the third case, he obviously has all the information he needs about the question, i.e., his uncertainty is zero.

3.6.2 Information Rate

A most useful application of the ideas discussed above is to the determination of the amount of information that can be extracted about a signal waveform. Consider a waveform $v(t) = s(t) + n(t)$ confined to frequencies below B c/s. Here $s(t)$ is the desired signal and $n(t)$ is noise. According to the sampling theorem (Section 1.5), the waveform can be completely characterized by sampling at time intervals of $1/2B$. This implies that, if no noise is present, a set of perfect measurements made of the amplitude of $v(t)$ at intervals of $1/2B$ will result in complete information (zero uncertainty) about the signal. However, there is always a certain amount of noise superposed on the signal and/or in the measuring instrument, and therefore a measurement of $v(t)$ at any instant of time will produce only an approximation of $s(t)$ and not its exact value. For example, if the noise has zero mean, is Gaussian, and its variance σ_n^2 is known to the observer, then a single measurement at time t_1, yielding a value v_o, will enable the observer to conclude that $s(t_1)$ is between $v_o - 3.54\,\sigma_n$ and $v_o + 3.54\,\sigma_n$ with a 99.96 percent probability of being correct. If we are satisfied to call a 99.96 percent probability a certainty, then we can say that, with a zero-mean Gaussian noise of known average power σ_n^2, it is possible through a measurement to confine the value of $v(t_1)$ within an interval of width $7.08\,\sigma_n$.

On this basis, we can choose a *quantization interval* $\Delta v = 7.08\,\sigma_n$, and divide up the scale of possible values of $v(t)$ into M intervals of width Δv, where $M\,\Delta v$ is the maximum range of possible values. Unless a particular $v(t)$ sample falls very near the boundary between quantization intervals it can be inferred from a measurement of the waveform at time t that the true value of $v(t)$ is within a specified interval, with a very high probability that the inference is correct.

Having set up this quantization of the signal, we now say that a measurement at any instant of time t_1 determines which of the M quantization intervals contains the true value of $v(t)$. If it is assumed that all M intervals are equally probable before the measurement is made, then according to (3.85), a gain of $\log_2 M$ bits of information has resulted from the measurement. Since sampling takes place every $1/2B$ s, then the maximum possible rate of acquisition of information about $v(t)$ (denoted by dI/dt) is $(\log_2 M)$ bits/sample multiplied by $(2B)$ samples/s, or

$$\frac{dI}{dt} = 2B \log_2 M \text{ bit/s} \tag{3.89}$$

We have so far been dealing explicitly with a Gaussian noise, because it is possible with such a noise to specify a probability of correctly deciding in which quantization interval $v(t)$ lies. For other types of noise this probability also exists, but may not be so easy to specify. However, the relation (3.89), usually known as the *Hartley law*,† is general and does not require that the noise be Gaussian.

An important result of Shannon's investigations is the *mean power theorem*, giving the maximum possible average rate of transmission of information over a communication channel perturbed by stationary white Gaussian noise. If the transmitted signal itself is a sample function of a stationary Gaussian random process, with uniform spectral density and mean power σ_n^2, the noise has mean power σ_n^2, and the *bandpass bandwidth*‡ of the channel in cps is $2B$; the maximum average rate of information transmission in bits per second is

$$\left[\frac{dI}{dt}\right]_{\max} = B \log_2 \left[1 + \frac{\sigma_s^2}{\sigma_n^2}\right] \tag{3.90}$$

The proof of (3.90) will be briefly sketched.§ First, the assumption of a white Gaussian signal is tantamount to an assumption that the signal has the maximum possible average *randomness* or *uncertainty* as defined by the

---

† Hartley, 1928.

‡ *Bandpass bandwidth* refers here to the actual spread between highest and lowest frequencies in a bandpass channel. In a low pass channel, the same designation refers to the spread from the negative of the highest frequency passed to the highest frequency passed itself. Half of this quantity would in the latter case be called *lowpass bandwidth*. This same point is mentioned in less detail on page 60, the second footnote.

§ See Woodward, 1953, Sections 3.2–3.8, especially Section 3.6, pp. 56–58, for elaboration on this material.

integral analog of (3.78)† before it has been measured at the receiver, relative to all other signals with mean power σ_s^2. This implies that, if one wishes to transmit a given number of information bits and has available a maximum average power, the way to transmit this information at the fastest possible rate (i.e., maximum uncertainty prior to reception implies maximum information acquired through reception) is to code it in such a way that it becomes a sample function of a Gaussian random process from the point of view of the receiver.

If both signal and noise are Gaussian, then it can be shown‡ that the average uncertainty in a single sample point of the pure noise waveform is $\frac{1}{2}\log_2(2\pi e\sigma_n^2)$ while that of a sample point of the signal-plus-noise waveform is $\frac{1}{2}\log_2[2\pi e(\sigma_s^2 + \sigma_n^2)]$. The information transmitted from sender to receiver is defined as

$$\log_2 \frac{p((s+n)/s_t)}{p(s_t)}$$

where s_t is the amplitude of the transmitted signal, $(s + n)$ is the received signal-plus-noise amplitude, $p(s_t)$ is the PDF for s_t and $p((s + n)/s_t)$ is the PDF for $s + n$, conditional on s_t having been sent. Multiplying this logarithm by the PDF's for s and n and integrating over all space results in an expression for the average information acquired in reading a single sample point of the signal. This average is $[\frac{1}{2}\log_2(\sigma_s^2 + \sigma_n^2) - \frac{1}{2}\log_2(\sigma_n^2)]$.

If the bandpass bandwidth is $2B$, then both signal and noise can be completely characterized by samples taken at a rate $(2B)$ per second (see Section 1.5). But the noise and the signal-plus-noise, both being white and limited by a bandpass bandwidth $2B$, have statistically independent values at time intervals of $(1/2B)$ (see Section 3.1.2).§ It can be shown|| that the

† That is,

$$I = -\int_{-\infty}^{\infty} [dx\, p(x)] \log_2 [p(x)\, dx]$$

where $p(x)\,dx$ is the probability that the signal $x(t)$ has a value between x and $x + dx$. The presence of the quantity $\log_2(dx)$ has generated a great deal of argument concerning the validity of this expression. This can be settled by noting that a given value of dx implies a given accuracy in determination of x, and setting the required accuracy at some fixed value by setting $dx = 1$.

‡ See Woodward, 1953, Section 3.6, pp. 56–58.

§ The ACF has zeros at intervals $1/2B$. Therefore, waveform values separated by $1/2B$ are uncorrelated. With Gaussian functions, this implies that they are statistically independent. This point was mentioned in Section 2.4.4.2.

|| The demonstration is as follows:

$$\begin{aligned} I &= -\int \cdots \int dx_1, \ldots, dx_M\, p(x_1, \ldots, x_M) \log_2 p(x_1, \ldots, x_M) \\ &= -\int \cdots \int dx_1, \ldots, dx_M\, p(x_1) \ldots p(x_M) \ldots \\ &\qquad [\log_2 p(x_1) + \log_2 p(x_2) + \ldots + \log_2 p(x_M)] \end{aligned}$$

average uncertainty in a set of samples at $1/2B$ intervals in a waveform of duration T with bandpass bandwidth $2B$ is $(2BT)$ multiplied by the uncertainty in a single sample. Thus the average information transmitted over the channel is $2BT[\frac{1}{2}\log_2(\sigma_s^2 + \sigma_n^2) - \frac{1}{2}\log_2(\sigma_n^2)]$, implying that the average rate of transmission is $B \log_2 (1 + \sigma_s^2/\sigma_n^2)$, as given by (3.90).

The maximum attainable information rate on the channel, defined as *channel capacity*, is that given by (3.90). However, attainment of such a high rate in practice would require *ideal coding*, i.e., coding of the transmitted signal such that it simulates white Gaussian noise at the receiver. The principal utility of (3.90) is that it provides a theoretical upper bound on the rate of information transmission over the channel.

REFERENCES

General

Cramer, 1955; pp. 114–116, 120, 183

Davenport and Root, 1958; Chapter 8

Freeman, 1958; Chapter 8

IRE, PGIT, June, 1960

Lawson and Uhlenbeck, 1950; Chapter 3, especially Sections 3.5, 3.7; also pp. 64–68

Middleton, 1960; Chapters 7, 8, 9

Parzen, 1960; Chapter 8, Section 5, pp. 371–378

Optimization of SNR and Matched Filters

Davenport and Root, 1958; Chapter 11, Section 11.7, pp. 244–247

IRE, PGIT, June, 1960

Middleton, 1960; Chapter 16, especially pp. 714–721

Schwartz, Bennett, and Stein, 1966; Section 2.4, pp. 63–68

Turin, 1960

Woodward, 1953; Chapter 4, especially Sections 4.3 through 4.9

Noise through Rectifiers and other Nonlinear Devices

Davenport and Root, 1958; Chapters 12, 13; also Section 9.5, pp. 189–199

Emerson, 1953

Kac and Siegert, 1947

$$= -\int dx_1\, p(x_1)\log_2 p(x_1) - \int dx_2\, p(x_2)\log_2 p(x_2) - \dots$$

$$- \int dx_M\, p(x_M)\log_2 p(x_M), \qquad (\text{if } x_1, \dots, x_M \text{ are statistically independent})$$

$$= I_1 + I_2 + \dots + I_M$$

If $I_1 = I_2 = \dots = I_M$, and $N = 2BT$, then $I = 2BTI_1$.

Middleton, 1960; Chapter 5
Rice, 1944–45; Part IV, pp. 247–294
Siegert, 1954
Terman, 1947; pp. 548–568

Physics of Noise in Electrical and Electronic Devices

Bennett, 1960
Davenport and Root, 1958; Chapter 7, pp. 112–144 (on shot noise)
Freeman, 1958; Chapter 4
Lawson and Uhlenbeck, 1950; Chapter 4; especially Sections 4.1 and 4.2, pp. 64–68; also Section 4.5, pp. 76–79 (experimental results)
Lee, 1960; Chapter 6, Section 1, pp. 177–186
Rice, 1944–45; Part I, pp. 145–161 (on shot noise)
Schwartz, 1959; Chapter 5
Smullin and Haus, 1959
Van der Ziel, 1954

Practical Implications of Central Limit Theorem

Cramer, 1955; pp. 114–116, 120, 183
Davenport and Root, 1958; pp. 81–84
Lawson and Uhlenbeck, 1950; pp. 46–47, also Section 3.6, pp. 50–53, Section 6.2, pp. 125–127
Middleton, 1960; p. 363, pp. 364–366

Information Theory

Brillouin, 1953
Brillouin, 1956; pp. 1–4, pp. 114–117; Chapters 12, 13, and 14
Cherry, 1961
Fano, 1961
Feinstein, 1958
Gabor, 1946
Goldman, 1956
Hartley, 1928
Middleton, 1960; Chapter 6
Nyquist, 1928
Raisbeck, 1964
Reza, 1961
Schwartz, 1963
Shannon, 1948
Shannon and Weaver, 1949
Woodward, 1953; pp. 21–25, pp. 56–58

4

STATISTICAL THEORY OF OPTIMUM SIGNAL PROCESSING

The present chapter covers some aspects of the elementary theory of optimum signal processing against noise, particularly Gaussian noise. More explicitly, it deals with the probabilistic approach to optimum design of detection and measurement systems. Many systems to be dealt with in later chapters fall under one of these two categories. Detection and measurement are basic constituents of many radio systems. The theory to be discussed in this chapter is therefore of great importance to our subject, and will appear repeatedly in connection with specific applications.†

4.1 DETECTION

One of the most important applications of noise theory is in the analysis and design of systems for detection of signals in noisy backgrounds. In particular, if it is decided beforehand what constitutes a "good system," i.e., what criterion of system merit is to be used, the theory will show what processing techniques applied to the noisy waveform will optimize system performance. In this section, we will concern ourselves with the most elementary class of detection systems, those which attempt to answer the question "Is the signal present or isn't it?" A useful performance criterion for such a system is

† Only a few of the simplest topics in the statistical theory of optimum signal processing are treated here. For elaboration on this vast subject area, the reader is referred to the bibliography at the end of the chapter.

detection probability, that is, the probability of successful determination of signal presence or absence. A more general criterion, of which detection probability is a special case, is the *average loss* or *average risk*. In using this criterion, a different risk or loss is associated with each of the two possible decisions on signal presence or absence.

The discipline used to analyze systems from the point of view of detection probability or average loss is known as *statistical decision theory*.† Consideration here will be limited to the theory of binary decision systems, that is, those whose task is to make a decision between only two alternatives. This theory applies not only to detection of the presence of a signal in a noisy background, but also to digital communication systems which code all intelligence in terms of *marks* and *spaces*.‡ Thus the range of applicability of the theory of binary decision systems is extensive.

4.1.1 Error Probabilities

If the purpose of the system is to make a binary decision, it is desirable that system performance should be measured in terms of the number of correct decisions in a large number of independent trials. This measure, for practical purposes, is equivalent to the "probability of a correct decision."

In making a decision between two alternative events a and b, the system must obviously either decide correctly or incorrectly. Thus the probability of an error plus the probability of a correct decision must add up to a probability of unity, or certainty. Therefore, the basic measure of system performance is usually chosen as the probability of error P_e, from which the probability of a correct decision can be computed (by subtracting P_e from unity).

There are two types of errors, that in which the alternative a is the true one and the system decides b, and that in which b is true and the system decides a. The first will be denoted by P_{ab} and the second by P_{ba}.

Let the system input be $X^{(i)}$ and let its output be X. For simplicity, let X represent a pair of numbers§ representing the cosine and sine components of the output voltage of a receiver (X_1 and X_2, respectively). Suppose we divide the entire range of values of X_1 and X_2 into two regions R_a and R_b as shown in Figure 4-1.

Let the system be designed so that, in the absence of noise, it will decide that event a has occurred if the output is in R_a and that b has occurred if the

† See Wald, 1950, Middleton and Van Meter, 1955, Middleton, 1960, Deutsch, 1965, Chapter 13, and Selin, 1965.

‡ See Chapter 12, Section 12.1.

§ In general, $X^{(i)}$ and X need not be single numbers or pairs of numbers. They may represent for example a set of N numbers denoting values of a waveform at N instants of time $t_1, \ldots, t_N$.

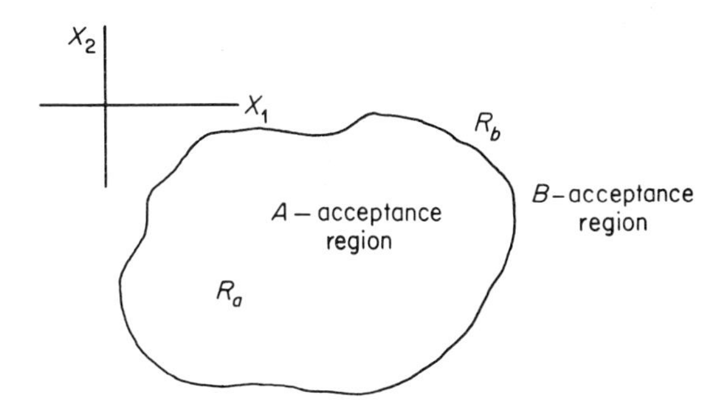

Figure 4-1 Two-dimensional decision space.

output is in R_b.† The conditional error probability, P_{ab}, is the probability that the output is in R_b if a has occurred, and P_{ba} is the probability that the output is in R_a if b has occurred, i.e.,

$$P_{ab} = \iint_{R_b} dX_1\, dX_2\, p_a(X_1, X_2) \tag{4.1}$$

$$P_{ba} = \iint_{R_a} dX_1\, dX_2\, p_b(X_1, X_2) \tag{4.2}$$

where $p_a(X_1, X_2)$ is the joint probability density function for X_1 and X_2 if a occurs and $p_b(X_1, X_2)$ is its counterpart if b occurs.

If the designer knows the statistics of the noise in the system for events a and b, that is, if he knows the probability density functions $p_a(X_1, X_2)$ and $p_b(X_1, X_2)$, he can (at least in principle) calculate conditional probabilities for the two types of errors, using (4.1) and (4.2), for any choice of R_a and R_b.

4.1.2 Optimum Decision Systems

The expressions (4.1) and (4.2) deal with conditional probabilities for the two possible types of errors, but they do not provide a criterion for over-all optimization of a decision system. This requires an over-all system measure of "goodness" which must be maximized or, equivalently, a measure of "badness" which must be minimized.

In general, choosing a larger R_a and thereby reducing the size of R_b will decrease the conditional error probability P_{ab} but will increase P_{ba}. Enlarging R_b, on the other hand, will increase P_{ab} and decrease P_{ba}. It seems reasonable that if P_{ab} and P_{ba} are combined in some fashion to produce an over-all measure of good performance, one will arrive at a choice of R_a and R_b which

† The decision can be made either automatically or by a human observer. The important point is that the decision rule, once defined, be precisely observed for every decision.

will maximize this measure. There are many possible ways of setting up such a measure. Certain measures, however, have become somewhat traditional. These measures seem to be consistent with the way in which the designer would approach the problem without mathematics, i.e., by common sense.

Before discussing the criteria for system optimization, we will specialize to the case where one attempts to detect the presence of a signal obscured by noise. In this case, the event b can be taken to represent the presence of a signal and the event a is "noise alone," or "no signal is present."

We will designate the conditional error probability P_{ab} as *false alarm probability* and P_{ba} as *false rest probability*.† This nomenclature is convenient and self explanatory.

The criteria to be defined will be applied in this discussion to the signal detection case, but the definitions can easily be extended to include any binary decision problem.

The three criteria that are somewhat classical are:

1. The ideal observer criterion.
2. The minimum average loss criterion.
3. The Neyman-Pearson criterion.

The first of these, the ideal observer criterion, considers the system optimized when the over-all probability of success is maximized, or, equivalently, the over-all probability of error is minimized.

The second, the minimum average loss criterion, assigns a *loss* to each of the two possible types of errors, defines the *average loss* as the sum of the products of the loss for a particular type of error and the probability of occurrence of that type of error, and minimizes the average loss so defined.

The third, the Neyman-Pearson criterion, minimizes the false rest probability for some fixed value of false alarm probability decided upon in advance.

4.1.2.1 Ideal Observer Criterion‡

To find the optimum decision rule with the ideal observer criterion, we will use a method following closely a development given by Davenport and Root.§ This approach does not require explicit use of the calculus of variations.

The total probability of error P_e is defined by

$$P_e = P^{(a)}P_{ab} + P^{(b)}P_{ba} \tag{4.3}$$

where $P^{(b)}$ is the probability of signal presence and $P^{(a)}$ that of absence of a signal, or "noise alone."

† This is sometimes called *false dismissal probability*.

‡ Lawson and Uhlenbeck, 1950, Section 2.5, pp. 167–173.

§ Davenport and Root, 1958, Sections 14.2 and 14.3.

Using (4.1) and (4.2) for conditional error probabilities P_{ab} and P_{ba}, we have

$$P_e = P^{(a)} \iint_{R_b} dX_1\, dX_2\, p_a(X_1, X_2) + P^{(b)} \iint_{R_a} dX_1\, dX_2\, p_b(X_1, X_2) \tag{4.4}$$

Since $R_a + R_b = R$, the entire space of X_1 and X_2, an integral over R_a is equivalent to the integral of the same integrand over R minus the integral over R_b. Thus

$$\begin{aligned} P_e &= P^{(b)} \int_R dX_1\, dX_2\, p_b(X_1, X_2) \\ &\quad + \int_{R_b} dX_1\, dX_2 \{P^{(a)} p_a(X_1, X_2) - P^{(b)} p_b(X_1, X_2)\} \\ &= P^{(b)} + P^{(a)} \iint_{R_b} dX_1\, dX_2\, p_a(X_1, X_2) \left[1 - \frac{P^{(b)} p_b(X_1, X_2)}{P^{(a)} p_a(X_1, X_2)}\right] \end{aligned} \tag{4.5}$$

The integral

$$\int_R p_b(X_1, X_2)\, dX_1\, dX_2$$

is the probability that X_1 and X_2 fall somewhere in R. This event is certain and therefore the integral is equal to unity.

The function $P^{(b)} p_b(X_1, X_2)/P^{(a)} p_a(X_1, X_2)$, the ratio of the PDF for output (X_1, X_2) with event b to its counterpart with event a, is known as the *likelihood ratio* and will be denoted by $\Lambda(X_1, X_2)$.

Suppose an alternative output space region R_b' is chosen as the b-acceptance region.† The difference between the error probability P_e' obtained with the choice of b-acceptance region R_b' and the error probability P_e obtained from (4.5) is

$$P_e' - P_e = P^{(a)} \iint_{(R'_b - R_b)} dX_1\, dX_2\, p_a(X_1, X_2)[1 - \Lambda(X_1, X_2)] \tag{4.6}$$

where $(R_b' - R_b)$ is the portion of region R_b' which is not a part of R_b.

If $\Lambda(X_1, X_2)$, which must be positive by definition, is greater than unity everywhere in the region $(R_b' - R_b)$, then the integral in (4.6) is negative, because $P^{(a)} p_a(X_1, X_2)$ must be positive. If $\Lambda(X_1, X_2)$ is less than unity everywhere in $(R_b' - R_b)$ then the integral is positive. Therefore, we can conclude that

$$\begin{aligned} P_e > P_e' \quad &\text{if } \Lambda(X_1, X_2) > 1 \quad \text{everywhere in } (R_b' - R_b) \\ P_e < P_e' \quad &\text{if } \Lambda(X_1, X_{2,}) < 1 \quad \text{everywhere in } (R_b' - R_b) \end{aligned} \tag{4.7}$$

If we choose R_b such that $\Lambda(X_1, X_2)$ is less than unity everywhere in the region $(R_b' - R_b)$ for all possible choices of R_b', then from (4.7) we can con-

† By *b-acceptance region* here we mean the region in the output space which will result in a decision that b rather than a is the true event that has occurred. The *a-acceptance region* is the part of the output space R which is not in the b-acceptance region.

clude that the error probability P_e obtained with R_b is smaller than the error probability P_e' obtained with *any* other choice of the b-acceptance region. This is equivalent to the statement that this choice of R_b is optimum, since it gives a lower over-all error probability than any other b-acceptance region we might have chosen.

Since we have assumed that $\Lambda(X_1, X_2)$ is less than unity throughout $(R_b' - R_b)$ for all possible R_b', then this holds true if R_b' is the entire output space R. But $R - R_b$ is the a-acceptance region R_a. Therefore the optimum choice of a- and b-acceptance regions R_a and R_b is

$$\begin{aligned} R_a &= \text{all values of } X_1 \text{ and } X_2 \text{ such that } \Lambda(X_1, X_2) < 1 \\ R_b &= \text{all values of } X_1 \text{ and } X_2 \text{ such that } \Lambda(X_1, X_2) > 1 \end{aligned} \tag{4.8}$$

The line in the (X_1, X_2)-plane dividing the space between R_a and R_b is the line on which $\Lambda(X_1, X_2)$ is equal to unity.

Thus the optimum processing scheme for signal detection, from the point of view of the ideal observer criterion, is one which computes the likelihood ratio $\Lambda(X_1, X_2)$ from the receiver output voltage components X_1 and X_2 and presents it to a decision device (either a human observer or an electronic circuit) which will call "signal present" when the computed $\Lambda(X_1, X_2)$ exceeds unity and "no signal" when $\Lambda(X_1, X_2)$ is less than unity. This processing scheme, if it can be mechanized, will provide the minimum over-all probability of error in deciding whether or not a signal is present.

4.1.2.2 Minimum Average Loss Criterion

The minimum average loss† criterion assigns a *loss* L_{ab} to a false alarm error and a *loss* L_{ba} to a false rest error. The *average loss* L is defined as the sum of the products of losses assigned to the two types of errors and the probabilities of occurrence of these errors. Thus

$$L = L_{ab}P^{(a)}P_{ab} + L_{ba}P^{(b)}P_{ba} \tag{4.9}$$

The optimum system is that which makes its decisions in such a manner as to minimize L. Comparison of (4.9) with (4.3) shows that the minimum average loss criterion degenerates into the ideal observer criterion when $L_{ab} = L_{ba} = 1$; $L = P_e$. The ideal observer criterion, then, is a special case of the minimum average loss criterion, applicable to the case where equal importance can be assigned to false rest and false alarm probabilities. If the designer has some reason to consider one type of error more disastrous than the other type, he uses the minimum average loss criterion, assigning a greater loss to that error which he considers to be the most serious.

For example, in land mine detection, false alarms may result in inconve-

† In some treatments the terms "risk" or "cost" are used instead of "loss."

nience or wasted time in stopping to attempt to destroy indicated but non-existent mines, but false rests may cost a soldier's life. The loss assigned to false rest errors, from the humanitarian viewpoint, should be much greater than that assigned to false alarms. However, from the viewpoint of long range military strategy, it may be decided to minimize the time required to traverse the minefield, even at a sacrifice in lives, in order to save more lives at some later date. In any case, the relative loss assignment to the two types of errors is a strategic decision.† Once this decision is made, the system designer can take over.

The minimum average loss criterion, by a simple extension of the reasoning leading from (4.3) to (4.8), results in the rule

$$\begin{aligned} R_a &= \text{all values of } X_1 \text{ and } X_2 \text{ such that} \\ &\qquad \Lambda(X_1, X_2) < \frac{L_{ab}}{L_{ba}} \\ R_b &= \text{all values of } X_1 \text{ and } X_2 \text{ such that} \\ &\qquad \Lambda(X_1, X_2) > \frac{L_{ab}}{L_{ba}} \end{aligned} \tag{4.10}$$

Thus the computation of the likelihood ratio is still the optimum processing scheme, but the detection threshold, instead of being unity as with the ideal observer criterion, is equal to the ratio of false alarm loss to false rest loss. Thus if the loss associated with false rests is regarded as high compared to that associated with false alarms, then the threshold L_{ab}/L_{ba} is set low so that R_a becomes a very small region. This renders false rests highly unlikely while it enhances the probability of false alarms.

4.1.2.3 The Neyman-Pearson Criterion

In using the Neyman-Pearson criterion, false rest probability P_{ba} is minimized for a given fixed false alarm probability P_{ab}, i.e.,

$$P_{ba} = P^{(b)} \iint_{R_a} dX_1\, dX_2\, p_b(X_1, X_2) \tag{4.11}$$

is minimized, with the constraint

$$P_{ab} = P^{(a)} \iint_{R_b} dX_1\, dX_2\, p_a(X_1, X_2) = K \tag{4.12}$$

where K is the chosen value of false alarm probability.

† It is interesting to note that decisions based on human value judgments are required somewhere in the process. Decision theory places these decisions at an earlier stage and then relies on mathematics to complete the design. One should always bear in mind in applying this discipline to problems that a human value judgment is always buried in the assumptions.

From (4.12) and the fact that $R = R_a + R_b$, we have

$$P_{ab} = P^{(a)} \iint_{R_b} dX_1\, dX_2\, p_a(X_1, X_2) = P^{(a)} \iint_{R} dX_1\, dX_2\, p_a(X_1, X_2) - P^{(a)} \iint_{R_a} dX_1\, dX_2\, p_a(X_1, X_2) = K \tag{4.13}$$

But the integral of $p_a(X_1, X_2)$ over the entire space R must be equal to unity, and therefore, from (4.13)

$$P^{(a)} \iint_{R_a} p_a(X_1, X_2)\, dX_1\, dX_2 = P^{(a)} - K \tag{4.14}$$

Adding $(1/\lambda)\, P_{ba}$ (where λ is a constant) to both sides of (4.14) results in

$$\iint_{R_a} dX_1\, dX_2\, P^{(a)} p_a(X_1, X_2) \left[1 - \frac{1}{\lambda} \Lambda(X_1, X_2)\right] + \frac{1}{\lambda} \iint_{R_a} dX_1\, dX_2\, P^{(b)} p_b(X_1, X_2) = P^{(a)} - K \tag{4.15}$$

Combining (4.11) and (4.15),

$$P_{ba} = \lambda(P^{(a)} - K) + \iint_{R_a} dX_1\, dX_2\, P^{(a)} p_a(X_1, X_2)[\Lambda(X_1, X_2) - \lambda] \tag{4.16}$$

Since $P^{(a)} p_a(X_1, X_2)$ must be everywhere nonnegative, the minimum value of P_{ba} for a given λ, $P^{(a)}$, and K will occur when R_a is chosen to be the entire region wherein $[\Lambda(X_1, X_2) - \lambda]$ is negative. Also, it can be shown that there is an optimum value of λ, i.e., a value which minimizes P_{ba} for a given choice of K.†

Without going further into the details of the optimization of the decision rule, it can be seen that the optimum acceptance region of the X_1, X_2 space for the presence of a signal is the region in which the likelihood ratio exceeds a certain value. This is the important point here, because combining this conclusion with the results of Sections 4.1.2.1 and 4.1.2.2 demonstrates that the likelihood ratio computer is the optimum signal processor for any of the three probabilistic optimization criteria that have been discussed here (i.e. ideal observer, minimum average loss, or Neyman-Pearson).

4.1.3 The Likelihood Ratio Detector

It is apparent from the preceding discussions that the *optimum decision system* is different for various detection criteria. However, as remarked above, the difference between the ideal observer, minimum average loss, and Neyman-Pearson criteria lies in the choice of threshold, and not in the basic

† Neyman and Pearson, 1928.

means of processing the output. In all three cases, the processing operation to be performed on the output is the computation of the likelihood ratio. This computation having been accomplished, the designer then decides which of the three criteria he wishes to use and chooses a threshold for the likelihood ratio accordingly. If this ratio is above the threshold, he calls "signal present"; if below the threshold, the decision is "noise alone."

In many detection systems, the factors which determine the criterion to use and the values of such quantities as "relative loss," "a priori probability of signal presence," or "maximum allowable false alarm probability" are highly arbitrary. Once a threshold has been set, a knowledge of these parameters is presupposed. However, it is quite conceivable that the best choice of values of the parameters may change during the game, due to changes in environment or strategy. In this case, the threshold on the likelihood ratio should be left arbitrary, to be chosen according to the whims of the operator.

Since the likelihood computer seems to be part of the optimum detection system in all cases, it would seem that it should be used regardless of the choice of threshold. Therefore, in an optimum *nondecision system*, i.e., one with arbitrary threshold, the likelihood ratio or some increasing function of the likelihood ratio is computed and displayed to the operator, who decides where to set the threshold on the basis of the problem at hand. Such a nondecision system can be converted into a decision system by setting a threshold and feeding the output into automatic deciding circuits.

It is natural to ask why the likelihood ratio has properties which enhance the ability to detect the presence of a signal and noise or to distinguish between two possible signals in noise. To show this in a simple way, we will restrict attention to the case of a decision between "signal present" and "noise alone," bearing in mind that the argument holds for other binary decision problems. For convenience, we will change the notation, using s to indicate signal presence and n to indicate noise alone. The symbols s and n will replace b and a, respectively, in all probabilities and probability densities. We redefine the likelihood ratio in this notation as follows:

$$\Lambda(X_1, X_2) = \frac{P^{(s)}p_s(X_1, X_2)}{P^{(n)}p_n(X_1, X_2)} \tag{4.17}$$

By Bayes theorem (see Eq. (2.3) in Section 2.1.2), the joint probability that the signal is present *and* the output is between X_1, X_2 and $(X_1 + dX_1)$, $(X_2 + dX_2)$ is equal to the probability that the signal is present multiplied by the conditional probability of the given output if the signal is present. This joint probability is also equal to the probability of an output in the interval $X_1 - (X_1 + dX_1)$, $X_2 - (X_2 + dX_2)$ multiplied by the conditional probability of signal presence if the given output is observed. In symbols,

$$P^{(s)}p_s(X_1, X_2)\, dX_1, dX_2 = p_2(X_1, X_2)P(s/X_1, X_2)\, dX_1\, dX_2 \tag{4.18}$$

where $p_2(X_1, X_2)\, dX_1\, dX_2$ is the probability of an output in the region $X_1 -$

$(X_1 + dX_1)$, $X_2 - (X_2 + dX_2)$ regardless of signal presence or absence and $P(s/X_1, X_2)$ is the conditional probability of signal presence given the output (X_1, X_2).

An analogous statement can be made regarding the noise alone case, i.e.,

$$P^{(n)}p_n(X_1, X_2)\, dX_1\, dX_2 = p_2(X_1, X_2)P(n/X_1, X_2)\, dX_1\, dX_2 \tag{4.19}$$

where $P(n/X_1, X_2)$ is the conditional probability of noise alone given the output (X_1, X_2).

Combining (4.18) and (4.19) and computing the likelihood ratio according to (4.17), we obtain

$$\Lambda(X_1, X_2) = \frac{P(s/X_1, X_2)}{P(n/X_1, X_2)} \tag{4.20}$$

Thus we are actually computing the ratio of the probability of signal presence to that of signal absence, given the output (X_1, X_2). Since the operator knows with certainty from his observation that the output is (X_1, X_2), he uses the output and his knowledge of the statistics of the interfering noise to calculate the relative probabilities of signal presence and absence. If $\Lambda(X_1, X_2)$ is well above unity, this means that signal presence is much more likely than signal absence, justifying a positive decision on the presence of the signal.

All of the above arguments on decision systems and likelihood detectors can easily be generalized to outputs of any number of dimensions. Typically one might use N samples of the output waveform, where N can have any value. The optimum processing of the waveform, using these N sample values $X_1, \ldots, X_N$ is the computation of the likelihood ratio

$$\Lambda(X_1, \ldots, X_N) = \frac{P^{(s)}p_s(X_1, \ldots, X_N)}{P^{(n)}p_n(X_1, \ldots, X_N)} \tag{4.21}$$

where $p_s(X_1, \ldots, X_N)\, dX_1, \ldots, dX_N$ and $p_n(X_1, \ldots, X_N)\, dX_1, \ldots, dX_N$ are the "signal-plus-noise" and "noise alone" probabilities, respectively, that the set of output samples have values between $X_1, \ldots, X_N$ and $(X_1 + dX_1), \ldots, (X_N + dX_N)$.

4.1.3.1 The Likelihood Detector for Gaussian Noise

In general, the likelihood ratio detector is a nonlinear device, and might be extremely complicated. If the interfering noise is Gaussian, however, the computation of $\Lambda(X_1, \ldots, X_N)$ is particularly simple and, in fact, linear. This will be shown below.

Suppose that the output is sampled each Δt s and the values of the samples are labeled $v_1, \ldots, v_N$. The output voltage waveform $v(t)$ is a superposition of a signal $s(t)$ and a noise $n(t)$. If the noise is Gaussian with mean zero and standard deviation σ_n and the samples are assumed to be statistically independent, the PDF $p_s(v_1, \ldots, v_N)$ is given by the following equation. (See Eq. (3.1) and consider the generalization of (3.5) to N statistically independent Gaussian random variables $X_1, \ldots, X_N$ with nonzero mean values.)

$$p_s(v_1, \ldots, v_N) = \frac{1}{(2\pi\sigma_n^2)^{N/2}} e^{-(v_1-s_1)^2/2\sigma_n^2} \ldots e^{-(v_N-s_N)^2/2\sigma_n^2} \tag{4.22}$$

where $s_1, \ldots, s_N$ are the signal contributions at the sample times $t_1, \ldots, t_N$.

The "noise alone" PDF is

$$p_n(v_1, \ldots, v_N) = \frac{1}{(2\pi\sigma_n{}^2)^{N/2}} e^{-v_1{}^2/2\sigma_n{}^2} \ldots e^{-v_N{}^2/2\sigma_n{}^2} \tag{4.23}$$

From (4.21), (4.22), and (4.23), the likelihood ratio is given by

$$\Lambda(v_1, \ldots, v_N) = \frac{P^{(s)}}{P^{(n)}} e^{\left[\frac{1}{2\sigma_n{}^2}\left\{2\sum_{k=1}^{N} v_k s_k - \sum_{k=1}^{N} s_k{}^2\right\}\right]} \tag{4.24}$$

Since the natural logarithm of the likelihood ratio is an increasing function of the ratio itself, it is permissible to compute the logarithm and set a threshold on it for optimum detection. Thus, the processing computation which should be performed on the output samples $v_1, \ldots, v_N$ is

$$U = \ln(v_1, \ldots, v_N) = \ln\left[\frac{P^{(s)}}{P^{(n)}}\right] - \frac{1}{2\sigma_n^2}\sum_{k=1}^{N} s_k^2 + \frac{1}{\sigma_n^2}\sum_{k=1}^{N} v_k s_k \tag{4.25}$$

It can be concluded from (4.25) that the optimum processing of the signal samples $v_1, \ldots, v_N$ is to compute the sum $\sum_{k=1}^{N} s_k v_k$ and set a threshold on it, the threshold setting being governed by the applicable detection criterion.

To instrument this processing computation, it would be necessary to multiply each sample value v_k by a number s_k which is the value of the signal $s(t)$ expected at time t_k. This means that an a priori knowledge of the signal waveform is required in order that reference values s_k can be constructed to form the products $s_k v_k$. It is not necessary that the absolute amplitude of the signal be known, only its waveshape, which can be normalized to any desired absolute amplitude.

The processing device which computes $\sum_{k=1}^{N} s_k v_k$ will be recognized as a sampling crosscorrelator.† Thus, it can be concluded that with Gaussian noise and a complete a priori knowledge of the signal waveform the optimum processing device using the N independent sample values $v_1, \ldots, v_N$ is a sampling crosscorrelator, i.e., a system which computes the sum of products of samples of the actual observed waveform $v(t)$ and samples of a reference signal $s(t)$ generated within the system itself.

---

† See Sections 2.4.4.5 and 3.2.1.

It was previously shown (Section 3.2.1) that the processing device which maximizes the output signal-to-noise ratio is a continuous crosscorrelator or "matched filter," provided the interfering noise is white over a frequency band large compared with that of the signal. Thus, from the signal-to-noise ratio point of view, the continuous crosscorrelator is the optimum processor against white noise, whether Gaussian or otherwise. We have shown that with a probabilistic criterion of system merit, and using N statistically independent samples of the output waveform, the optimum processor against Gaussian noise, regardless of its spectrum, is a sampling crosscorrelator.†

We now consider the question of when the output noise samples can be considered statistically independent. In the case of a Gaussian noise, it was shown in Section 3.1.2 that two samples of a noise waveform are statistically independent if they are uncorrelated, i.e., separated by a time interval τ such that the autocorrelation function $\rho(\tau)$ vanishes for that time interval. Thus, if we examine the noise autocorrelation function and find a sampling interval Δt such that $\rho(\Delta t) = 0$, we can sample at time instants Δt apart and assure statistical independence between samples.

If noise is bandlimited white with central angular frequency ω_0 and bandpass bandwidth $2B$ in c/s, its NACF, according to (2.50), is

$$\rho(\tau) = \cos \omega_0 \tau \operatorname{sinc} (2\pi B \tau) \tag{4.26}$$

This function has zeros at the zeros of the function $\sin (2\pi B \tau)$, other than that at $\tau = 0$, i.e., at the values of τ equal to $n/2B$ where $n = \pm 1, \pm 2, \pm 3, \ldots$. Thus for this type of noise, the samples will be statistically independent if they are chosen at intervals equal to the reciprocal of the bandpass bandwidth of the noise in cycles per second. For example, if the noise bandwidth is 1 MHz the sampling interval should be 1 μs. A bandwidth of 10MHz dictates a sampling interval of $\frac{1}{10}$ μs. As the bandwidth approaches infinity, the sampling crosscorrelator approaches a continuous crosscorrelator.

According to the sampling theorem (Section 1.5), a time function of duration T s and whose highest frequency is B c/s can be completely specified by $2BT$ sample values taken at equal intervals.

It follows that sampling at the zeros of the autocorrelation envelope, i.e., every $1/2B$ s, completely specifies the noise contribution to the output waveform. Since the signal contribution is assumed to be known, this means that the entire waveform is specified by sampling each $1/2B$ s. Thus a continuous crosscorrelator will be an excellent approximation to the true optimum processor, and can be used in its place very effectively, if the

† There is a subtle difference between the crosscorrelator discussed here and that of Section 3.2.1. The latter produces an output in the same frequency region as the input. The device discussed in the present section translates the input frequency downward to the region near dc. The output amplitudes are the same, however.

interfering agent approximates bandlimited white Gaussian noise of bandwidth $2B$.†

4.1.3.2 Detection of a Signal with Unknown Parameters in Gaussian Noise

In the theory discussed in 4.1.3.1, it is assumed that the properties of the signal are known well enough to construct a reference waveform simulating it. However, in most applications there is at least some a priori uncertainty concerning signal properties. For example, the amplitude and phase of a sinusoid or the occurrence time of a pulse can rarely be assumed to be known at the receiving point.

Some alternative procedures for dealing with this situation will be discussed below. First a *searching crosscorrelator* can be used. The crosscorrelator can be designed with variable reference signal parameters and these parameters can be varied slowly as the crosscorrelation process is performed. Since a long integration time is required for reliable performance over each small range of the signal parameter in question, this process may require an enormous expenditure of time.

Another possible procedure is to use a multiple channel system, i.e., a bank of crosscorrelators each designed for a small range of values of the uncertain parameters, such that the signal is fed simultaneously into all correlators. This process will require less time than the searching crosscorrelator. The price of the time reduction is equipment complexity. For example, suppose a signal parameter x has an initial uncertainty range Δx, and a 90 percent detection probability is desired. Suppose further that an integration time of 10 s would be required to attain this detection probability with a given estimated input signal level and the known noise level. Assume that the range Δx can be divided into 100 equal subintervals, such that a bank of

---

† From (1.70), where $T = N/2B$, the continuous crosscorrelator output can be written in the form

$$\int_0^T dt'\, s(t')v(t')$$
$$= \sum_{m,n=1}^{N} v\left(\frac{m}{2B}\right) s\left(\frac{n}{2B}\right) \int_0^{N/2B} dt' \operatorname{sinc}\left[2\pi B\left(t' - \frac{m}{2B}\right)\right] \operatorname{sinc}\left[2\pi B\left(t' - \frac{n}{2B}\right)\right]$$

If the sinc functions are expressed in terms of their Fourier transforms and the order of integration is reversed, the above summation is proportional to

$$\sum_{m,n=1}^{N} v\left(\frac{m}{2B}\right) s\left(\frac{n}{2B}\right) \operatorname{sinc}\left[2\pi B(m-n)\right]$$

But $\operatorname{sinc}[2\pi B(m-n)]$ vanishes unless $m = n$; therefore the integral is proportional to

$$\sum_{n=1}^{N} v\left(\frac{n}{2B}\right) s\left(\frac{n}{2B}\right)$$

100 crosscorrelators can be designed for these subintervals, each one having a 90 percent probability of detecting the signal within 10 s. A searching correlator switching from one subinterval to the next at time intervals of 10 s could require up to 1000 s to detect the signal. Although the bank of 100 correlators would perform the same function in only 10 s, this time advantage would be bought at a high price, since the construction of 100 correlators is a cumbersome procedure, and in applications where equipment size and weight are important, and microminiaturized components are unavailable, such an assembly may be too bulky to be practical.

In principle, it would be possible to use maximum likelihood estimator theory (see Section 4.2) to dictate the design of a device to optimally measure the unknown variable. This measured value of the parameter could then be used in the crosscorrelator reference signal.

Another possible procedure is that of designing the correlator to average over the unknown parameters. This is dictated by the theory if it is generalized to include an initial uncertainty in signal parameters. The likelihood ratio resulting from this generalization of the theory is

$$\Lambda(v_1, \ldots, v_N) = \frac{\int \cdots \int d\beta_1 \ldots d\beta_K P^{(s)}(\beta_1, \ldots, \beta_K) p_s(v_1 \ldots, v_N/\beta_1, \ldots, \beta_K)}{P^{(n)} p_n(v_1, \ldots, v_N)} \tag{4.27}$$

where $P^{(s)}(\beta_1, \ldots, \beta_K)$ now represents the probability of the presence of a signal with parameters $\beta_1, \ldots, \beta_K$ and $p_s(v_1, \ldots, v_N/\beta_1, \ldots, \beta_K)\, dv_1, \ldots, dv_N$ is the conditional probability that the signal produces an output between $v_1, \ldots, v_N$ and $(v_1 + dv_1) \ldots, (v_N + dv_N)$ if it has parameter values $\beta_1, \ldots, \beta_K$.

The specialization of (4.27) to the case of Gaussian noise results in a processor output

$$U(v_1, \ldots, v_N) = \ln \int \cdots \int_{\boldsymbol{\beta}\text{-space}} d\boldsymbol{\beta} \frac{P^{(s)}(\boldsymbol{\beta})}{P^{(n)}} e^{\left[\sum_{k=1}^{N} \frac{v_k s_k(\boldsymbol{\beta})}{\sigma_n^2} - \sum_{k=1}^{N} \frac{s_k^2(\boldsymbol{\beta})}{2\sigma_n^2}\right]} \tag{4.28}$$

where the vector $\boldsymbol{\beta}$ now represents the set of parameters $\beta_1, \ldots, \beta_K$.

Note that the optimum processing filter dictated by (4.28) is again a sampling crosscorrelator, but this time the theory tells us that it should:

1. Take the exponential of the computed crosscorrelation function

$$\frac{1}{\sigma_n^2} \sum_{k=1}^{N} v_k s_k(\boldsymbol{\beta})$$

for each value of every unknown parameter and multiply it by the exponential of

$$-\frac{1}{2\sigma_n^2} \sum_{k=1}^{N} s_k^2(\boldsymbol{\beta})$$

2. Weight the product resulting from step (1) with the probability density for the given values of the unknown parameters.

3. Integrate the product resulting from steps (1) and (2) over all possible values of all unknown parameters.

The operations (1), (2), and (3) are seen to involve, in a complicated way, a search for the values of the unknown parameters. The difference between this idealized procedure and conventional search technique is that, with the latter, one does not take the exponential of the crosscorrelator output. The usual search technique, then, is equivalent to a series expansion of the exponential of (4.28), retention of only the linear term, and averaging over unknown parameters. As will become apparent in later discussions of specific applications, under small SNR conditions, this technique is for practical purposes equivalent to the theoretical optimum detection technique dictated by (4.28).

Logically, what we are doing when we process according to (4.28) is testing the two alternative hypotheses (1) "a signal $s(t)$ is present with specific parameter values $\boldsymbol{\beta}$," and (2) "such a signal is not present," performing a likelihood ratio analysis for this specific set of values of the components of $\boldsymbol{\beta}$, and averaging over all possible combinations of the parameters $\boldsymbol{\beta}$. If this is to be done on a continuous basis, it must carry the assumption that the statistics of all signal and noise processes remain the same during the entire processing period, and it must maintain the parameter values effectively constant during each computation of the crosscorrelation function $\sum_k v_k s_k(\boldsymbol{\beta})$, i.e., sweep these parameters so slowly that they remain effectively constant over the duration of the signal. This, like the search technique, implies a requirement of *time*, a commodity which is not always readily available.

4.2 MEASUREMENT

The process of measurement is in general more complex than that of detection, since it involves the attempt to answer the question, "What is the value of a particular variable?" In either case, the decision to be made is no longer binary, as in the detection problems discussed in Section 4.1. It is a decision between many alternatives, and therefore in the information-theoretic sense is an attempt to acquire more information than is asked for in a detection process. In general, if the variable to be measured has the *true value* x_t, the instrument makes a transformation on x_t which can be denoted in general by $x_i = f(x_t)$, where x_i symbolizes the *indicated value* of x_t. The error is defined as

$$\epsilon = x_i - x_t = f(x_t) - x_t \tag{4.29}$$

Two general types of errors exist: (1) systematic errors and (2) random errors. A systematic error is not statistical in nature. It is simply a bias existing on the measurement that is completely independent of any noise which might be present. An example is a scale using a spring that obeys Hooke's law $W = k_1 x$ for loads up to 100 lbs. and obeys a law $W = k_2\sqrt{x}$ beyond 100 lbs., where W is weight and x is elongation. If the designer calibrates the scale as if Hooke's law were obeyed above 100 lbs., then if k_1 is unity and $k_2 = 10$ the indicated weight of a 150-lb. object is 225 lbs. This represents a large error due to the failure of the scale designer to account for departures from Hooke's law. If he had made the necessary measurements before designing the scale, the error could have been eliminated by proper calibration.

Systematic errors, then, are due to failure to compensate for biasing of the measurement due to *known* physical effects. If the error is known, then at least in principle, it can always be biased out in calibrations. "Known" is a key word here, because if an effect is unknown to the designer or observer of an instrument, then by definition it becomes dependent on random variables *from his point of view*.† The resulting error is then a *random error*. Even with our scale example, if the designer had no way of knowing of the deviation from Hooke's law, then he could not compensate it out, and *strictly from his point of view*, it must be considered as a kind of random error or noise.

In general, random errors are of two types, those with finite mean value and those with zero mean value. We will denote these error types as Class I and Class II, as follows:

Class I:

$$\langle \epsilon \rangle = \langle f(x_t) - x_t \rangle \neq 0 \tag{4.30}$$

Class II:

$$\langle \epsilon \rangle = 0, \text{ but } \langle \epsilon^2 \rangle = \langle [f(x_t) - x_t]^2 \rangle \neq 0$$

Class I random errors cannot be eliminated by the familiar technique of simply averaging over many samples from the ensemble of the random variables involved. No matter how much averaging is done, the residual mean error $\langle \epsilon \rangle$ will always remain. To deal with this situation, one can in principle measure the average error and subtract it out, leaving only a Class II error.

An example of a Class I random error is a case where an electrical signal y is accompanied by additive noise n, and the sum is driven through a square law device. Thus $x_t = y^2$ and the indicated output, if calibration neglects the presence of noise, is

$$x_i = f(x_t) = (y + n)^2 = y^2 + 2yn + n^2 \tag{4.31}$$

† The idea of a subjective probability, i.e., a probability from the point of view of a particular person, is invoked in many places throughout this book. Its use reflects the author's philosophy concerning the use of probability concepts in engineering analysis. The probability idea is useful in situations where a certain degree of ignorance prevails. It need not be universal ignorance, but only individual ignorance. If one thinks of it this way, it is easy to understand the phrase "probability from his point of view."

The observer now thinks he is reading y^2, while he is actually reading the quantity on the right-hand side of (4.31). If $\langle n \rangle = 0$, the average error is

$$\langle \epsilon \rangle = \langle (y^2 + 2yn + n^2) - y^2 \rangle = \langle n^2 \rangle \tag{4.32}$$

This is a Class I random error, and it cannot be averaged out. However, if $\langle n^2 \rangle$ is accurately known to the instrument designer, it can be electronically subtracted out after squaring such that the transformation performed by the instrument is

$$x_i = f(x_t) = (y + n)^2 - \langle n^2 \rangle \tag{4.33}$$

The average error is now

$$\langle \epsilon \rangle = \langle y^2 + 2yn + n^2 - \langle n^2 \rangle - y^2 \rangle = 0 \tag{4.34}$$

and the instrument contains only Class II errors, which can be reduced by averaging.

The root mean square error in this case is (recalling that $\langle n \rangle = 0$)

$$\epsilon_{\text{rms}} = \sqrt{\langle \epsilon^2 \rangle} = \sqrt{4y^2\langle n^2 \rangle + \langle n^4 \rangle + 4y\langle n^3 \rangle - \langle n^2 \rangle^2} \tag{4.35}$$

while that in the case where $\langle n^2 \rangle$ is not subtracted out is

$$\epsilon_{\text{rms}} = \sqrt{4y^2\langle n^2 \rangle + 4y\langle n^3 \rangle + \langle n^4 \rangle} \tag{4.36}$$

4.2.1 Optimum Measurement

We have discussed the optimization of the process of detection of a signal in noise. We now ask whether there is an analogous optimization theory for the process of measurement of a variable in noise. The answer is in the affirmative, with modifications in the definition of "optimum."

The optimum process of measurement of a single variable β can be defined as that process producing a measured value β_m as near as possible to the true value β_t. To approach the problem from the point of view of the party making the measurement, whom we will call A, let us recognize that A does not know the true value of the variable, for if he did there would obviously be no need to measure it. He does, however, have some a priori knowledge of the possible range of values of β, e.g., suppose he knows it to be confined to a region between a minimum value β_1 and a maximum value β_2. Designating $(\beta_2 - \beta_1)$ as $\Delta\beta$, we can define an a priori probability density function for β, from A's point of view, as follows:

$$\begin{aligned} p(\beta) &= \frac{1}{\Delta\beta}, \qquad \beta_1 \leq \beta \leq \beta_2 \\ &= 0, \qquad \beta < \beta_1;\ \beta > \beta_2 \end{aligned} \tag{4.37}$$

Now suppose A designs his measuring instrument, based on his knowledge of a physical principle involving the variable β. He takes a set of N readings of the output of his instrument, which we will denote by $x_1, \ldots, x_N$.

Bayes' theorem (Eq. (2.3)) tells us that

$$p(\beta)p(x_1, \ldots, x_N/\beta) = p(x_1, \ldots, x_N)p(\beta/x_1, \ldots, x_N) \tag{4.38}$$

where $p(x_1, \ldots, x_N/\beta)$ represents the PDF for observation of the N outputs $x_1, \ldots, x_N$, conditional on the designated value of β, $p(x_1, \ldots, x_N)$ is the PDF for observation of $x_1, \ldots, x_N$ regardless of the value of β, and $p(\beta/x_1, \ldots, x_N)$ is the PDF for β conditional on observation of $x_1, \ldots, x_N$.

We will make a special point of the fact that (4.38) is a "mental equation" written by our observer A, i.e., it is a subjective statement reflecting A's state of ignorance about the variable β. Another observer who knew more about the true value of β would write another version of (4.38). Also, exclusively from A's point of view, all that is known about β prior to the measurement is its a priori PDF given by (4.37). The function $p(\beta/x_1, \ldots, x_N)$ is the observer's PDF for β *after* he has made the measurement and is known as the a posteriori probability density function for β.

Now strictly speaking, the a posteriori PDF for β represents all the knowledge of β that has been made possible through the measurement. Using (4.37) in (4.38) we note that A could write, after observing the instrument outputs $x_1, \ldots, x_N$

$$\begin{aligned} p(\beta/x_1, \ldots, x_N) &= \frac{1}{\Delta\beta}\frac{p(x_1, \ldots, x_N/\beta)}{p(x_1, \ldots, x_N)}, && \beta_1 \leq \beta \leq \beta_2 \\ &= 0, && \beta < \beta_1; \beta > \beta_2 \end{aligned} \tag{4.39}$$

That is, using (4.39) he could write the PDF for β from the instrument outputs $x_1, \ldots, x_N$, provided he knows the statistics of the noise in his instrument, as given by the conditional PDF $p(x_1, \ldots, x_N/\beta)$.† This would constitute the maximum use of the information about β that could be acquired through the measurement.

However, in most practical measurement problems, a complete knowledge of the a posteriori distribution of β is not necessary; it is required only that a decision be made about the value of β as a result of the measurement. In this connection, the most reasonable quantity to ask for is the "most probable value of β," i.e., that value which maximizes the a posteriori probability given by (4.39). It must be noted, however, that this value is only the most probable from the observer's point of view after he has made the measurement. In this sense, the computation of the most probable value of β is the "optimum" processing that can be done on the instrument output readings $x_1, \ldots, x_N$, if it is postulated that a definite decision must be made about the value of β. In making such a decision, one is throwing away information about β contained in a complete display of $p(\beta/x_1, \ldots, x_N)$, but ordinarily this information is superfluous from the practical point of view. A decision would be required eventually, and if a human observer were to make it from

† After making the measurement, A knows with certainty that the values $x_1, \ldots, x_N$ have been observed; therefore from his point of view, $p(x_1, \ldots, x_N)\, dx_1 \ldots dx_N = 1$.

a display of the a posteriori PDF, he would in many cases choose the most probable value.

To find the most probable value of β, usually known as the *maximum likelihood estimator* of β (since it is the estimated value of β that maximizes $(1/\Delta\beta)p(x_1, \ldots, x_N/\beta)$, which can be designated as the *likelihood function*), we differentiate (4.39) with respect to β (noting that the derivative of a step function is an impulse function) and set the result equal to zero, as follows:

$$\frac{\partial p}{\partial \beta}(\beta/x_1, \ldots, x_N) = 0 = \frac{1}{\Delta\beta}\frac{\partial p}{\partial \beta}(x_1, \ldots, x_N/\beta) + p(x_1, \ldots, x_N/\beta)\frac{1}{\Delta\beta}\cdot[\delta(\beta - \beta_1) - \delta(\beta - \beta_2)], \quad \beta_1 \leq \beta \leq \beta_2 \tag{4.40}$$

The second term vanishes identically except at the end points β_1 and β_2. Outside the interval $(\beta_1 \leq \beta \leq \beta_2)$, the a posteriori probability must be negligibly small, since no measurement can place a possible value of β outside the interval in which it was known to be located before the measurement was made. Also, it can be assumed that $p(x_1, \ldots, x_N/\beta)$ is negligibly small at the end points, because, with the aid of (4.39), this assumption becomes tantamount to the statement that the possible range of values of β has been reduced by the measurement (i.e., the a posteriori PDF $p(\beta/x_1, \ldots, x_N)$ is negligibly small at the end points). A measurement which does not fulfill this assumption has not reduced the uncertainty about the value of β, and consequently such a measurement would be worthless.

If the above considerations are accounted for in (4.40), the optimum processing of the instrument output readings $x_1, \ldots, x_N$ is the computation dictated by the condition

$$\frac{\partial p}{\partial \beta}(x_1, \ldots, x_N/\beta) = 0 \tag{4.41}$$

The result (4.41) is not critically dependent on the chosen a priori probability for β given by (4.37) provided the function $p(\beta)$ is essentially flat over a region whose width is greater than that of $p(\beta/x_1, \ldots, x_N)$. This will generally be true in practice, since as remarked above, a measurement in which it does not apply would not reduce the uncertainty in the value of the parameter being measured and hence would be of little value.

A very simple and useful example of the application of (4.41) is the case where $x_1, \ldots, x_N$ are independent samples from a Gaussian random function with variance σ_n^2, where the mean values x_i are functions of β. In this case, (4.41) takes the form

$$\frac{\partial}{\partial \beta}\left[e\left[-\frac{1}{2\sigma_n^2}\sum_{i=1}^{N}[x_i - \langle x_i(\beta)\rangle]^2\right]\right] = -\frac{1}{\sigma_n^2}\,e\left[\frac{1}{2\sigma_n^2}\sum_{i=1}^{N}[x_i - \langle x_i(\beta)\rangle]^2\right] \cdot\left\{\sum_{i=1}^{N}[x_i - \langle x_i(\beta)\rangle]\frac{\partial\langle x_i\rangle}{\partial \beta}\right\} = 0 \tag{4.42}$$

The condition (4.42) is equivalent to

$$\sum_{i=1}^{N} x_i \frac{\partial\langle x_i(\beta)\rangle}{\partial\beta} = \sum_{i=1}^{N} \langle x_i(\beta)\rangle \frac{\partial\langle x_i(\beta)\rangle}{\partial\beta} \tag{4.43}$$

4.2.2 Measurement of a Static Variable

One particular specialization of (4.43) leads to a fact well-known to everyone who has ever made measurements, namely that random errors in measurement of a static variable† can be reduced by simply taking the arithmetic mean of a large number of independent readings of the instrument output. The more readings, the more accurate the results. It might be asked whether this averaging out of random errors is truly optimum, or whether some other method of combining the output readings would produce better results.

The answer is that if the error-producing agent is additive Gaussian noise with zero mean, computation of the arithmetic mean is optimum in the sense that it is the maximum likelihood estimator of the variable, as discussed above.

If each independent sample of the instrument output consists of the true value of the variable, assumed to be constant in time, plus a sample of Gaussian noise, then the maximum likelihood estimator is given by (4.43), where $\langle x_i\rangle = \beta$. In this case (4.43) degenerates into

$$\frac{1}{N}\sum_{i=1}^{N} x_i = \beta \tag{4.44}$$

Thus (4.44) tells us that the time-honored procedure of taking an arithmetic mean of as many independent readings as possible is, theoretically, the optimum processing (in the maximum likelihood estimator sense) that can be done to enhance the accuracy of measurement of a variable in additive Gaussian noise. It is not necessarily optimum with other forms of interference.

We can evaluate the performance of the processing method indicated by (4.44) in the presence of additive noise by determining: (1) the rms error versus the number of samples N, where error is defined as the difference between true and measured value of β, or (2) the probability that the measured value of β will differ from the true value by no more than some specified amount, say 1 percent or 10 percent or a given number of units of the variable.

The rms error, with N samples, is

$$\epsilon_N = \sqrt{\langle(\beta_m - \beta_t)^2\rangle} = \sqrt{\langle\beta_m^2\rangle - \beta_t^2} \tag{4.45}$$

where β_m and β_t are measured and true values of β, respectively.‡ Note

† A variable that does not change its average value over a long period of time, say hours or days.

‡ Note that $x_i = \beta_t$ by assumption. Therefore,

$$\langle\beta_m\rangle = \frac{1}{N}\sum_{i=1}^{N}\langle x_i\rangle = \beta_t \quad \text{and} \quad \langle(\beta_m - \beta_t)^2\rangle = \langle\beta_m^2 - \beta_t^2\rangle = \langle\beta_m^2\rangle - \beta_t^2$$

that ϵ_N is equivalent to the standard deviation of the random variable β.

Combining (4.44) and (4.45), we obtain

$$\epsilon_N = \sqrt{\frac{1}{N^2} \sum_{i,j=1}^{N} \langle(\beta_t + n_i)(\beta_t + n_j)\rangle - \beta_t^2} = \sqrt{\frac{1}{N^2} \sum_{j=1}^{N} \langle n_j^2 \rangle} = \frac{\sigma_n}{\sqrt{N}} \tag{4.46}$$

where n_i refers to the noise sample at $t = t_i$, the assumed statistical independence of n_i and n_j and the assumption of zero mean assure the vanishing of $\langle n_i n_j \rangle$ for $i \neq j$, and σ_n^2 is the mean-square noise $\langle n_i^2 \rangle$, which is the same from sample to sample because the noise has been assumed stationary.

The result (4.46) tells us the rather well-known fact that the rms error decreases as the square root of the number of readings. Note that it contains no assumption about the statistics of the noise except that it is additive with β, has zero mean, and the samples are independent.

If we prefer a probabilistic criterion of system performance to the rms error, we can assume the noise to be Gaussian, in which case the probability that the sum (4.44) falls within a region from $\beta_t - (\delta\beta/2)$ to $\beta_t + (\delta\beta/2)$ is

$$\begin{aligned} p(\delta\beta) &= \text{Prob}\left\{\beta_t - \frac{\delta\beta}{2} \leq \beta_m \leq \beta_t + \frac{\delta\beta}{2}\right\} \\ &= \frac{\sqrt{N}}{\sqrt{2\pi\sigma_n^2}} \int_{\beta_t - (\delta\beta/2)}^{\beta_t + (\delta\beta/2)} e^{\left[-\frac{(x-\beta_t)^2}{2(\sigma_n{}^2/N)}\right]} dx = \text{erf}\left[\frac{\delta\beta}{2\sigma_n}\sqrt{\frac{N}{2}}\right] \\ &= \text{erf}\left[\frac{\delta\beta}{2\sqrt{2}\,\epsilon_N}\right] \end{aligned} \tag{4.47}$$

where we have used (4.46), the fact that the sum of Gaussian random variables is a Gaussian random variable (see Section 3.1.3) and the additional fact that the standard deviation of the sum β in (4.44) is $\sigma_n/\sqrt{N}$, as implied by (4.45).† The error function erf (x) was defined in Section 3.1.1 as

$$\frac{2}{\sqrt{\pi}} \int_0^x dx'\, e^{-x'^2}$$

and is extensively tabulated (see the footnote below Eq. (3.2)).

The procedure in evaluating the accuracy of the measurement is to plot $P(\delta\beta)$ versus $[(\delta\beta/2\sigma_n)\sqrt{N/2}]$ (see Figure 4-2), using tables of the error function, and to find the point on the curve corresponding to some value of $P(\delta\beta)$ that we are willing to consider as certainty, e.g., $P(\delta\beta) = 0.9, 0.99$, or 0.999 (the less conservative among us might be willing to regard a 75 percent probability as a state of certainty). Having decided upon such a value of $P(\delta\beta)$ and designating the corresponding value of $\delta\beta$ by $\delta\beta_0$, we can then regard our measurement as accurate to within $\delta\beta_0$, where

$$P(\delta\beta_0) = \text{erf}\left[\frac{\delta\beta_0}{2\sqrt{2}\,\epsilon_N}\right] \tag{4.48}$$

† Strictly speaking, (4.47) should be averaged over β_t, but since the result turns out to be independent of β_t, this step need not be shown.

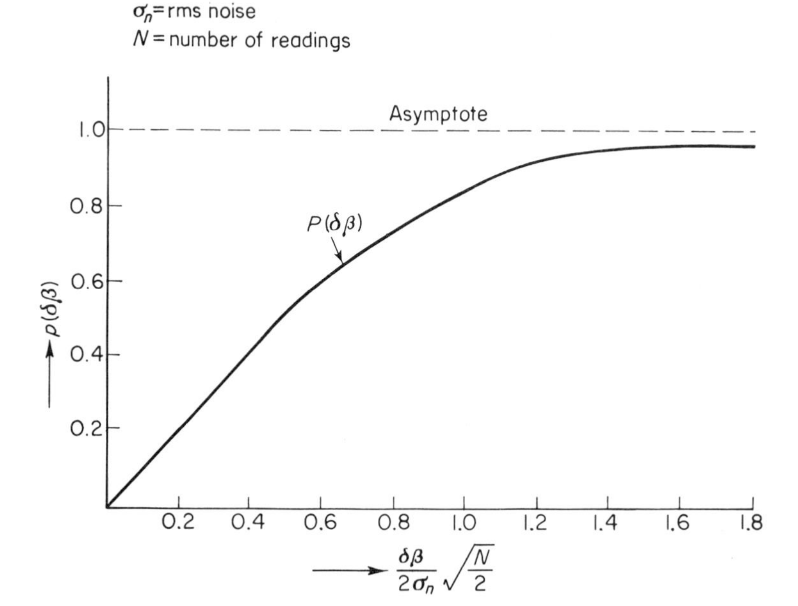

Figure 4-2 Probability that measured variable falls between $[\beta_t - (\delta\beta/2)]$ and $[\beta_t + (\delta\beta/2)]$.

If, for example, the chosen value of $P(\delta\beta_0)$ is 0.999, then (4.48) gives us

$$\delta\beta_0 = (6.56)\epsilon_N \tag{4.49.a}$$

For a few other values of $P(\delta\beta_0)$, we have

$$\delta\beta_0 = 5.14\,\epsilon_N, \qquad P(\delta\beta_0) = 0.99 \tag{4.49.b}$$

$$\delta\beta_0 = 3.28\,\epsilon_N, \qquad P(\delta\beta_0) = 0.90 \tag{4.49.c}$$

$$\delta\beta_0 = 2.30\,\epsilon_N, \qquad P(\delta\beta_0) = 0.75 \tag{4.49.d}$$

Thus with Gaussian noise, the rms error ϵ_N is linearly related to the quantity $(\delta\beta_0)$, the range of values within which the true value of the measured variable β is 99.9 percent (or, if desired, 99 percent, or 90 percent, or 75 percent) certain to be located.

4.2.3 Dynamic Measurement

Throughout the remainder of this book, we will often be discussing, directly or indirectly, dynamic measurement processes. These are defined as processes which involve measurement of rapidly changing variables and consequently must be carried out within a short period of time.

The theory of static measurements, as developed in 4.2.1, will be the foundation for discussions of dynamic measurement processes. The important difference between the two types of process is that in the case of the former the number of independent samples of the time dependent variable $\beta(t)$ that can be summed, and hence the attainable accuracy, is unlimited, while for the latter, the maximum number is limited by the period during which the value of the variable can be considered effectively constant.

To illustrate these points in the simplest manner we will confine attention to the additive Gaussian noise case discussed above. In this case, the samples must be taken at least τ_c time units apart, where τ_c is the noise correlation time (see Section 2.4.4.1) or, equivalently, the reciprocal of twice the highest noise frequency or the bandpass bandwidth of the noise, $2B_n$. If our variable $\beta(t)$ remains effectively constant for only ΔT time units, then we can apply the static measurement averaging process discussed above, but are limited to summation of $(2B_n \, \Delta T)$ independent samples in order to measure the effectively static value of β during this interval. The statement that β remains effectively constant only over time ΔT is equivalent to the statement that B_s, the lowpass bandwidth of $\beta(t)$, is the reciprocal of $\kappa \, \Delta T$, where κ is a constant. The precise value of κ depends on the shape of the signal spectrum,† but it should usually differ from unity by less than an order of magnitude.

The net result of the above reasoning is as follows: The maximum number of independent values of $\beta(t)$ that can be summed during an interval in which $\beta(t)$ has an effectively constant value (or, equivalently, the maximum number of independent samples N) is $\kappa(B_n/B_s)$, a specified fraction of the ratio of the bandwidth of the interfering noise to that of $\beta(t)$. In symbols,

$$N_{\max} = \kappa \frac{B_n}{B_s} \tag{4.50}$$

where $N_{\max}$ is the maximum number of independent samples of $\beta(t)$.

Let us consider the implications of (4.50). Combining (4.46) and (4.50) we have

$$\epsilon_{\min} = \frac{\sigma_n}{\kappa} \frac{B_s}{B_n} \tag{4.51}$$

---

† For example, by a simple analysis we can determine the percentage change in a signal $s(t)$ during the time interval 0 to t, given by

$$\left| \frac{s(t) - s(0)}{s(0)} \right| \times 10^2$$

where $s(t)$ has a Fourier transform that is flat between $\omega = -\pi B_s$ and $\omega = \pi B_s$. This percentage change is $|\operatorname{sinc}(\pi B_s t) - 1| \times 10^2$. If $(\pi B_s t)$ is sufficiently small, the first two terms of the power series expansion of sinc $(\pi B_s t)$ are adequate to describe the percentage change, which turns out to be roughly equal to $|(\pi B_s t)^2/6 \times 10^2| + 0((\pi B_s t)^4)$. Suppose we define ΔT as the time during which the change is less than 10 percent. Then in this case, if terms $0((\pi B_s t)^4)$ can be neglected, κ is about 4. If the change is 5 percent, then κ is about 6.

where $\epsilon_{\min}$ is the minimum attainable value of the rms error ϵ_N. But $\epsilon_{\min}$ is related (through (4.48) and Figure 4-2) to the accuracy of measurement in the probabilistic sense. Consequently, (4.51) can be considered as a limitation imposed on the attainable measurement accuracy by the presence of noise whose bandwidth is not overwhelmingly large compared to that of the time-varying quantity one is attempting to measure. For a sufficiently narrow band $\beta(t)$, i.e., one which is effectively constant over a long period, and an extremely wide band noise, the accuracy is not seriously limited.

Note that we have not explicitly mentioned the process of electronically averaging independent samples of a dynamic variable $\beta(t)$. To accomplish the equivalent operation with an RC integrator, we can design the integrator with a time constant T_c such that an integration is effectively carried out over the period ΔT, i.e., all past inputs are "forgotten" every ΔT time units. In accordance with the arguments presented above, this implies an integrator for which

$$T_c = RC \simeq \Delta T = \frac{\kappa}{B_s} \tag{4.52}$$

A longer time constant will "smear" out the information contained in the time variations of the signal, while a shorter one will not take full advantage of all independent samples available during the "static" period ΔT.

In a qualitative sense, the result (4.52) is that which simple frequency domain reasoning would also have provided, in a case where the signal spectrum is such that $\kappa \simeq 1$. The filter bandwidth is the reciprocal of the time constant, and hence if the noise has a wider band than the signal, the highest signal-to-noise ratio would naturally be achieved by using a filter with bandwidth equal to that of the signal. This elementary line of reasoning is thus seen to be essentially equivalent to that based on the statistical concept of averaging out random errors in a measurement.

To carry the argument further, note that a filter with bandwidth B_s will be ineffective in noise suppression if B_s is greater that κB_n. A glance at (4.50) shows us that in this case, the maximum number of independent samples available for integration is less than unity. Thus no enhancement of signal-to-noise ratio is attainable through the technique of averaging out random errors.

Another question arises when one attempts electronic integration of a signal in noise, that of sampling of the input to the integrator. If the input waveform is sampled periodically at intervals of $\tau_c = 1/2B_n$, where $B_n > B_s$, as postulated above, then no information about the waveform has been lost, and the integration is equivalent to that of a continuous input waveform. To illustrate the point, we compare the integrals

$$\mathscr{I}_1(t) = \int_0^t s(t - t')\, e^{-t'/T_c}\, dt' \tag{4.53}$$

and

$$\mathscr{I}_2(t) = \tau_c \int_0^t s(t - t') e^{-t'/T_c} \sum_{n=1}^{N} \delta(t - t' - n\tau_c)\, dt' \tag{4.54}$$

If $s(t)$ is a square pulse, from 0 to t, then the results of integration of (4.53) and (4.54), respectively, are

$$\mathscr{I}_1(t) = T_c(1 - e^{-t/T_c}) \tag{4.55}$$

and

$$\mathscr{I}_2(t) = \tau_c \sum_{n=1}^{N} e^{-(t-n\tau_c)/T_c} = \tau_c e^{-t/T_c} \left\{ \sum_{n=0}^{\infty} [e^{\tau_c/T_c}]^n - \sum_{n=N+1}^{\infty} [e^{\tau_c/T_c}]^n - 1 \right\} \tag{4.56}$$

With the aid of the series expansion for $1/(1 - x)$,† we obtain, provided that $\tau_c/T_c \simeq B_s/B_n \ll 1$ (aided by the fact that $N\tau_c = t$),

$$\mathscr{I}_2(t) \simeq T_c(1 - e^{-t/T_c}) \tag{4.57}$$

which is equivalent to (4.55). Thus if

$$\frac{B_s}{B_n} \ll 1$$

and if the integrator has a (lowpass) bandwidth B_s, then the effect of integrating the continuous input waveform is essentially the same as that of integrating a waveform sampled at intervals of $1/2B_n$.

REFERENCES

Detection Theory, Estimation Theory, Theory of Errors in Measurement

Beers, 1953

Davenport and Root, 1958; Chapter 14

Deutsch, 1965; especially Chapter 9

Dwork, 1950

Hancock and Wintz, 1966

Helstrom, 1960

Kotelnikov, 1960

Middleton, 1960; Part 4

Middleton and Van Meter, 1955

Middleton and Van Meter, 1956

Selin, 1965

Swerling, 1959

Topping, 1955

Wainstein and Zubakov, 1962

Woodward, 1953; Chapter 4

---

† $1/(1 - x) = \sum_{n=0}^{\infty} x^n$, if $|x| < 1$.

5

ANALYSIS OF RADIO SYSTEMS

The applications of statistical communication theory to radio technology are so extensive that no hope exists of covering all of them within a few chapters of a single book. The best we can hope to do is to discuss some of the simplest ideas and to briefly suggest extensions of these to more abstract situations. Before undertaking this, it is desirable to first discuss some of the physical and engineering ideas and facts basic to radio system analysis. The present chapter is included to provide some of the elementary ideas in radio wave propagation, antennas, and other important radio engineering topics, in order to facilitate the reading of later chapters in which these topics are used in discussions of radio system applications of statistical communication theory.

This chapter is peripheral to the mainstream of the book. It can be bypassed without a serious effect on continuity, under the assumption that the reader has some prior familiarity with elementary aspects of radio engineering. Those who do not are advised to read the chapter in order to acquire a cursory glance at this vast subject, then eventually to supplement their reading with more detailed treatments to be found in standard textbooks on various aspects of radio science and radio engineering, some of which are cited in the reference list at the end of the chapter. In order to keep the length of the chapter within reasonable bounds (in view of its peripheral nature), it was necessary to present most of the material without mathematical derivations or extensive background discussions.

The basic radio system, as it will be defined here, consists of a transmitter generating waves at a frequency between 3 kc/s and 100,000 Mc/s,† a medium through which the wave travels, and a receiver that intercepts the wave and extracts information from it. The medium might include the earth, the ionosphere, flying targets as in radar, environmental appendages such as mountains, or, in some radio situations, a portion of outer space.

5.1 RADIO WAVES IN INFINITE MEDIA

In the general discussion of waves in Section 1.3 it was noted that the behavior of radio waves is governed theoretically by the Maxwell equations. The wave equation (1. 29) applied to electric and magnetic fields can be developed from these equations. The reader is referred to the literature for a demonstration of this fact.‡ For present purposes, it suffices to say that the wave equation itself can be regarded as basic and many important facts about radio wave propagation can be deduced from it.

The discussion in Section 1.3.1 applies to the propagation of a plane wave in a linear, homogeneous, isotropic medium of infinite extent. The simple analysis presented here is based on solution of (1.29), where the complex amplitude is assumed to be a function of only one coordinate. In the present discussion the complex amplitude referred to is an electric or magnetic field. Either field obeys (1.29) in an infinite medium, where *propagation constant* k at frequency $\omega/2\pi$ is ω/v, v being the velocity of the waves in the medium, i.e.,

$$k = \frac{\omega}{v} \tag{5. 1}$$

$$v = c\sqrt{\frac{\epsilon_0 \mu_0}{\epsilon \mu}}$$

where c is the velocity of light in free space (3×10^8 m/s), ϵ_0 and μ_0 are, respectively, the permittivity and magnetic permeability of free space, ϵ/ϵ_0 is the permittivity of the medium relative to that of free space, and μ/μ_0 is the magnetic permeability of the medium relative to that of free space.

In the MKS system of units, $\mu_0 = 4\pi(10^{-7})$ H/m and $\epsilon_0 = (1/36\pi)(10^{-9})$ F/m. Also in all media of concern here, $\mu = \mu_0$. Therefore variations in refractive index are determined entirely by variations in ϵ.

Some media of interest to radio propagation (e.g., moist soil and sea water) are *lossy*, i.e., they have an electrical conductivity σ that is large compared to ($\epsilon\omega$), the latter when multiplied by the electric field being referred

---

† We exclude *extremely low frequencies* (ELF), i.e., frequencies below 3 kc/s.

‡ See Jordan, 1950, Chapter 5, Part 1, especially pp. 113–114.

to as *displacement current density*. The *conduction current density* is $\mathbf{J} = \sigma\mathbf{E}$, where $\mathbf{E}$ is the electric field. The *lossiness* of a medium is usually characterized by its ratio of conduction current to displacement current. Displacement current involves no dissipation of energy, being merely a local oscillation of an electric field, such as takes place in the dielectric of a capacitor excited by an ac source. Conduction current, on the other hand, involves a coupling of the electric field with free electrons or ions in the medium, usually involving a progressive loss of energy by the propagating wave. The latter shows itself mathematically in the complex representation of the Maxwell equations† through the fact that the dielectric coefficient of the medium turns out to be complex and of the form

$$\epsilon_c = \epsilon + j\frac{\sigma}{\omega} \tag{5.2}$$

From (5.1), (5.2), and the assumption of a highly lossy medium, that is, a medium in which $\sigma/\omega\epsilon \gg 1$, it follows that

$$k = \frac{\omega}{v} = \omega\sqrt{\mu\epsilon_c} \simeq \sqrt{\frac{\omega\mu\sigma}{2}}(1 + j) \tag{5.3}$$

The electric field vector $\mathbf{E}$ or magnetic field vector $\mathbf{H}$ of a wave travelling through the lossy medium in the positive z-direction has the form (where $\mathbf{i}_x$, $\mathbf{i}_y$, and $\mathbf{i}_z$ are unit vectors along the x-, y-, and z-axes)

$$\begin{matrix} \mathbf{E} & \mathbf{i}_x(E_0) \\ \simeq & \\ \mathbf{H} & \mathbf{i}_y\left(\dfrac{k}{\omega\mu}E_0\right) \end{matrix} \quad \ldots\; e^{j(z/\delta)-(z/\delta)} \tag{5.4}$$

where E_0 is the amplitude of the electric field, assumed polarized in the x-direction,‡ and $\delta = \sqrt{2/\omega\mu\sigma}$. This is known as the *skin depth*, this nomenclature expressing the fact that the wave amplitude is attenuated in the amount (e^{-1}) as it travels a distance δ; hence the skin depth is a measure of the depth of penetration into a lossy medium. Note that a phase shift of $\pi/2$ also takes place as the wave travels a distance $(\pi/2)\delta$, and note also that skin depth is inversely proportional to the square root of frequency. For this reason, a very lossy medium favors low frequency waves.

In slightly lossy medium, i.e., where $\sigma/\omega\epsilon \ll 1$, the fields take the form§

$$\begin{matrix} \mathbf{E} & \mathbf{i}_x(E_0) \\ \simeq & \\ \mathbf{H} & \mathbf{i}_y\left(\dfrac{k}{\omega\mu}E_0\right) \end{matrix} \quad \ldots\; e^{j\omega\sqrt{\mu\epsilon}z}\, e^{-(\sigma\sqrt{\mu}/2\sqrt{\epsilon})z} \tag{5.5}$$

† See, Jordan, 1950, pp. 124–128.

‡ That is, the oscillating electric field is entirely in the x-direction.

§ The expansion of $\sqrt{\epsilon_c} = \sqrt{\epsilon(1 + j(\sigma/\omega\epsilon))}$ in powers of $(\sigma/\omega\epsilon)$ yields $\sqrt{\epsilon}(1 + j(\sigma/2\omega\epsilon) + 0((\sigma/\omega\epsilon)^2)$, from which (5.5) follows.

The *wave impedance* of the medium Z_w, consisting of a real *wave resistance* R_w and an imaginary *wave reactance* X_w, is defined as follows:

$$Z_w = \left|\frac{E_x}{H_y}\right| = \sqrt{\frac{\mu}{\epsilon}} = R_w + jX_w \tag{5.6}$$

and is analogous to a circuit impedance in the sense that its reciprocal is a measure of current (displacement plus conduction) flow in response to a given applied voltage, the latter being proportional to the magnitude of the electric field associated with the propagating wave.

The power flow in a plane wave is given by the real part of the component of the *Poynting vector* **P** in the direction of propagation; that is, with the aid of (5.6),

$$\text{power flow} = \mathbf{i}_z P = \tfrac{1}{2}\,\text{Re}\,\{\mathbf{E} \times \mathbf{H}^*\} = \frac{|E_0|^2 R_w}{2|Z_w|^2} \tag{5.7}$$

where $\mathbf{i}_z$ is the z-directed unit vector.

The quantity defined in (5.7) is actually the power flowing through unit area, or the *intensity* of the wave. For a medium that is nearly lossless, the wave impedance is almost purely real and according to (5.5) and (5.6) is approximately equal to $\sqrt{\mu/\epsilon}$. In this case

$$\text{power flow} = \sqrt{\frac{\epsilon}{\mu}}\,\frac{|E_0|^2}{2} \tag{5.8}$$

The wave impedance of free space will be denoted by Z_{fs}. In free space, $Z_{fs} = \sqrt{\mu/\epsilon} = 377$ ohms.

The ideas developed in the present section are applicable only to a wave propagating in an infinite medium in the *far zone* of its source, i.e., many wavelengths away. This is not a serious restriction, since virtually all radio applications of concern to us will involve far zone fields.

5.2 THE RADIO SPECTRUM

The conventional frequency classification of radio systems is as given in Table 5-1. (Note that kHz, MHz, and GHz denote kilocycles per second, megacycles per second, and kilomegacycles per second (or gigacycles per second), respectively, or in more modern terms, kilohertz, megahertz, and gigahertz.)

As indicated in Table 5-1, each band of radio frequencies has its important practical distinguishing characteristics. Some of these are due to the ratio of the dimensions of equipment or of objects within the propagation path relative to the wavelength of the radiation. For example, a mountain is entirely opaque to SHF radiation, having dimensions that are enormous relative to wavelengths of the order of centimeters. At the lower end of the

Table 5-1 Frequency Classification of Radio Systems

Frequency range	*Wavelength range*	*Designation*	*Special features*
3–30 kHz	10–100 km	Very Low Frequency VLF	Waveguide mode transmission over long distances; low fading; antennas large and inefficient
30–300 kHz	1–10 km	Low Frequency LF	Ground and sky wave, but ground wave predominates; small amount of fading due to sky waves; large antennas; high atmospheric noise
300 kHz–3 MHz	100 m–1 km	Medium Frequency MF	Ground and sky wave transmission, more fading and less atmospheric noise than VLF; quite reliable for communication
3–30 MHz	10–100 m	High Frequency HF	Sky wave transmission; fading often severe; atmospheric noise less important than fading; widely used in military communication
30 MHz–300 MHz	1–10 m	Very High Frequency VHF	Sky wave transmission; fading is a problem; atmospheric noise is low; antenna structures required are smaller than HF, larger than UHF.
300 MHz–3 GHz	10 cm–1 m	Ultra-High Frequency UHF (low end of microwave band)	Line-of-sight transmission; small antennas with substantial gain and directivity; fading is a problem; atmospheric noise not too important; extraterrestrial noise of some importance
3–30 GHz	1–10 cm	Super High Frequency SHF (intermediate and high end of microwave band)	Can penetrate ionosphere; high directivity and gain with very small antenna structures; extraterrestrial noise very important; used in radar and microwave line-of-sight communication, space communication
30–300 GHz	1 mm–1 cm	Extremely High Frequency EHF (millimeter wave band)	Exhibits advantages of SHF to a greater degree; extremely high directivity and gain with small antenna structures; equipment tolerances very critical; power sources are a problem; oxygen and water vapor absorption; extraterrestrial noise high; can be used in radar, microwave line-of-sight, space communication

VLF band, however, where wavelengths are of the order of 100 km, it is merely a diffracting obstacle and radiation fields are only slightly perturbed by its presence.

Reasons for some of the most important characteristics distinguishing one frequency band from another as mentioned in Table 5-1 will become apparent in the discussions of antennas, receivers, power limitations, noise sources, and propagation effects in Sections 5.3, 5.4, and 5.5.

5.3 ANTENNAS

There are a number of features of transmitting and receiving antennas that have important effects on the performance of radio systems. Steps can be taken toward circumventing the need for special processing of signals by designing the antenna configuration in optimal fashion.

In discussing radio antennas, we assume that reciprocity† holds in all cases. What is meant is as follows: "If voltage V impressed at a point a on antenna A produces a current I at a point b on antenna B, then the same voltage V applied at b will produce the current I at a." The important consequence of this is that an antenna will have the same properties when used for transmitting as when used for receiving, for example, gain, bandwidth, beam pattern, efficiency, etc. Thus we can reverse the roles of transmitting and receiving antennas and will not change the transmission between the two antennas. Note that this property is dependent on the propagation medium as well as the antennas. It can justifiably be assumed in nearly all terrestrial radio applications.

We will now consider some of the antenna design features that are important to the optimum design of a radio system.

First, in any specific system application, there is always a requirement of angular coverage. If the transmitter or receiver must cover a solid angle Ω in free space‡ then the required size of the antenna effective aperture area A_e (see Section 1.3.2) is

$$A_e = \frac{\lambda^2}{\Omega} \tag{5.9}$$

where λ is wavelength. A larger aperture than that given by (5.9) for a particular wavelength will result in inadequate angular coverage. A smaller

† See Jordan, 1950, pp. 327–337, or Kerr, 1951, especially Appendix A, pp. 693–698.

‡ Equation (5.9) must be appropriately modified to account for the presence of the earth. The fields at a point in space due to an antenna with a specified current distribution can be calculated with account taken of the earth's presence. From this calculation the beam pattern and hence the angular coverage can be determined. An effective area can then be calculated from (5.9). It will not be the same as the effective area of the same antenna in free space.

aperture will result in coverage of superfluous angular regions, thus (in the case of a transmitting antenna) wasting power that should properly be transmitted only into the region of interest. The latter is usually a more serious problem since at the lower end of the radio spectrum upper limits exist on the aperture areas that can be attained in practical antenna structures. For aperture areas that are many square wavelengths, the aperture area is some fixed fraction of the physical antenna area, about 0.5 in the case of a parabolic dish and ranging (roughly) between 0.2 and 0.8 for other standard antenna types.† For antennas that are small compared to wavelength (dipoles), the aperture area is proportional to the square of the wavelength.

The directivity requirement is often a stimulus to the use of the highest frequencies that are practical in a particular application. It follows from (5.9) that a halving of wavelength, or, equivalently, a doubling of frequency, with fixed aperture area will decrease the solid angle of coverage by a factor of 4.

Another important antenna feature is the ability to match the characteristic impedance of the transmission line feeding the antenna to the input impedance of the antenna (see Appendix I), As radio frequencies become very low (e.g., $\approx$ 100 kHz), the real part of the input antenna impedance, known as *radiation resistance* (denoted by R_{rad}), becomes extremely small. For short ($\ll\lambda$) dipoles,‡ R_{rad} is proportional to $(\ell/\lambda)^2$ where ℓ is the antenna length. To be more precise, for a short electric dipole with uniform current,§ the theoretical radiation resistance is

$$R_{\text{rad}} = 80\pi^2\left(\frac{\ell}{\lambda}\right)^2 \tag{5.10}$$

Equation (5.10) illustrates the reason for difficulty in constructing an efficiently radiating antenna if (ℓ/λ) is appreciably less than unity. For example, if $(\ell/\lambda) \simeq .001$, then R_{rad} is about .0008 ohms, so small that the inductance coils used to match the feed line to the reactive part of the input impedance (which may be very high, e.g., $\simeq$ 100–1000 ohms) cannot be constructed with such a small resistance. The resistance necessarily incurred may be tens or hundreds of ohms. This results in a mismatch and a reduction in the power fed to the antenna relative to that realized with the perfect match condition. This power reduction is given in decibels (see Eq. (I-2)) by

$$\text{power reduction} = 10 \log_{10}\frac{4}{(1 + (R_s/R_{\text{rad}}))^2} \quad \text{dB} \tag{5.11}$$

† See ITT, 1956, pp. 750–751.

‡ The short dipole is usually a good approximation for antennas of practical size at frequencies below VHF. Radiation resistance is defined as the ratio of the power radiated into all space to the current squared on the antenna. The former is the integral of the Poynting vector of the radiated fields over the unit sphere surrounding the antenna.

§ See Jordan, 1950, Section 10.03 and 10.04, pp. 309–312.

where it is assumed that the reactive part of the input impedance (but not the radiation resistance), has been matched out and where R_s is the resistance of the power source and feed line.

The loss in efficiency given by (5.11) is sometimes an important consideration with low and very low frequency antennas.

5.4 THE RADIO RECEIVER

Two types of radio receivers in use are the tuned RF and the superheterodyne. Most modern applications involve superheterodyne receivers, which are the more sensitive of the two (see Section 6.8). Consideration here will therefore be specialized to superheterodyne receivers, but the principles of statistical processing are the same for tuned RF receivers.

A block diagram of the superheterodyne radio receiver is shown in Figure 5-1.

The important feature of this type of receiver is the mixing operation. The local oscillator (L.O.) generates a sinusoidal signal of frequency ($f_R + f_I$), where f_R is the radio frequency (RF) and $f_I \ll f_R$. This L.O. signal is *mixed* (i.e., effectively multiplied) with the incoming RF signal, resulting in a superposition of two signals, one having the *sum* frequency ($2f_R + f_I$), the other having the *difference* frequency f_I. The mixer output is then fed into a bandpass filter centered at f_I, with a bandwidth wide enough to pass the intelligence in the incoming signal, but sufficiently narrow to exclude the sum frequency component, whose frequency is very high compared to f_I.

The frequency f_I is called the *intermediate frequency* (IF). The IF signal preserves the RF signal envelope, and the mixing operation is effectively

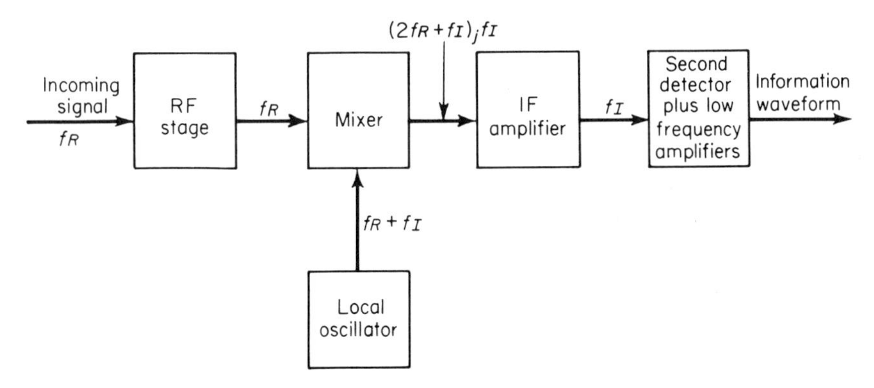

Figure 5-1 The superheterodyne receiver.

linear, acting only to shift the signal's spectrum downward without distorting its shape. For this reason, the additive Gaussian property of the receiver noise that accompanies radio signals at RF (see Section 3.4) is preserved at IF. This is fortunate for the noise analyst, because it implies that the simple assumptions about signals and noise that apply at RF are not affected by heterodyning.

In an *amplitude modulation* (AM) receiver, the IF signal enters the *second detector*, which is a rectifier usually of the linear or square-law variety, and if it is a linear rectifier, may be either half wave or full wave. (See Section 3.3.)

In an *angle modulation* receiver, i.e., one employing *frequency modulation* (FM) or *phase modulation* (PM), the IF signal is fed into a *discriminator*, which converts variations in frequency or phase into variations in amplitude.

In a receiver handling analog transmissions the output of the second detector or discriminator is generally fed into a final stage of low frequency (audio or video) filtering, which both amplifies the signal and strips off the IF components in the rectifier or discriminator output, leaving only the information bearing (audio or video) frequencies.

In some cases (e.g., certain classes of digital communication systems) information about the IF waveform, such as frequency or phase, must be preserved. As will be explained in Chapter 6, a synchronous detector, which is essentially equivalent to a matched filter or crosscorrelator, is used as the second detector. The output of the synchronous detector is fed into deciding circuits, where an automatic decision is made as to what kind of signal was transmitted.

Perhaps the most important feature of a radio receiver is its sensitivity, usually measured in terms of its threshold signal power, that is, the smallest received signal power that will permit the receiver to perform its assigned task, e.g., detection of a radar echo or extraction of intelligence from a transmitted signal. The sensitivity is of course dependent on the noise level and bandwidth.

Other features of receivers are also important. One such feature is the *dynamic range*, defined as the range of received signal strengths over which its prerectifier response is linear. For extremely small signal strength, even linear rectifiers behave like square law rectifiers and the response is nonlinear.† For extremely large signal strength, various components exhibit limiting characteristics and the receiver goes into saturation, i.e., further increases in input signal strength have no effect on output signal strength. The range between these two extremes of behavior is the dynamic range.

The gain and frequency selectivity are also key design parameters. The selectivity and gain are essentially those of the IF amplifiers in a superheterodyne receiver. There are tradeoffs between selectivity, sensitivity, and fidelity. Extreme selectivity implies small bandwidth and therefore low noise and high sensitivity. However, it may also suppress important frequency

† See Terman, 1955, pp. 558–559.

components at the edges of the signal spectrum and thereby reduce fidelity. High selectivity facilitates high amplifier gain, since it is easier to provide high gain over a narrow frequency band than over a wide band. There are in fact, rough rules of thumb in amplifier design fixing the gain-bandwidth product.†

5.5 ANALYSIS OF A RADIO SYSTEM

To analyze a radio system it is necessary to specify the relative positions of the transmitter and receiver, the purpose of the transmission (e.g., voice communication, telemetry, etc.), the properties of the transmission medium, the power available at the transmitter, and the sensitivity of the receiver. Given these items, one can formulate a performance criterion and determine analytically how well the system does its work.

In the present section, some of the problems involved in analysis of a radio system will be delineated. It is only fair to say, however, that many problems arise in such analysis that could not possibly be covered here. In fact, as new radio applications appear, completely new problems come into being. It is hoped that the reader will gain from these pages a modicum of insight into the way in which one might begin an analysis of a radio system. Only further study and experience can teach the fine points of such analysis.

It will be assumed in what follows that the reciprocity theorem holds (see Section 5.3). This assumption is rigorously justified except in those cases where the effect of the earth's magnetic field on the propagation characteristics of the ionosphere is of major importance.‡ In rough analysis of many radio propagation problems, the magnetic field can be ignored, or its effects can be considered in a way that preserves reciprocity.

Because of reciprocity, we can regard propagation over a path from point A to point B as behaving in the same way as propagation over the same path from B to A. We can also regard antennas as having the same characteristics (gain, effective aperture area, etc.) when used for reception as for transmission.

5.5.1 The Beacon Equation in Infinite Free Space

We will begin the analysis with an extreme idealization from which we can later build up more realistic models. Consider a radio transmitter operating in infinite free space. If its antenna radiates average§ power P_T into space, the average§ power received at a point a distance r from the trans-

† See Terman, 1955, pp. 288–292 and pp. 414–416.

‡ See Davies, 1965, Section 5.4.6, pp. 253–254.

§ "Average" here refers to a time average over a time interval T that contains an enormous number of RF cycles.

mitter and at angle θ, ϕ relative to the transmitter is

$$P_R = \frac{P_T G_T A_{eR}}{4\pi r^2} g_T^2(\theta, \phi) \tag{5.12}$$

where A_{eR} is the effective aperture area of the receiving element, G_T is the peak gain of the transmitting antenna relative to an isotropic radiator† with the same total radiated power, and $g_T^2(\theta, \phi)$ is the angular relative power distribution or *beam pattern* of the transmitting antenna. By virtue of the relation

$$G = \frac{4\pi A_e}{\lambda^2} \tag{5.13}$$

(where G and A_e are peak gain and effective aperture area of an antenna, respectively),‡ it is possible to express (5.12) in the form

$$P_R = \frac{P_T A_{eT} A_{eR}}{\lambda^2 r^2} g_T(\theta, \phi)\, g_R(\theta', \phi') \tag{5.14}$$

where the subscripts T and R refer in all cases to transmitting and receiving, respectively, and where the primes on θ' and ϕ' in g_R indicate that transmitting and receiving antenna orientation angles are in different coordinate systems.

Due to the possibility of transmitter antenna mismatch and transmission line losses, the total radiated power is in general a fraction $1/L_T$ of the total generator power P_G. Also, the signal power at the input to the second detector is a fraction $1/L_A$ of that at the receiving antenna terminals, again due to transmission line losses and mismatch. Including these losses, and using (3.62) as a representation of the basic receiver noise (exclusive of noise figure), the SNR at the second detector input, in decibels, is

$$\begin{aligned} \text{SNR}_{\text{dB}}^{(i)} = 39 &+ (P_G)_{\text{dBw}} - (L_T)_{\text{dB}} - (L_R)_{\text{dB}} + 20 \log_{10} f_{\text{kHz}} \\ &+ 10 \log_{10}(A_{eT})_{\text{m}^2} + 10 \log_{10}(A_{eR})_{\text{m}^2} - 20 \log_{10} r_{\text{m}} \\ &- 10 \log_{10} B_{\text{kHz}} - 10 \log_{10}(T_{\text{abs}}) - (F_n)_{\text{dB}} \\ &+ 10 \log_{10} [g_T(\theta, \phi)] + 10 \log_{10} [g_R(\theta', \phi')] \end{aligned} \tag{5.15}$$

where the subscripts dB, dBw, kHz, m, and m², refer, respectively, to decibels, decibels above 1 watt, kiloHertz, meters, and square meters, and where the quantities B (bandwidth), F_n(noise figure) and T_{abs} (absolute temperature) are discussed in Section 5.5.3.1.

In most cases the quantity calculated through (5.15) would provide a rather optimistic estimate of the SNR of a terrestrial radio system. However,

† That is, one which radiates with equal intensity in all directions.

‡ Note that (5.13) and (5.9) are equivalent because the gain G, by definition, is the ratio of 4π, the solid angle of the unit sphere, to Ω, the solid angle covered by the antenna beam.

it is a convenient starting point in analyzing terrestrial systems and is directly applicable to some space radio systems, The additional features that degrade performance in terrestrial systems such as ground path loss and atmospheric noise can be subtracted from (5.15) to improve the realism of the SNR estimate.

The performance of a radio system can be evaluated in an infinite variety of ways, depending on the contemplated application. For example, detection systems may be evaluated in terms of decision theory criteria such as the detection probability or average loss, and digital communication systems are usually evaluated in terms of bit error probability or message reception probability.

Most of the noise in the prerectification stages of a radio receiver is Gaussian. It will become apparent in later discussions that with Gaussian input noise all of the standard probabilistic criteria of performance result in measures of system quality that are increasing functions of the SNR at the input to the second detector. Throughout the present chapter, where the precise applications of the radio systems discussed will not be specified, $SNR^{(i)}$ will be regarded as the measure of system quality. In later chapters, where functional applications are treated, it will be shown how specific increases or decreases in detection probability, communication system reliability, or other probabilistic measures of performance can be determined from a given increase or decrease in $SNR^{(i)}$.

In the section to follow, we will investigate the effect on $SNR^{(i)}$ of changes in the various parameters of (5.15) and other parameters to be added as the realism of the model is increased.

5.5.2 Signal Power Limitations

The limitations in received signal power arise through the following mechanisms.

1. Upper limits on transmitter output power.
2. Limits on antenna efficiency and directivity.
3. Ground losses.
4. Atmospheric attenuation.
5. Losses within the receiver.

5.5.2.1 Transmitter Power Limits

The upper limit on transmitter power takes the form of a limit on average power with or without a limit on peak power. Some power sources, particularly certain microwave sources, have a peak power limit. This can be a serious problem in radar, particularly if it is desired to pack a large amount of power into a single pulse.

If transmitter power were unlimited, radio systems could overcome any amount of internal and atmospheric receiver noise, atmospheric attenuation, and effects of the presence of the earth, and could operate at any range. It might be said that the brute force method of improving range is to increase transmitter power.

It is important in analyzing radio systems to determine the limits in both peak and average transmitter power, In many applications, particularly those wherein the interfering agent is additive bandlimited white Gaussian noise, the important parameter is the ratio of total signal energy to noise-per-unit bandwidth, regardless of how the signal energy is distributed in time. (See Sections 6.2 and 6.3.) Thus the theory seems to tell us that the instantaneous signal power is not significant. However, as a practical matter, any pattern of power transmission that exceeds the peak power limit instantaneously is precluded, even though it does not violate the average power or total energy limit throughout the period of transmission.

Note that we have defined $\mathrm{SNR}_{\mathrm{dB}}^{(i)}$ in (5.15) as a ratio of time averaged signal power to the mean noise power. The average is over a time interval T. Then, if the instantaneous transmitter power is $P_{Gi}(t)$, the power $(P_G)_{\mathrm{dBw}}$ in (5.15) actually represents the quantity

$$\frac{1}{T}\int_0^T dt'\, P_{Gi}(t')$$

This quantity has a maximum possible value, which is the *average* power limit. The limit on the peak instantaneous power $P_{Gi}(t)$ which is not explicitly shown in (5.15) must be accounted for in analyzing a system.

5.5.2.2 Limits on Antenna Efficiency and Directivity

It was pointed out in Section 5.3 that a limit exists on the ability to match the power source to the antenna input impedance at low frequencies. This limit (Eq. (5.11)) takes many different mathematical forms depending on the antenna structure used. However, in general it can be said that it is difficult to build an antenna with an efficiency greater than a few percent whose key dimensions are appreciably smaller than a wavelength. This implies that an efficiency problem exists at VLF, LF, and over an appreciable part of the MF band, where wavelengths are of the order of hundreds of meters or kilometers. To couple appreciable energy to free space at those frequencies, enormous antenna structures are required (e.g., wires stretched for miles across valleys to be used as VLF antennas).

The efficiency limit referred to above concerns the total power radiated into space. The limit on the ability to direct the beam into a selected part of space may be an equally severe problem at VLF, LF, and MF, as mentioned in Section 5.3.

The efficiency losses in both transmitting and receiving antennas may be lumped into the set of terms $[-(L_T)_{dB} - (L_R)_{dB}]$ in (5.15). The limits on transmitting and receiving antenna directivity can be accounted for in (5.15) through the practical upper limits on the set of terms $[10 \log_{10} A_{eT} + 10 \log_{10} A_{eR}]$.

5.5.2.3 Ground Losses

Ground loss in terrestrial radio wave propagation is the attenuation of the *ground wave*, or *surface wave* (i.e., the radio wave propagated along the surface of the earth), due to the fact that the earth is not a perfect conductor. A set of plots of ground-wave field strength as a function of transmission distance with typical earth conductivity values and for various frequencies is shown in Figure 5-2. These curves are taken from Terman (1955). The basic source of curves like these is a classical (1936) paper by K. A. Norton, reporting the results of theoretical calculations of the factors influencing ground wave radio propagation over a wide range of frequencies, based on Sommerfeld's original work on propagation over a lossy half-space. The

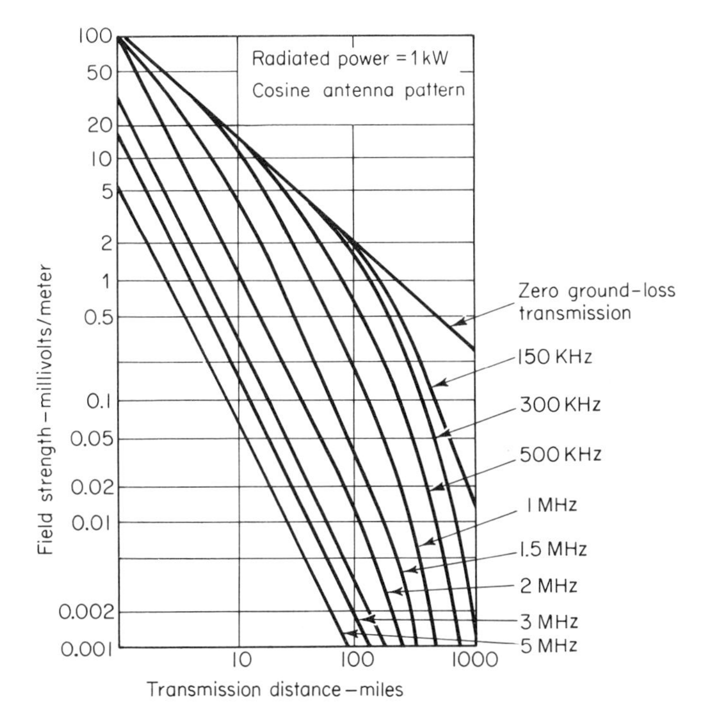

Figure 5-2 Ground-wave field strength as a function of distance. From Terman, 1955, Figure 22-3, p. 807.

general features of these results have been well substantiated experimentally and the results have appeared in various forms in textbooks and handbooks.†

The ground wave is essentially vertically polarized but with a slight forward tilt in its E-vector due to the continuous flow of energy into the earth along the propagation path. The loss depends on the constitutive parameters of the earth in a mathematically complicated way, and shows somewhat different functional dependence on these parameters in different frequency regions.

VLF propagation involves the ground wave influenced by the ionosphere. The ground wave can be used as a mode of short range propagation at LF and MF. At HF and above, its attenuation is too severe for effective use; thus terrestrial HF radio communication relies on waves reflected from the ionosphere (see Section 5.5.4.5), or *sky waves.*

5.5.2.4 Atmospheric Attenuation

Another limitation on radio range in systems operating entirely or partially‡ within the earth's atmosphere is atmospheric attenuation. Such attenuation may be due to natural constituents of the atmosphere, irregularities along the transmission path, or weather conditions. For example, SHF radar transmissions must contend with oxygen absorption, which has a resonant peak at about 5 mm (60 GHz), water vapor absorption, with a resonant peak at about 1.33 cm (22.4 GHz), fog and cloud droplets and raindrops (above 10 GHz, where raindrop radius becomes comparable to wavelength).

The way in which atmospheric attenuation appears in (5.12) is through the insertion of an exponential factor $e^{-2\alpha r}$, where α is the field strength attenuation in nepers per unit of distance. We can convert the attenuation α into decibels per unit of distance by noting that $10 \log_{10} e^{-2\alpha r} = -(20(\log_{10} e)\alpha)\, r = -2\gamma r$, where $\gamma = 4.343\alpha$ = field strength in decibels per unit of distance. The power attenuation expressed in decibels per unit of distance is 8.686α.

5.5.2.5 Receiver Losses

Receiver losses are those arising from lossy or mismatched transmission lines or waveguides between the antenna and the receiver circuitry or those in the circuitry itself. Such losses, which in a well-designed receiver should

† See Kerr, 1951, Sections 2.6, 2.7, 2.8, pp. 58–86; Alpert, 1960, Chapter VI, No. 22, pp. 191–203; Terman, 1955, pp. 803–808; ITT, 1956, pp. 714–715. Some results of the generalization accounting for earth curvature are reported in Alpert, 1960, pp. 191–203. K. A. Norton accounted for earth curvature in some of his early work (1941). The results discussed here are for a flat earth approximation.

‡ This includes earth-to-space or space-to-earth transmission where only part of the propagation takes place within the atmosphere.

be no higher than 2 or 3 dB, should be accounted for in evaluating the SNR at the receiver. If operation takes place near the threshold signal region, losses of 2 or 3 dB can be significant. These losses can be lumped into the term $-(L_R)_{\text{dB}}$ in (5.15).

5.5.3 Additive Noise Sources

There are various noise sources in radio systems whose outputs are additive with the transmitted signal. Since the generation of noise by these sources is statistically independent of the mechanism of generation of the signal, the noise is statistically independent of the signal. The features of additivity and statistical independence with respect to the signal act to simplify the analysis of signals perturbed by noise from these sources.

The important additive noise sources are:

1. Internal receiver noise.
2. Atmospheric noise.
3. Extraterrestrial noise.
4. Man-made noise and interference.

Each of these will be briefly discussed below.

5.5.3.1 Receiver Noise–Noise Figure

The thermal and shot noise within the receiver is Gaussian and white over the receiver passband. (See Section 3.4.) The effect of this noise is contained in the three terms $[-10 \log_{10} B_{\text{kHz}} - 10 \log_{10} T_{\text{abs}} - (F_n)_{\text{dB}}]$ in (5.15). The first of these terms, that associated with receiver bandwidth, is fixed by the specific application. It is determined by the amount of information to be transmitted and the time allowed for the transmission. For example, TV transmission may require a bandwidth of the order of 6 to 10MHz, i.e., $-10 \log_{10} B_{\text{kHz}}$ is between -40 and -38. Voice transmission with amplitude modulation requires between 3 and 4 kHz, i.e., $-10 \log_{10} B_{\text{kHz}}$ is between -6 and -5. In general, the greater the information rate, the greater the required bandwidth, and the more noise must be allowed into the receiver.

The effect of the term $-10 \log_{10} T_{\text{abs}}$ can be reduced by cooling the receiver. Such cooling is actually involved in the operation of maser amplifiers.† In ordinary radio situations not involving the use of masers it is customary to assume T_{abs} to be about 300°K, i.e., $-10 \log_{10} T_{\text{abs}} \simeq -25$.

The noise figure F_n varies greatly in different types of radio receivers. In general, it increases with radio frequency unless modern techniques of low noise amplification are used to offset the high noise figures of receivers

† See L. S. Nergaard, "Modern Low-Noise Devices," in Berkowitz, 1965, pp. 432–470, and the reference list following the article.

operating at microwave frequencies. These noise figures are between about 10 and 17 dB. Noise figures as low as 3 to 6 dB are attainable with old-fashioned design techniques below UHF. With such devices as masers and parametric amplifiers, effective noise figures as low as 1.5 or 2 dB can be achieved in some parts of the radio spectrum.

Much of this book will be devoted to the subject of processing techniques designed to reduce the effects of internal receiver noise. Obviously, the use of low noise amplifiers minimizes the need for such processing. However, there are many applications in which, for various reasons, processing techniques are valuable as a substitute for, or a supplement to, the use of these low noise amplification techniques.

The noise figure of a radio receiver is a sufficiently important concept to warrant elaboration here. In general, the *noise figure* of a group of linear electrical networks in cascade is a measure of the amount of noise generated by the cascade. It is defined as the ratio of the total noise power in the output of the cascade to the noise power which would be present in the output if the cascade generated no noise of its own but simply passed the input noise with a gain or loss. Like all power ratios, it is most conveniently expressed in decibels.

To formalize the definition, we consider a network designated as No. 1, with gain $G^{(1)}$. If its input is noise with total power $P_{Ni}^{(1)}$, the noise figure $F^{(1)}$ is defined as follows:

$$F^{(1)} = 1 + \frac{\Delta P_N^{(1)}}{G^{(1)} P_{Ni}^{(1)}} \tag{5.16}$$

where $\Delta P_N^{(1)}$ is the amount of noise generated within the network. Clearly, if the network generates no noise, $\Delta P_N^{(1)} = 0$, the noise output power is $G^{(1)} P_{Ni}^{(1)}$, and the noise figure is unity, or, equivalently, zero dB.

Consider a second network in cascade with No. 1, whose parameters are denoted with superscript (2). The input to network No. 2 is $G^{(1)} P_{Ni}^{(1)} + \Delta P_N^{(2)}$. The noise figure of the cascade of networks 1 and 2 is

$$F^{(1),(2)} = \frac{G^{(2)} G^{(1)} P_{Ni}^{(1)} + G^{(2)} \Delta P_N^{(1)} + \Delta P_N^{(2)}}{G^{(2)} G^{(1)} P_{Ni}^{(1)}} = F^{(1)} F^{(2)} \tag{5.17}$$

where

$$F^{(2)} = G^{(2)} + \frac{\Delta P_N^{(2)}}{G^{(1)} P_{Ni}^{(1)} + \Delta P_N^{(1)}} = 1 + \frac{\Delta P_N^{(2)}}{G^{(1)} G^{(2)} P_{Ni}^{(1)} F^{(1)}}$$

The derivation of (5.17) from (5.16) can be extended to include any number of linear networks in cascade. If subscript (j) applies to the jth network in the cascade and there are a total of M networks, then

$$F^{(1),(2),\ldots,(M)} = F^{(1)} F^{(2)} \ldots F^{(M)} \tag{5.18}$$

where

$$F^{(j)} = 1 + \frac{\Delta P_N^{(j)}}{P_{Ni}^{(1)} G^{(1)} G^{(2)} G^{(3)} \ldots G^{(j)} F^{(1)} F^{(2)} \ldots F^{(j-1)}}$$

The total noise at the output of the cascade, according to (5.18), is the product of the noise input to the first stage and the noise figure of the cascade. This applies to cascades of any number of stages, so long as all stages are linear. There is a very useful consequence of (5.17). If most of the noise is introduced in the second stage of a cascade, then high gain amplification in the first stage can reduce the over-all noise figure of the cascade. This is apparent from the definition of $F^{(2)}$ in (5.17), which shows that, if $G^{(1)}$ is made sufficiently large compared to

$$\frac{\Delta P_N^{(2)}}{P_{Ni}^{(1)}}$$

then $F^{(2)}$ can be made as near as desired to unity. There is one possible deterrent to this. It may be that an increased gain in stage 1 cannot be accomplished without an increase in $F^{(1)}$. In such a case, it may be impossible to reduce the over-all noise figure $F^{(1)}F^{(2)}$ even with a reduction in $F^{(2)}$, since $F^{(1)}$ will increase in any attempt to do this.

The use of high gain amplification to reduce noise figure can be applied to a cascade of any number of stages, as is evidenced by (5.18). The important point is that the amplification must be applied in stages prior to the stage where most of the noise is generated. A common application is in a superheterodyne radio receiver, where most of the noise is introduced in the mixer and IF stages. An RF amplifier with sufficiently high gain can reduce the noise figure significantly.

5.5.3.2 Atmospheric Noise

In addition to internal noise, there is atmospheric noise over a wide range of radio frequencies. It arises from electrical discharges in the atmosphere, e.g., intercloud lightning discharges and cloud-to-ground lightning flashes associated with thunderstorms in various parts of the world. Such noise tends to decrease with frequency, being most prominent at VLF and diminished to nearly negligible levels at VHF and above. A plot of median atmospheric noise level (i.e., field strength at the receiving antenna in microvolts/meter as a function of frequency under various conditions taken from the ITT Handbook† is shown in Figure 5-3.

It is evident from the figure that atmospheric noise is not spectrally flat, but for rough analysis purposes, it can be considered as such over a typical receiver passband. It can be determined how wide the receiver band must be before the bandlimited white noise model becomes a poor approximation. This can be done by examining the curve of Figure 5-3, and setting up a criterion for the maximum allowable receiver bandwidth for validity of

---

† Figure 5-3 is a slightly modified rendition of parts of Figure 1 on p. 763 of ITT, 1956. The use of the curves to calculate noise power levels with specific receiving antennas is facilitated by referring to the instructions below Figure 1 on p. 763 of the Handbook.

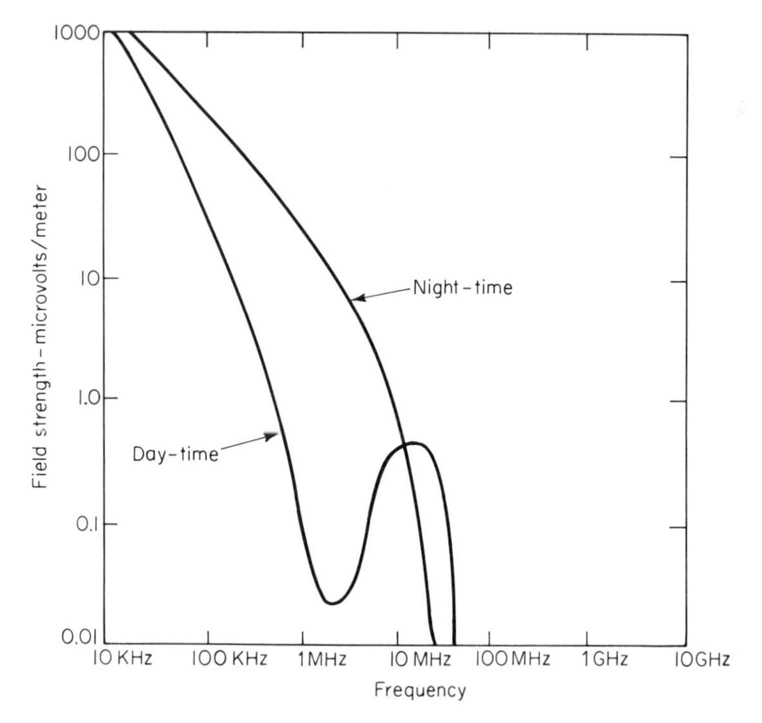

Figure 5-3 Median atmospheric noise level as a function of frequency. Reproduced by permission from *Reference Data for Radio Engineers*, copyright © 1956 by International Telephone and Telegraph Corporation, p. 763.

the bandlimited white noise approximation. This may be, for example, the band over which the average atmospheric noise level does not change by more than 10 percent.

Except for the highly impulsive noise due to nearby or overhead thunderstorms, most atmospheric noise can be considered Gaussian, because it arises from a superposition of outputs of many statistically independent sources (see Section 3.5). Thus it can be regarded as statistically equivalent to internal noise (bandlimited white, Gaussian, additive) within the receiver passband, and can be analyzed as such.

5.5.3.3 Extraterrestrial Noise

At higher frequencies (e.g., UHF, SHF), where atmospheric noise is negligible, one encounters another disturbance in the form of *extraterrestrial* or *galactic noise*, e.g., that arising from certain galaxies. Such noise also exists at lower frequencies, but in terrestrial radio applications it can usually be ignored, because it is absorbed by the ionosphere, which becomes effectively opaque to radio transmissions normally incident on it at frequencies below a few megacycles per second. In space applications, it could become a problem even at lower radio frequencies.

Galactic noise can become especially troublesome at SHF. In particular, it is the limiting factor on the performance of supersensitive radar receivers, e.g., those which use masers or other types of low noise amplifiers and thereby reduce the level of internal noise. Average galactic noise in dBw for a dipole antenna is shown as a function of frequency in Figure 5-4 (from ITT Handbook, 1956). Like atmospheric noise, extraterrestrial noise can be regarded as Gaussian because it is essentially a superposition of noise transmissions from a large number of independent sources (see Section 3.5).

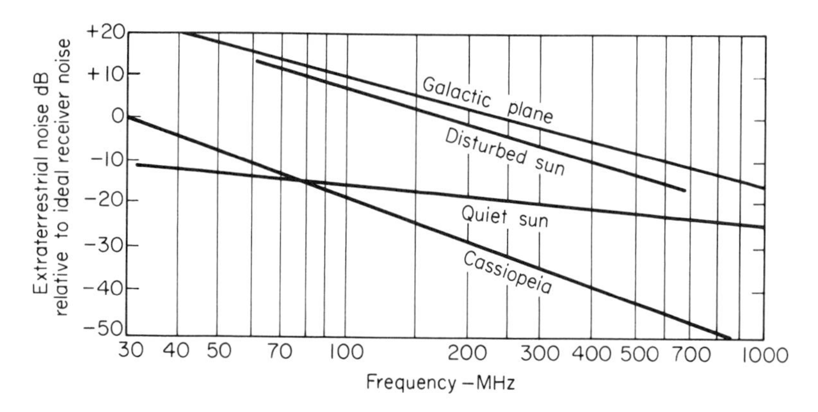

Figure 5-4 Extraterrestrial noise as a function of frequency. Reproduced by permission from *Reference Data for Radio Engineers*, copyright © 1956 by International Telephone and Telegraph Corporation, p. 764.

5.5.3.4 Man-Made Noise and Interference

In any terrestrial environment, there are many man-made sources of unwanted radio signals, Some of these signals have the characteristics of impulse noise (e.g., ignition noise at low radio frequencies). Others are essentially modulated CW tones such as those that arise from nearby radio transmissions. Power lines and vibrating machinery are sometimes sources of radio noise. Such noise sources will obviously be most prevalent in heavily populated areas such as cities and will present no serious problem in the open country.

Very little can be said in general about man-made noise sources. They cannot be classified as natural phenomena in the usual sense of this phrase. They depend not on the state of nature, but rather on the state of civilization within the radio environment. If transmissions from enough such sources are simultaneously active, then because of the central limit theorem their aggregate will behave like Gaussian noise and can be treated as such in analysis. If only a few discrete sources of man-made noise are present, and they are of an impulsive character or are pure CW tones, their statistics will be totally different from that of Gaussian noise. Very little general theory exists on these kinds of disturbance, and each case must be treated individually.

A type of man-made noise that is of great importance in military radio systems is enemy jamming. A great body of literature exists on jamming and methods of dealing with it, but most of it carries a government security classification and cannot be discussed in open literature.

5.5.3.5 Total Additive Noise

It is true under nearly any conditions that can be imagined that each of the noise sources discussed in Sections 5.5.3.1 through 5.5.3.4 are mutually independent, and consequently add incoherently.† Moreover, each of the types of noise discussed in the preceding sections has a Gaussian component because at least a part of it arises from superposition of a large number of independent sources. Atmospheric noise may have impulsive components and man-made noise may have both impulsive components and narrow bands of energy that we will call *discrete frequency components* or *tones.* Both of these components are generally statistically independent of the corresponding Gaussian components because they arise from different sources than does the Gaussian noise.

The noise in the receiver arising from the superposition of the Gaussian parts of the internal, atmospheric, extraterrestrial and man-made noises is itself a Gaussian noise (see Section 3.1 and Eq. (3.10)) with mean-square value at the receiver input equal to

$$\sigma_{gn}^2 = (\sigma_{gi}^2 + \sigma_{ga}^2 + \sigma_{ge}^2 + \sigma_{gm}^2) \tag{5.19}$$

where the subscript g on all σ's refers to *Gaussian component*, the subscripts i, a, e, and m refer to internal, atmospheric, extraterrestrial, and man-made, respectively. The mean internal noise power σ_{gi}^2 is $kT_{\text{abs}}\, B_R$, and this power is distributed uniformly over the receiver bandwidth B_R, so that its spectral

† See Section 3.4.2, text below Eq. (3.66). The idea of "incoherent" addition of signals is based on the fact that the mean power in a sum of N voltage signals v_k is proportional to

$$\left\langle \left(\sum_{k=1}^{N} v_k \right)^2 \right\rangle = \sum_{k=1}^{N} \langle v_k^2 \rangle + \sum_{\substack{j,k=1 \\ j \neq k}}^{N}{}' \langle v_j v_k \rangle.$$

The addition is said to be incoherent if the jth and kth signals are uncorrelated, i.e., if $\langle v_j v_k \rangle = 0$ for all j, k. In this case, the total power is the sum of the power levels of the individual signals.

density is kT_{abs}. The spectral densities of σ_{ga}^2 and σ_{ge}^2 can be obtained from the curves of Figures 5-2 and 5-3, respectively. The man-made noise power σ_{gm}^2 is critically dependent on the environment and nothing further will be said of it here.

To account for non-Gaussian noise in analysis of an actual radio system, it would be necessary to first study the environment and determine the statistical properties of the noise disturbances that may be present. Interference from other radio transmissions, for example, would take the form of small bands of energy around their central frequencies. Impulse noise due to nearby thunderstorms could be represented analytically by impulse functions or short pulses occurring in some random sequence determined by the characteristics of the storm. The mean power in the superposition of the major types of noise disturbance could be represented generally by

$$\sigma_n^2 = \sigma_g^2 + \sigma_I^2 + \sigma_d^2 \tag{5.20}$$

where σ_g, the Gaussian contribution, is defined in (5.19), σ_I^2, the impulsive contribution, is principally due to nearby storms but could also arise from man-made sources, and σ_d^2, the discrete frequency contribution, consists of narrow bands of energy due to various radio transmissions.

The spectral density of the total noise is

$$G_n(\omega) = kT_{\text{abs}}[G_{ga}(\omega) + G_{ge}(\omega) + G_{gm}(\omega)] + G_I(\omega) + G_d(\omega) \tag{5.21}$$

where each $G(\omega)$ is a spectral density corresponding to one of the σ^2 quantities defined in (5.19) and (5.20).

The important quantity to be determined from the point of view of radio system analysis is the total additive noise power at the input to the second detector of the receiver, to be denoted by σ_{nT}^2. This is obtained from (5.21) by the transformation

$$\sigma_{nT}^2 = \frac{1}{2\pi}\int_{-\infty}^{\infty} d\omega\, G_n(\omega)\,|F(\omega)|^2 N_F(\omega) \tag{5.22}$$

where $F(\omega)$ is the frequency response function of the portion of the receiver prior to the second detector, and $N_F(\omega)$ is the receiver noise figure, which may be frequency dependent within the receiver passband.

5.5.4 Propagation Effects

There are effects on radio waves due to the presence of the earth and atmosphere that either enhance or disturb radio system performance. Some of those which result in power losses (e.g., ground loss factor) and additive noise (e.g., atmospheric noise) have already been discussed. There are others not yet mentioned that result in losses or multiplicative noise.

Some propagation phenomena that are important in radio system analysis will be discussed below.

5.5.4.1 Field Strength and Polarization

When discussing radio wave propagation, it is best to think in terms of the field strength of the received radio wave before evaluating its power. In propagation theory, the quantity usually calculated is the field strength at a point in space due to a given transmitting antenna with a particular current distribution. The instantaneous received field strength E_R, in volts per meter, is proportional to the instantaneous current I_T on the transmitting antenna. Assuming the receiving point to be in the far zone of the transmitter, as is nearly always the case, the received wave is approximately a plane wave and the time averaged received power flow is given by (5.8), i.e.,

$$\text{power flow} = \frac{1}{2}\frac{\epsilon}{\mu}\overline{|E_R|^2} = \frac{1}{2}\frac{\epsilon}{\mu}\left|\frac{C}{r}g_T(\theta, \phi)\right|^2\overline{I_T^2} = \frac{K}{r^2}\overline{P_T}\left|g_T(\theta, \phi)\right|^2 \tag{5.23}$$

where $(C/r)g_T(\theta, \phi)$ is the proportionality factor between E_R and I_T. It is indicated that the time averaged transmitted power $\overline{P_T}$ is proportional to $\overline{I_T^2}$, and $K|g_T|^2/r^2$ is the proportionality factor between the average received power flow and the average transmitted power, K being a constant. Multiplication of (5.23) by the effective aperture area of the receiving antenna results in an expression equivalent to (5.12) with a number of parameters lumped into the factor K.

The polarization of the radio wave is defined as the direction of its electric field. When adding electric fields due to two waves, one must remember that these fields are vectors. Consider two such fields in an x, y-coordinate plane. The resultant field is the vector superposition of the two fields. Suppose one field has amplitude E_1, oscillates at angular frequency ω, and is directed along the x-axis. The other has amplitude E_2, is oscillating at the same frequency as the first but with a phase differing by $\Delta\psi$ from that of the first, and is directed at an angle θ relative to the x-axis. The average power flow associated with the resultant wave field is, from (5.8),

$$\text{power flow} = \frac{1}{2}\frac{\epsilon}{\mu}\{[E_1 \cos \omega t + E_2 \cos (\omega t + \Delta\psi)]$$

$$[E_1 \cos \omega t + E_2 \cos (\omega t + \Delta\psi)]\} = \frac{1}{4}\frac{\epsilon}{\mu}\{E_1^2 + E_2^2 + 2E_1E_2 \cos \theta \cos \psi\} \tag{5.24}$$

The oscillation patterns corresponding to various phase differences and space angle differences can be complicated in some cases.

Many radio antennas at frequencies below SHF are either vertical straight wires which respond only to the vertically polarized component of the incoming wave or horizontal loops or horizontal straight wires which respond to the horizontally polarized component of the wave. The received power referred to in the beacon equation (5.12), then, is the power associated only with that polarization component to which the antenna is sensitive.

The two basic types of polarization in radio waves are linear and elliptical. A linearly polarized wave is one whose **E**-field oscillates in one given direction, usually the vertical or horizontal. By designing transmitting antennas to send out vertically or horizontally polarized waves and receiving antennas to respond to whatever polarization was transmitted, the designer maximizes the transfer of power between transmitter and receiver with respect to polarization.

If horizontally and vertically polarized waves with $\pm 90^\circ$ phase difference are superposed, the result is *elliptical polarization* with major and minor axes of the "polarization ellipse" in the vertical and horizontal directions.† In the discussion leading to (5.24), this is the case where $\theta = \pi/2$, $\Delta\psi = \pi/2$.

The oscillation due to such an arrangement,‡ is a field vector rotating at the oscillation frequency $\omega/2\pi$ and tracing out an ellipse of eccentricity $|E_1/E_2|$. The x and y resultant field components, as can easily be shown from the arguments above, are

$$\begin{aligned} E_x &= E_1 \cos \omega t \\ E_y &= -E_2 \sin \omega t \end{aligned} \tag{5.25}$$

The average power flow in this case (since $\cos\theta \cos\psi = 0$ in (5.24)) is proportional to $(E_1^2 + E_2^2)$. If E_1 and E_2 are equal, then the total power is twice that in a single polarization component, and the tip of the rotating field vector traces out a circle. This case is known as *circular polarization.*

Antennas consisting of combinations of vertical and horizontal elements with proper relative phase can be designed to transmit and receive elliptically polarized waves.

5.5.4.2 Wave Interference

Wave interference is an effect that can be important in propagation between two points whose height is comparable to or greater than a wavelength. This condition is met in air-to-air or air-to-ground communication at MF, HF, VHF, or UHF, and in ground-to-ground communication between antennas at HF, VHF, and UHF.§

---

† This is discussed in many texts. See ITT, pp. 666–668; Jordan, 1950, pp. 123–124, 418–420; Davies, 1965, pp. 84–86; and Terman, 1955, pp. 817, 831 (the latter two references are in the context of ionospheric radio propagation).

‡ See any of the references cited above, in particular, Jordan.

§ In lower frequency regions, ground antenna heights or even aircraft altitudes are small fractions of wavelength, therefore path lengths of direct and reflected waves have negligible phase difference. In higher frequency regions, irregularities in the propagation path would cause the direct and reflected waves to add incoherently, and the interference would not manifest itself.

The transmitting antenna radiates in a range of directions determined by its radiation pattern. One wave path is directly to the receiving antenna, the other is indirectly to the antenna through reflection from the ground. These two paths are known as direct and reflected waves (or rays), respectively. The total current induced in the receiving antenna is the superposition of the two currents induced by the electric fields of the direct and reflected rays. (The superposition of the two waves is called the *space wave*.) Whether this interference between the two waves is constructive or destructive depends on the relative path length (in wavelengths) of the two waves. It is constructive if the phase difference is $2n\pi$, destructive if it is $(2n + 1)\pi$ (where n is an integer), and partially constructive if the phase difference is other than $2n\pi$ or $(2n + 1)\pi$.

The geometry of radio interference is shown in Figure 5-5. It is assumed that the transmission distance r is small compared to earth radius, so that a flat earth approximation is valid. It is also assumed that transmitting and receiving antenna heights h_T and h_R, respectively, are small compared to r, and r is large in terms of wavelength; that is, reception is in the far zone of the transmitter. The total induced current is easily shown (using the law of cosines) to be proportional to

$$I=\sqrt{[g_T(\theta_D)g_R(\theta'_D)]^2+[g_T(\theta_G)g_R(\theta'_G)]^2+2g_T(\theta_D)g_R(\theta'_D)g_T(\theta_G)g_R(\theta'_G)\cos\frac{2\pi\delta}{\lambda}+\Psi_G} \tag{5.26}$$

where $g_T(\theta)$ and $g_R(\theta')$ are, respectively, the transmitting and receiving antenna patterns in the vertical plane, subscripts D and G refer to *direct wave* and *ground reflected wave*, respectively, ψ_G is the phase shift due to reflection, and δ is the path difference between direct and reflected waves, given by

$$\delta = (d_D + d_R) - \sqrt{d_D^2 + d_R^2 - 2d_D d_R \cos(\pi - 2d_G)} \simeq \frac{(h_T + h_R)}{\sin \alpha_G}\left\{1 - \left[1 - \frac{4h_T h_R \alpha_G^2}{2(h_T + h_R)^2}\right]\right\} \simeq \frac{2h_T h_R}{r} \tag{5.27}$$

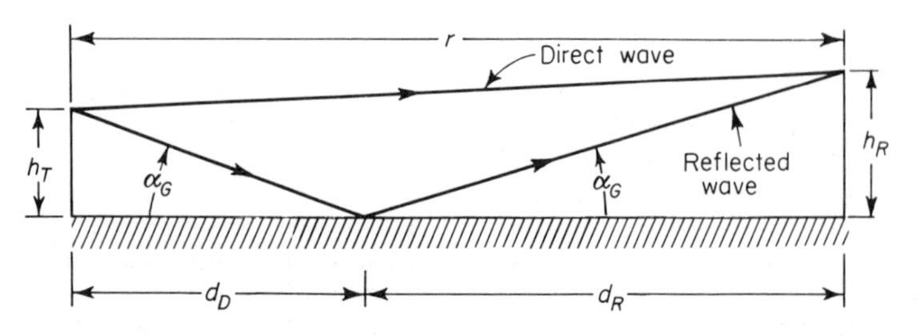

Figure 5-5 Wave interference geometry.

where α_G is the angle of reflection relative to the horizontal, h_T and h_R are transmitting and receiving antenna elevations, respectively, and d_D and d_R are distance between ground reflection point and transmitter and receiver, respectively. The parameters in (5.26) and (5.27) are illustrated in Figure5-5.

Equation (5.26) can easily be extended to take account of earth curvature, but the result is rather cumbersome.† The approximation (5.27) is adequate to cover many cases of practical interest.

5.5.4.3 Atmospheric Refraction

Because of the stratification in density of the earth's atmosphere, some degree of refraction occurs in propagating radio waves within the atmosphere. This is illustrated in Figure 5-6. This effect is of particular importance in navigational devices where angle measurement is involved. It introduces an error into such measurements resulting from the discrepancy between the true and apparent source of a received radio wave.

Refraction usually has a beneficial effect on long distance terrestrial radio transmission. It enables radio waves to be "bent" around the horizon.

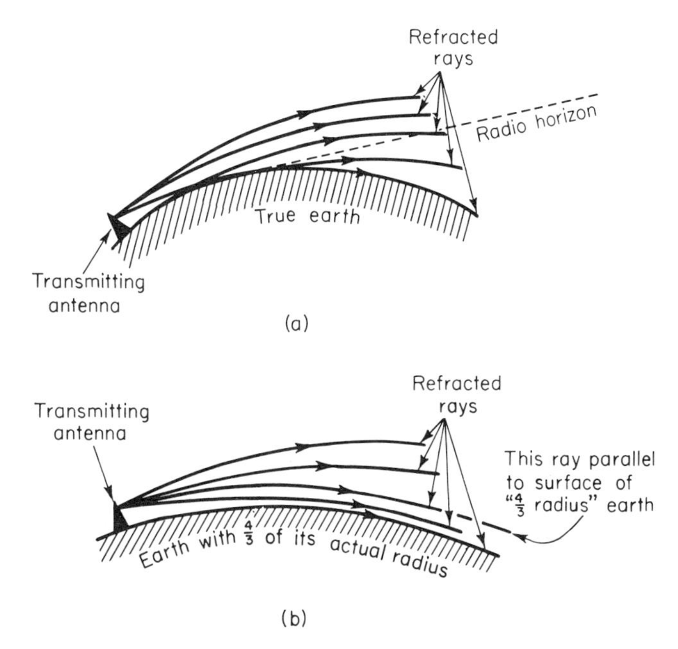

Figure 5-6 Atmospheric refraction.

† See Kerr, 1951, Section 2.13, pp. 113–122 (by W. T. Fishback) or Alpert, 1960, Chapter VI, No. 23, pp. 208–218.

This is equivalent to an effective increase in the earth's radius with respect to radio transmissions. Elaborate stratified atmosphere theories exist that allow determination of the effective radio horizon for any given frequency.† A convenient "rule of thumb" often used for rough estimates of the effect of refraction on radio transmission is the assumption of an earth's radius that is four-thirds of the actual earth radius.‡ This rule is roughly applicable over much of the radio spectrum, but it should be noted that it will not necessarily give a highly accurate estimate under any set of conditions, but represents a sort of rough average over a range of probable conditions. In any given situation, atmospheric refraction may differ widely from that corresponding to the $\frac{4}{3}$ earth radius approximation. A knowledge of local meteorological conditions would be required to determine its effects for specific radio problems.

5.5.4.4 Diffraction around the Earth

The earth represents a diffracting obstacle to radio waves, especially to those at the lower end of the spectrum. However, even at UHF and SHF, some power is diffracted around the horizon. In the *diffraction zone*, immediately below the effective radio horizon,§ there are a series of power maxima and minima, just as are observed on the far side of a diffraction grating. This diffraction pattern can be calculated from the theory of propagation of radio waves around the curved earth.|| The spacing between maxima and minima is an increasing function of wavelength, as would be expected from elementary considerations.

The implication of this is that radio transmission can take place beyond the effective radio horizon through the diffraction mechanism under conditions where the receiver is near a power maximum and not near a minimum. This condition becomes more difficult to maintain continuously as frequency increases, because the positions of maxima and minima may shift with time due to temporal changes in atmospheric refraction.

5.5.4.5 Ionospheric Effects

The ionosphere is a region between (roughly) 40 and 300 mi above the earth considered as a collection of layers of ionized gas. See Table 5-2 for a delineation of the ionospheric layers and their important properties.

The effect of the ionosphere on radio propagation can be determined

---

† See Kerr, 1951, Chapter 2.

‡ Kerr, 1951, p. 6.

§ "Effective" refers to the horizon after accounting for refraction, e.g., the "$\frac{4}{3}$ earth radius horizon."

|| For some of the results of such calculations, see Alpert, 1960, pp. 191–203. Also see Kerr, 1951, Section 2.1.2, pp. 109–112, for discussion of some of the theory behind evaluation of the fields in the diffraction region.

Table 5-2 Ionospheric Layers[a]

Ionospheric layer	*Approximate range of heights*	*Properties*
D-layer	50–90 km	Present only in daytime; reflects VLF and LF; absorbs MF and attenuates HF
E-layer	105–115 km	Reflects HF; important in short and medium range ($\leq$ 1500 km); daytime HF propagation and long-range ($\leq$ 150 km) nighttime propagation modes
F_1-layer	175–250 km	Attenuates (but does not usually reflect) long-range HF "sky waves" on their way to and from F_2 layer; generally absent at night; consequently signal strength of long-range HF is higher at night
F_2-layer	250–400 km	Reflects HF; important in long-range HF propagation

[a] Summarized from ITT, 1956, pp. 718–719.

from the theory of wave propagation through plasma media.† The ionospheric layers are plasma media, composed of free electrons that oscillate in response to the electric field of an incoming radio wave. The effect of this oscillation of electrons is that the plasma becomes an artificial propagation medium, with effective permittivity and conductivity different from those of free space. The effective changes in these parameters in response to a plane wave can be calculated from standard theory combining the Maxwell equations with the equations of motion of the electrons.‡ In the general case where the earth's magnetic field is accounted for, the Appleton-Hartree magneto-ionic theory§ is used. If the effects of the earth's magnetic field are neglected|| the results are:

† The general theory, not necessarily in the special context of ionospheric wave propagation, can be found in many texts on electromagnetic theory or plasma physics; for example, see Stratton, 1941, pp. 327–330; Holt and Haskell, 1965, Sections 11.1, 11.2, Chapter 13.

‡ See Alpert, 1960, Appendix I, pp. 377–379.

§ The theory was derived independently by many workers but these names are often associated with it. See Davies, 1964, pp. 63–99.

|| The effect of the earth's magnetic field is to transform the ionosphere into an anisotropic medium. An incoming plane-polarized wave becomes a pair of circularly polarized waves, one with right-handed and the other with left-handed circular polarization. These are called the *ordinary* and *extraordinary* waves. The medium presents different refractive indices to these two waves; hence their propagation properties are different.

$$\frac{\epsilon}{\epsilon_0} \simeq 1 - \frac{n_e q_e^2}{m_e(\nu_c^2 + [2\pi f]^2)} \tag{5.28.a}$$

$$\sigma \simeq \frac{n_e q_e^2 \nu_c}{m_e(\nu_c^2 + [2\pi f]^2)\epsilon_0} \tag{5.28.b}$$

where

f ≡ frequency in hertz
ν_c ≡ number of collisions per second, or *collision frequency*
n_e ≡ number of electrons per cubic meter, or *electron density*
q_e ≡ electron charge in coulombs
m_e ≡ electron mass in kilograms
ϵ_0 ≡ permittivity of free space = 3.85×10^{-12} F/m
σ ≡ effective conductivity in mhos per meter
$\frac{\epsilon}{\epsilon_0}$ = effective permittivity relative to that of free space

A little manipulation of Eqs. (5.28) leads to expressions for the effective refractive index and the reflection coefficient of the ionospheric layers. If $\nu_c \ll 2\pi f$, as is the case throughout much of the ionosphere, the refractive index N_r is the square root of ϵ/ϵ_0, given approximately by

$$N_r \simeq \sqrt{1 - \frac{81 n_e}{f^2}} \tag{5.29}$$

If a layer of the ionosphere is considered to be a sharply bounded homogeneous medium, a wave incident on its lower surface from free space will obey Snell's law of refraction. If the angle of incidence (defined as the angle between the vertical and the direction of the incoming wave) is θ_i, Snell's law is

$$\sin \theta_i = N_r \sin \theta_r \tag{5.30}$$

where θ_r is the angle of refraction (the angle between the vertical and the direction of the refracted wave). Total reflection occurs when $\theta_r = 90°$, in which case we have, from (5.29) and (5.30),

$$\sqrt{1 - \frac{81 n_e}{f^2}} = \sin \theta_i \tag{5.31}$$

Waves at vertical incidence on the ionosphere will be reflected at all frequencies below a *critical frequency* f_c, defined as that frequency corresponding to the condition $\theta_i = 0$. From (5.31), f_c is given by

$$f_c \simeq 9\sqrt{n_e} \tag{5.32}$$

It follows from (5.31) and (5.32) that the *maximum usable frequency* (MUF), defined as the highest frequency of waves that will be returned to earth by ionospheric reflection, is

$$\text{MUF} \simeq \frac{9 n_e}{|\cos \theta_i|} = f_c |\sec \theta_i| \tag{5.33}$$

Equations (5.32) and (5.33) are convenient rules of thumb for estimating the highest frequency that can be used for sky wave transmission, but they do not tell the whole story. In practice, (5.29) must be modified to account for ionospheric collisions, the earth's magnetic fields, and the fact that ionospheric layers do not have a homogeneous electron density, but rather a continuous positive variation of n_e with height until a maximum n_e is reached, after which n_e decreases with height. The wave propagates through the lower part of a layer, is attenuated therein, and finally reaches a value of n_e where it is totally reflected. In general, the higher the frequency, the smaller the ionospheric attenuation; hence the further the wave will propagate before it is reflected back.

These effects are quite complicated, and a great deal of theoretical and experimental research has been done to determine precisely what happens to radio waves in the ionosphere. It has been found that its structure has very pronounced geographical, diurnal, and seasonal variations and a long term variation due to the sunspot cycle.

From (5.31) and electron density versus ionosphere height curves (*electron density profiles*),† it is possible to determine a height at which n_e reaches a value consistent with (5.33) for a given frequency f and a given angle θ_i and hence a height at which waves of that frequency and that angle will be reflected. By simple geometry (using a flat earth model) this information can be used to find a transmission distance at which waves of frequency f will be received. It is worth noting that the daytime behavior of sky wave transmission differs from the nighttime behavior. E-layer and sometimes F-layer reflection generally occur at night when the D-layer is not present, the latter being the principal mechanism for daytime sky wave transmission at VLF and LF and an absorber for MF and HF waves. It is also worth noting that *skip distances* exist between sky wave reception points, such that no reception occurs except over a narrow range determined by the frequency and current ionospheric characteristics.

For detailed information on the ionosphere and sky wave radio transmission, the reader is referred to standard papers, textbooks, and handbooks listed at the end of this chapter. The brief discussion presented here is meant only to familiarize the reader with a few extremely simple ideas that often arise in discussions of terrestrial radio systems.

Consider now the way in which the ionosphere affects radio wave propagation in various frequency regions, beginning with VLF.

At VLF, terrestrial propagation takes place as if the waves were confined within a spherical shell bounded by two nearly perfect conductors, the earth and the ionosphere. The theory by which this phenomenon is analyzed is

† Such curves, together with discussion of the theory behind them, can be found in Alpert, 1960, Chapter III, pp. 117–158; Davies, 1965, Section 3.3.3, pp. 124–133.

known as the *waveguide mode theory* of VLF radio propagation.† Analysis based on this theory shows that the region between earth and ionosphere is like a waveguide for VLF waves, allowing the "guided" waves to propagate over enormous terrestrial distances. For this reason, and because atmospheric irregularities are small compared to a VLF wavelength and tend to average out in the propagation, thus reducing problems of rapid fading, the VLF region is a useful part of the radio spectrum for reliable long-distance transmission.

The waveguide mode theory, together with extensive empirical evidence, shows that VLF propagation from a vertical short electric dipole antenna at the earth's surface is governed by an expression of the approximate form,

$$|E_R| \simeq \frac{11.9}{\lambda}\frac{M_I}{r}\frac{\sqrt{r/\lambda}}{h_T/\lambda}\sqrt{\frac{r/R_e}{\sin(r/R_e)}}\,e^{-2\alpha r} \tag{5.34}$$

where

$|E_R| \equiv$ magnitude of received field strength in volts per meter
M_I = dipole current moment, i.e., current $\times$ length in ampere-meters
h_T = effective antenna height in meters
λ = wavelength in meters
α = attenuation constant in nepers per kilometer dependent on wavelength and on earth and ionosphere parameters
R_e = earth radius in kilometers $\simeq$ 6400
r = distance from transmitter to receiver in kilometers

Noting that the total power radiated into space by a short dipole is $40\pi^2 M_I^2/\lambda^2$,‡ squaring (5.34), and invoking (5.8), we can determine that the power (in watts) received at range r is

$$P_r = \frac{0.473P_T}{h_T\lambda r}\left[\frac{r/R_e}{\sin(r/R_e)}\right]e^{-2\alpha r} \tag{5.35}$$

where P_T is the total power radiated into space in watts, r and R_e are now expressed in meters and α in nepers per meter.

Note that the variation of received power with key parameters is somewhat different for VLF propagation over the earth than that in free space. Comparison of (5.35) with (5.14) indicates that if $r/R_e \ll 1$ (flat earth approximation, valid for short range transmission), the received power varies approximately as $(1/\lambda r)\,e^{-2\alpha r}$ and not as $1/\lambda^2 r^2$ as in free space. The power attenuation factor $e^{-2\alpha r}$ has a variation with r that is functionally equivalent to that of atmospheric attenuation at SHF (see Section 5.5.2.4), although the mechanism of attenuation is somewhat different in these two cases.

---

† See Davies, 1965, Chapter 9, pp. 393–441.

‡ Kraus, 1950, pp. 136–137, Eqs. (5.53)–(5.56).

At LF and MF, the ionosphere begins to be sufficiently transparent to radio waves to render unreliable the waveguide mode picture used at VLF. The waveguide becomes increasingly more "leaky" as the frequency increases above $\simeq$ 100 kHz. The ionosphere's effect can best be viewed in terms of reflection of plane waves, or sky wave transmission. In the region from 100 kHz to 3 MKz, transmission is primarily by way of the ground wave. To calculate fields at these frequencies, one uses the Sommerfeld theory of propagation in free space accounting for the presence of the earth but neglecting that of the ionosphere. This is the theoretical model that leads to the curves of Figure 5-2. In the 3 to 30 MHz (HF) region, the ground wave attenuation is extremely severe (see Figure 5-2) and the radio transmission is accomplished with the sky wave.

The maximum usable frequency, or MUF as it is usually called, is an important factor in determining how the ionosphere affects various frequency bands. The MUF is a random quantity, but rough average values can be estimated with reasonable accuracy for a given path and at given times of year and day. The average MUF varies from the 2 to 12 MHz neighborhood for short paths (less than 100 km) to nearly 50 MHz for long paths (about 3000 km). Thus ionospheric sky wave transmission at short distances must take place in the low end of the HF band or the high end of the MF band, while long distance sky wave transmission is feasible throughout the HF band and the lower part of the VHF band.

5.5.4.6 Tropospheric and Ionospheric Scatter

Over the horizon VHF, UHF, and SHF transmissions were observed in the early days of radio under conditions where neither atmospheric refraction nor diffraction theory would explain them. A body of theoretical and experimental knowledge† has developed which, to many workers in the field, indicates that these transmissions are attributable to some sort of scattering mechanism. The waves may be scattered from layers in the troposphere or ionosphere, or from small spatial gradients in the constitutive parameters of the troposphere or ionosphere, or "blobs," as they are often called. Theories differ on the exact mechanism of this so-called *scatter propagation*. One of the most popular theories is that due to Booker and Gordon‡ which attributes the phenomenon to "blob" scattering. But

† See *Proc. IRE*, special issue on "Scatter Propagation," October, 1955. This material is now somewhat classical, but the fundamental ideas of scatter propagation are covered in the papers contained in this issue.

‡ Booker and Gordon, 1950.

since this theory does not explain all observations, other theories have arisen, and certain questions on the physics of scatter propagation cannot be regarded as completely settled.

In the practical sense, however, it has been established that the use of highly directive antennas and high power sources will result in some very reliable transhorizon radio communication systems. These are known as *troposcatter* or *ionoscatter* systems, depending on whether the *common volume*† of the transmitting and receiving antenna beams is in the troposphere or ionosphere.

Troposcatter systems are operated at frequencies between about 40 MHz and 4 GHz, covering much of the VHF and UHF bands. Ionoscatter systems use frequencies between 20 MHz and 60 to 70 MHz, covering primarily the lower part of the VHF band. Transmission distances for typical ionoscatter systems are between 600 and 1200 mi. Those for troposcatter are a few hundred miles beyond the radio horizon. To obtain long troposcatter ranges, it becomes desirable to elevate the antenna as much as possible. For antenna elevations attainable in practice, about 400 mi is the approximate limit for ultrareliable troposcatter transmissions.

If receiver and antenna parameters and transmission distance are held constant, the power received in a troposcatter system decreases quite slowly with increasing frequency. In ionoscatter systems, on the other hand, the power decrease with frequency is much more pronounced.

An extremely important feature of all kinds of scatter systems is the fading they exhibit. This will be discussed in Section 5.5.4.7.

5.5.4.7 Multiplicative Noise—Multipath and Fading

Discussions of noise in Section 5.5.3 were limited to the kind of noise that adds to the signal. Another kind of noise, that which multiplies the signal and produces such effects as fading, is an important source of difficulty in some classes of radio transmission.

One physical mechanism by which multiplicative noise arises is that of multiple transmission modes, usually known as *multimode* or *multipath*. This phenomenon arises wherever the medium will concurrently support two or more totally different propagation paths, e.g., a ground wave and a sky wave, or a single hop reflected from different ionospheric heights. The transmitted signal in this case is propagated over a number of paths. These transmissions arrive at the receiver with different time delays because of differences in the electrical lengths of the various paths.

† That is, the volume in space occupied simultaneously by both beams.

The signals from the various paths add together, sometimes constructively and sometimes destructively. As the properties of the propagation medium vary slowly with time, the resulting variation of the relative path lengths produces a fluctuation in the signal amplitude and phase, known as multipath fading. If the paths are sufficiently separated in electrical length, the signals on different paths arrive during different time intervals. This leads to the simultaneous reception of signals that were transmitted at different times, a disturbing effect in many communication systems.†

A mathematical model showing how multipath affects radio signals will now be constructed. We begin by noting that a transmitted radio signal $s_T(t)$ can be represented in terms of its Fourier transform $S_T(\omega)$,

$$s_T(t) = \frac{1}{2\pi} \int_{-\infty}^{\infty} d\omega \, S_T(\omega) \, e^{j\omega t} \tag{5.36}$$

In the case of propagation over a single path, the transmitted signal undergoes a time delay τ_1 equal to the ratio of the electrical length of the path to the wave velocity and also undergoes amplitude changes and phase shifts which may be frequency selective because propagation effects are in general frequency dependent. The selectivity effect can be represented by a frequency function $M_1(\omega)$. The received signal in the single path case takes the form

$$[s_R(t)]_1 = \frac{1}{2\pi} \int_{-\infty}^{\infty} d\omega \, M_1(\omega) S_T(\omega) \, e^{j\omega(t-\tau_1)} \tag{5.37}$$

If there are N separate paths, each with a *time delay* τ_k and a *path response function* $M_k(\omega)$, the multipath signal is a superposition of the signals travelling these paths, and has the form

$$s_R(t) = \sum_{k=1}^{N} [s_R(t)]_k \tag{5.38}$$

where

$$[s_R(t)]_k \equiv \frac{1}{2\pi} \int_{-\infty}^{\infty} d\omega \, M_k(\omega) S_T(\omega) \, e^{j\omega(t-\tau_k)}$$

or, equivalently,

$$s_R(t) = \frac{1}{2\pi} \int_{-\infty}^{\infty} d\omega \, S_T(\omega) M(\omega) \, e^{j\omega t} \tag{5.39}$$

where

$$M(\omega) \equiv \sum_{k=1}^{N} M_k(\omega) \, e^{-j\omega\tau_k}$$

The representation (5.39) implies that the general effect of a multipath

† See the discussion in Section 12.4 on intersymbol interference in digital communication systems.

medium on a transmitted signal is equivalent to that of a linear filter with impulse response

$$m(t) = \sum_{k=1}^{N} m_k(t - \tau_k) \tag{5.40}$$

where $m_k(t)$ = inverse Fourier transform of $M_k(\omega)$.

The precise effects on the transmitted signal depend on certain features of the medium's frequency response function $M(\omega)$. To determine these effects, we first observe that the transmitted signal consists of a pair of narrow bands around $+\omega_c$ and $-\omega_c$ where ω_c is the carrier frequency. The signal information is contained in the shape of these bands. We write

$$S_T(\omega) = S_T(\omega - \omega_c) + S_T(\omega + \omega_c) \tag{5.41}$$

where $S_T(\omega)$ is the spectrum of the waveform containing the signal information.

To facilitate analysis, we will use a very simple multipath model in which the effect is nonselective and amplitude differences between paths are negligible. Then $M_k(\omega)$ is a constant which can be called M_0. We also assume that $S_T(\omega)$ is flat between $\omega \mp \omega_c = -\pi B_s$ and $+\pi B_s$ and takes on a value S_T between these limits. Then from (5.39), (5.40), and (5.41)

$$s_R(t) = 2M_0 S_T B_s \sum_{k=1}^{N} \cos \omega_c(t - \tau_k) \operatorname{sinc}(\pi B_s(t - \tau_k)) \tag{5.42}$$

There is an alternative way of writing (5.42), in which $s_R(t)$ is recognized as a narrowband noise process, as follows:

$$s_R(t) = x_{SR}(t) \cos \omega_c t + y_{SR}(t) \sin \omega_c t \tag{5.43}$$

where

$$x_{SR}(t) = 2M_0 S_T B_s \sum_{k=1}^{N} \cos \omega_c \tau_k \operatorname{sinc}(\pi B_s(t - \tau_k))$$

$$y_{SR}(t) = 2M_0 S_T B_s \sum_{k=1}^{N} \sin \omega_c \tau_k \operatorname{sinc}(\pi B_s(t - \tau_k))$$

Both $x_{SR}(t)$ and $y_{SR}(t)$ are slowly varying compared to cos $(\omega_c t)$ and sin $(\omega_c t)$. The path delays, from the receiver's viewpoint, are random variables. In certain propagation mechanisms there is some correlation between paths separated by small delays, because they may be generated by the same physical mechanism. For example, two paths arising from reflection from two adjacent parts of the same ionospheric layer or tropospheric "blob" are likely to be somewhat correlated. However, there is a *correlation time*, i.e., a time delay separation τ_c beyond which paths can be considered uncorrelated and even statistically independent. If there are enough paths and the total *multipath spread* (the total spread in delay times over all the paths) is sufficiently large, then $x_{SR}(t)$ and $y_{SR}(t)$ can both be considered as superpositions of a number of independent random variables (because the time variation is slow compared to cos $\omega_c t$ and sin $\omega_c t$, the functions $x_{SR}(t)$ and $y_{SR}(t)$ can be regarded as constant over many *RF* periods). If the number is at least as high as 6 or 7, the central limit theorem discussed in Section 3.5 applies, and according to Eq. (3.73) of that section, both x_{SR} and y_{SR}have approximately Gaussian statistics. The amplitude $\sqrt{x_{SR}^2 + y_{SR}^2}$ has the Rayleigh PDF (Eq. (3.40)) and the phase $\tan^{-1}(-y_{SR}/x_{SR})$ has a uniform PDF.†

The real-world implication of the above arguments is that the signal amplitude and phase both fluctuate at rates that are slow compared to cos $\omega_c t$ or sin $\omega_c t$. This "fading" fluctuation is not known at the receiver and must be treated as a random function. Measurements show that much of the fading in radio transmissions has a Rayleigh amplitude distribution, evidence of the correctness of the theoretical multipath model in explaining the observed effects.‡

The assumption of flat signal spectrum in the analysis is not really basic to the conclusions. The general features of the results would hold for any signal spectrum that is narrow compared to the radio frequency. A somewhat basic assumption is that of nearly equal amplitudes for all the paths. If some small number of paths (less than about 4) have amplitudes overwhelmingly larger than the rest, then the effect will be somewhat similar to the action of wave interference (described in Section 5.5.4.2) on a moving receiver. If the relative path lengths fluctuate in time, the resulting fading amplitude will not have Rayleigh statistics. The fading statistics will in fact be highly dependent on the statistics of the delay time τ_k. If the amplitude of a single path is large compared to the others, or if the number of paths add coherently, then we have the equivalent of a strong steady sinusoidal signal superposed on the Rayleigh fading. The amplitude PDF in this case is described by the Rice distribution (3.46) and the phase PDF§ is somewhat more complicated than in the zero signal case. The steady signal part of the received waveform is known as the *specular fading component.* Most fading that occurs in actual radio systems is a superposition of Rayleigh and specular components.

Terrestrial propagation modes for which fading often presents severe problems are (1) troposcatter and ionoscatter, where the multipath effect may arise through scattering from different "blobs," (2) HF sky wave transmission, where signals reflected from different parts of the ionosphere may be received simultaneously and the ionospheric properties may be varying with time, and (3) line-of-sight UHF and microwave transmission, where fading sometimes arises through statistical irregularities in the refractive index of the atmosphere.

---

† To establish the uniform phase PDF, we would express $p_2(x_n, y_n)$ in (3.36) in terms of $|v_n|$ and ϕ_n as given by (3.38) and (3.39), then integrate from $|v_n| = 0$ to $|v_n| = \infty$.

‡ It should be noted that the Rayleigh PDF could also be predicted from other theoretical models.

§ See Davenport and Root, 1958, pp. 166–167.

5.5.5 Space Radio Transmission

Most of the discussions in Sections 5.5.1 through 5.5.4 have been more or less focussed on the effects encountered in strictly terrestrial radio transmissions. In modern space technology, radio is used extensively for communication and for detection, location, and tracking of space vehicles. It is important to include in our discussion certain special features of space radio transmissions.

There are a number of possible modes of transmission in space applications. The three basic modes are: (1) ground-to-space, (2) space-to-ground, and (3) space-to-space. Modes (1) and (2) both involve transmission through the troposphere and the ionosphere in the case where ordinary earth satellites are the transmitting or receiving terminals, and through other ionized regions in deep space (such as the Van Allen Belt) as well in the case of deep-space vehicles. Mode (3) may or may not involve ionospheric transmission. A radio link between two satellites both at ionospheric altitudes will obviously involve such transmission, while a link between two deep-space vehicles will not. It is rather interesting that in some respects the most glamorous of space radio links, the deep-space-to-deep-space link, is the least complicated, provided sufficiently high frequencies are used. In that mode, propagation takes place through a medium nearly equivalent to infinite free space, unhampered by obstacles and ionospheric effects.

If we review the material presented earlier in this chapter, we will find that the considerations of Sections 5.1 through 5.4 are as applicable to space radio as to terrestrial radio. The same applies to many of the discussions of Section 5.5, in particular those of 5.5.1. The beacon equation (5.14) and also Eq. (5.15) (the expression for SNR in dB derived from (5.14)) are more directly applicable to space transmission than to terrestrial transmission. Many of the perturbing effects of the earth and its environment, which reduce the direct utility of Eqs. (5.14) and (5.15), are absent in space transmission modes. Some of the signal power limitations treated in Section 5.5.2 are still present in space radio, except for the obvious absence of ground losses. Atmospheric attenuation will still be present in a space-to-ground or ground-to-space link. However, for these modes we must also consider the attenuation of the radio wave in traversing the ionosphere. Referring to Section 5.5.4.5, we note the fundamental difference in the role of the ionosphere (which acts as a high pass filter) in entirely terrestrial modes and in modes involving propagation between earth and space. In the former, to exploit the ionosphere as a reflector, one uses frequencies well *below* the critical frequencies of its layers. In the latter, one uses frequencies well *above* the critical frequencies of these layers in order to penetrate them with a minimum of attenuation and phase shift. This dictates the use of extremely high frequencies, primarily SHF, in such modes. Fortunately, another consideration also dictates the use of these frequencies, that of minimization of size and weight of equipment, a prime factor in space vehicles. Antennas can be extremely small and still provide adequate directivity at SHF, as pointed out in Section 5.3.

On the negative side, the use of increasingly higher frequencies in order to more effectively reduce ionospheric attenuation will increase the attenuation of the wave in traversing the lower atmosphere, particularly in regions near oxygen and water vapor absorption peaks. (See Section 5.5.2.4.) This is a factor that sometimes must be accounted for in choosing a frequency for this type of radio link.

All of the additive noise sources referred to in Section 5.5.3 are present to some degree in space transmissions. Man-made noise and interference generated on earth can act effectively on the ground station in a link between space and ground. For such noise to directly affect the space terminal, it would almost be required at the present stage of space technology that it be deliberate and intentional. This will no longer be true when space vehicle traffic becomes sufficiently dense to produce an active radio environment in space, similar to that which now exists in the regions where ordinary aircraft fly.

Extraterrestrial noise is a somewhat greater problem in space than in purely terrestrial radio systems. One reason for this is the absence of the ionospheric shielding of such noise that protects terrestrial radio systems. However, before making too large a point of this, one should note from the discussion above that links between space and ground employ frequencies well above the spectral region in which the ionosphere would be a very effective shield. Hence systems operating entirely below the ionosphere *at these same frequencies* would be subject to nearly the same amount of extraterrestrial noise as would space-to-ground and ground-to-space links.

Referring now to the propagation effects discussed in Section 5.5.4, most of these have meaning only in the context of transmission in an earth environment. There are some, however, that do have some pertinence to space radio. One is the topic discussed in 5.5.4.7, multipath and fading. At sufficiently high frequencies to allow virtually loss-free ionospheric penetration, the phase perturbing effects exhibited by the ionosphere are minimized and hence fading and multipath are not usually serious problems at these frequencies. However, in such space modes as satellite-to-earth or earth-to-satellite communication at VHF, random irregularities in the ionosphere will perturb the phase front of a passing radio wave in such a manner as to produce scintillations in the received signal. Fading induced by these effects can be a serious problem in cases where frequencies as low as VHF are employed in space-ground radio transmission. The reader is referred to the literature for further discussion of fading due to ionospheric scintillations.†

† See Aarons, Barron, and Castelli, 1958; Aarons, 1963, Chapter 3, pp. 38–134; Davies, 1965, Section 5.4.4, pp. 248–253.

Another propagation effect of special importance in radio links between space and ground is atmospheric refraction (Section 5.5.4.3). Slight errors in angular measurements of space vehicles hundreds or thousands of miles from earth can be extremely serious. In accurate angular measurement and tracking work, it is important to know (or be able to measure) rather precisely the degree of refraction in the ionosphere and the lower atmosphere in order to correct for it. Random errors in angular measurement also may occur due to ionospheric scintillations, but as remarked above, the use of sufficiently high frequencies minimizes that effect.

Space radio has become such a vast technology that there is a great deal of literature now available on the subject. The reader interested in pursuing it further is referred to the bibliography at the end of this chapter.

REFERENCES

Radio Engineering, Textbooks and Handbooks (*Transmitters, Receivers, Propagation*)

Henney, 1959
ITT, 1956
Seely, 1950
Terman, 1955

Electromagnetic Theory, Radio Wave Propagation, Ionosphere

Alpert, 1963
Burrows and Attwood, 1949
Davies, 1965
ITT, 1965; Chapter 24, pp. 710–758
Jordan, 1950 (electromagnetic radiation slanted toward radio systems)
Kerr, 1951 (microwave propagation)
Ramo and Whinnery, 1953 (electromagnetic theory with radio systems "flavor")
Terman, 1955; Chapter 22, pp. 803–863

Antennas

ITT, 1956; Chapter 23, pp. 662–710
Kraus, 1950
Silver, 1949 (microwave antennas)
Terman, 1955; Chapter 23, pp. 864–934

Radio Noise

See reference list on physical sources of noise in Chapter 3 for references on internal radio noise. On external radio noise, see:

Davies, 1965; Section 7.5, pp. 310–315
Bennett, 1960; Chapter 6

ITT, 1956; Chapter 25, pp. 762–777
Menzel, 1960

Space Radio Transmission

Most of these references are in the context of space communications:

Balakrishnan, 1963
Brown, 1962
Golomb, 1964
Maeda and Silver, 1965
Marsten, 1966
Stiltz, 1966
Tischer, 1965

6

PROCESSING OF RADIO SIGNALS FOR MAXIMUM SNR

The next three chapters are devoted to certain applications of statistical communication theory to the processing of signals entering a radio receiver. We choose to call the topic to be discussed in what follows *statistical processing*, to indicate that the signals are operated upon somewhat in accordance with optimization criteria indicated by statistical communication theory. In the present chapter, attention is confined to processing for maximum signal-to-noise ratio, where the interfering noise is assumed to be Gaussian.†

An underlying philosophical point permeates the discussions in these chapters. In many cases, the optimum processing techniques indicated theoretically are not significantly different from the time-honored procedures arrived at intuitively long before the theory was in use by engineers. This is not to imply that the theory is superfluous. Quite to the contrary, it provides the engineer with the knowledge of how close his intuitively designed systems are to the ultimate theoretical upper limits of performance. Moreover, it will show the limits of applicability of the intuitive approach and indicate under what conditions greater sophistication is required.

Suppose we ask the question, "What are the best filtering operations to perform on the incoming radio signal?" This question has meaning only if we define a performance criterion, that is, decide on the parameter to be used as a measure of quality of the system. Having decided on such a measure, we know that we want the filtering operations to maximize this measure.

† See Sections 5.5.3.1 through 5.5.3.5 for a discussion of the nature of additive noise in radio receivers.

The measure of quality that immediately comes to mind if we leave unspecified the kind of information to be extracted from the signal is the signal-to-noise ratio (SNR). In the case of additive Gaussian noise, it will become apparent in later discussions that measures of degradation (e.g., error probability in detection and digital communication systems, and mean square error in analog measurement systems) are monotonically decreasing functions of SNR, at least in cases where we deal with *linear* systems. For this reason, where the troublesome noise is additive and Gaussian, it is always desirable to maximize the SNR at the output of the linear stages of the receiver before attempting to extract information from the signal. Since this maximization will be accomplished through linear filtering if the noise is Gaussian,† and since most of the noise at the receiver input is narrowband Gaussian,‡ most of the present chapter will concern the subject of linear filtering of radio signals in narrowband Gaussian noise.

6.1 MATHEMATICAL REPRESENTATION OF A RADIO SIGNAL

A radio signal at the input to a receiver§ is a radio wave carrier modulated by some form of information or intelligence. The task of the receiver is to extract this information from the carrier. The carrier itself is no longer needed once the signal has traversed space and arrived at the receiver.

The radio signals to be discussed here will all be characterized by the fact that the bandwidth of the information "riding" on the carrier is very small compared to the carrier frequency itself. If the latter is denoted (in angular frequency units) by ω_c and the information bandwidth|| (also in angular frequency units) by $2\pi B_s$, then our assumption is

$$\omega_c \gg 2\pi B_s \tag{6.1}$$

The radio signal at the receiver input can always be represented in the form††

$$s_i(t) = x_i(t) \cos \omega_c t - y_i(t) \sin \omega_c t = a_i(t) \cos (\omega_c t + \psi_i(t)) \tag{6.2}$$

---

† See Section 3.2.1.

‡ See Section 3.3, especially Eq. (3.33), and Sections 5.5.3.1 through 5.5.3.5.

§ The "signal" here refers to the voltage across the receiving antenna's output terminals, i.e., at the point of entry into the receiver. This voltage, induced by the electric field incident on the antenna, is due to the arrival of the transmitted wave.

|| "Information bandwidth" refers here to the highest significant frequency in the information waveform or, equivalently, the lowpass bandwidth of the information waveform.

†† Note that $x_i(t)$ and $y_i(t)$ are not uniquely determined by $s_i(t)$. Since the zero reference level of time is arbitrary, one could as well write $s_i(t) = x_i(t) \cos \omega_c(t - t_1) - y_i(t) \sin \omega_c(t - t_1)$, in which case $x_i(t)$ and $y_i(t)$ would be different from the $x_i(t)$ and $y_i(t)$ of (6.2.).

A convenient mathematical tool for analyzing such signals as (6.2) is that of the *complex envelope*. This tool will be developed here and will be found to be very useful in subsequent analysis. Let us write (6.2) in the form

$$s_i(t) = \tfrac{1}{2}\{\hat{s}_i(t)\, e^{j\omega_c t} + \hat{s}_1^*(t)\, e^{-j\omega_c t}\} = \text{Re}\,\{\hat{s}_i(t)\, e^{j\omega_c t}\} \tag{6.3}$$

where

$$\hat{s}_i(t) = x_i(t) + jy_i(t)$$

The complex function $\hat{s}_i(t)$ will be called the *complex envelope* of $s_i(t)$. It is a slowly varying (compared to $e^{\pm j\omega_c t}$) function of time, and contains all of the signal information. One of its interesting and important properties is

$$\hat{s}_i(t) = a_i(t)\, e^{j\psi_i(t)} \tag{6.4}$$

where

$$a_i(t) = \sqrt{[x_i(t)]^2 + [y_i(t)]^2}$$

is the signal amplitude as defined through (6.2), and

$$\psi_i(t) = \tan^{-1}\left[\frac{y_i(t)}{x_i(t)}\right]$$

is the signal phase as defined through (6.2). Note that both $a_i(t)$ and $\psi_i(t)$ are real functions of time.

6.2 LINEAR FILTERING OF RADIO SIGNALS

In applying the notion of complex amplitude to linear filtering, we assume first that the linear bandpass filtering done on the signal at the RF stage of the receiver will have certain characteristics, as follows:

1. The filter will be tuned to a central frequency ω_0 very near or equal to the carrier frequency (a discrepancy between ω_0 and ω_c is usually due to the lack of precise knowledge of ω_c at the receiver), such that

$$|\omega_0 - \omega_c| \ll \omega_c, \omega_0 \tag{6.5}$$

2. The bandwidth† of the filter, denoted by B_f, will be small compared to ω_c or ω_0, i.e., as an analog of (6.1).

$$\omega_c \gg 2\pi B_f \tag{6.6}$$

from which it follows through (6.5) that

$$\omega_0 \gg 2\pi B_f \tag{6.7}$$

By steps equivalent to those used above for the signal, we can develop

† Although the filter under discussion here is a bandpass filter, its bandwidth B_f will be defined as a lowpass bandwidth in correspondence with the definition of B_s.

a complex amplitude model for the impulse response of the filter, denoted by $f(t)$. Changing s to f and changing the subscript i to the subscript f in (6.2), (6.3), and (6.4), we have

$$\begin{aligned} f(t) &= x_f(t)\cos\omega_0 t - y_f(t)\sin\omega_0 t = a_f(t)\cos(\omega_0 t + \psi_f(t)) \\ &= \operatorname{Re}\{\hat{f}(t)\,e^{j\omega_0 t}\} \end{aligned} \tag{6.8}$$

where

$$\begin{aligned} \hat{f}(t) &= x_f(t) + jy_f(t) = a_f(t)\,e^{j\psi_f(t)} \\ a_f(t) &\equiv \sqrt{[x_f(t)]^2 + [y_f(t)]^2} \\ \psi_f(t) &\equiv \tan^{-1}\left[\frac{y_f(t)}{x_f(t)}\right] \end{aligned}$$

The output of the filter with the input given by (6.2) and the impulse response given by (6.8) is, from (1.5),

$$\begin{aligned} s_o(t) &= \int_{-\infty}^{\infty} dt'\, s_i(t')f(t-t') = \int_{-\infty}^{\infty} dt'\, s_i(t-t')f(t') \\ &= \operatorname{Re}\{\hat{s}_o(t)\,e^{j\omega_c t}\} \end{aligned} \tag{6.9}$$

where $\hat{s}_o(t)$, the complex amplitude of the output, is given by

$$\begin{aligned} \hat{s}_o(t) &= \frac{1}{2}\int_{-\infty}^{\infty} dt'\, \hat{s}_i(t-t')\hat{f}(t')e^{j\Delta\omega t'} \\ &\quad + \frac{1}{2}\int_{-\infty}^{\infty} dt'\, \hat{s}_i(t-t')\hat{f}^*(t')\,e^{-j\Delta\omega t'}\,e^{-2j\omega_c t'} \\ &= a_o(t)e^{j\psi_o(t)} \end{aligned}$$

where $\Delta\omega \equiv \omega_0 - \omega_c$ and where $a_o(t)$ and $\psi_o(t)$ represent the amplitude and phase respectively of the output signal.

Because of the assumptions (6.1), (6.5), (6.6), and (6.7), the factor $\hat{s}_i(t-t')\hat{f}^*(t')e^{j\Delta\omega t'}$ in the second integral of (6.7) is substantially constant over many cycles of $e^{-2j\omega_c t}$ and the second integral is therefore negligibly small relative to the first. Then, to very good approximation,

$$\begin{aligned} \hat{s}_o(t) &\simeq \frac{1}{2}\int_{-\infty}^{\infty} dt'\, \hat{s}_i(t-t')\hat{f}(t')e^{j\Delta\omega t'} \\ &= e^{j\Delta\omega t}\left(\frac{1}{2}\int_{-\infty}^{\infty} dt''\, \hat{s}_i(t'')\hat{f}(t-t'')e^{-j\Delta\omega t''}\right) \end{aligned} \tag{6.10}$$

The expression (6.10) relates the complex envelopes of input signal, filter impulse response, and output signal in such a manner that the effect of the carrier frequency ω_c has been completely eliminated. Note that this elimination of the carrier's effect is only possible because of the special "narrowband" properties of radio information signals and filters designed to process them, as characterized by the assumptions (6.1), (6.5), (6.6), and (6.7). The effect of a discrepancy between the carrier frequency and the central or tuning frequency of the filter is accounted for very conveniently through the factor $e^{j\Delta\omega t}$ in (6.10).

We will now express the idea of the complex envelope in terms of Fourier transforms of signals and frequency response functions of filters. This is easily accomplished through application of (1.4) and the convolution theorem (1.5) to Eq. (6.10),

$$F\{\hat{s}_o(t)\} = \tfrac{1}{2} F\{\hat{s}_i(t)\} F\{\hat{f}(t)\, e^{j\Delta\omega t}\} \tag{6.11}$$

Denoting the Fourier transform of an arbitrary complex envelope $\hat{s}(t)$ by $\hat{S}(\omega)$, we can express (6.11) in the form

$$\frac{\hat{S}_o(\omega)}{\hat{S}_i(\omega)} = \frac{1}{2} \hat{F}(\omega - \Delta\omega) \tag{6.12}$$

where $\hat{F}(\omega)$ is the frequency response function corresponding to $\hat{f}(t)$. Note that $\hat{f}(t)$ is the complex envelope of the filter impulse response referred to the tuning frequency ω_0 and not the carrier frequency ω_c.

To relate Fourier transforms of complex envelopes to those of the real signals represented by these complex envelopes, we invoke (1.4) expressing it in the form

$$\begin{aligned} S_o(\omega) &= \int_{-\infty}^{\infty} dt'\, e^{-j\omega t'} \operatorname{Re}\{\hat{s}_o(t')\, e^{j\omega_c t'}\} = S_i(\omega) F(\omega) \\ &= \int_{-\infty}^{\infty} dt'\, e^{-j\omega t'} \operatorname{Re}\{\hat{s}_i(t')\, e^{j\omega_c t'}\} \int_{-\infty}^{\infty} dt''\, e^{-j\omega t''} \operatorname{Re}\{\hat{f}(t'')\, e^{j\omega_0 t''}\} \end{aligned} \tag{6.13}$$

or, equivalently,

$$\begin{aligned} &\tfrac{1}{2}\{\hat{S}_o(\omega - \omega_c) + \hat{S}_o^*(-[\omega + \omega_c])\} \\ &\qquad = \tfrac{1}{4}\{\hat{S}_i(\omega - \omega_c) + \hat{S}_i^*(-[\omega + \omega_c])\}\{\hat{F}(\omega + \omega_0) + \hat{F}^*(-[\omega + \omega_0])\} \end{aligned} \tag{6.14}$$

By virtue of the assumptions (6.1), (6.5), (6.6), and (6.7), there is no significant overlap in frequency space between the functions $\hat{S}_o(\omega - \omega_c)$, $\hat{S}_i(\omega - \omega_c)$, $\hat{F}(\omega - \omega_0)$ and the functions $\hat{S}_o(\omega + \omega_c)$, $\hat{S}_i^*(-[\omega + \omega_c])$, $\hat{F}(-[\omega + \omega_0])$. The latter have their peak magnitudes in the vicinity of $\omega \simeq -\omega_0, -\omega_c$, while the former have their peaks in the vicinity of $\omega \simeq \omega_0, \omega_c$. These regions are so widely separated under the narrowband assumption that the functions with argument $(\omega - \omega_0)$ or $(\omega - \omega_c)$ decay to negligible values far outside the region where those with argument $(\omega + \omega_0)$ or $(\omega + \omega_c)$ have significant magnitudes. Thus cross products between these two types of functions in (6.14) are negligible, and we have

$$\begin{aligned} \hat{S}_o(\omega - \omega_c) &\simeq \tfrac{1}{2} \hat{S}_i(\omega - \omega_c)\, \hat{F}(\omega - \omega_0) \\ \hat{S}_o^*(-[\omega + \omega_c]) &\simeq \tfrac{1}{2} \hat{S}_i^*(-[\omega + \omega_c]) \hat{F}^*(-[\omega + \omega_0]) \end{aligned} \tag{6.15}$$

Then

$$
\begin{aligned}
s_o(t) &= \frac{1}{4\pi}\int_{-\infty}^{\infty} d\omega'\, \hat{S}_o(\omega' - \omega_c)\, e^{j\omega' t} \\
&\quad + \frac{1}{4\pi}\int_{-\infty}^{\infty} d\omega'\, \hat{S}_o^*(-[\omega' + \omega_c])\, e^{j\omega' t} \\
&= \operatorname{Re}\left\{e^{j\omega_c t}\frac{1}{2\pi}\int_{-\infty}^{\infty} d\omega'\, \hat{S}_o(\omega')\, e^{j\omega' t}\right\}
\end{aligned}
\tag{6.16}
$$

or, equivalently,

$$
\begin{aligned}
\hat{s}_o(t) &= \frac{1}{2\pi}\int_{-\infty}^{\infty} d\omega'\, \hat{S}_o(\omega')\, e^{j\omega' t} \\
&= \frac{1}{4\pi}\int_{-\infty}^{\infty} d\omega'\, \hat{S}_i(\omega')\hat{F}(\omega' - \Delta\omega)\, e^{j\omega' t}
\end{aligned}
\tag{6.17}
$$

Since $s_i(t)$ and $s_o(t)$ both represent voltages or currents, the input and output signal power levels are proportional to the squares of $s_i(t)$ and $s_o(t)$, respectively, averaged over many cycles of the RF carrier.

From (6.2), (6.3), and (6.9), noting that time averages $(\overline{e^{\pm 2j\omega_c t}})$ are zero for practical purposes, we have

$$
\begin{aligned}
\overline{[s_i(t)]^2} &= \tfrac{1}{4}\,\overline{\left[\hat{s}_{\substack{i\\o}}(t)\, e^{j\omega_c t} + \hat{s}_{\substack{i\\o}}^*(t)\, e^{-j\omega_c t}\right]^2} \\
&= \tfrac{1}{4}\left\{\left[\hat{s}_{\substack{i\\o}}(t)\right]^2 e^{2j\omega_c t} + \left[\hat{s}_{\substack{i\\o}}^*(t)\right]^2 e^{-2j\omega_c t} + 2\left[\hat{s}_{\substack{i\\o}}(t)\,\hat{s}_{\substack{i\\o}}^*(t)\right]\right\} \\
&= \tfrac{1}{2}\,|\hat{s}_{\substack{i\\o}}(t)|^2 = \tfrac{1}{2}\left[a_{\substack{i\\o}}(t)\right]^2
\end{aligned}
\tag{6.18}
$$

Note that we have uncoupled the process of time averaging over the slow variations of $\hat{s}_{\substack{i\\o}}(t)$ from that of time averaging over the rapid variations of $e^{\pm 2j\omega_c t}$. This is perfectly valid here because of the virtually complete absence of any coherent relationship between those two fluctuations, which are not only on two widely separated time scales but whose frequencies are in general not commensurable.

We now turn to the problem of evaluating the effect of linear filters on noise.

To determine the noise power output of a linear filter, we assume that the noise bandwidth† B_n is substantially greater than the filter bandwidth B_f. This will be true in the radio applications to be considered here: The noise power can be conveniently determined without the use of complex envelopes. We therefore return to the real signal and filter representation to perform noise power calculations.

The mean noise output power at the time t is, in general,

$$
\langle [n_o(t)]^2 \rangle = \iint_{-\infty}^{\infty} dt'\, dt''\, \langle n_i(t - t')n(t - t'')\rangle f(t')f(t'') \tag{6.19}
$$

---

† Noise bandwidth B_n will be defined here in correspondence with B_s and B_f, i.e., the *lowpass* noise bandwidth or the frequency spacing between zero and the highest significant noise frequency. This is consistent with the definition in Section 3.2.1.

If we assume, as indicated above, that

$$B_n \gg B_f \tag{6.20}$$

then within the integrand of (6.14), we can use the approximation†

$$\langle n_i(t')n_i(t'+\tau)\rangle \simeq \frac{\sigma_n^2}{2B_n}\,\delta(\tau) \tag{6.21}$$

where σ_n^2 is the mean noise power and $\delta(\tau)$ is the unit impulse function.

It follows, through (6.6) and (6.14), that

$$\langle [n_o(t)]^2\rangle \simeq \frac{\sigma_n^2}{2B_n}\int_{-\infty}^{\infty} dt'\, f^2(t') = \frac{\sigma_n^2}{4B_n}\int_{-\infty}^{\infty} dt'\,[a_f(t')]^2 \tag{6.22}$$

We can also use frequency domain reasoning, and write, as in Section 3.2.1,

$$\langle [n_o(t)]^2\rangle = \frac{1}{2\pi}\int_{-\infty}^{\infty} d\omega'\, G_n(\omega')\,|F(\omega)|^2 = \frac{\sigma_n^2}{16\pi B_n}\int_{-\infty}^{\infty} d\omega'\,|\hat{F}(\omega')|^2 \tag{6.23}$$

By Parseval's theorem,‡ for an arbitrary time function $f(t)$,

$$\int_{-\infty}^{\infty} dt'\, f^2(t') = \frac{1}{2\pi}\int_{-\infty}^{\infty} d\omega'\,|F(\omega')|^2 \tag{6.24}$$

which demonstrates the agreement that must exist between the impulse response and frequency response approaches to the calculation of mean noise output power.

In all applications to be considered the filters will have passbands that are symmetrical about their central frequencies.

A convenient definition of the filter's *lowpass bandwidth* in this case is half of the integral of the absolute square of the complex envelope of the frequency response function, relative to its value at the central frequency. For a flat filter, this would be (by definition) the true bandwidth. For certain other filter shapes, it comes close to the bandwidth as defined in other ways, e.g., in terms of half-power points. For example, the magnitude squared of the frequency response function of an *RLC* bandpass filter,§ given by (1.26) or (1.28), peaks at $\omega_0 = 1/\sqrt{LC}$ and decays to half its peak value at a frequency $\omega_1 = \omega_0 \pm (R/2L)$ (in the series *RLC* case) or $\omega_1 = \omega_0 \pm (1/2RC)$ (in the parallel *RLC* case). In these particular cases, then, the half-power

---

† Note that (6.20) is equivalent to the assumption that the noise is completely flat over all frequency space from $\omega = -\infty$ to $\omega = +\infty$; that is, it is *white noise* in the strict sense of the term.

‡ See Korn and Korn, 1961, No. 4-11-4; p. 136. Equation (6.24), known classically as Parseval's theorem, can easily be derived by using (1.4) to express $F(\omega)$ as the Fourier transform of $f(t)$ in the integral on the right-hand side of (6.24), then invoking (1.6.f).

§ If the filters with the frequency response functions given by (1.26) and (1.28) are considered as *narrowband* and peaked at the resonant frequencies $\omega_0 = 1/LC$, then under the narrowband condition $|\omega - \omega_0| \ll \omega_0$, we have in (1.26)

point definition of bandwidth and this definition differ only by a factor $\pi/2$, which is not far from unity.†

The above definition of bandwidth, stated mathematically, is

$$B_f = \frac{1}{2}\left\{\frac{1}{2\pi\,|\hat{F}(0)|^2}\int_{-\infty}^{\infty} d\omega'\,|\hat{F}(\omega')|^2\right\} \tag{6.25}$$

Obviously, if the complex frequency response of the filter is flat from $\omega' = -B_f$ to $\omega' = B_f$, then (6.25) is an identity.

Using (6.23) and (6.25), the mean noise output power at any time t (the output power is independent of t because of the stationarity of the input noise) is

$$\langle [n_o(t)]^2 \rangle \simeq \frac{\sigma_n^2\,|\hat{F}(0)|^2 B_f}{4B_n} \tag{6.26}$$

The output SNR at time t, denoted by $\rho_o(t)$, can be calculated from (6.17), (6.18), and (6.26) as follows:

$$\rho_o(t) = \frac{[s_o(t)]^2}{\langle [n_o(t)]^2 \rangle} = \frac{B_n}{2\sigma_n^2 B_f\,|\hat{F}(0)|^2}\left|\frac{1}{2\pi}\int_{-\infty}^{\infty} d\omega'\,\hat{S}_i(\omega')\hat{F}(\omega' - \Delta\omega)\,e^{j\omega' t}\right|^2 \tag{6.27}$$

or, equivalently, using (6.10) in lieu of (6.17),

$$\rho_o(t) = \frac{B_n}{2\sigma_n^2 B_f\,|\hat{F}(0)|^2}\left|\int_{-\infty}^{\infty} dt'\,\hat{s}_i(t - t')\hat{f}(t')\,e^{j\Delta\omega t'}\right|^2 \tag{6.28}$$

6.3 COHERENT MATCHED FILTERING OR CROSSCORRELATION

If we have complete knowledge of the carrier frequency and the signal waveshape, including the phases of all frequency components in $s_o(t)$, we can

---

$$\left|\frac{j\omega C}{1 + j\omega RC + (j\omega)^2 LC}\right|^2 = \left|\frac{j\omega_0 C + j(\omega - \omega_0)C}{\{1 - [\omega_0 + (\omega - \omega_0)]^2 LC\} + j\{\omega_0 RC + (\omega - \omega_0)RC\}}\right|^2$$
$$\simeq \left|\frac{j\omega_0 C}{\{-2\omega_0(\omega - \omega_0)LC\} + j\omega_0 RC}\right|^2$$
$$= \left|\frac{1}{R}\,\frac{1}{1 + j\dfrac{(\omega - \omega_0)}{(R/2L)}}\right|^2 \propto \frac{1}{1 + \dfrac{(\omega - \omega_0)^2}{(R/2L)^2}}$$

and in (1.28)

$$\left|\frac{j\omega L}{1 + j\omega\dfrac{L}{R} + (j\omega)^2 LC}\right|^2 \simeq \left|\frac{1}{1 + j\dfrac{(\omega - \omega_0)}{(1/2RC)}}\right|^2 = \frac{1}{1 + \dfrac{(\omega - \omega_0)^2}{(1/2RC)^2}}$$

† Note that $\dfrac{1}{4\pi}\displaystyle\int_{-\infty}^{\infty}\frac{d\omega}{1 + \dfrac{(\omega - \omega_0)^2}{2\pi B}} = \frac{\pi}{2}B.$

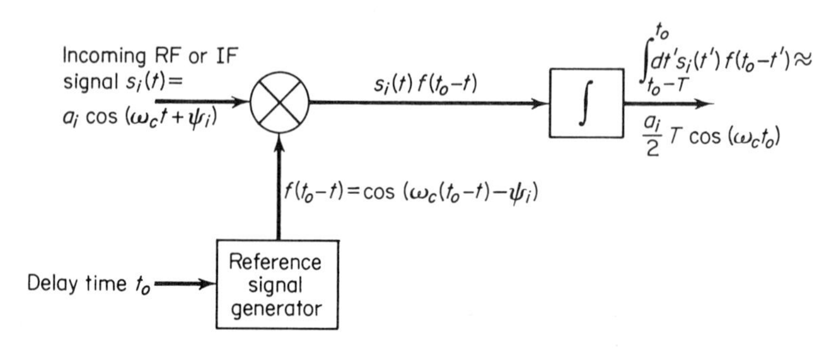

Figure 6-1 Coherent crosscorrelator or matched filter for radio signals.

in principle construct an ideal matched filter or crosscorrelator (illustrated for radio signals in Figure 6-1). As indicated in Section 3.2, this is the ultimate in linear filtering, in the sense that it leads to the highest possible SNR at the output of the filter.

To treat matched filtering in terms of complex envelopes, we first note that the impulse response of the matched filter, according to (3.31), is

$$f(t) = Ks_i(-t) \tag{6.29}$$

where K is a constant which can (and will) be set equal to unity without loss in generality. With the aid of (6.6) and the assumption that ω_0 is set exactly equal to ω_c, we have

$$f(t) = \text{Re}\,\{\hat{f}(t)\,e^{j\omega_0 t}\} = \text{Re}\,\{\hat{s}_i(-t)\,e^{j\omega_c t}\} = s_i(-t) \tag{6.30}$$

which implies that

$$\hat{f}(t) = \hat{s}_i(-t) \tag{6.31}$$

In the ideal case, the output of the matched filter should be read out at $t = 0$. From (6.10), (6.18), (6.30), and (6.31), the complex envelope of the output is

$$\begin{aligned} \hat{s}_o(0) = \frac{1}{2}\int_{-\infty}^{\infty} dt'\,\hat{s}_i(-t')\hat{s}_i^*(-t') &= \frac{1}{2}\int_{-\infty}^{\infty} dt'\,|\hat{s}_i(t')|^2 \\ &= \int_{-\infty}^{\infty} dt'\,s_i^2(t') \\ &= \frac{1}{2\pi}\int_{-\infty}^{\infty} d\omega'\,|S_i(\omega')|^2 = E_s \end{aligned} \tag{6.32}$$

where E_s is the total signal energy, as defined in Section 3.2.1 (Eq. (3.29)). From (6.18),†

† Note the difference between (6.33) and (6.18). Time averaging of $e^{\pm 2j\omega_c t}$ takes place in (6.18) but not in (6.33). The latter is the square of the instantaneous value of $s_o(t)$ at $t = 0$.

$$[s_o(0)]^2 = |\hat{s}_o(0) + \hat{s}_o^*(0)|^2 = [\hat{s}_o(0)]^2 + [(\hat{s}_o(0)^*]^2 + 2|\hat{s}_o(0)|^2 = 4E_s^2 \quad (6.33)$$

The SNR at $t = 0$, from (6.27) or (6.28) and (6.29), is

$$\rho_o(0) = \frac{2B_n E_s^2}{\sigma_n^2 B_f |\hat{F}(0)|^2} \quad (6.34)$$

But from (6.25)

$$B_f = \frac{1}{2|\hat{F}(0)|^2} \frac{1}{2\pi} \int_{-\infty}^{\infty} d\omega' \, |\hat{S}_i(\omega')|^2 = \frac{E_s}{|\hat{F}(0)|^2} \quad (6.35)$$

from which it follows that

$$\rho_o(0) = \frac{E_s}{\sigma_n^2/2B_n} \quad (6.36)$$

Note that (6.36) is equivalent to (3.29). This must be the case, since (3.29) was derived for a general signal waveshape, of which the narrowband waveshape characteristic of a radio signal is only a special case.

6.4 EFFECT OF PHASE UNCERTAINTY—NONCOHERENT CROSSCORRELATION

It is often true in radio situations that a reasonably good knowledge of the signal power spectrum exists at the receiver, but the phase as a function of frequency is virtually unknown.† In this case, coherent matched filtering is not effective, and we must modify our filtering techniques to account for the lack of a priori phase information.

To demonstrate the defects of coherent matched filtering in this event, we postulate a case where the signal power spectrum is completely known, and the filter is designed to match it. The phase at any given frequency, however, is completely unknown, and hence in designing a matched filter the receiver must guess at it and use the guess to set a value on the phase in his filter response function.

Implicit in the assumption of complete knowledge of the signal power spectrum is a knowledge of the carrier frequency and hence the ability to match the filter's central frequency to the carrier frequency. The complex envelope of the output signal at time t, according to (6.17) and the assumption of perfect matching of signal and filter *amplitude* (but not phase) at every

† For example, in communication modes involving ionospheric transmission (see Section 5.5.4.5), such as HF modes, the phase of a frequency component may be severely and unpredictably distorted as it traverses the ionospheric path, while the effect on its amplitude may be very slight. Hence, at the receiver, phase information is totally lacking although the amplitude spectrum may be reasonably well-known. The same often holds true with any other sort of randomly fluctuating path.

frequency, is

$$
\begin{aligned}
s_o(t) &= \frac{1}{4\pi}\int_{-\infty}^{\infty} d\omega' \, |S_i(\omega')|\,|F(\omega')|\, e^{j[\psi_i(\omega')+\psi_f(\omega')]}\, e^{j\omega' t} \\
&= \frac{1}{4\pi}\int_{-\infty}^{\infty} d\omega' \, |S_i(\omega')|^2\, e^{j\omega' t}\, e^{j[\psi_i(\omega')+\psi_f(\omega')]}
\end{aligned}
\tag{6.37}
$$

where $\psi_i(\omega)$ and $\psi_f(\omega)$ represent the phase of the signal's Fourier transform and that of the filter's frequency response function, respectively, at the receiver.

We now make an assumption which, although quite unrealistic in a practical situation, serves to illustrate the point about the effect of phase ignorance on attempts at matched filtering. The assumption is that the filter designer at the receiver knows the frequency variation of phase and attempts to design his filter phase to be equal and opposite to signal phase at all frequencies, but he makes a phase error that is uniform over the entire frequency spectrum, i.e.,

$$
\psi_i(\omega) + \psi_f(\omega) \equiv \Delta\psi(\omega) = \Delta\psi = \text{constant} \tag{6.38}
$$

Then at $t = 0$,

$$
\hat{s}_o(0) = e^{j\Delta\psi} E_s \tag{6.39}
$$

The output signal power at time t is proportional to†

$$
\begin{aligned}
[s_o(t)]^2 &= [\hat{s}_o(t)\, e^{j\omega_c t} + \hat{s}_o^*(t)\, e^{-j\omega_c t}]^2 \\
&= 2\,|\hat{s}_o(t)|^2 + [\hat{s}_o(t)]^2\, e^{2j\omega_c t} + [\hat{s}_o^*(t)]^2\, e^{-2j\omega_c t}
\end{aligned}
\tag{6.40}
$$

At time $t = 0$, (6.40), with the aid of (6.39), becomes

$$
[s_o(0)]^2 = 2E_s^2[1 + \cos(2\,\Delta\psi)] = 4E_s^2 \cos^2(\Delta\psi) \tag{6.41}
$$

From (6.27) or (6.28) and (6.41), we observe that the output SNR of the coherent matched filter is degraded in the amount specified by the equation

$$
(\rho_o)_{\Delta\psi} = (\rho_o)_0 \cos^2(\Delta\psi) \tag{6.42}
$$

where $(\rho_o)_{\Delta\psi}$ and $(\rho_o)_0$ are the output SNR's for finite $\Delta\psi$ and zero $\Delta\psi$ respectively. From (6.42), we see, for example, that a constant phase error of 90° would eliminate the signal output entirely.

It is apparent that the process of squaring the filter output followed by averaging *out of RF* fluctuations, i.e., taking the time averages $e^{\pm 2j\omega_c t}$, will produce a filter output consisting of the first term of (6.40). Since the influence of the phase error is contained in the second and third terms, the squaring and RF averaging operations produces an output independent of the phase error. But the squaring plus RF averaging process is the equivalent of taking the

† The same remarks apply to (6.40) as to (6.33) with respect to the relationship of these equations to (6.18).

squared envelope† of the signal output. If the square root is taken after squaring and RF averaging, then the process is that of computing the envelope† itself. To perform those operations, we feed the signal output through a *linear envelope rectifier*, or a *square-law envelope rectifier*, as described in Sections 3.3.2 and 3.3.3, respectively. We can use interchangeably the phrases "rectification," "extraction of the signal amplitude," "noncoherent filtering," "envelope detection," "noncoherent detection," or "noncoherent crosscorrelation" to describe this process (illustrated in Figure 6-2). The feature most pertinent to our discussion is the elimination of the effect of phase mismatch between the signal and the filter.

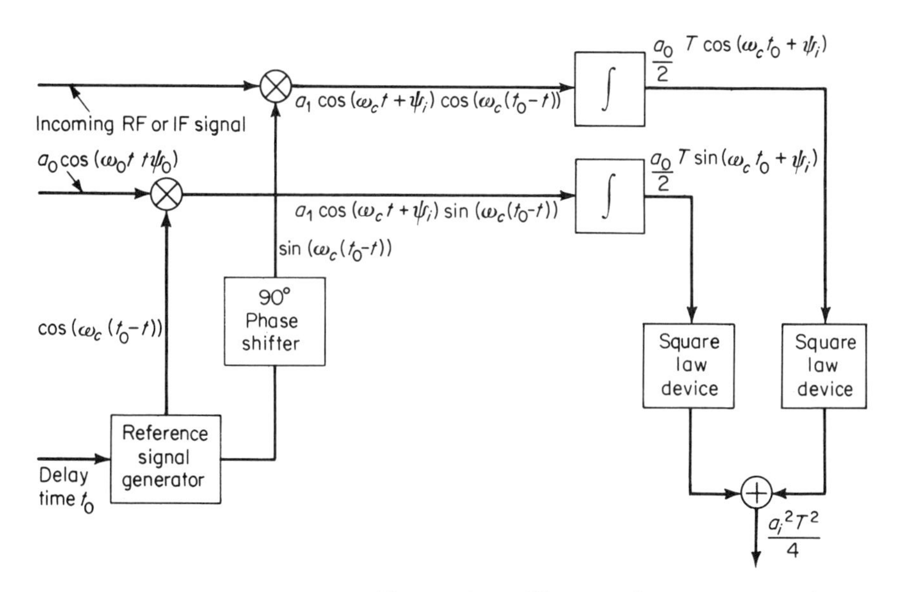

Figure 6-2 The noncoherent crosscorrelator.

It is prudent to ask what price must be paid for the advantage of elimination of the phase-matching requirement, since it is obvious that we have designed a filter with something less than complete a priori knowledge of the signal, and to do this without some sort of sacrifice would seem to be "getting something for nothing." It is clear that in eliminating the second and third terms of (6.40), we have lost some signal power relative to the case where these terms are included and are each equal to the first term, as would be true if there were no phase mismatch. This means that half the signal power, that is, 3dB, is lost when we average out the phase.

† This is the true envelope, not the complex envelope. The latter is a useful mathematical device, but a physical fiction, while the former is a physically meaningful entity.

Let us look at this signal power loss in a more precise manner. We first note that, if the signal-plus-noise at the input and at the output of the filter is denoted by $v_i(t)$ and $v_o(t)$, respectively, the filter output can be expressed in the form

$$v_o(t) = \int_{-\infty}^{\infty} dt'\, v_i(t') f(t - t') = s_o(t) + n_o(t) = \text{Re}\,\{\hat{v}_o(t)\, e^{j\omega_c t}\} \tag{6.43}$$

where $\hat{v}_o(t) = \hat{s}_o(t) + \hat{n}_o(t)$, and where $\hat{n}_o(t)$ is the complex envelope of the noise, given by

$$\hat{n}_o(t) = \frac{1}{2} \int_{-\infty}^{\infty} dt'\, \hat{n}_i(t)\, \hat{f}(t - t')$$

In analogy to (6.4), we can write (6.43) in the form

$$v_o(t) = \text{Re}\,\{\hat{v}_o(t)\, e^{j\omega_c t}\} = \text{Re}\,\{a_{v_o}(t)\, e^{j\psi_{v_o}(t)} \cdot e^{j\omega_c t}\} \tag{6.44}$$

where $a_{v_o}(t)$ and $\psi_{v_o}(t)$ are the amplitude and phase of $\hat{v}_o(t)$, respectively. From (6.43)

$$\begin{aligned} a_{v_o(t)} &= \sqrt{[a_{s_o}(t)]^2 + [a_{n_o}(t)]^2 + 2a_{s_o}(t)a_{n_o}(t) \cos(\psi_{s_o}(t) - \psi_{n_o}(t))} \\ \psi_{v_o}(t) &= \tan^{-1}\left[\frac{a_{s_o}(t) \sin \psi_{s_o}(t) + a_{n_o}(t) \sin \psi_{n_o}(t)}{a_{s_o}(t) \cos \psi_{s_o}(t) + a_{n_o}(t) \sin \psi_{n_o}(t)}\right] \end{aligned} \tag{6.45}$$

where $a_{s_o}(t)$ and $a_{n_o}(t)$ denote amplitude of output signal and noise, respectively, and $\psi_{s_o}(t)$ and $\psi_{n_o}(t)$ are the phase of output signal and noise, respectively.

We are forced to use more cumbersome subscripts on our amplitudes and phases here than in previous sections, because we must distinguish between input, output, signal, noise, and signal-plus-noise quantities.

The mean power P_o in the output of an envelope detector operating on $v_o(t)$ is proportional to the ensemble average of the square of $a_{v_o}(t)$. (The RF time averaging process has already been carried out; thus the only remaining averaging process is statistical ensemble averaging.)

$$\begin{aligned} P_o = \langle [a_{v_o}(t)]^2 \rangle = [a_{s_o}(t)]^2 + \langle [a_{n_o}(t)]^2 \rangle + 2a_{s_o}(t) \cos \psi_{s_o}(t) \langle x_{n_o}(t) \rangle \\ + 2a_{s_o}(t) \sin \psi_{s_o}(t) \langle y_{n_o}(t) \rangle \end{aligned} \tag{6.46}$$

where $x_{n_o}(t)$ and $y_{n_o}(t)$ are the inphase and quadrature noise components, equal, respectively, to $a_{n_o}(t) \cos \psi_{n_o}(t)$ and $a_{n_o}(t) \sin \psi_{n_o}(t)$. From the noise properties defined earlier (See Section 3.1), both $x_{n_o}(t)$ and $y_{n_o}(t)$ have zero-mean values. Thus P_o contains only a signal term $[a_{s_o}(t)]^2$ and a noise term $\langle [a_{n_o}(t)]^2 \rangle$.

The output SNR of the filtered and envelope-detected signal is

$$(\rho_o)_{\text{e.d.}} = (\rho_o)_{\text{n coh}} = \frac{[a_{s_o}(t)]^2}{\langle [a_{n_o}(t)]^2 \rangle} \tag{6.47}$$

where the subscript "e.d." refers to "envelope detected" and "n coh" refers to "noncoherent filtered." These two designations are equivalent and can be interchanged.

To evaluate in a simple way the loss incurred by using noncoherent as opposed to coherent matched filtering, we specialize to the case of a purely sinusoidal carrier signal with constant amplitude a_s and phase ψ_s (with respect to some arbitrary reference phase). For this case, we do not require the complex envelope formalism, and can proceed very simply with the real representations, as follows:

$$s_i(t) = a_{s_i} \cos(\omega_c t + \psi_{s_i}) \tag{6.48}$$

The impulse response of the filter is assumed to be correctly matched to the input frequency but to have a phase error $(\psi_f - \psi_{s_i}) = \Delta\psi$ with respect to the input phase. It is also assumed to have unit amplitude (the filter amplitude will cancel out in the SNR and hence is arbitrary) and to be activated from $t = 0$ to $t = T$ and deactivated when $t < 0$ and $t > T$, where it is assumed that $T \gg 1/\omega_c$.

From (6.8), (6.9), and (6.48), we have

$$\begin{aligned} s_o(t) &= a_{s_i} \int_0^T dt' \cos(\omega_c t + \psi_{s_i}) \cos(\omega_c(t - t') + \psi_f) \\ &= x_{s_o}(t) \cos \omega_c t + y_{s_o}(t) \sin \omega_c t \\ &= a_{s_o}(t) \cos(\omega_c t + \psi_{s_o}(t)) \end{aligned} \tag{6.49}$$

where

$$x_{s_o}(t) \equiv a_{s_i} \int_0^T dt' \cos(\omega_c t' + \psi_{s_i}) \cos(\omega_c t' - \psi_f)$$

$$y_{s_o}(t) \equiv a_{s_i} \int_0^T dt' \cos(\omega_c t' + \psi_{s_i}) \sin(\omega_c t' - \psi_f)$$

The noise output waveform is

$$\begin{aligned} n_o(t) &= \int_0^T dt' \, n_i(t') \cos(\omega_c(t - t') + \psi_f) \\ &= x_{n_o}(t) \cos \omega_c t + y_{n_o}(t) \sin \omega_c t \end{aligned} \tag{6.50}$$

where

$$x_{n_o}(t) \equiv \int_0^T dt' \, n_i(t') \cos(\omega_c t' - \psi_f)$$

$$y_{n_o}(t) \equiv \int_0^T dt' \, n_i(t') \sin(\omega_c t' - \psi_f)$$

If we are to do coherent matched filtering, we assume that $\Delta\psi = \psi_{s_i} + \psi_f = 0$ and read out the filter output at $t = 0$. The resulting approximate signal output is

$$s_o(0) = x_{s_o}(0) \simeq a_{s_i} \frac{T}{2} \tag{6.51}$$

The mean-square noise output, with the aid of (6.21), is

$$\langle [n_o(0)]^2 \rangle = \langle [x_{n_o}(0)]^2 \rangle$$
$$\simeq \int_0^T \int_0^T dt'\, dt''\, \langle n_i(t') n_i(t'') \rangle \cos(\omega_c t' - \psi_f) \cos(\omega_c t'' - \psi_f) \simeq \frac{\sigma_n^2 T}{4B_n} \tag{6.52}$$

where as usual, a bandlimited white noise with the property $B_n \gg 1/T$ has been assumed.

The output SNR with coherent matched filtering is

$$(\rho_o)_{\text{coh}} = \frac{[s_o(0)]^2}{\langle [n_o(0)]^2 \rangle} = \frac{(a_{s_i}T/2)^2}{\sigma_n^2 T/4B_n} = \frac{E_s}{(\sigma_n^2/2B_n)} = 2\rho_i B_n T \tag{6.53}$$

where $\rho_i \equiv a_{s_i}{}^2/2\sigma_n^2$ = input SNR and in this case the total signal energy is equal to $a_{s_i}{}^2 T/2$.

The SNR given by (6.53) is consistent with (6.36) and with (3.30). With noncoherent filtering, we can obtain the output SNR from (6.47), (6.49), and (6.50), resulting in

$$\begin{aligned}(\rho_o)_{\text{n coh}} &= \frac{[x_{s_o}(t)]^2 + [y_{s_o}(t)]^2}{\langle [x_{n_o}(t)]^2 \rangle + \langle [y_{n_o}(t)]^2 \rangle} \\ &= \frac{(a_{s_i}T/2)^2 (\cos^2(\Delta\psi) + \sin^2(\Delta\psi))}{(\sigma_n^2 T/4B_n) + (\sigma_n^2 T/4B_n)} \\ &= \frac{1}{2}\frac{E_s}{(\sigma_n^2/2B_n)} = \rho_i B_n T = \frac{1}{2}(\rho_o)_{\text{coh}}\end{aligned} \tag{6.54}$$

Comparison of (6.53) and (6.54) shows that the use of noncoherent crosscorrelation in lieu of coherent crosscorrelation on a sinusoidal carrier signal results in a loss of $\frac{1}{2}$ or 3 dB in SNR. The loss results from the process of averaging out the RF carrier phase.

For a carrier with signal information the loss is different from 3 dB, the magnitude of the difference depending on signal spectrum shape and degree of filter matching. However, in most cases we can regard the incoming signal as a sinusoid whose amplitude remains constant over a time interval about equal to a fraction of the reciprocal bandwidth of the signal information. The sinusoidal signal representation, then, can be used as an analysis tool as long as T, the filter integration time, is short compared to the reciprocal information bandwidth.†

6.5 EFFECTS OF FREQUENCY UNCERTAINTY

In an attempt at matched filtering (coherent or noncoherent), suppose that we should slightly mismatch the frequency of the reference signal and

† For example, if we were to divide a voice signal (bandwidth of about 3–4 kHz) into short time-slots and apply noncoherent matched-filtering to each time slot, we could

that of the incoming radio signal. This would not affect the mean output noise power, provided the input noise bandwidth were sufficiently large compared with the frequency error. If the noise bandwidth is large compared with the filter bandwidth and the frequency error is at most comparable in magnitude to the filter bandwidth, then shifting the frequency to which the filter is tuned will not affect the total noise passed by the filter. These conditions are met in a large class of applications.†

Suppose the filter is matched to the incoming signal in every parameter but central frequency. To analyze this situation, we will return to the complex envelope formalism. The complex envelope of the frequency response function of the filter takes the form

$$\hat{F}(\omega) = \hat{S}_i^*(\omega) \tag{6.55}$$

or, equivalently, its impulse response complex envelope has a form similar to (6.31), but with the stipulation that $\Delta\omega$ in (6.9) is not zero. Then, from (6.10), we have for the signal output at $t = 0$,

$$\hat{s}_o(0) = \frac{1}{2}\int_{-\infty}^{\infty} dt'\, |\hat{s}_i(t')|^2\, e^{-j\Delta\omega t'} = \frac{1}{2}\int_{-\infty}^{\infty} dt'\, |a_i(t')|^2\, e^{-j\Delta\omega t'} \tag{6.56}$$

Using the frequency representation of input signal and filter characteristic, we obtain, from (6.17) and (6.55),

$$\hat{s}_o(0) = \frac{1}{4\pi}\int_{-\infty}^{\infty} d\omega' \hat{S}_i(\omega')\hat{S}_i^*(\omega' - \Delta\omega) \tag{6.57}$$

The precise numerical value of the SNR degradation due to frequency mismatch depends on the shape of the input signal and hence on that of the filter impulse response or frequency response. To illustrate this degradation in a simple way, we again invoke the sinusoidal carrier signal and filter models of Section 6.4. To develop the ideas, we need only assume, as in Section 6.3, that $a_i(t)$ in Eq. (6.56) is a constant equal to a_i from 0 to T and zero outside that region. For analytical convenience we assume the filter amplitude to be a_i, although as remarked previously, this assumption will not affect the results.

The complex envelope of the signal output at $t = 0$, as given by (6.56), is

$$\hat{s}_o(0) = \frac{a_i^2}{2}\int_0^T dt'\, e^{-j\Delta\omega t'} = \frac{a_i^2}{2}\, e^{-j\Delta\omega T/2}\, \text{sinc}\left(\frac{\Delta\omega T}{2}\right) \tag{6.58}$$

---

realize nearly the full theoretical gain if the time-slots were no longer than about a third of a millisecond. Some degradation could be expected with longer time-slots.

† As indicated in Sections 5.5.3.1 and 5.5.3.2, internal receiver noise is spectrally flat over most of the radio spectrum and atmospheric noise changes very little over frequency regions of a few MHz. Hence any narrowband signalling, where bandwidths are of the order of kHz (e.g., voice) or even a few MHz (e.g., TV), would have this property for practical purposes.

For a sinusoidal carrier input, the output SNR of a coherent filter with a frequency mismatch $\Delta\omega$ but perfect phase matching, according to an extension of arguments in Section 6.3 to include the effect of the frequency mismatch, is

$$(\rho_o)_{\mathrm{coh}} = (2\rho_i B_n T)\left[\operatorname{sinc}\left(\frac{\Delta\omega T}{2}\right)\right]^2 \tag{6.59}$$

In most cases where frequency mismatch might exist, phase knowledge would be lacking, and noncoherent crosscorrelation would be used.

Using the arguments of Section 6.4 with the additional feature of a frequency mismatch, we obtain

$$(\rho_o)_{\mathrm{n\ coh}} = \rho_i B_n T\left[\operatorname{sinc}\left(\frac{\Delta\omega T}{2}\right)\right]^2 \tag{6.60}$$

Thus the effect of frequency mismatch is to introduce a factor sinc $[(\Delta\omega T/2)]^2$ on the output SNR. This can be interpreted by considering the integration stage of the crosscorrelation process as a stage of lowpass filtering. The bandwidth of this lowpass filter, by the definition (6.25), is

$$B_f = \frac{1}{4\pi}\int_{-\infty}^{\infty} d(\Delta\omega)\left|\operatorname{sinc}\left(\frac{\Delta\omega T}{2}\right)\right|^2 = \frac{1}{2T} \tag{6.61}$$

The bandwidth of the integrator, considered as a lowpass filter, is one-half of the reciprocal integration time. If the bandwidth $(2\pi B_f)$ were defined as the frequency spacing between $\Delta\omega = 0$ and the first zero of sinc $(\Delta\omega T/2)$, then the substitution for (6.61) would be

$$B_f = \frac{1}{T} \tag{6.62}$$

which differs from (6.61) by a factor of 2. Regardless of the precise definition of bandwidth, the reciprocal relationship between integration time and bandwidth still holds, with changes only in the proportionality constant.

If we think of the integrator as a lowpass filter, then we can conclude that a frequency mismatch is tolerable provided the filter bandwidth is considerably greater than $\Delta\omega$. If we think in terms of integration, then the same idea can be conveyed by the statement that the integration time must be small compared with $1/\Delta\omega$ in order that frequency mismatch will not be disastrous.

6.6 EFFECTS OF WAVESHAPE PARAMETER UNCERTAINTY

The general problem of parameter uncertainty in matched filtering is quite involved, but there are certain special cases that are of interest in radio applications and can be simply analyzed.

One of these is the problem of matching to a purely sinusoidal carrier signal of finite duration (i.e., an RF pulse) whose radio frequency is precisely known but whose exact position and duration are not known. To apply matched filtering, we would construct a reference sinusoid of the same radio frequency and phase (or we would filter noncoherently if phase is unknown), and would attempt to position it on the time scale in such a way that it is synchronized with the incoming pulse. If this is not done, and no overlap exists between incoming and reference pulse, then obviously the product of the two vanishes, and the signal output of the filter is zero.

As long as *some* overlap exists between input and reference pulse, a signal output will appear. Determination of the filter as a function of the degree of overlap is a straightforward but tedious problem, which is detailed in Appendix II. Some of the results are shown graphically in Figures 6-3 and 6-4.

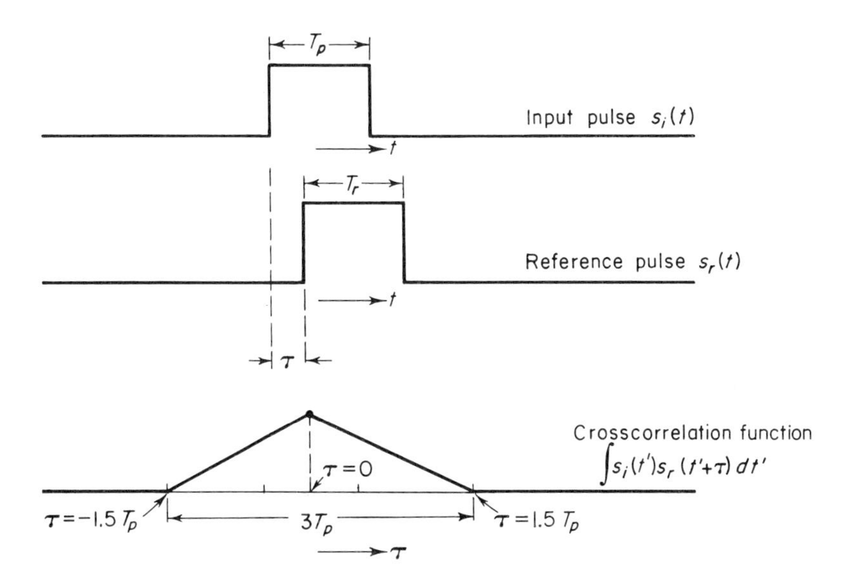

Figure 6-3 Crosscorrelation of pulses—input and reference pulse durations are equal.

Another common case of parameter uncertainty is that in which the incoming signal's central frequency is known, but the slow variations in its amplitude and phase are unknown.† An approach in this case is to generate

† This would be true for any analog transmission, e.g., voice, telemetry of measured variables to ground from a space vehicle, etc. It would also be true for random channels, wherein the fades distort both the amplitude and phase of the signal in a way which the receiver cannot know precisely.

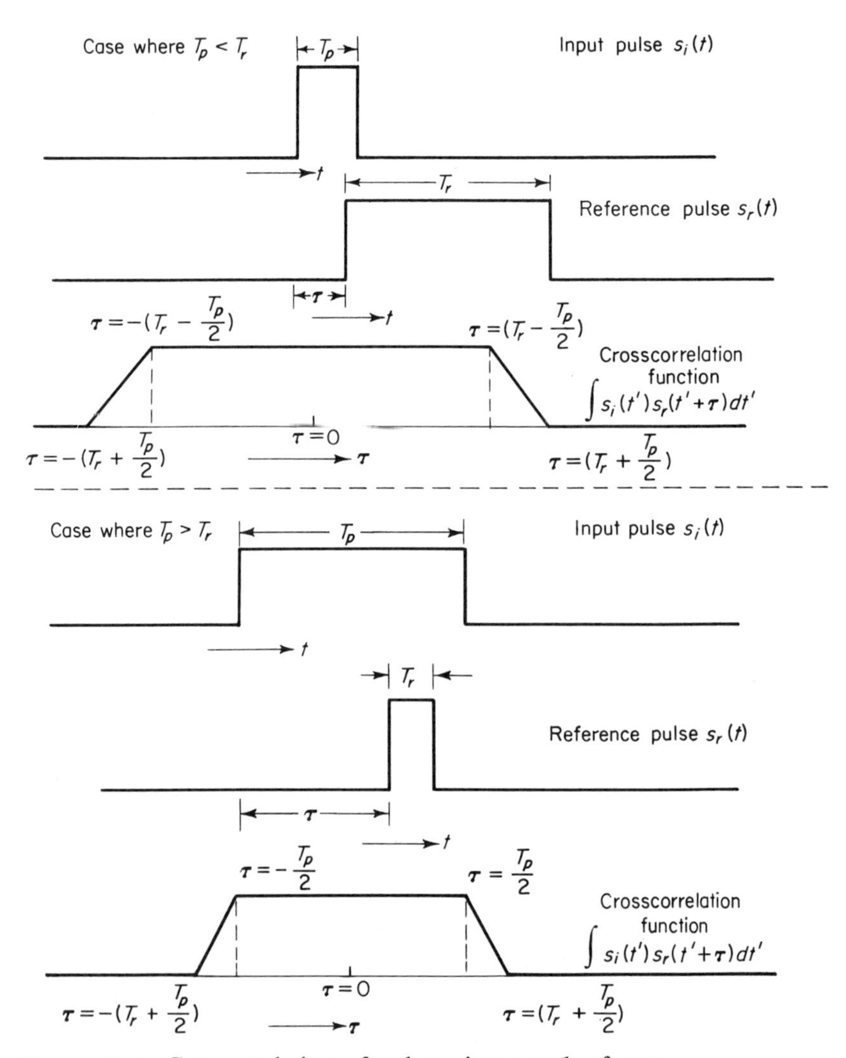

Figure 6-4 Crosscorrelation of pulses—input and reference pulse durations are unequal.

as a reference signal a pure sinusoid with frequency equal to the central frequency of the incoming radio signal. How would we determine the resulting SNR degradation (relative to perfect matched filtering)?

We represent the complex envelope of the incoming signal in the form

$$\begin{aligned} \hat{s}_i(t) &= \bar{a}_i e^{j\psi_i} + \Delta\hat{s}_i(t), \qquad -T \leq t \leq 0 \\ &= 0, \qquad\qquad\qquad\qquad t < -T, \quad t > 0 \end{aligned} \tag{6.63}$$

where $\bar{a}_i e^{j\psi_i}$ represents the mean value of the signal's complex envelope and where $\Delta\hat{s}_i(t)$ is the deviation from the mean due to the slow amplitude variations, unknown at the receiver.

The reference signal, whose frequency is the same as that of the incoming signal, has the complex envelope

$$\hat{f}(t) = K\bar{a}_i e^{-j\psi_i} = K[\hat{s}_i^*(-t) - \Delta\hat{s}_i^*(-t)], \qquad -T \leq t \leq 0$$
$$= 0, \qquad t < -T, \quad t > 0 \tag{6.64}$$

where K is an arbitrarily chosen constant.

If the RF phase were known, we could attempt to use coherent matched filtering. The complex envelope of the coherent filter output, as obtained from a generalization of (6.32), that accounts for the mismatch between incoming and reference signals, is

$$\hat{s}_o(0) = K\left\{\int_{-T}^{0} dt'\,|\hat{s}_i(t')|^2 - \int_{-T}^{0} dt'\,\hat{s}_i(t')\,\Delta\hat{s}_i^*(t')\right\}$$
$$= K\left\{\int_{-T}^{0} dt'\,|\hat{s}_i(t')|^2 - \bar{a}_i e^{j\psi_i}\int_{-T}^{0} dt'\,\Delta s_i^*(t') - \int_{-T}^{0} dt'\,|\Delta\hat{s}_i(t')|^2\right\} \tag{6.65}$$

The signal fluctuation $\Delta\hat{s}_i(t)$ can be regarded as a sample function of a random process. From the viewpoint of the receiver, who doesn't know its parameters, it is just that. Moreover, for purposes of this purely illustrative analysis, we can regard it as a *stationary* random process over a time interval sufficiently short so that the signal statistics remain roughly constant. It will be assumed that T is short enough to allow the assumption that $\Delta\hat{s}_i(t)$ is stationary during the time interval $-T \leq t \leq 0$, and long enough so that the ensemble average of $\Delta\hat{s}_i(t)$ can be approximated by

$$\Delta\hat{s}_i \simeq \frac{1}{T}\int_{-T}^{0} dt'\,\Delta\hat{s}_i(t') = 0 \tag{6.66}$$

By taking the absolute square of (6.65), ensemble averaging the quantities involving $\Delta\hat{s}_i(t)$, invoking (6.66), and recognizing that the mean-square noise output is still given by (6.26) in the present case, we obtain for the output SNR of the filter

$$(\rho_o)_{\text{coh}} \simeq (2\rho_i B_n T)L \tag{6.67}$$

where $(2\rho_i B_n T)$ is recognized as the SNR of an ideal coherent matched filter (see Eq. (6.53)) and L is a degradation factor given by

$$L = \left\{1 - \frac{\int_{-T}^{0} dt'\,\langle|\Delta s_i(t')|^2\rangle}{E_s} + \frac{\int_{-T}^{0}\int_{-T}^{0} dt'\,dt''\,\langle|\Delta\hat{s}_i(t')|^2\,|\Delta\hat{s}_i(t'')|^2\rangle}{4E_s^2}\right\} \tag{6.68}$$

Under the assumption that $\Delta\hat{s}_i(t)$ is stationary,

$$\langle|\Delta\hat{s}_i(t)|^2\rangle = \sigma_s^2 \quad \text{(independent of time)} \tag{6.69}$$

Under a further assumption that $s_i(t)$ has Gaussian statistics,†

$$\langle|\Delta\hat{s}_i(t')|^2\,|\Delta\hat{s}_i(t'')|^2\rangle = \langle|\Delta\hat{s}_i(t')|^2\rangle\langle|\Delta\hat{s}_i(t')|^2\rangle + |\langle\Delta\hat{s}_i(t')\,\Delta\hat{s}_i^*(t'')\rangle|^2 = \sigma_s^4[1 + \rho_s(t' - t'')] \tag{6.70}$$

where $\rho_s(\tau)$ is the NACF of $\Delta\hat{s}_i(t)$.

From (6.67), (6.68), (6.69), and (6.70), we obtain

$$(\rho_o)_{\mathrm{coh}} \simeq (2\rho_i B_n T)\left\{1 - \frac{\sigma_s^2 T}{E_s} + \frac{\sigma_s^4}{4E_s^2}T^2 + \int_{-T}^{0}\int_{-T}^{0} dt'\,dt''\,\rho_s(t' - t'')\right\} \tag{6.71}$$

Computation of $(\rho_o)_{\mathrm{coh}}$ from (6.71), of course, requires further specialization of $\Delta\hat{s}_i(t)$. Without proceeding further, we can conclude from (6.71) that the output SNR can be greatly different from that of a coherent filter that is precisely matched to the incoming signal.

6.7 BANDPASS FILTERING

As pointed out in Section 6.3, the matched filter is actually a very special type of bandpass filter centered at the expected carrier frequency of the incoming signal. Although it is the theoretical optimum linear filter, our intuition should tell us that other types of bandpass filter, if centered at the expected carrier frequency, would also enhance the SNR, *not as much*, of course, as would the matched filter. In many systems, the construction of an ideal matched filter would be impractical. It is important to know the degradation incurred by using an easily constructed bandpass filter in lieu of a matched filter.

As a special case of bandpass filtering, consider the extension of matched filtering (coherent or noncoherent) where the integration process is carried out with a time constant T_c. If the time constant is long compared with the reciprocal of the carrier frequency, then the contributions to the integrand involving oscillations in the vicinity of twice the carrier frequency will still be negligible. We assume that the signal is a sinusoid. The signal and reference duration (and hence the signal and noise integration time) is T.

If the integration time T is infinite, this case is equivalent to the *RLC* bandpass filter. To see this, we refer to (1.25) and (1.27) and to the footnote below Eq. (6.24) in Section 6.2. The complex envelopes of $f(t)$ for series and parallel *RLC* circuits as given in (1.25) and (1.27) are proportional to $e^{(-R/2L)t}$

---

† See Appendix III, Eq. (III. 9) and the references cited below Eqs. (III. 8) through (III. 10).

and $e^{-t/2RC}$, respectively. The quantities $(2L/R)$ and $(2RC)$ play the role of time constant in series and parallel RLC circuits respectively.

Results obtained in Sections 6.1 through 6.6 are for filters with integrators whose time constants are effectively infinite. The same analytical procedures applied to the case of a finite time constant T_c, with all assumed conditions otherwise equivalent, will modify these results as follows:
(*Note:* Equations are numbered according to their counterparts for the case of infinite time constant (e.g., $(\widetilde{6.49})$ represents the counterpart of (6.49) for finite T_c) with the exception of $(\widetilde{6.58})'$, whose counterpart does not appear elsewhere.)

$$s_o(t) \simeq \frac{a_{s_i} T_c}{2}(1 - e^{-T/T_c}) \tag{$\widetilde{6.49}$}$$

$$[n_o(t)]^2 = [n_o(0)]^2 \simeq \frac{\sigma_n^2}{8B_n} T_c(1 - e^{-2T/T_c}) \tag{$\widetilde{6.52}$}$$

$$(\rho_o)_{\substack{\text{coh}\\ \text{n coh}}} = \frac{(2)}{(1)} \rho_i B_n T_c \frac{(1 - e^{-T/T_c})}{(1 + e^{-T/T_c})} \tag{$\widetilde{6.54}$}$$

$$\hat{s}_o(0) = \frac{a_{s_i}^2 T_c}{2} \frac{[1 - e^{-j\Delta\omega T} e^{-T/T_c}]}{(1 + j\Delta\omega T_c)} \tag{$\widetilde{6.58}$}$$

$$|\hat{s}_o(0)|^2 = \left(\frac{a_{s_i}^2 T_c}{2}\right)^2 \frac{1}{[1 + (\Delta\omega T_c)^2]} \{1 + e^{-2T/T_c} - 2e^{-T/T_c} \cos(\Delta\omega T)\} \tag{$\widetilde{6.58}'$}$$

$$(\rho_o)_{\substack{\text{coh}\\ \text{n coh}}} = \frac{(2)}{(1)} \rho_i B_n T \frac{1}{[1 + (\Delta\omega T_c)^2]} \left\{\frac{1 + e^{-2T/T_c} - 2e^{-T/T_c} \cos(\Delta\omega T)}{1 - e^{-2T/T_c}}\right\} \tag{$\widetilde{6.59}$, $\widetilde{6.60}$}$$

In general, from examination of the results given above, we can conclude that a finite time constant T_c with a finite integration time T and a frequency mismatch $\Delta\omega$ has the effect of modifying the SNR by a factor,

$$g(T_c, \Delta\omega) = \frac{\dfrac{T_c}{T}\left\{\dfrac{1}{[1 + (\Delta\omega T_c)^2]}\right\}\{1 + e^{-2T/T_c} - 2e^{-T/T_c}\cos(\Delta\omega T)\}}{\left[\operatorname{sinc}\left(\left(\dfrac{\Delta\omega T_c}{2}\right)\dfrac{T}{T_c}\right)\right]^2 [1 - e^{-2T/T_c}]} \tag{6.72}$$

which degenerates into

$$g(T_c, 0) = \frac{T_c}{T} \frac{[1 - e^{-T/T_c}]}{[1 + e^{-T/T_c}]} \tag{6.73}$$

for perfect frequency matching, i.e., for $\Delta\omega = 0$.

Let us consider the approximate forms assumed by (6.72) and (6.73) if the time constant is (1) small and (2) large compared to the integration time T.

(1) $T_c \ll T$ (i.e., $e^{-T/T_c} \ll 1$)

$$g(T_c, \Delta\omega) \simeq \frac{T_c}{T} \frac{1}{[1 + (\Delta\omega T_c)^2]\left[\operatorname{sinc}\left(\left(\dfrac{\Delta\omega T_c}{2}\right)\dfrac{T}{T_c}\right)\right]^2} \tag{6.74}$$

$$g(T_c, 0) \simeq \frac{T_c}{T} \tag{6.75}$$

(2) $T_c > T$ (In this case, expand e^{-T/T_c} in a power series and retain only the terms linear and square in T/T_c.)

$$g(T_c, \Delta\omega) \simeq \frac{\dfrac{2T_c^2}{T^2}\left\{1 - \dfrac{T}{T_c}\sin^2\left(\dfrac{\Delta\omega T}{2}\right) + \dfrac{T^2}{4T_c^2}\left(1 - \dfrac{T}{2T_c}\right)^2\right\}}{[1 + (\Delta\omega T_c)^2]\left[\text{sinc}\left(\dfrac{\Delta\omega T}{2}\right)\right]^2\left[1 - \dfrac{T}{T_c}\right]} \tag{6.76}$$

$$g(T_c, 0) \simeq \frac{1}{2}\frac{\left(1 - \dfrac{T}{2T_c}\right)^2}{\left(1 - \dfrac{T}{T_c}\right)} \tag{6.77}$$

It is apparent from (6.60) and (6.75) that the output SNR with perfect frequency matching is proportional to the time constant T_c if the integration time is very long compared to the time constant. Increasing the integration time will not significantly enhance SNR once it has attained a value much greater than T_c. This reflects the limiting of the effective integration time provided by a finite time constant. Also, since T_c can be regarded as the reciprocal bandwidth, the output SNR can be regarded as inversely proportional to bandwidth, which is consistent with all of our past results.

If the time constant is much greater than the integration time, even with perfect frequency matching, Eqs. (6.60) and (6.77) tell us that the output SNR is roughly proportional to the integration time, is depressed by approximately 3dB from its value in the ideal case, and is affected only slightly by increases in integration time as long as these increases retain the condition $T \ll T_c$. A filter that integrates over a period much shorter than its time constant cannot perform as a true integrator and hence cannot provide the kind of SNR improvement that one usually expects from the integration process. Plots of $g(T_c, \Delta\omega)$ versus (T/T_c) for various values of $(\Delta\omega T_c)$ are shown in Figure 6-5(a). From these curves, we conclude that $g(T_c, \Delta\omega)$: (1) decreases monotonically with T/T_c, for frequency mismatch small compared to reciprocal time constant, (2) decreases to a minimum, then increases with T/T_c for mismatch about equal to reciprocal time constant, and (3) becomes highly oscillatory when the mismatch greatly exceeds the reciprocal time constant. Curves showing the ratio of the output SNR with a finite integration time to its value with an infinite integration time, plotted against (T/T_c) with small mismatch are shown in Figure 6-5. These curves show a general tendency for this ratio to increase with (T/T_c) until a plateau is reached, then increase only very slowly with further increases in (T/T_c), asymptotically approaching unity as (T/T_c) becomes infinite. Thus, once T has reached a value about three times T_c, an attempt to enhance SNR through further increases in integration time is of little avail.

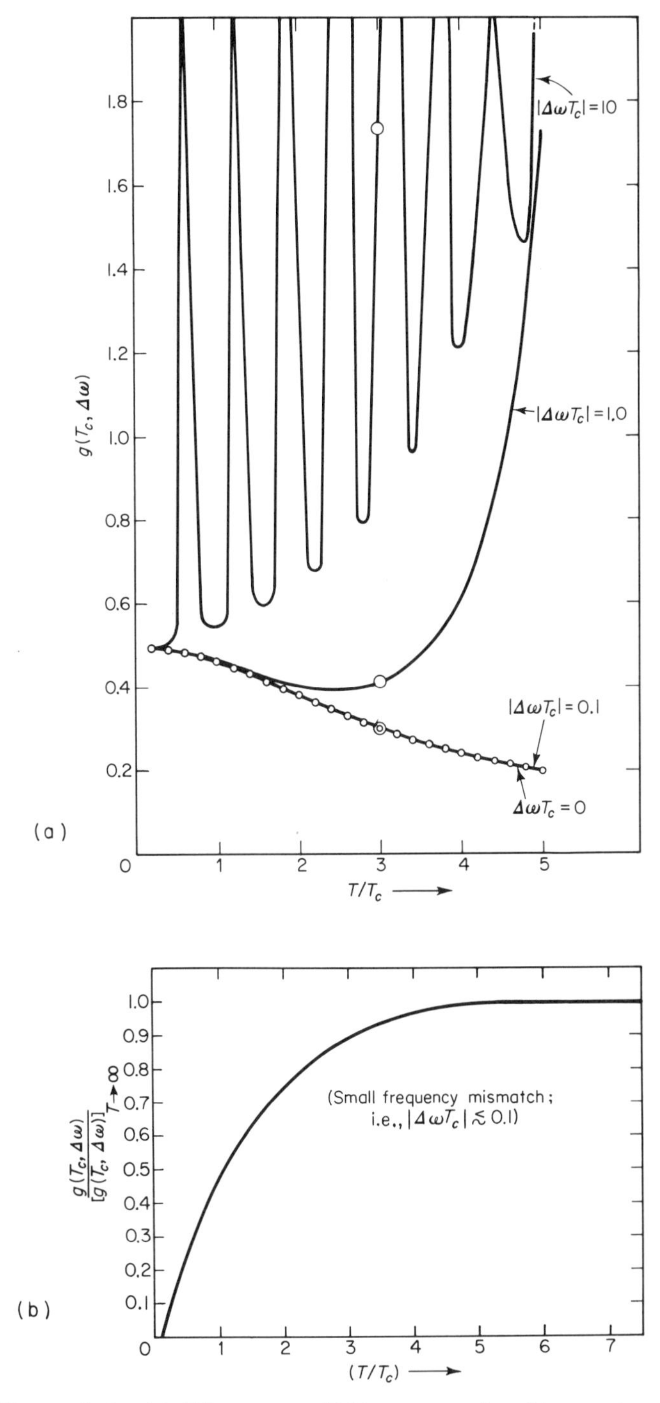

Figure 6-5 (a) Filter output SNR versus ratio of integration time to time constant. (b) Effect of finite integration time on output SNR.

6.8 OFFSET CORRELATION AND HETERODYNING

In all of the discussions of Sections 6.1.1 through 6.6, the optimum or near-optimum processing filter for radio signals was assumed to be centered on the expected carrier frequency. As a practical matter, it is sometimes more convenient to work at a frequency other than the expected carrier frequency. This is the basis of schemes known as *offset correlation* or a special case usually known as *heterodyning*, and is also the basis of the SNR enhancement properties of the superheterodyne receiver.

We can confine the discussion of heterodyning to pure sinusoidal input signals with perfect a priori knowledge of frequency, phase, and occurrence time. The arguments will then carry over approximately to noncoherent filtering and to realistic information-bearing RF signals. Modifications to the idealized results are identical to those discussed in the previous sections.

The operation of the superheterodyne receiver is briefly described in Section 5.4 and illustrated in Figure 5-1.

The input to the mixing stage (actually a summation, squaring, and integration process that produces at its output the product of the local oscillator signal and the incoming signal) is the sum of the incoming signal-plus-noise, denoted by $v_i(t)$, and the local oscillator signal, denoted by $v_m(t)$. The waveform $v_i(t)$ consists of the signal $s_i(t)$ and its accompanying noise $n_i(t)$. Expressing each of these in terms of their complex envelopes, we have, for the input to the mixing stage,

$$[v_i(t) + v_m(t)] = \tfrac{1}{2}\{[\hat{s}_i(t) + \hat{s}_m e^{j\Delta\omega t}]\, e^{j\omega_c t} + [\hat{s}_i^*(t) + \hat{s}_m^* e^{-j\Delta\omega t}]\, e^{-j\omega_c t} + [\hat{n}_i(t)e^{j\omega_c t} + \hat{n}_i^*(t)e^{-j\omega_c t}]\} \tag{6.78}$$

We now feed $[\hat{v}_i(t) + \hat{v}_m(t)]$ into a square-law detector followed by a bandpass filter centered on the frequency $\Delta\omega$, assumed for mathematical convenience to have a flat impulse response from 0 to T.

The output of this *IF filter* or *IF amplifier*† is

$$v_o(T) = \text{Re}\left\{\int_0^T dt'\, (\widehat{[v_o(t') + v_m(t')]^2})\, e^{-j\Delta\omega(t-t')}\right\} \tag{6.79}$$

where $(\widehat{[v_i(t) + v_m(t)]^2})$ denotes the complex envelope of the squared output.

The complex envelope of $[v_i(t) + v_m(t)]^2$ is

† In practical systems, the IF filter has a gain and is used for signal amplification prior to rectification.

$$
\begin{aligned}
(\overbrace{[v_i(t) + v_m(t)]^2}^{\wedge}) = \tfrac{1}{4}\{&2\,|\hat{s}_i(t) + \hat{s}_m\, e^{j\Delta\omega t}\,|^2 \\
&+ 2\,\mathrm{Re}\,\{[\hat{s}_i(t) + \hat{s}_m\, e^{j\Delta\omega t}]^2\, e^{2j\omega_c t}\} \\
&+ 2\,\mathrm{Re}\,[(\hat{s}_i(t) + \hat{s}_m\, e^{j\Delta\omega t})\hat{n}_i(t)\, e^{2j\omega_c t}] \\
&+ 2\,\mathrm{Re}\,[(\hat{s}_i(t) + \hat{s}_m\, e^{j\Delta\omega t})\,\hat{n}_i^*(t)] \\
&+ 2\,|\hat{n}_i(t)\,|^2 + 2\,\mathrm{Re}\,\{[\hat{n}_i(t)]^2\, e^{2j\omega_c t}\}
\end{aligned} \tag{6.80}
$$

We will now make a simplifying assumption that generally holds in practical superheterodyne receivers. The amplitude of the local oscillator signal is assumed large compared to that of the incoming signal, i.e.,

$$|\hat{s}_m\,| \gg |\,\hat{s}_i(t)\,| \tag{6.81}$$

It is also assumed that

$$T \gg \frac{1}{\omega_c} \tag{6.82}$$

The IF filter always has a passband that is narrow compared to ω_c. It is also narrow compared to $\Delta\omega$. Then both the $e^{\pm 2j\omega_c t}$ terms and the dc terms in (6.80) will be effectively filtered out. The surviving portion of the IF filter output after rejection of these terms is

$$v_o(T) \simeq \mathrm{Re}\,\{e^{j\Delta\omega t}\,\hat{v}_o(T)\} \tag{6.83}$$

where

$$
\begin{aligned}
\hat{v}_o(T) = &\int_0^T dt'\,\{|\,\hat{s}_i(t') + \hat{s}_m e^{j\Delta\omega t'}\,|^2\, e^{-j\Delta\omega t'} \\
&+ [(\hat{s}_i(t') + \hat{s}_m e^{j\Delta\omega t'})\hat{n}_i^*(t')e^{-j\Delta\omega t'}] + |\,\hat{n}_i(t')\,|^2 e^{-j\Delta\omega t'}\} \\
= &\int_0^T dt'\,|\,\hat{s}_i(t')\,|^2\, e^{-j\Delta\omega t'} + \int_0^T dt'\,|\,\hat{s}_m\,|^2\, e^{j\Delta\omega t'} + \int_0^T dt'\;\hat{s}_m^*\hat{s}_i(t') \\
&+ 2\int_0^T dt'\;\hat{s}_i(t')\hat{n}_i^*(t')e^{-j\Delta\omega t'} + 2\int_0^T dt'\;s_m^*\,\hat{n}_i(t') \\
&+ 2\int_0^T dt'\,|\,\hat{n}_i(t')\,|^2 \simeq 2\left\{\int_0^T dt'\;\hat{s}_m^*\hat{s}_i(t') + \int_0^T dt'\;\hat{s}_m^*\hat{n}_i(t')\right. \\
&\left. + \int_0^T dt'\,|\,\hat{n}_i(t')\,|^2\right\}
\end{aligned}
$$

Even if the incoming signal amplitude is comparable to the noise level, or even well below it, it is usually possible to set the local oscillator signal at a very high level, such that, with the aid of (6.72), we have

$$\left|\int_0^T dt'\,[n_i(t')]^2\right| \ll \left|\int_0^T dt'\;\hat{s}_m^*\hat{s}_i(t')\right| \tag{6.84}$$

$$\left|\int_0^T dt'\,[n_i(t')]^2\right| \ll \left|\int_0^T dt'\;\hat{s}_m^*\hat{n}_i(t')\right| \tag{6.85}$$

with the result

$$v_o(T) = \mathrm{Re}\left\{\left(\hat{s}_m^*\left[\int_0^T dt'\,\hat{s}_i(t') + \int_0^T dt'\,\hat{n}_i(t')\right]\right)e^{j\Delta\omega t}\right\} \tag{6.86}$$

Except for a proportionality constant and a frequency shift from ω_c to $\Delta\omega$ the output (6.86) has the same form as the signal-plus-noise output of a filter operating at the RF stage. Thus, although the mixing operation is inherently nonlinear, involving square-law detection as it does, it can be rendered equivalent to a purely linear operation by choosing the IF sufficiently high to allow narrowband IF filtering and making the local oscillator signal strong enough to eliminate the effect of the nonlinear noise term in the IF filter output.

The entire heterodyning process, then, can be considered simply as a downward frequency shift, and the functional output form of the output SNR is unaffected by the shift. An important reason for heterodyning, in fact, is to bring the signal frequency down from RF to a region in which it is easier to operate on it electronically. The increased sensitivity inherent in the superheterodyne receiver is at least partially due to the integration or lowpass filtering in the last stage of the process. In principle (ignoring the differences in physical properties of devices operating in different frequency regions), the same sensitivity increase could be attained at RF, but as a practical matter highly sensitive RF amplifiers are usually more difficult to construct than IF amplifiers operating at lower frequencies, so that heterodyning is usually worthwhile from this point of view. A key parameter is the amplifier bandwidth (equivalently, reciprocal integration time), which for maximum sensitivity should be made as small as possible consistent with passage of significant signal information. The smallest attainable bandwidth is in general some given fraction of the central frequency. Suppose the RF is 10 GHz and the smallest attainable bandwidth is 10 percent of the central frequency. If the signal is heterodyned down to an IF of (say) 100 MHz, a 10 percent IF bandwidth is 10 MHz. If an information band about 10 MHz wide is to be passed, then amplifying at RF provides a minimum bandwidth of 1 GHz. If this entire 1 GHz passband is open, there is a 990 MHz portion of it that passes noise but no additional signal energy. Thus, the reduction in bandwidth made possible by superheterodyne operation would seem to enhance sensitivity very substantially in this case.

6.9 THE AUTOCORRELATOR

The crosscorrelator technique, although optimum in the case where the signal waveshape is known, is not necessarily practical otherwise. A useful reference signal can be constructed only if some reasonable degree of knowledge of the signal characteristics exists. There are many radio situations, however, where only very meager knowledge is available at the receiver, such that any attempt to construct a reference signal based on guesswork could be futile.† If the guess were wrong, the reference signal could be completely different from the actual received signal, nullifying any possible SNR improvement, and possibly even degrading the quality of the signal relative to the case where it was left unprocessed. In principle, of course, a flexible reference signal could be constructed and multiple channels or a search technique could be used to find the correct combination of unknown parameters. However, in cases of extreme ignorance of signal properties, the required time or complexity could be prohibitive.

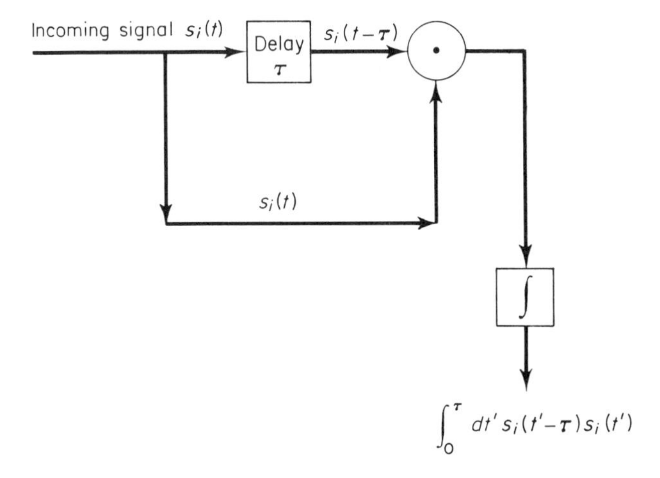

Figure 6-6 The autocorrelator.

A method of circumventing these difficulties is provided by autocorrelation techniques (see Figure 6-6), that is, the use of a device to compute the autocorrelation function of the signal. Such a device is known as an *autocorrelator*. Note that in the absence of noise, autocorrelation performs an operation similar to crosscorrelation, i.e., multiplication of the incoming signal by a delayed replica of itself followed by integration of the product. Moreover, under certain conditions, the delayed replica in the autocorrelation case is the actual signal itself, and certainly no artificially constructed reference signal could duplicate the true signal as well as this.

There is a price to be paid for this *high fidelity* reference signal. First, the feasibility of an autocorrelator generally requires one of two conditions,

† See Section 10.3.2.

namely (1) the signal is periodic with fundamental frequency known at the receiver and the correlator delay time τ is set at or near a multiple of the fundamental period, or (2) the periodicity is either nonexistent or unknown to the receiver, in which case the delay time must be set to zero. The second case is merely that of a square-law rectifier followed by an integrator.

The output of a crosscorrelator whose input signal and noise are $s_i(t)$ and $n_i(t)$, respectively, whose delay time is τ, and whose integration time is T, is

$$v_o(\tau; T) = v_{ss}(\tau; T) + v_{sn}(\tau; T) + v_{ns}(\tau; T) + v_{nn}(\tau; T) \tag{6.87}$$

where

$$v_{ss}(\tau; T) \equiv \int_0^T dt'\, s_i(t')s_i(t' + \tau)$$

$$v_{sn}(\tau; T) \equiv \int_0^T dt'\, s_i(t')n_i(t' + \tau)$$

$$v_{ns}(\tau; T) \equiv \int_0^T dt'\, n_i(t')s_i(t' + \tau)$$

$$v_{nn}(\tau; T) \equiv \int_0^T dt'\, n_i(t')n_i(t' + \tau)$$

The signal term $v_{ss}(\tau)$ has precisely the same form as the signal output of a crosscorrelator with a perfectly matched reference signal. The first of the two signal-noise crossterms $v_{sn}(\tau)$ is identical to the output noise term of a crosscorrelator. The noise-noise crossterm $v_{nn}(\tau)$ is unique to an autocorrelator and does not appear in the output of a crosscorrelator.

The details of calculation of the output SNR of an autocorrelator are somewhat cumbersome and are therefore relegated to an appendix (see Appendix III). The results of the calculations in Appendix III are presented below.

If the autocorrelator is designed in such a manner that the delay time is a multiple of the RF signal period, $T \gg \tau$, and noise bandwidth is far in excess of signal bandwidth, then $\rho_s(\tau) = 1$, and from (III. 13)

$$\frac{\rho_o}{(\rho_o)_{\text{coh}}} \simeq \frac{\rho_i}{[1 + 2B_nT|\hat{\rho}_n(\tau)|^2] + \rho_i[1 + 2B_nT\{2\,\text{Re}\,(\hat{\rho}_n^*(\tau)\}} \tag{6.88}$$

where B_n is the lowpass noise bandwidth and where

$$\hat{\rho}_n(\tau) \equiv \frac{\langle \hat{n}_i(t')\hat{n}_i^*(t' + \tau)\rangle}{\langle|\hat{n}_i(t')|^2\rangle}$$

The symbol $\hat{n}_i(t)$, of course, denotes the complex envelope of $n_i(t)$. If $\rho_i \ll 1$,

$$\frac{\rho_o}{(\rho_o)_{\text{coh}}} \simeq \frac{\rho_i}{1 + 2B_nT|\hat{\rho}_n(\tau)|^2} \tag{6.89}$$

If $\rho_i \gg 1$,

$$\frac{\rho_o}{(\rho_o)_{\text{coh}}} \simeq \frac{1}{1 + 4B_nT\,\text{Re}\,\{\hat{\rho}_n^*(\tau)\}} \tag{6.90}$$

Note that the autocorrelator simulates the measurement of the sum of: (1) autocorrelation function of the signal, (2) the crosscorrelation function between signal and noise, and (3) the autocorrelation function of noise, all with a given delay time τ. We can expect, then, that the signal-noise crosscorrelation term should be small for all values of τ, provided the integration time T is large enough for the time integration to simulate an infinite time averaging process. The condition

$$T \gg \frac{1}{B_n} \tag{6.91}$$

is required, since the reciprocal noise bandwidth is a measure of the *correlation time* of the noise. If the condition (6.81) holds, the integration will be over a large number of cycles of noise and the integral will be small.

The noise-noise term in the autocorrelator output, if the condition (6.81) holds, will simulate the ACF of the noise. This term will be close to unity if τ is much less than the noise correlation time, and small compared to unity if τ is much greater than the correlation time. Thus, for example,

$$|\hat{\rho}_n(\tau)| = 1, \qquad \text{if } \tau = 0 \tag{6.92.a}$$

$$|\hat{\rho}_n(\tau)| \simeq 1, \qquad \text{if } \tau \ll \frac{1}{B_n} \tag{6.92.b}$$

$$|\hat{\rho}_n(\tau)| \simeq 0.3 \text{ to } 0.7, \qquad \text{if } \tau \simeq \frac{1}{B_n} \tag{6.92.c}$$

$$|\hat{\rho}_n(\tau)| \ll 1, \qquad \text{if } \tau \gg \frac{1}{B_n} \tag{6.92.d}$$

In the special case where a square-law rectifier followed by a lowpass filter is used, the delay time τ is zero. The noise waveforms corresponding to t and $t + \tau$ are identical and therefore completely correlated (as is implied by (6.92.a)). In this case, the ratio of the output SNR to that of the coherent crosscorrelator, obtained from (6.88), is

$$\frac{\rho_o}{(\rho_o)_{\text{coh}}} \simeq \frac{2\rho_o}{1 + B_nT + 2\rho_o[1 + 4B_nT]} \tag{6.93}$$

In using this square-law technique, one circumvents the requirement that the delay time be set equal to a multiple of the signal period. However, the price to be paid for this is the increased noise due to the large value of the noise-noise output. This output would be vanishingly small in an autocorrelator whose delay time was far in excess of the noise correlation time. Thus the square-law scheme should be used only when ignorance of the signal period is so great that it is not possible to choose the proper delay time in a true autocorrelator. Recall (see Section 6.4) the case where we must substitute noncoherent for coherent filtering because of ignorance of the signal phase, and thereby sacrifice 3 dB of SNR. The case presently under discussion is analogous. We again lose SNR because we must substitute square-law detection for autocorrelation due to our a priori ignorance of a signal parameter, this time the frequency.†

6.10 PROCESSING AGAINST FADING—DIVERSITY

The phenomenon of fading of radio signals was discussed in Section 5.5.4.7. The presence of fading in the incoming radio signal changes the "ground rules" on processing for maximum SNR. If the fading is extremely rapid, such that its frequencies are comparable to those of the information contained in the received signal, then processing against it becomes a complicated matter. We will consider this case as beyond the scope of this book. If the fading is sufficiently slow to regard the amplitude level as constant during many cycles of the information-bearing signal, then the problem is considerably simplified. We will limit ourselves to the latter case in the discussions to follow.

There are a number of methods of processing against slow fading. Many of these methods are variations of a general class of antifading techniques known as *diversity* or *diversity reception.* This is not to say that some of the modern methods of dealing with fading, for example adaptive techniques,‡ are not markedly different from standard diversity techniques. However, it is often true that an adaptive technique can be considered philosophically as roughly equivalent to one or other class of diversity, in terms of the time, bandwidth, or equipment complexity required to attain a given level of antifading improvement.

The basic idea behind diversity reception is that a *choice between* or *addition of* two or more *independently* fading radio signals will tend to average out fluctuations and thereby reduce the deleterious effects of fading. To

---

† In radar applications (see Chapters 9 and 10) frequency uncertainty may be introduced through unknown target Doppler shifts. If in addition, the transmission medium has random fluctuations sufficient to distort the frequency, there may be considerable frequency uncertainty at the receiver. In such cases, the square-law technique would be in order. In general, autocorrelation techniques might be used in passive detection applications (see Section 10.3.2) where one is looking for hidden periodicities in an apparently random signal. See the somewhat classical paper by Lee, Cheatham, and Wiesner (1950) for a discussion of this.

‡ See Schwartz, Bennett, and Stein, 1966, pp. 395, 396, 400–402, or P. A. Bello, J. Ekstrom, and D. Chesler, 1962.

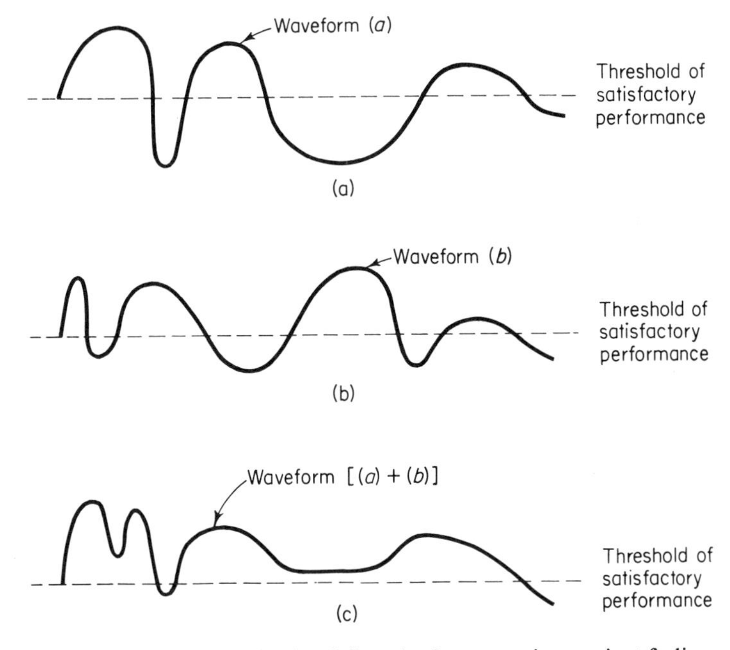

Figure 6-7 The basis of diversity in processing against fading. Note that (*b*) is above threshold during a portion of the time that (*a*) is below threshold and that waveform [(*a*) + (*b*)] is below the threshold for a smaller fraction of the time than either (*a*) or (*b*).

show in an extremely elementary manner why this is true, we consider the two random amplitude waveforms shown in Figure 6-7 (a) and Figure 6-7(b). Suppose the fluctuations of these two waveforms (which could represent amplitudes of received radio signals) are mutually independent, that is, suppose no causal connection exists between the fluctuations of (a) and those of (b). In both (a) and (b), a threshold level, which might be the receiver noise level, is shown by a horizontal dashed line. During periods when either (a) or (b) is in a deep fade, i.e., well below threshold, it is quite likely that the other waveform will be above threshold. If the amplitude level of both waveforms are monitored, and the highest is chosen for use, there is a much better chance for consistent operation above threshold than if one were to rely entirely on either (a) or (b). This technique is known as *dual* (*or 2nd order*) *selection diversity*. It can be extended to any number of independently fading signals, say N, in which case it is known as *N-fold* (*or Nth order*) *selection diversity*.

Another class of diversity techniques is known as *diversity combining*. *Dual* (*or 2nd order*) *combining* is illustrated in (c) of Figure 6-7, in which the sum of (a) and (b) is shown to be above threshold† more often than either (a) or (b).

6.10.1 Classes of Diversity

Within the two general classifications of selection and combining, some of the common classes of diversity used in radio practice are:

1. Time diversity.
2. Frequency diversity.
3. Space diversity.
4. Angle diversity.
5. Polarization diversity.
6. Combinations of 1, 2, 3, 4, and 5.

The signals used in time diversity are one or more repetitions of the transmitted signal separated in time. The time separation between them must be large enough so that their fading envelopes and phases are essentially uncorrelated. This means in effect that the separation time between repetitions must be at least of the order of the reciprocal bandwidth of the fading. An advantage of time diversity over other diversity schemes is that it can be accomplished with a single antenna and receiver. An obvious disadvantage is the long expenditure of time required to accomplish it.

Frequency diversity makes use of two or more signals separated from each other in frequency by at least the *coherent bandwidth*, defined as the minimum frequency separation required in order that the fading associated with the different signals be essentially uncorrelated. This type of diversity can also be accomplished with a single antenna provided the antenna is sufficiently wide-band to handle all of the signal frequencies. Frequency diversity circumvents the requirement for large available time and substitutes a requirement for large bandwidth.

Space diversity involves the use of two or more receiving antennas. The antennas are spaced so that the signal waves entering them will have travelled paths sufficiently different to ensure that their fading envelopes are mutually uncorrelated. Angle diversity may use the same antenna assembly for all the diversity channels, but uses two or more subantennas (e.g., feeds) responding to waves with different angles of arrival. Again, the angular separation

---

† The threshold for (c) is $\sqrt{2}$ times that for (a) or (b), under the assumptions that the noise is strictly additive with the signal amplitude and the noise levels in signal channels (a) and (b) are equal. The two noise powers add incoherently. Therefore the noise power associated with (c) is twice that of (a) or (b). The amplitude, being the square root of power, is $\sqrt{2}$ times that of the noise on (a) or (b).

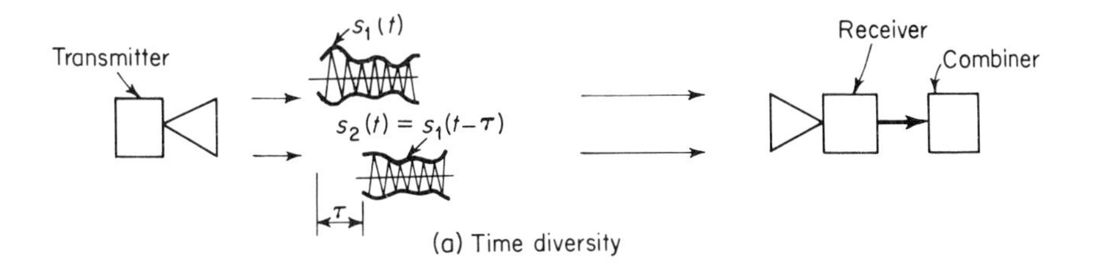

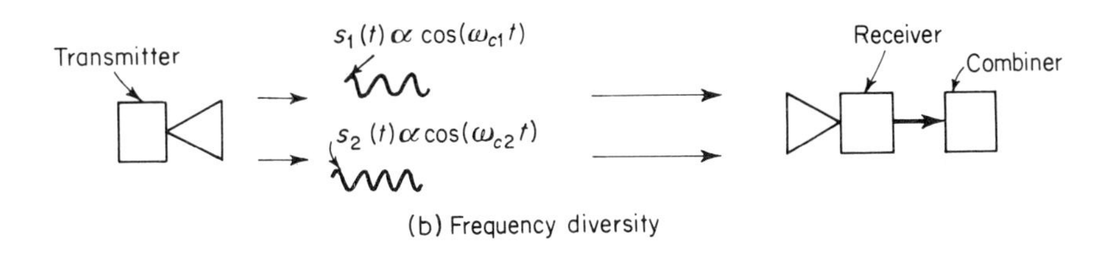

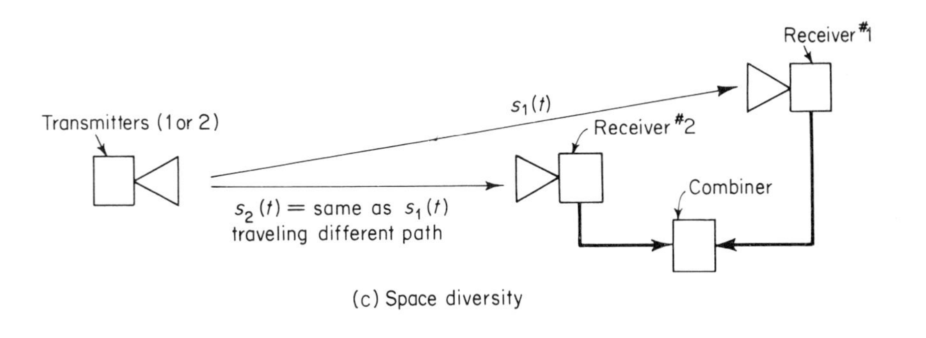

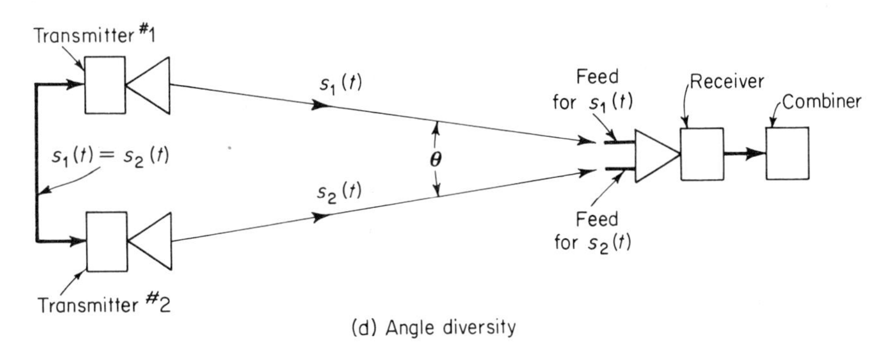

Figure 6-8 Four classes of two-fold diversity.

between paths must assure that the fading waveforms are mutually uncorrelated. Complexity is substituted for time and bandwidth as the commodity required to accomplish space or angle diversity.

Polarization diversity is less commonly used than the other classes of diversity cited above. Its feasibility is based on the fact that radio waves propagating with different polarizations (e.g., horizontal and vertical) undergo different propagation effects (see Section 5.5.4.2) and hence are, in general, mutually uncorrelated or weakly correlated. For this reason, two or more transmission systems with different polarizations and carrying the same message may be used as a diversity link.

The first four of the classes of diversity listed above are illustrated in Figure 6-8(a), (b), (c), and (d).

6.10.2 Analysis of Diversity

It is possible to analyze diversity reception by the techniques of statistical communication theory without specifications of the class of diversity (i.e., time, frequency, space, or angle).

As pointed out in Section 5.5.4.7, the Rayleigh distribution of amplitudes is a good model for a wide class of fading situations met in practice. Attention will be confined to Rayleigh fading in the analysis to follow. In that part of the analysis dealing only with SNR, i.e., not involving calculation of probabilities, it is not necessary to specify the amplitude PDF. The SNR analysis is therefore applicable to fading with other than Rayleigh amplitude statistics.

The analysis of diversity schemes for the Rayleigh fading model (strictly from the viewpoint of maximizing SNR) will be briefly discussed below. It is noted here that well-known analytical treatments of diversity reception by Brennan,† Pierce and Stein,‡ Baghdady,§ Schwartz, Bennett, and Stein,|| and others†† will take the reader much more deeply into the subject than the treatment to follow. The latter is intended to be only a brief introductory summary. Further discussions bearing on diversity will appear in Chapters 10 and 12 in the contexts of radar and communication systems, respectively.

The criterion of performance used in the analysis will be the SNR of the combined signals as a function of the SNR's of the individual diversity branches. To formulate the model, which includes selection diversity as a

---

† Brennan, 1959.

‡ Pierce and Stein, 1960.

§ Baghdady, 1961, Chapter 7.

|| Schwartz, Bennett, and Stein, 1966, Part III, Chapters 10 and 11.

†† See the reference list on diversity at the end of this chapter. The treatment presented in this book draws heavily upon some of the references cited.

special type of combining, let the input signal in the kth branch consist of a signal $s_k(t)$ and an additive noise $n_k(t)$. The signal is a product of a waveform $v(t)$ containing the carrier and the information and common to all diversity branches, a time-varying amplitude $f_k(t)$ (considered as a sample function of a stationary random process) characterizing the fading in the kth diversity branch and a branch gain g_k artificially introduced at the receiver. The fading is assumed to be slowly varying compared to the modulated carrier $v(t)$, a reasonable assumption over a wide range of practical situations.

The time scale of interest for the analysis is an interval 0–T which is long compared to the period of the highest significant frequency in the signal information but short enough to preclude any significant time variation in the fading amplitude. The numerator of the SNR will be defined as the signal power averaged over this time interval and also averaged over the ensemble of all possible fading amplitudes. The denominator is the noise power time averaged over the interval and also ensemble averaged. The SNR of the kth branch, then, is defined as follows:

$$\rho_k = \frac{\dfrac{g_k^2}{T} \displaystyle\int_0^T dt'\, v^2(t) \langle f_k^2(t) \rangle}{\dfrac{g_k^2}{T} \displaystyle\int_0^T dt\, \langle n_k^2(t) \rangle} = \frac{\langle f_k^2 \rangle}{\langle n_k^2 \rangle} \left(\frac{1}{T} \int_0^T dt\, v^2(t) \right) \tag{6.94}$$

where we have accounted for the stationarity of both $n_k(t)$ and $f_k(t)$ and the constancy of $f_k(t)$ during the averaging interval, and where real rather than complex envelope representations have been used.

An Nth order linear diversity combining scheme, by definition, is characterized by signal and noise waveforms

$$s(t) = \sum_{k=1}^{N} g_k f_k(t) v(t) \tag{6.95}$$

$$n(t) = \sum_{k=1}^{N} g_k n_k(t) \tag{6.96}$$

where the branch gains g_k have different forms with different combining schemes.

The general expression for the SNR of the combiner output is

$$\rho(N) = \frac{\displaystyle\sum_{k_1, k_2 = 1}^{N} g_{k1} g_{k2} \left(\frac{1}{T} \int_0^T v^2(t)\, dt \right) \langle f_{k1} f_{k2} \rangle}{\displaystyle\sum_{k_1, k_2 = 1}^{N} g_{k1} g_{k2} \langle n_{k1} n_{k2} \rangle} \tag{6.97}$$

If it is assumed that: (1) the noise waveforms n_k have zero means and are mutually uncorrelated, i.e., $\langle n_{k1}\, n_{k2} \rangle = 0$, and (2) the fading waveforms are uncorrelated but have finite means, i.e., $\langle f_{k1} f_{k2} \rangle = \langle f_{k1} \rangle \langle f_{k2} \rangle$, then (6.97) takes the form

$$\rho(N) = \overline{v^2}\,\frac{\left(\sum_{k=1}^{N} g_k\langle f_k\rangle\right)^2}{\sum_{k=1}^{N} g_k^2\langle n_k^2\rangle} \tag{6.98}$$

where $\overline{v^2} \equiv \frac{1}{T}\int_0^T v^2(t)\,dt$.

6.10.3 Diversity Combining

The performance of diversity schemes is strongly influenced by the method of combining the signals from the various branches. There are three widely used types of combining, as follows:

1. Optimal selection combining, in which the SNR's of the various branches are monitored and the branch with the highest SNR is chosen for use while the other branches are deactivated.
2. Equal gain combining, in which the signals in the various branches are summed with the same weighting coefficient or "gain" on each branch. This type of combining can be either coherent or noncoherent, i.e., the voltages or the powers in the diversity branches can be added.
3. Optimal ratio combining, in which individual branch SNR's are monitored and the relative branch gains are weighted according to the relative SNR's before summation in such a way that the maximum attainable SNR is realized at the output. Like equal gain combining, optimal ratio combining can also be coherent or noncoherent.

These three combining techniques will be discussed below.

6.10.3.1 Optimal Selection Combining

The SNR for optimal selection combining is obviously that of the individual branch with the highest SNR. It is not really necessary to use (6.98) to calculate the SNR in this case. To show that (6.98) is applicable to optimal selection combining, however, we note that the use of this type of combining implies that $g_k = 0$ for all k except that of the selected branch which we will call the jth branch. In this case, (6.98) takes the form

$$\rho(N) = \overline{v^2}\,\frac{g_j^2\langle f_j\rangle^2}{g_j^2\langle n_j\rangle^2} = \rho_j \tag{6.99}$$

where ρ_j was defined in (6.94) as the SNR of the jth branch.

6.10.3.2 Equal Gain Combining

Equal gain combining is characterized by the conditions $g_k = g =$ constant for all k,

$$\rho(N) = \overline{v^2}\,\frac{\left(\sum_{k=1}^{N}\langle f_k\rangle\right)^2}{\sum_{k=1}^{N}\langle n_k^2\rangle} \tag{6.100}$$

If $\langle f_k\rangle$ and $\langle n_k^2\rangle$ are the same in all diversity branches, and equal to $\langle f\rangle$ and σ_n^2, respectively, then (6.100) takes the form

$$\rho(N) = \frac{\overline{v^2}N^2\langle f\rangle^2}{N\sigma_n^2} = N\rho(1) \tag{6.101}$$

where $\rho(1) = \overline{v^2}\langle f\rangle^2/\sigma_n^2 =$ SNR for a single branch.

The conclusion to be drawn from (6.101) is that the SNR gain due to N-fold diversity with equivalent fading and noise in each branch is increased over that of a nondiversity system by a factor N. This is a substantial improvement for large N, and (6.101) shows that increasing the order of diversity by a factor N increases the SNR also by a factor N.

6.10.3.3 Maximal Ratio Combining

In maximal ratio combining, the optimum gains of the diversity branches are first found theoretically in terms of the relative SNR's of the branches. In the system itself, the SNR's of all branches are monitored and the relative gains set in accordance with measured values of SNR.

Under the assumptions basic to our model, it can be shown† that the highest combined output SNR is attained when the branch gains g_k are all proportional to the ratios $\langle f_k\rangle/\langle n_k^2\rangle$. In this case, according to (6.98) and (6.94),

$$\rho(N) = \overline{v^2}\,\frac{\left(\sum_{k=1}^{N}\frac{\langle f_k\rangle^2}{\langle n_k^2\rangle}\right)^2}{\sum_{k=1}^{N}\frac{\langle f_k\rangle^2}{\langle n_k^2\rangle}} = \sum_{k=1}^{N}\overline{v^2}\,\frac{\langle f_k\rangle^2}{\langle n_k^2\rangle} = \sum_{k=1}^{N}\rho_k \tag{6.102}$$

Equation (6.102) tells us that the SNR for maximal ratio combining is the sum of SNR's for the individual branches. In the case where all branch SNR's are equal, it follows from (6.102) that

$$\rho(N) = N\rho(1) \tag{6.103}$$

Equation (6.104) is identical to (6.101), which applies to equal gain combining. It is obvious that the optimal branch weightings would all be equal in this case; hence optimal ratio combining must degenerate into equal gain combining if the branch SNR's are equal.

---

† See Schwartz, Bennett, and Stein, 1966, Section 10.5, pp. 440–442, or Brennan, 1959.

6.11 PROCESSING AGAINST NON-GAUSSIAN NOISE

It was mentioned in Section 5.5.3 that certain types of non-Gaussian noise, e.g., impulse noise and *tone interference* (or *coherent signal interference*), are often important in radio systems. The methodology used in dealing with such noise constitutes an interesting and important topic in radio system analysis. Discussions of non-Gaussian noise and the methods of processing against it will not be included in this book. The reader should consult some of the references cited at the end of this chapter under "Non-Gaussian Noise" for information on impulse noise and tone interference.

REFERENCES

Matched Filtering or Crosscorrelation Detection of Radio Signals

See the reference lists in Chapters 2 and 3, particularly those on statistical analysis of waveforms, etc. in Chapter 2 and those on optimization of SNR and matched filters in Chapter 3. Most of them apply to these topics in the context of radio signals.

Two papers not previously referenced are of some interest in connection with these topics:

Huggins and Middleton, 1955

Smith, 1951

Autocorrelation Detection

Bendat, 1958; Chapters 6, 7

Lee, 1960; Chapter 12, especially Section 1, pp. 288–290

Lee, Cheatham, and Wiesner, 1950

Diversity

Baghdady, 1961; Chapter 7

Barrow, 1962

Barrow, 1963

Bello and Nelin, March 1962

Brennan, 1959

Grandlund and Sichak, 1961

Pierce and Stein, 1960

Schwartz, Bennett, and Stein, 1966; Chapters 10, 11

Staras, 1956

Stein and Jones, 1967; Chapter 17

Turin, 1961

Turin, 1962

Non-Gaussian Noise (e.g., Tone and Impulse Noise)

Baghdady, 1961; Chapter 19, Section V, "Reception in the Presence of Additive Disturbances—Coherent-Signal Interference," pp. 483–508.

Baghdady, 1961; Chapter 19, Section VI, "Impulse Noise and its Suppression," pp. 508–521

Jones, Pierce, and Stein, 1964; Section 2.9.1, pp. 214–215 (impulse noise)

R. Lerner; Chapter 10, Section VI of Baghdady, 1961, "Signalling in the Presence of Impulse Noise," pp. 263–277

Splitt, 1966

7

THEORY OF DETECTION OF RADIO SIGNALS IN GAUSSIAN NOISE

The present chapter will be devoted to the detection of radio signals corrupted by additive Gaussian noise. Previous discussions (see Section 3.4 and Section 5.5.3) have indicated that this type of noise is a major source of difficulty in radio systems. It is not the *only* type of noise that is troublesome, as pointed out in Section 5.5.3. It is, however, the type to which the communication theorists have usually addressed themselves (to some extent because it is the simplest type of noise to analyze). Hence the theory is highly developed for additive Gaussian noise and developed only in a fragmentary way for other types of noise.

7.1 DETECTION OF SIGNAL PRESENCE

7.1.1 Introduction

The theory of optimum detection of signal presence in Gaussian noise was treated in Section 4.1. In the discussion to follow, the material covered in Section 4.1 will be specialized to the case of detection of a signal entering a radio receiver in a background of noise assumed to be bandlimited white and Gaussian. Both internal receiver noise and external noise with Gaussian statistics may be present. The signal-plus-noise model, then, is the same as that of Chapter 6 with the added specification of a finite time interval $t = 0$ to $t = T_s$ for the signal. This does not really restrict the generality of the discussion, since T_s is arbitrary.

The general signal model to be used, a specialization of (6.2) to the case of a finite signal interval, is

$$s_i(t\,;\boldsymbol{\beta}) = a_i(t\,;\boldsymbol{\beta})\cos\left(\omega_c(\boldsymbol{\beta})t + \psi_i(t\,;\boldsymbol{\beta})\right), \qquad 0 \leq t \leq T_s \tag{7.1}$$

where the multidimensional vector $\boldsymbol{\beta}$ represents the set of parameters that are a priori unknown at the receiver and hence must be averaged out in the processing, and where the functional dependence of s_i, a_i, and ψ_i on both time and on the elements of $\boldsymbol{\beta}$ and the functional dependence of ω_c on $\boldsymbol{\beta}$ are explicitly indicated through the notation. As in Chapter 6, the signal is narrowband; hence $a_i(t\,;\boldsymbol{\beta})$ and $\psi_i(t\,;\boldsymbol{\beta})$ are slowly varying compared with $\cos\omega_c t$ or $\sin\omega_c t$.

It was shown in Section 4.1 that optimum detection of a signal $s_i(t)$ in Gaussian noise is accomplished by a device that computes the likelihood ratio, which in this case, has the form given by Eq. (4.28). The sums within the exponentials of (4.28) are equivalent to integrals in the case of band-limited white noise whose bandpass bandwidth is $2B_n$ and whose mean power is σ_n^2, as shown in the footnote at the end of Section 4.1.3.1.

$$\sum_{n=1}^{N} \frac{v_n s_{in}(\boldsymbol{\beta})}{\sigma_n^2} \simeq \frac{2B_n}{\sigma_n^2}\int_0^{T_s} dt'\, v_i(t')s_i(t'\,;\boldsymbol{\beta}) \tag{7.2}$$

$$\sum_{n=1}^{N} \frac{s_{in}^2(\boldsymbol{\beta})}{2\sigma_n^2} \simeq \frac{B_n}{\sigma_n^2}\int_0^{T_s} dt'\, s_i^2(t'\,;\boldsymbol{\beta}) \tag{7.3}$$

that is, Eqs. (7.2) and (7.3) represent a coherent matched filtering operation, as discussed in Section 6.3.

Using the theory developed in Section 4.1 we arrive at a general expression for the likelihood ratio (except for a proportionality constant) which is the quantity that the theory tells us should be computed in the optimum processing operation

$$\Lambda = \int \cdots \int d\boldsymbol{\beta}\, p(\boldsymbol{\beta}) e^{-\frac{B_n}{2\sigma_n^2}\int_0^{T_s} dt'[a_i(t';\boldsymbol{\beta})]^2 - \frac{B_n}{2\sigma_n^2}\int_0^{T_s} dt'\{[a_i(t';\boldsymbol{\beta})]^2 \cos[2\omega_c t' + 2\psi_i(t';\boldsymbol{\beta})]\}} \cdots \left\{e^{\frac{B_n}{\sigma_n^2}\int_0^{T_s} dt' v_i(t') a_i(t';\boldsymbol{\beta})\cos(\omega_c t' + \psi_i(t';\boldsymbol{\beta}))}\right\} \tag{7.4}$$

where it is assumed that the observation time is the entire signal duration T_s.

We now make the assumption that the signal duration T_s is large compared with the RF period, often true in practice. Thus

$$T_s \gg \frac{1}{\omega_c} \tag{7.5}$$

from which it follows that the integral in (7.4) that involves $\cos(2\omega_c t + 2\psi_i(t;\boldsymbol{\beta}))$ will effectively vanish.

An information-bearing radio signal can be regarded as a constant amplitude sine wave over a time interval that is short compared with the reciprocal of B_s, where B_s is the bandwidth of the information waveform, known as the *information bandwidth*.† Detection theory is most easily applied when the signal is a sine wave with constant amplitude and phase; consequently, in this section we will treat the incoming radio signal as such. This is tantamount to an assumption that the observation time T is somewhat smaller than or at most comparable to the reciprocal information bandwidth.

We now change the integration time in (7.4) from T_s, the signal duration, to a shorter time T during which the signal can be considered as a sine wave with constant amplitude a_i and constant phase ψ_i.

Then, with the aid of (7.5), we obtain

$$\Lambda = \int \cdots \int d\boldsymbol{\beta}\, p(\boldsymbol{\beta}) e^{-\frac{B_n}{2\sigma_n^2}(a_i^2 T) + \frac{B_n T}{\sigma_n^2}(a_i \overline{v_i(t) \cos(\omega_c t + \psi_{si}))}} \tag{7.6}$$

where the line over the symbol refers to time averaging over the interval 0 to T, i.e.,

$$\overline{v_i(t) \cos(\omega_c t + \psi_i)} \equiv \frac{1}{T} \int_0^T dt' v_i(t') \cos(\omega_c t' + \psi_i)$$

7.1.2 Coherent Detection

Roughly speaking, *coherent detection* denotes the technique of crosscorrelating the incoming radio signal with a reference signal of the same radio frequency and phase as the incoming signal. The reference signal should be as close to an exact replica of the incoming signal as possible. The feasibility of this technique in its idealized form requires complete knowledge of signal parameters (except possibly the amplitude; see Section 7.1.5), such that the reference signal constructed at the receiver can precisely duplicate the incoming signal. In this case, all components of the vector $\boldsymbol{\beta}$ are assumed to be precisely known. If the nth component of $\boldsymbol{\beta}$ is denoted by β_n, then the joint probability density function of the β_n's is

$$p(\boldsymbol{\beta}) = p(\beta_1, \beta_2, \ldots, \beta_M) = \delta(\beta_1 - \beta_{10})\delta(\beta_2 - \beta_{20}) \ldots \delta(\beta_M - \beta_{M0}) \tag{7.7}$$

where $\boldsymbol{\beta}_0 = (\beta_{10}, \beta_{20}, \ldots, \beta_{M0})$ now represents the set of a priori known values of $\beta_1, \beta_2, \ldots, \beta_M$. If (7.7) holds, then coherent detection is the theoretically optimum detection technique (see Section 4.1). The parameters β_1, β_2, and β_3 will represent signal amplitude a_i, phase ψ_i, and carrier frequency ω_c in what follows. Then

$$p(\boldsymbol{\beta})\, d\boldsymbol{\beta} = p(a_i, \psi_i, \omega_c)\, da_i\, d\psi_i\, d\omega_c \tag{7.8}$$

It is assumed in general that a_i, ψ_i, and ω_c, even if unknown to the observ-

† We can also call B_s the *signal bandwidth*.

er, are statistically independent from his point of view, i.e.,

$$p(\boldsymbol{\beta})\,d\boldsymbol{\beta} = p(a_i)\,da_i\,p(\psi_i)\,d\psi_i\,p(\omega_c)\,d\omega_c \tag{7.9}$$

Consider first the highly idealized case where a_i, ψ_i, and ω_c are known by the observer to be precisely equal to a_{i0}, ψ_{i0}, and ω_{c0} respectively. Then, from (7.7) and (7.9),

$$p(\boldsymbol{\beta}) = \delta(a_i - a_{i0})\,\delta(\psi_i - \psi_{i0})\,\delta(\omega_c - \omega_{c0}) \tag{7.10}$$

From (4.28), (7.6), and (7.10),

$$\Lambda \simeq e^{-\frac{a^2{}_{i0}B_nT}{2\sigma_n{}^2} + \frac{B_na_{i0}}{\sigma_n{}^2}\int_0^T dt'v_i(t')\cos(\omega_ct' + \psi_{i0})} \tag{7.11}$$

In implementing the optimum detector, we should compute a quantity proportional to the logarithm of the likelihood ratio and set a threshold on it. The processor that computes the quantity $\log\Lambda$ is a perfectly valid optimum detector, because $\log\Lambda$ is a monotonic increasing function of Λ, and because we are not at the moment concerned about the exact value of the threshold. The quantity U, on which the detection threshold will be set, is

$$U = U_s + U_n = K\int_0^T dt'\,v_i(t')\cos(\omega_ct' + \psi_r) = Ka_i\int_0^T dt'\cos(\omega_ct' + \psi_i)$$
$$\cos(\omega_ct' + \psi_r) + K\int_0^T dt'n_i(t')\cos(\omega_ct' + \psi_r) \tag{7.11}$$

where a_r and ψ_r are the reference signal's amplitude and phase and K is a proportionality constant to account for the fact that the exact signal amplitude is not generally known at the receiver and the reference signal's amplitude may be greatly different from that of the incoming signal. The constant K does not affect performance, except for its role in determination of the optimum threshold. For a fixed threshold value, error probabilities will turn out to be dependent on SNR at the processor output, and K, being common to both signal and noise, cancels out in the SNR. This point was mentioned in Section 6.3 in connection with matched filtering. Because of its irrelevance to the final results, we will set $K = 1$ throughout the remaining discussions.

The probabilities of error in attempting to detect the signal are given in general by (4.11) and (4.12). Equation (4.11) gives the false rest probability and (4.12) gives the false alarm probability. The operation represented by (7.11) may be considered as linear; hence the noise at the integrator output, like that at the input, is Gaussian.† The conditional PDF's $p(U/s)$ and $p(U/n)$, representing the PDF for an output U *with* a signal present and *without* a signal present, respectively, are both Gaussian functions. Therefore, from (4.11) and (4.12) the false rest and false alarm probabilities are

† See Sections 3.1.3 and 3.1.4.

$$P_{fr} = P^{(s)} \int_{-\infty}^{U_T} \frac{dU}{\sqrt{2\pi\sigma_u^2}} e^{-(U-U_s)^2/2\sigma_u^2} = \frac{1}{2} P^{(s)} \left\{1 + \text{erf}\left(\frac{U_T - U_s}{\sqrt{2\sigma_u^2}}\right)\right\} \tag{7.13}$$

$$P_{fa} = P^{(n)} \int_{U_T}^{\infty} \frac{dU}{\sqrt{2\pi\sigma_u^2}} e^{-U^2/2\sigma_u^2} = \frac{1}{2} P^{(n)} \left\{1 - \text{erf}\left(\frac{U_T}{2\sigma_u^2}\right)\right\} \tag{7.14}$$

where $P^{(s)}$ and $P^{(n)}$ are a priori probabilities for "signal present" and "noise alone," respectively, U_T is the *decision threshold*, i.e., the threshold level on U separating the "signal present" and "noise alone" decision regions, σ_u^2 is the mean-square deviation of U from its mean (due to noise) and U_s is the signal component of U.

To determine U_s and σ_u^2, we note from (7.12) that the signal part of U is approximately $a_i T/2$ and the mean noise power output is equivalent to that given by (6.19). The noise output has zero mean (because the input noise has zero mean and the mean of an integral of the form $\int dt' f(t') n_i(t - t')$ is the integral of the mean) and is uncorrelated with the signal. Thus, the mean output of the filter is equivalent to the signal output U_s. Now,

$$\sigma_u^2 = \langle \overline{U^2} \rangle - \langle \overline{U} \rangle^2 = \langle (U_s - U_n)^2 \rangle - U_s^2 = \langle U_n^2 \rangle = \frac{\sigma_n^2 T}{4B_n} \tag{7.15}$$

$$U_s = \frac{a_i T}{2} \tag{7.16}$$

From (6.46), (7.15), and (7.16),

$$\frac{U_s}{\sqrt{2\sigma_u^2}} = \sqrt{\rho_i B_n T} = \sqrt{\frac{(\rho_0)_{\text{coh}}}{2}} \tag{7.17}$$

that is, $U_s/\sqrt{2\sigma_u^2}$, the parameter that essentially determines the error probabilities for a given threshold level, is proportional to the square root of the matched filter output SNR. To go one step further, we write

$$U_T = K_T U_s \tag{7.18}$$

i.e., the threshold level is chosen to be a constant parameter K_T multiplied by the signal level. It follows from (7.13), (7.14), (7.17), and (7.18) that the *conditional error probabilities*† are

$$P_{frc} = \frac{P_{fr}}{P^{(s)}} = \frac{1}{2}\{1 - \text{erf}\,[(1 - K_T\sqrt{\rho_i B_n T})]\} \tag{7.19}$$

$$P_{fac} = \frac{P_{fa}}{P^{(n)}} = \frac{1}{2}\{1 - \text{erf}\,(K_T\sqrt{\rho_i B_n T})\} \tag{7.20}$$

The technique of detection discussed above can be called *coherent cross-correlation detection, coherent filter detection* or merely *coherent detection.*

† That is, the false rest error probability conditional on the presence of a signal and the false alarm error probability conditional on the absence of a signal.

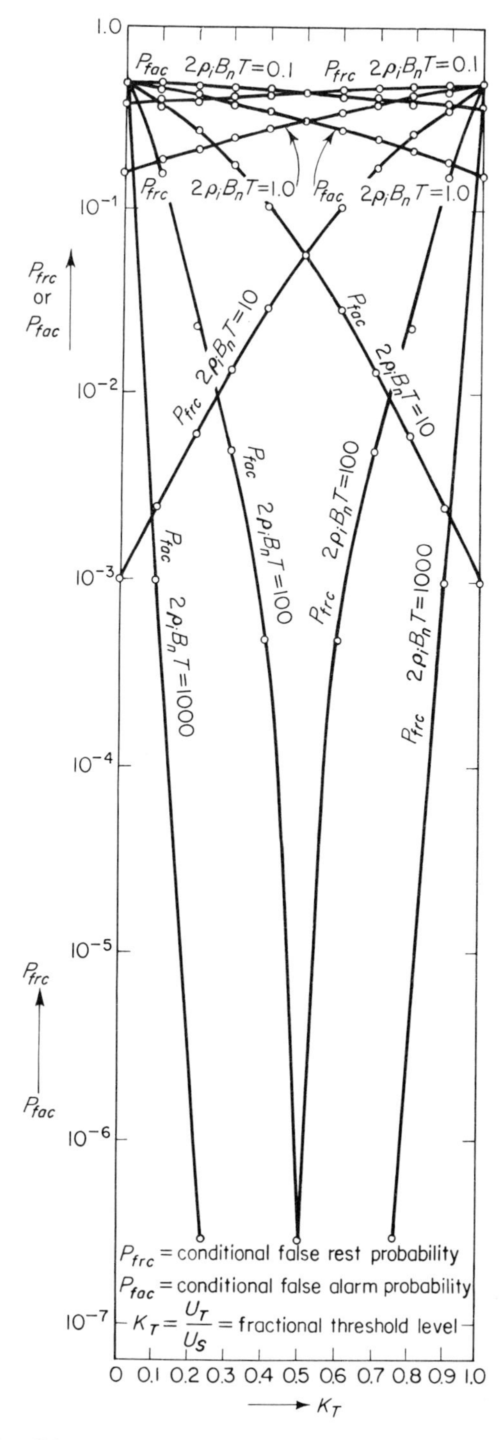

Figure 7-1 Conditional false rest and false alarm probabilities versus threshold—coherent detection.

Note that the filtering technique involved is equivalent to coherent matched filtering, as discussed in Section 6.3.

The interpretation of (7.19) and (7.20) is quite simple, aided by the curves of Figure 7-1. These curves show conditional false rest and false alarm probabilities (P_{frc} and P_{fac}, respectively) for the coherent detector as functions of K_T for fixed values of $2\rho_i B_n T$. As noted above, the parameter $2\rho_i B_n T$ represents the output SNR of the crosscorrelation filter in voltage or current units. If this SNR is very low (Case (a)) both P_{frc} and P_{fac} are high, as would be expected. The chance of a false alarm naturally decreases as the threshold is raised while that of a false rest increases. In case (b), where the SNR is high, the same qualitative behavior is observed, but both P_{frc} and P_{fac} are generally lower and much less sensitive to the threshold level within the range of thresholds shown on the curves. Case (c), where $2\rho_i B_n T$ is unity, shows behavior intermediate between the low SNR and high SNR cases, as would be expected.

In Figure 7-2, plots of P_{frc} and P_{fac} against $2\rho_i B_n T$ for fixed K_T are shown. As would be expected, both conditional error probabilities decrease with increases in $2\rho_i B_n T$, but both the absolute and relative rates of decrease are very sensitive to the choice of threshold level.

The case considered above is highly idealized because complete a priori knowledge of signal parameters is rare in most radio applications. However, the curves shown in Figure 7-1 and 7-2 represent a theoretically optimum condition. They provide us with a standard, corresponding to the best that can be done, by which more realistic situations may be evaluated. In what follows, as was done in Chapter 6 for evaluation of SNR, we will consider the effect on error probabilities of increases in the realism of both the radio signal and the state of a priori knowledge at the receiver. We will do this in gradual steps, often aided by intuition where rigorous analytical methods are prohibitively cumbersome or where extrapolation from the simple theory is adequate to provide an understanding of the phenomenon under discussion.

7.1.3 Effects of Phase Uncertainty—Noncoherent Detection†

Consider the case where signal phase is completely unknown to the observer. From his point of view, the phase can be regarded as randomly distributed, i.e., Eq. (7.10) is replaced by

$$p(\boldsymbol{\beta}) = \delta(a_i - a_{i0})\delta(\omega_c - \omega_{c0})\frac{1}{2\pi}, \qquad 0 \leq \psi_i \leq 2\pi \tag{7.21}$$

and (7.11) is replaced by

† This is often designated as *envelope detection.*

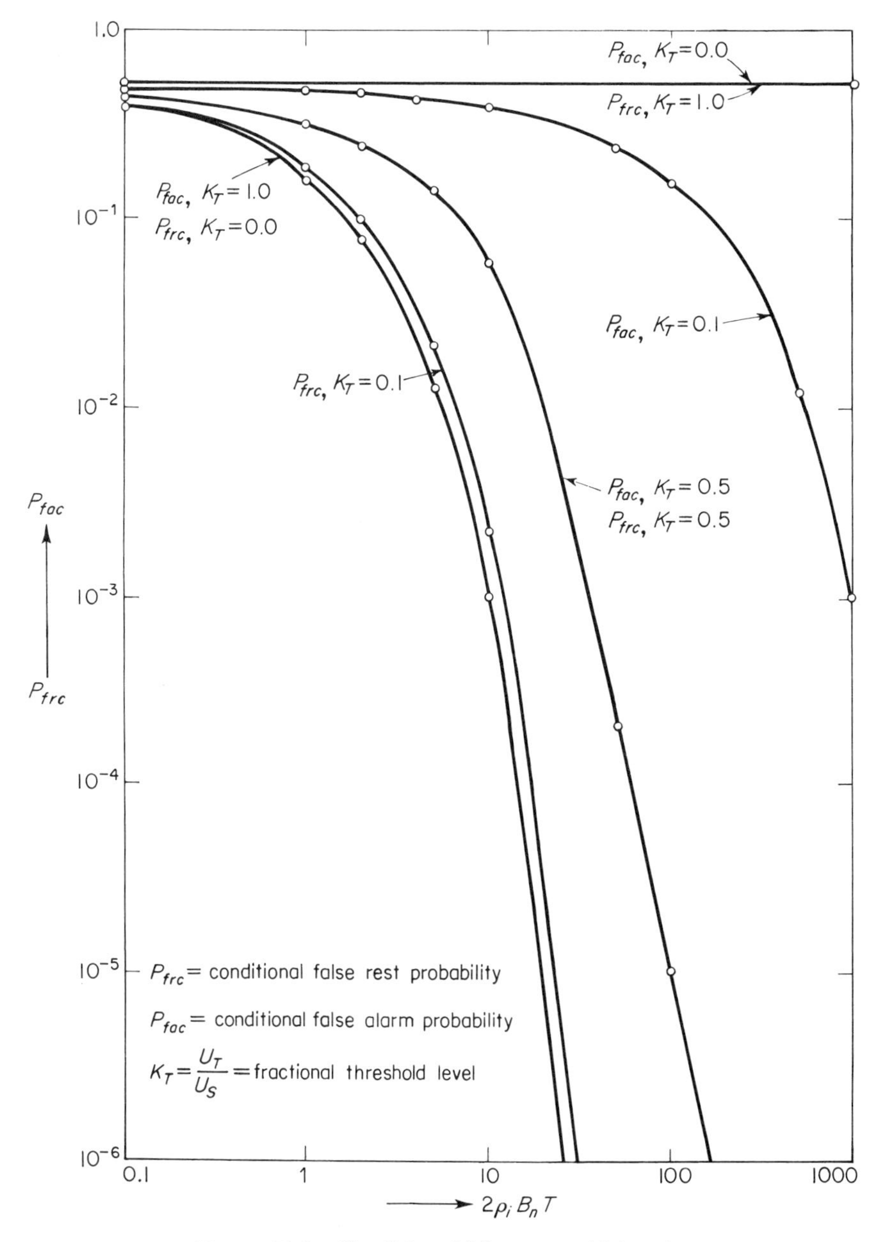

Figure 7-2 Conditional false rest and false alarm probabilities versus output-SNR—coherent detection.

$$\Lambda \simeq \frac{e^{-a^2_{i0}B_nT/2\sigma_n^2}}{2\pi} \int_0^{2\pi} d\psi_i \, e^{a_{i0}B_n/\sigma_n^2} \sqrt{\left[\left(\int_0^T dt' \, v_i(t') \cos \omega_{c0}t'\right)^2 + \left(\int_0^T dt' \, v_i(t') \sin \omega_{c0}t'\right)^2\right]} \cos (\psi_i - \alpha) \tag{7.22}$$

where

$$\alpha \equiv \tan^{-1} \left\{ \frac{\int_0^T dt' \, v_i(t') \sin \omega_{c0}t'}{\int_0^T dt' \, v_i(t') \cos \omega_{c0}t'} \right\}$$

By virtue of the relation†

$$I_0(x) = \frac{1}{2\pi} \int_0^{2\pi} d\psi \, e^{x \cos \psi} \tag{3.45}$$

where $I_0(x)$ is the zero order modified Bessel function, Eq. (7.22) takes the form

$$\Lambda \simeq e^{-a^2_{i0}B_nT/2\sigma_n^2} \, I_0\left(\frac{a_{i0}B_n}{\sigma_n^2} \sqrt{[v_c(T)]^2 + [v_s(T)]^2}\right) \tag{7.23}$$

where

$$v_c(T) \equiv \int_0^T dt' \, v_i(t') \cos \omega_{c0}t'$$

$$v_s(T) \equiv \int_0^T dt' \, v_i(t') \sin \omega_{c0}t'$$

The quantity on which a threshold is set is

$$\ln \Lambda \simeq -\frac{a_{i0}^2 B_n T_s}{2\sigma_n^2} + \ln I_0\left(\frac{a_{i0}B_n}{\sigma_n^2} \sqrt{[v_c(T)]^2 + [v_s(T)]^2}\right) \tag{7.24}$$

The modified Bessel function $I_0(x)$ is a monotonic increasing function of x,‡ hence the processing operation dictated by the theory is to compute the argument of the modified Bessel function and set a threshold on it. Thus the quantity to be computed is

$$U = K\sqrt{[v_c(T)]^2 + [v_s(T)]^2} \tag{7.25}$$

where K is a constant.

The result (7.25) tells us that the optimum detection procedure in the absence of a priori phase information consists of crosscorrelation of the input signal with two sinusoids in quadrature followed by squaring and summing of the correlator outputs. The summing of the squares of the two reference sinusoids to eliminate the need for phase information has been designated as *noncoherent crosscorrelation* or *noncoherent matched filtering* in Section 6.1.

† See Abramowitz and Stegun, 1965, No. 9.6.16, p. 376.

‡ See Abramowitz and Stegun, 1965, Figure 9.7, p. 374.

To calculate error probabilities, we recognize the signal-plus-noise input to the cosine and sine filters as a narrowband Gaussian process with the representation (3.33) where the time reference is such that $\psi_i = 0$ and where $a_s = a_i$, $\omega_c = \omega_{c0}$. The resulting filter outputs are approximately

$$v_c(T) \simeq \frac{a_{i0}T}{2} + \frac{1}{2}\int_0^T dt'\, x_n(t') \tag{7.26.a}$$

$$v_s(T) \simeq \frac{1}{2}\int_0^T dt'\, y_n(t') \tag{7.26.b}$$

where the assumption (7.5) applied to the case where $T_s = T$ has effectively cancelled out integrals of sine-cosine products and integrals involving radio frequency sinusoids.

Each integral in (7.26.a) and (7.26.b) is a Gaussian random variable.† One property of these integrals, brought about by the statistical independence of $x_n(t)$ and $y_n(t)$ and the fact that both of these quantities have zero mean, is

$$\langle v_c(T)v_s(T)\rangle = \frac{a_{i0}T}{4}\int_0^T dt'\langle y_n(t')\rangle + \frac{1}{4}\int_0^T dt' \int_0^T dt''\langle x_n(t')y_n(t'')\rangle = 0 \tag{7.27}$$

Since $v_c(T)$ and $v_s(T)$ are Gaussian, it follows from (7.27) that they are statistically independent.‡

The joint PDF of v_c and v_s (the functional dependence of v_c and v_s on T will no longer be explicitly indicated) is

$$p(v_c, v_s) = \frac{1}{\sqrt{2\pi\sigma_{nc}^2}}\, e^{-\left(v_c - \frac{a_{i0}T}{2}\right)^2/2\sigma^2_{nc}} \frac{1}{\sqrt{2\pi\sigma_{ns}^2}}\, e^{-v_s^2/2\sigma^2_{ns}} \tag{7.28}$$

where

$$\sigma_{nc}^2 \equiv \left\langle \left[v_c - \frac{a_{i0}T}{2}\right]^2 \right\rangle$$

$$\sigma_{ns}^2 \equiv \langle v_s^2 \rangle$$

A convenient transformation of variables can be invoked at this point in order to simplify the notation, as follows:

$$X \equiv \frac{v_c}{\sigma_{nc}}, \qquad \overline{X} \equiv \frac{a_{i0}T}{2\sigma_{nc}}$$
$$Y \equiv \frac{v_s}{\sigma_{ns}} \tag{7.29}$$

Equation (7.27) in terms of the new variables defined in (7.28) now takes the form

$$p(X, Y)\frac{1}{2\pi}\, e^{-\frac{1}{2}(X - \overline{X})^2 - \frac{1}{2}Y^2} \tag{7.30}$$

† See Section 3.1.4.

‡ Uncorrelated *Gaussian* processes are statistically independent. See Section 3.1.2.

The error probability can be calculated by transforming the variables X, Y, analogous to rectangular coordinates, to a pair of polar coordinates R, θ, as follows:

$$\begin{aligned} R &= \sqrt{X^2 + Y^2} \\ \theta &= \tan^{-1}\left(\frac{Y}{X}\right) \end{aligned} \tag{7.31}$$

and integrating over θ exactly as is done in Section 3.3, proceeding from Eq. (3.45). This results in a PDF exactly equivalent to that of Eq. (3.46).†

The quantities σ_{nc} and σ_{ns}, with the aid of (7.5) and (6.21) are given by

$$\sigma_{nc}^2 = \frac{1}{4}\int_0^T \int_0^T dt'\, dt'' \langle x_n(t')x_n(t'')\rangle \simeq \frac{\sigma_n^2 T}{8B_n} \tag{7.32.a}$$

$$\sigma_{ns}^2 = \frac{1}{4}\int_0^T \int_0^T dt'\, dt'' \langle y_n(t')y_n(t'')\rangle \simeq \frac{\sigma_n^2 T}{8B_n} \tag{7.32.b}$$

The conditional error probabilities are:

$$P_{frc} = \int_0^{K_T\sqrt{\rho_i B_n T}} dR\, R\, e^{-\frac{1}{2}(R^2 + 4\rho_i B_n T)} I_0(R\sqrt{4\rho_i B_n T}) \tag{7.33.a}$$

$$P_{fac} = \int_{K_T\sqrt{\rho_i B_n T}}^{\infty} dR\, e^{-R^2/2} R = e^{-K_T{}^2 \rho_i B_n T/2} \tag{7.33.b}$$

where $K_T\sqrt{\rho_i B_n T}$ is the SNR corresponding to the detection threshold.

The integral of (7.33.a) is a special case of *Q-function,*

$$Q(y, b) = \int_b^\infty dx\, x\, e^{-(x^2 + y^2)/2} I_0(xy) \tag{7.34}$$

tabulated by Marcum.‡ Equations (7.33.a) and (7.33.b) assume the forms§

$$\begin{aligned} P_{frc} &= Q(\sqrt{4\rho_i B_n T}, 0) - Q(\sqrt{4\rho_i B_n T}, K_T\sqrt{\sigma_i B_n T}) \\ &= 1 - Q(\sqrt{4\rho_i B_n T}, K_T\sqrt{\rho_i B_n T}) \end{aligned} \tag{7.35.a}$$

$$P_{fac} = e^{-K_T{}^2 \rho_i B_n T/2} \tag{7.35.b}$$

Plots of P_{frc} and P_{fac} as functions of K_T for various values of $(2\rho_i B_n T)$ and plots of P_{frc} and P_{fac} as functions of $(2\rho_i B_n T)$ for various values of K_T, are shown in Figures 7-3 and 7-4, respectively. These sets of curves are the analogs, for noncoherent detection, of the curves of Figures 7-1 and 7-2, applicable to coherent detection.

Comparison of Figures 7-3 and 7-4 with Figures 7-1 and 7-2 shows the variation of P_{frc} and P_{fac} with K_T and $(2\rho_i B_n T)$ with noncoherent detection

---

† This is often called the *Rice distribution*, after S. O. Rice (see Rice, 1945, Section 3.10, pp. 104–107 (pp. 236–239 of Wax, 1954)).

‡ Marcum, 1950; Marcum and Swerling, 1960.

§ A special property of the Q-function is:

$$Q(y, 0) \equiv \int_0^\infty dx\, x\, e^{-(x^2 + y^2)/2} I_0(x\, y) = 1$$

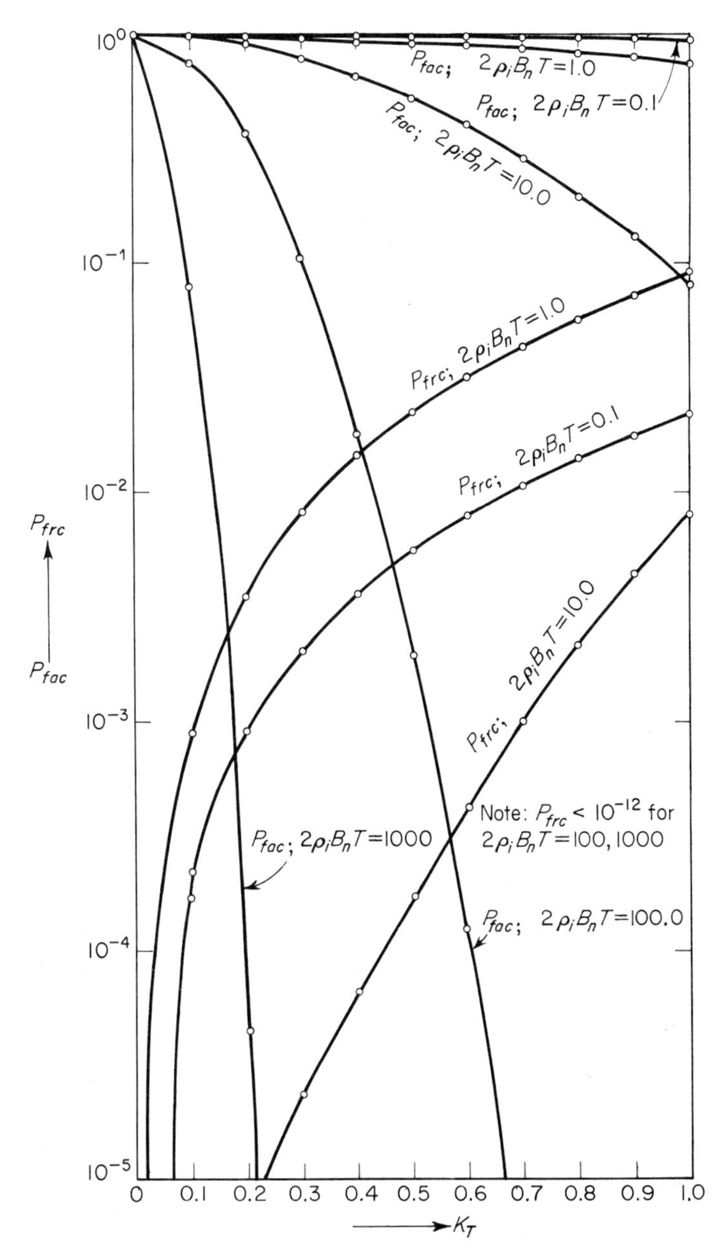

Figure 7-3 Conditional false rest and false alarm probabilities versus threshold level—noncoherent detection.

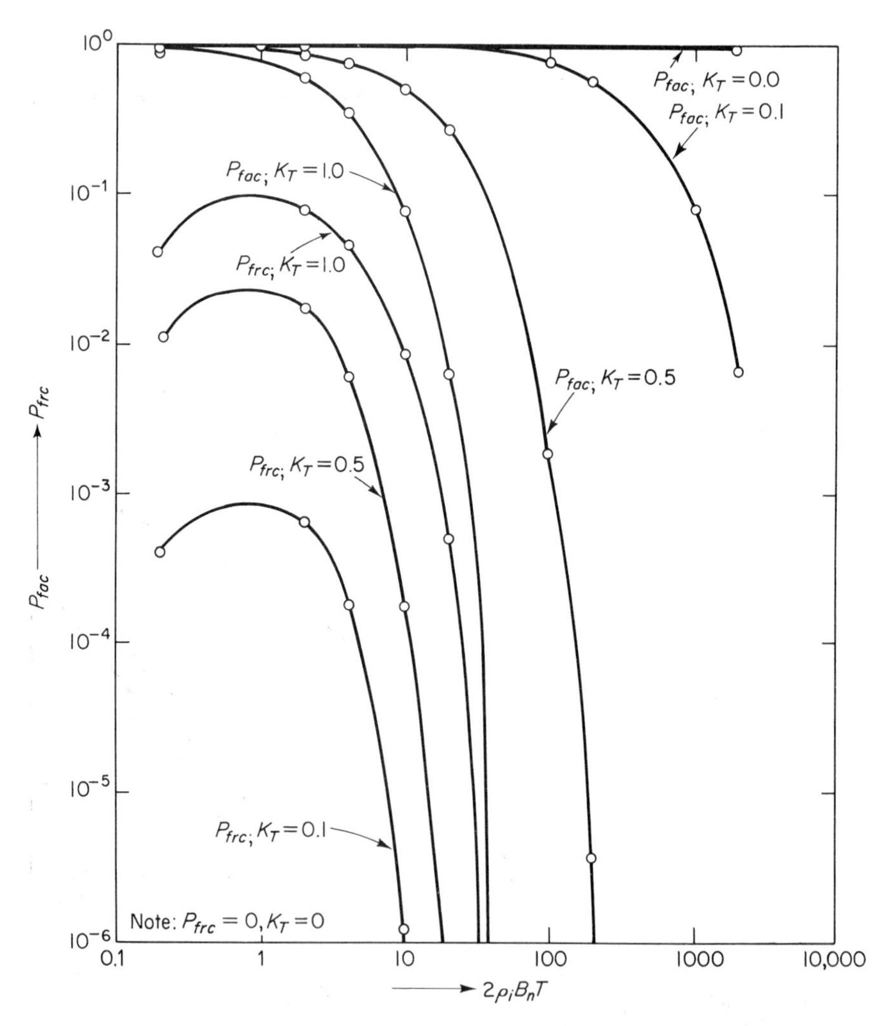

Figure 7-4 Conditional false rest and false alarm probabilities *versus* output SNR—noncoherent detection.

to be qualitatively similar to the analogous error probability behavior with coherent detection. However, the coherent case is generally superior to the noncoherent case. The reason for this becomes apparent if one relates the error probabilities to the SNR at the output of the correlator. Since the use of a noncoherent detector degrades the output SNR, it seems reasonable that a given error probability should be attainable coherently at an SNR below that required for noncoherent detection.

A comparison of Figures 7-3 and 7-4 with Figures 7-1 and 7-2 reveals some interesting differences between coherent and noncoherent detection. First, note that both false rest and false alarm probabilities always *decrease* with increasing SNR, except for the noncoherent false rest probability in the case where $2\rho_i B_n T \leq 1$. The decrease with increasing SNR is to be expected, since intuition tells us that an error becomes less likely as the SNR increases. The apparently anomalous case, where the noncoherent false rest probability *increases* with SNR, is due to the choice of decision rule.

To explain this apparent anomaly, we note that the decision threshold in both coherent and noncoherent cases is chosen to be a fixed fraction of the signal level. In the coherent case, a very small SNR would imply a decision threshold very near zero. Since the noise can swing either positively or negatively around its mean value (the mean value is the signal level), positive and negative swings of a given magnitude are equally probable. As the SNR approaches zero, both the mean value and the decision threshold become negligibly small compared to RMS noise. The output PDF in this case approaches that corresponding to noise alone. Thus false rests and false alarms become about equally probable and both attain the "pure chance level," i.e., $P_{frc} = P_{fac} = \frac{1}{2}$. According to Figures 7-1 and 7-2, this is what happens as SNR approaches zero.

In the noncoherent case, the situation for small SNR is radically different. In this case, the output PDF is that of *amplitude*, which obviously can never swing negative. As the SNR decreases to very small values, the decision threshold, being proportional to signal level, becomes negligibly small compared to the noise level. Eventually, when SNR becomes sufficiently small, the threshold is so low that the chance of the output amplitude falling below it becomes negligible. Thus we can expect the false rest probability to approach zero as SNR approaches zero. As SNR attains larger values, this probability must revert to normal behavior, that is, it must decrease with increasing SNR. Figures 7-3 and 7-4 show this to be the case. The false rest probability increases to a maximum at $2\rho_i B_n T \simeq 1$, then decreases with further increases in SNR. Note that this is strictly due to the choice of decision threshold as a fixed fraction of signal level. If the threshold were a fixed number independent of signal level, the false rest probability would decrease with SNR throughout the entire range of possible SNR.

Before closing this discussion, a further note is in order concerning the two limiting cases of low SNR and high SNR in noncoherent detection. Referring to (7.24), note that $\ln I_0(x)$ can be expanded in a power series in x, retaining only the first few terms in the case $x \ll 1$. For $x \gg 1$, the asymptotic form of $I_0(x)$ for large x can be used. The results are†

† See Abramowitz and Stegun, 1965, No. 9.6.12, p. 375, and No. 9.7.1, p. 377.

$$\ln I_0(x) \simeq \ln\left(\frac{e^x}{\sqrt{2\pi x}}\right) = x - \frac{1}{2}\ln(2\pi x) \simeq x \tag{7.36}$$

if $x \gg 1$, and

$$\ln I_0(x) = \ln\left[1 + \frac{x^2}{4} + 0(x^4)\right] \simeq \frac{x^2}{4} + 0(x^4) \tag{7.37}$$

if $x \ll 1$.

The results (7.36) and (7.37) would seem to dictate the use of a square-law rectifier for optimum detection at low SNR and a linear rectifier at high SNR. The detection performance is not highly sensitive to the type of rectifier used. The important part of the operation is the computation of x, which is equivalent to the quantity U of (7.25). Whether one uses U itself or its square makes little difference in the quality of the detection process.

7.1.4 Effects of Frequency Uncertainty

Consider the uncertain phase case covered in Section 7.1.3 with the added feature that the frequency ω_c is not precisely known to the observer. From the observer's point of view, then, frequency is characterized by a PDF $p(\omega_c)$ other than that used in (7.21).

7.1.4.1 The Likelihood Function with Unknown Frequency

The likelihood function is obtained from a slight generalization of (7.23) to account for the process of averaging over frequency, as follows:

$$\Lambda \simeq \int_0^\infty d\omega_c\, p(\omega_c)\, e^{-a_{i_0}B_nT/2\sigma_n^2}\, I_0\left(\frac{a_{i_0}B_n}{\sigma_n^2}\int_0^T\int_0^T dt'\, dt'' v_i(t')v_i(t'')\cos\omega_c(t'-t'')\right) \tag{7.38}$$

If we use the expansion (7.24), and assume the input SNR to be sufficiently small to neglect terms $0(x^4)$, the only term of Λ involving the processing of the incoming signal is proportional to

$$\begin{aligned} \Lambda_s &\simeq \int_0^\infty d\omega_c\, p(\omega_c)\int_0^T\int_0^T dt'\, dt'' v_i(t')v_i(t'')\cos\omega_c(t'-t'') \\ &= \int_0^T\int_0^T dt'\, dt''\, v_i(t')v_i(t'')\langle\cos\omega_c(t'-t'')\rangle \end{aligned} \tag{7.39}$$

Now suppose the observer has no a priori knowledge of ω_c except that it is somewhere between $\omega_{c0} - \Delta\omega_c/2$ and $\omega_{c0} + \Delta\omega_c/2$; then

$$\begin{aligned} \langle\cos\omega_c(t'-t'')\rangle &= \frac{1}{\Delta\omega_c}\int_{\omega_{c0}-\frac{\Delta\omega_c}{2}}^{\omega_{c0}+\frac{\Delta\omega_c}{2}} \cos\omega_c(t'-t'')\,d\omega_c \\ &= \cos\omega_{c0}(t'-t'')\operatorname{sinc}\left(\frac{\Delta\omega_c}{2}(t'-t'')\right) \end{aligned} \tag{7.40}$$

Now (7.39) is substituted into (7.38).

The effect of the function $\operatorname{sinc}\left(\frac{\Delta\omega_c}{2}(t' - t'')\right)$ is to remove contributions to the integral from regions in the t', t''-plane in which $|t' - t''|$ is appreciably greater than $2/|\Delta\omega_c|$. The diagram of Figure 7.5 illustrates this point.

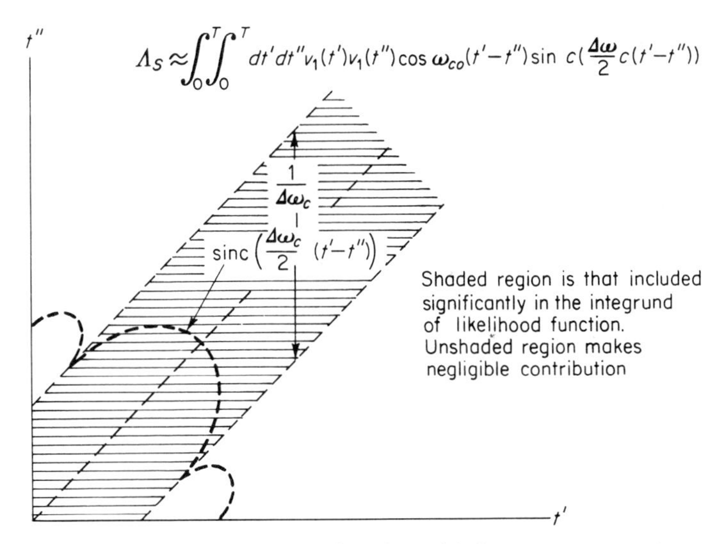

Figure 7-5 Likelihood function with frequency uncertainty.

In the extreme case where the size of the frequency uncertainty region $\Delta\omega_c$ is extremely large, then for practical purposes,

$$\operatorname{sinc}\left(\frac{\Delta\omega_c}{2}(t' - t'')\right) \simeq \pi\delta\left(\frac{\Delta\omega_c}{2}(t' - t'')\right) = \frac{2\pi}{\Delta\omega_c}\,\delta(t' - t'') \qquad (7.41)$$

and consequently,

$$\Lambda_s \simeq \frac{2\pi}{\Delta\omega_c}\int_0^T dt'\,[v_i(t')]^2 \qquad (7.42)$$

The result (7.42) tells us that, with extremely large a priori uncertainty in carrier frequency, the optimum detection scheme is to square the incoming signal and integrate *after* squaring. This is not too surprising when one considers the inefficiency of crosscorrelating the incoming signal with a reference sinusoid in the case where it is necessary to take a completely random guess at the proper frequency to use in the reference signal. Integration will not improve performance if the guess is markedly wrong, and might even degrade it. Therefore, it is advisable to square first, which in the case of low noise would be equivalent to crosscorrelation with a perfect reference signal. We can then draw on the noise reducing benefits of integration with the assurance that the reference signal has the correct frequency. The price to be paid for this assurance is an additional noise term, as indicated in Section 6.9 (See Eq. (6.93)). The problem of calculating error probabilities with the processing operation indicated in (7.42), a special case of the more general problem of square-law detection followed by video or audio filtering (see Section 3.1.4), has been treated by Kac and Siegert† and Emerson,‡ and is discussed by Davenport and Root.§ It will not be discussed further here. In certain limiting special cases, the problem degenerates into an extremely simple one. For example, if the bandwidth of the integrator is extremely small compared to the input noise bandwidth, the output statistics do not deviate substantially from Gaussian (see Section 3.5). In one such case, that where $\int_0^T dt'\, v^2(t')$ is computed and T (which corresponds roughly to reciprocal integrator bandwidth) far exceeds the reciprocal noise bandwidth, the PDF of the output could be very roughly approximated by||

$$p(v_0) \simeq \frac{1}{2\pi_u^2}\, e^{-(v_0-\langle v_0\rangle)^2/2\sigma_u^2} \tag{7.43}$$

where

$$\langle v_0\rangle = \sigma_n^2 T(1+\rho_i)$$
$$u = \sigma_n^2 \sqrt{\frac{T}{2B_n}}\sqrt{1+2\rho_i}$$

and where σ_n is the rms input noise, ρ_i the input SNR, and B_n the noise bandwidth.

In this case, the error probabilities would be approximated, again very roughly, by (7.13) and (7.14), with approximate changes in parameter nomenclature.

If the bandwidth of the integrator is very large compared with the input noise bandwidth, then the output PDF is roughly that of the square-law envelope detector, as given by (3.50). In this case, the error probabilities could be computed from (7.33.a) and (7.33.b), where the signal-plus-noise and noise alone PDF's entering into the integrands of (7.33.a) and (7.33.b), respectively, are given, with appropriate parameter changes, by (3.50) and by the form (3.50) would take with $a_s = 0$. This case is not of interest to us

† Kac and Siegert, 1947.

‡ Emerson, 1953.

§ Davenport and Root, 1958, Section 9.5, pp. 189–193.

|| To determine $\langle v_0\rangle$ and σ_u in (7.43) we calculate the mean and mean-square deviation of $\int_0^T dt'\, v^2(t')$, invoking (6.21) and (III.9) in the calculation of the latter. The calculation is easier if we use complex envelope concepts, noting that $v^2(t) = \{Re[(\hat{s}(t)+\hat{n}(t))e^{j\omega_c t}]\}^2$, $\langle \hat{n}(t')n^*(t'')\rangle \approx (\sigma_n^2/B_n)\delta(t'-t'')$, and $\langle|\hat{n}(t)|^2\rangle = 2\sigma_n^2$. See Section 6.2.

and will not be discussed further. The assumption behind it implies that no appreciable integration takes place after squaring, which in turn implies no processing operations designed to reduce noise. In practice, in a low SNR situation (i.e., near detection threshold), it would be undesirable to square the signal and then provide no noise reduction feature afterward. The squaring operation would only enhance the noise and effectively reduce the SNR.†

7.1.4.2 The Spectrum Analyzer

The squaring and integrating technique discussed above applies to a case where frequency uncertainty is quite large. Another method of dealing with frequency uncertainties of all magnitudes naturally suggests itself. Returning to (7.39), we note that an optimum detection technique for the case of a uniform PDF in frequency is to perform a noncoherent crosscorrelation process for a frequency ω_c and slowly search through all possible values of ω_c. An alternative process would be to construct a bank of noncoherent crosscorrelators to operate concurrently, each covering a frequency band of width $\Delta\omega_c/N$.

The single searching correlator scheme and the multiple correlator scheme are both special methods of implementing a spectrum analyzer. The conventional spectrum analyzer, of course, is designed to determine the power spectrum of a signal. In attempting to detect a signal whose spectral shape we do not know by means of frequency searching correlators or multiple correlators, we are doing nothing more than performing a spectral analysis of the signal using optimal filtering techniques (i.e., matched filtering or crosscorrelation), with a view toward locating the signal in frequency space. This point will be elaborated upon in Section 8.7.

To show that the single searching correlator procedure and the multiple correlator procedure are essentially equivalent in the theoretical sense, we represent the former by a quasi-static model; that is, the reference signal is centered at a given frequency ω_1 and the correlator operates for a period T_1, registers an output for that period, then operates at ω_2 for a period T_2, at ω_3 for a period T_3, etc. This is a crude mathematical model of a continuous

† In Section 6.9, we allude to the operation of square-law detection followed by integration. This is a special case of autocorrelation detection. In Eq. (6.89), the output SNR ρ_o is given as a function of the input SNR ρ_i for $\rho_i \ll 1$. Noting that $(\rho_o)_{\text{coh}}$ as used in (6.89) is proportional to ρ_i, it follows that ρ_o is proportional to ρ_i^2 for small input SNR. Thus, if ρ_i were very small, then ρ_o, being proportional to its square, would decrease rapidly with further decreases in ρ_i. This is known as *small signal suppression* in detectors, and it occurs also in detector types other than square-law. See also Davenport and Root, 1958, pp. 265–267.

frequency-searching crosscorrelation filter (which we could also call a sweeping spectrum analyzer).

Extrapolation from the results of an analysis based on such a model will provide an estimate of the quantitative behavior of a continuously searching crosscorrelator.

If the quasi-static searching process is carried out in such a manner that the intervals T_1, T_2, . . . etc., are greater than the noise correlation time, the correlator outputs for different intervals are uncorrelated and therefore are statistically independent if the noise is Gaussian (see Section 3.1.2). The noise correlation time is equivalent to the reciprocal of the (lowpass) noise bandwidth B_n. The intervals T_1, T_2 are assumed all equal and have the value T_r. The condition for approximate statistical independence is

$$B_n T_r \gtrsim 1 \tag{7.44}$$

or preferably

$$B_n T_r \gg 1 \tag{7.44$'$}$$

The integration time is roughly the reciprocal of the filter bandwidth B_f. Thus we can write (7.44) in the form

$$B_n T_r \simeq \frac{B_n}{B_f} \gtrsim 1 \tag{7.45}$$

In particular, suppose the filter is matched to frequency ω_1 for an interval $0 - t_1$ and uses a reference signal $\cos \omega_1 t$ during that interval. The filter produces a noise output $n_o(t_1)$ at the end of the interval. Then it is matched to frequency ω_2 for the interval $t_1 - t_2$, using a reference signal $\cos \omega_2 t$, and produces a noise output $n_o(t_2)$ at the end of that interval. The CCF between these two outputs is

$$\begin{aligned} R_o(t_1, t_2) &= \langle n_o(t_1) n_o(t_2) \rangle \\ &= \int_{t_1}^{t_2} \int_0^{t_1} dt'\, dt''\, \langle n_i(t') n_i(t'') \rangle \cos(\omega_1 t') \cos(\omega_2 t') \end{aligned} \tag{7.46}$$

If $t_1 = T_r$, $(t_2 - t_1) = T_r$, and the condition (7.44) holds (or particularly if the stronger condition (7.44)$'$ holds) then we can regard the noise inputs as approximately uncorrelated and it follows, since $n_i(t)$ has zero mean, that

$$R_o(t_1, t_2) = \int_{t_1}^{t_2} \int_0^{t_1} dt'\, dt''\, \langle n_i(t') \rangle \langle n_i(t'') \rangle \cos(\omega_1 t') \cos(\omega_2 t'') \simeq 0 \tag{7.47}$$

If the multiple correlator technique is used, then the noise *inputs* to the filters are identical and therefore perfectly correlated, but we can show that the noise *outputs* are mutually uncorrelated. In this case, we assume (7.44)$'$ and invoke its implication (6.21). Then (7.46) is replaced by

$$
\begin{aligned}
R_o(T_r, T_r) &= \int_0^{T_r} \int_0^{T_r} dt'\, dt''\, \langle n_i(t')n_i(t'')\rangle \cos(\omega_1 t') \cos(\omega_2 t'') \\
&= \frac{\sigma_n^2}{4B_n} \int_0^{T_r} dt' \,[\cos(\omega_1 - \omega_2)t' + \cos(\omega_1 + \omega_2)t'] \\
&= \frac{\sigma_n^2 T_r}{4B_n} \{\text{sinc}\,((\omega_1 - \omega_2)T_r) + \text{sinc}\,((\omega_1 + \omega_2)T_r)\} \\
&\simeq \frac{\sigma_n^2 T_r}{4B_n} \text{sinc}\,((\omega_1 - \omega_2)T_r)
\end{aligned} \tag{7.48}
$$

since $(\omega_1 + \omega_2) \gg (\omega_1 - \omega_2)$. The implication of (7.48) is that

$$R_o(T_r, T_r) \simeq 0 \quad \text{if} \quad (\omega_1 - \omega_2)T_r \gg \pi \tag{7.48)'$$

Equation (7.48)′ tells us that the CCF between the two correlator outputs corresponding to reference frequencies ω_1 and ω_2 becomes vanishingly small at intervals greatly in excess of $\omega_1 - \omega_2 = \pi/T_r$. This corresponds to a cycle per second frequency interval that is about $1/2T_r$ or $B_f/2$. Since the interfering noise is assumed to be Gaussian, if the two outputs differing in frequency by $1/2T_r$ are uncorrelated, they are also statistically independent.† Thus if we choose our reference signal frequencies at intervals $1/2T_r$, the outputs will be approximately independent.

From the above arguments, we conclude that the quasi-static search and the multiple correlator approaches both give approximately statistically independent outputs with different reference signal frequencies, provided (1) in the multiple correlator case, the frequencies are separated by an interval well above half the reciprocal of the correlator integration time, or equivalently, well above half the correlator bandwidth, and (2) in the case of the searching correlator, the search rate is sufficiently slow to guarantee the condition (7.44)′, or at least the weaker condition (7.44).

To calculate the error probabilities for the multiple correlator or searching correlator processes, we must first choose a decision rule. This is a more complicated matter here than in cases previously discussed, and the choice of decision rules is far from unique.

Suppose we are interested *only* in detecting the signal's presence, i.e., we don't care whether we correctly determine its frequency. In this case, we can choose a threshold $U_T = K_T\sqrt{\rho_i B_n T}$, decide "signal present" when at least one of the filter outputs is above U_T, and "no signal" if all outputs fall below U_T. The conditional false rest probability, then, is the conditional probability that, *if* a signal is present *somewhere* within the range of frequencies covered, all outputs are below U_T. This is

$$P_{frc} = \prod_{k=1}^{N} (P_{frc})_k \tag{7.49}$$

where $(P_{frc})_k$ is the conditional false rest probability for the case where the

† See Section 3.1.2, below Eq. (3.5).

signal is passed through the kth filter, i.e., the probability that the kth filter output falls below U_T if a signal is present somewhere within the passbands of the entire bank of filters.

The false alarm probability is the probability that, *if* no signal is present, at least one output is above U_T. This is equivalent to unity minus the conditional probability that all outputs are below U_T, if no signal is present anywhere; that is,

$$P_{fac} = 1 - \prod_{k=1}^{N} [1 - (P_{fac})_k] \tag{7.50}$$

where $(P_{fac})_k$ is the conditional false alarm probability for the kth filter channel.

Note that, according to (7.49), even if the output of a channel far from the true signal frequency were to be above U_T, we would call "signal present." Such an event could be caused by noise, i.e., there is a finite chance that noise in one of the filters could swing the output above U_T even with negligible signal energy within the passband of that filter.

Because a priori knowledge of phase is highly unlikely in a case where frequency is a priori unknown, we assume noncoherent detection, with the central frequencies of the filter passbands being separated by intervals of $W = 2\pi/T_r$ and the true signal frequency differing from the central frequency of the jth filter by an amount $\Delta\omega$. We then modify (7.35.a) and (7.35.b) to account for the frequency displacement, the modification being provided by (6.60). Then (7.46) and (7.47) assume the forms

$$\begin{aligned} P_{frc} = \prod_{k=1}^{(WT/2\pi)} \Bigg\{ & Q\left(\sqrt{4\rho_i B_n T \left| \operatorname{sinc} \frac{\Delta\omega T}{2} + \pi(k-j)\right|^2}, 0\right) \\ & - Q\left(\sqrt{4\rho_i B_n T \left| \operatorname{sinc} \frac{\Delta\omega T}{2} + \pi(k-j)\right|^2}, K_T\sqrt{\rho_i B_n T}\right)\Bigg\} \end{aligned} \tag{7.51.a}$$

$$P_{fac} = 1 - \prod_{k=1}^{(WT/2\pi)} \{1 - e^{-K_T^2 \rho_i B_n T/2}\} = 1 - \{1 - e^{-K_T^2 \rho_i B_n T/2}\}^{(WT/2\pi)} \tag{7.51.b}$$

Note that if the jth filter were tuned exactly to the true signal frequency (i.e., $\Delta\omega = 0$), then the kth filter output SNR would be proportional to $|\operatorname{sinc}(\pi(k-j))|^2$. But $\operatorname{sinc}(\pi(k-j)) = 1$ for $k = j$ and zero for $k \neq j$. In this case, we have

$$\begin{aligned} P_{frc} &= \{Q(0,0) - Q(0, K_T\sqrt{\rho_i B_n T})\}^{(WT/2\pi)-1} \ldots \\ & \quad \{Q(\sqrt{4\rho_i B_n T}, 0) - Q(\sqrt{4\rho_i B_n T}, \sqrt{\rho_i B_n T})\} \\ &= \{1 - Q(0, k_T\sqrt{\rho_i B_n T})\}^{(WT/2\pi)-1}\{1 - Q(\sqrt{4\rho_i B_n T}, \sqrt{\rho_i B_n T})\} \end{aligned} \tag{7.52}$$

(See footnote under Eq. (7.35).)

Figure 7-6 shows plots of P_{frc} and P_{fac} against $(W/2\pi)$ with K_T and $2\rho_i B_n T$ as parameters, in the case $\delta\omega = 0$ (i.e., Eqs. (7.51) and (7.52)).

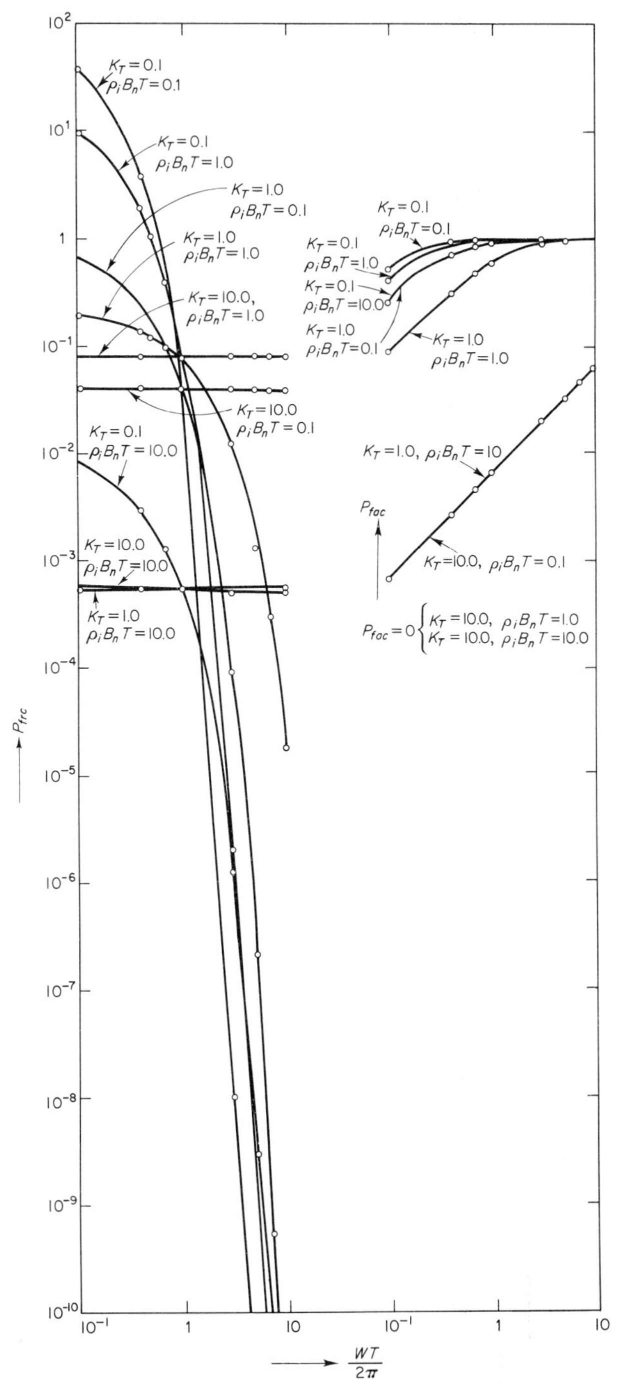

Figure 7-6 Conditional false rest and false alarm probabilities versus WT product—multiple or searching correlators with frequency uncertainty.

Comparison of these curves with those of Figures 7-3 and 7-4 shows that detection performance is degraded quite substantially from the noncoherent detection case where frequency is a priori known. It is also evident that the degradation increases with $\Delta\omega$, as would be expected.

7.1.5 Effect of Amplitude Uncertainty

It is evident that an uncertainty in the amplitude of an incoming sinusoidal signal affects the design of the optimum crosscorrelation detector only with respect to the detection threshold setting. This is true so long as the signal can be regarded as a sinusoid. In choosing the waveshape of the reference signal, it is only necessary that it match the incoming signal in frequency (and in the coherent detector case, also in phase); its amplitude is irrelevant. To convince ourselves of this, we examine the error probabilities for both coherent and noncoherent cases (Eqs. (7.13), (7.14), (7.35.a) and (7.35.b)). In both cases, the error probabilities are functions of the output SNR of the integrator and the detection threshold level. The amplitude of the reference signal, denoted by a_R, appears in both the signal output and noise output power in a scale factor a_R^2. This factor cancels out in the SNR and consequently does not affect the error probabilities, provided the detection threshold is chosen to be proportional to SNR rather than absolute signal level.

It is only when the radio signal is representable as a sine wave that these remarks are strictly valid. Over a long enough time scale, wherein the time variations of the RF or IF signal amplitude are substantial, it would be desirable to know these time variations and incorporate this knowledge into the design of the reference signal. In the case of pure detection (as opposed to waveform information extraction), it would be possible to match the reference signal in frequency using an arbitrary constant amplitude and still improve performance, although lack of information about the amplitude variations would degrade the attainable gain. This was discussed in Section 6.6 from the pure SNR point of view.

7.2 BINARY SIGNAL DETECTION

The problem of detecting the presence of a radio signal in Gaussian noise is a special case of that of distinguishing between two possible signals in Gaussian noise. The analysis in Section 4.1, up to the point in Section 4.1.3 where we specialize to the signal presence detection case, is applicable to this problem. In applying that discussion to radio signals, we will assume, as in Section 7.1, that the two signals are both narrowband RF functions whose amplitudes and phases are slowly varying compared with the carrier and whose central frequencies may be different.

The solution of this problem, known as binary detection, is vital in the analysis of binary digital communication systems, wherein one of two possible signals is transmitted. At the receiver, it must be decided which of the two signals was actually transmitted. These systems will be discussed in detail in Chapter 12.

From an extension of (4.17) and its consequences to the binary signal case, we can deduce that the optimum processing scheme is a likelihood ratio computer, which computes the ratio

$$\Lambda = \frac{\int d\boldsymbol{\beta}_1\, p(\boldsymbol{\beta}_1)\, e^{-\frac{1}{2\sigma_n^2}\sum_{i=1}^{N}[v-s_1(\boldsymbol{\beta}_1)]^2}}{\int d\boldsymbol{\beta}_2\, p(\boldsymbol{\beta}_2)\, e^{-\frac{1}{2\sigma_n^2}\sum_{i=1}^{N}[v-s_2(\boldsymbol{\beta}_2)]^2}} \simeq \frac{\int d\boldsymbol{\beta}_1\, p(\boldsymbol{\beta})\, e^{-\frac{1}{2\sigma_n^2}\int_0^T dt'[v(t')-s_1(t';\boldsymbol{\beta}_1)]^2}}{\int d\boldsymbol{\beta}_2\, p(\boldsymbol{\beta})\, e^{-\frac{1}{2\sigma_n^2}\int_0^T dt'[v(t')-s_2(t';\boldsymbol{\beta}_2)]^2}} \tag{7.53}$$

where $v(t)$, and $s_1(t\,;\boldsymbol{\beta}_1)$ and $s_2(t_2\,;\boldsymbol{\beta})$ are, respectively, the waveform actually received and the two desired signals. The sums can be regarded as equivalent to the 0 to T time integrals, by virtue of arguments presented in Section 4.1.3.1 (in the footnote below Eq. (4.26)). The vectors $\boldsymbol{\beta}_1$ and $\boldsymbol{\beta}_2$ represent the unknown parameters in $s_1(t\,;\boldsymbol{\beta}_1)$ and $s_2(t\,;\boldsymbol{\beta}_2)$, respectively, and $p(\boldsymbol{\beta}_1)$ and $p(\boldsymbol{\beta}_2)$ are PDF's for $\boldsymbol{\beta}_1$ and $\boldsymbol{\beta}_2$, respectively. The assumed signals are:

$$s_1(t) = a_1(t)\cos(\omega_1 t + \psi_1(t)) \tag{7.54.a}$$

$$s_2(t) = a_2(t)\cos(\omega_2 t + \psi_2(t)) \tag{7.54.b}$$

In this case,

$$\boldsymbol{\beta}_1 = (\mathbf{a}_1, \omega_1, \boldsymbol{\psi}_1), p(\boldsymbol{\beta}_1) = p(\mathbf{a}_1, \omega_1, \boldsymbol{\psi}_1) \tag{7.55.a}$$

$$\boldsymbol{\beta}_2 = (\mathbf{a}_2, \omega_2, \boldsymbol{\psi}_2), p(\boldsymbol{\beta}_2) = p(\mathbf{a}_2, \omega_2, \boldsymbol{\psi}_2) \tag{7.55.b}$$

where the vectors, $\mathbf{a}_1$, $\boldsymbol{\psi}_1$ and $\mathbf{a}_2$, $\boldsymbol{\psi}_2$ represent sets of sample values of the functions $a_1(t)$, $\psi_1(t)$, $a_2(t)$, and $\psi_2(t)$, and where the signal durations are long compared with the RF periods corresponding to ω_1 and ω_2, so that integrals whose integrands contain $\cos 2\omega_{1,2}t$ and $\sin 2\omega_{1,2}t$ vanish.

Using (7.55.a) and (7.55.b) in (7.53), we obtain

$$\Lambda = \frac{\int\cdots\int d\mathbf{a}_1\, d\omega_1\, d\boldsymbol{\psi}_1\, p(\mathbf{a}_1, \omega_1, \boldsymbol{\psi}_1)\, e^{\frac{1}{\sigma_n^2}\int_0^T dt' v(t')a_1(t')\cos(\omega_1 t'+\psi_1(t'))}\, e^{-\frac{1}{2\sigma_n^2}\int_0^T dt' a_1^2(t')}}{\int\cdots\int d\mathbf{a}_2\, d\omega_2\, d\boldsymbol{\psi}_2\, p(\mathbf{a}_2, \omega_2, \boldsymbol{\psi}_2)\, e^{\frac{1}{\sigma_n^2}\int_0^T dt' v(t')a_2(t')\cos(\omega_2 t'+\psi_2(t'))}\, e^{-\frac{1}{2\sigma_n^2}\int_0^T dt' a_2^2(t')}} \tag{7.56}$$

It is assumed that the components of $\mathbf{a}_1$ and $\mathbf{a}_2$ are statistically independent of those of $\boldsymbol{\psi}_1$ and $\boldsymbol{\psi}_2$ and both $\mathbf{a}_{1,2}$, and $\boldsymbol{\psi}_{1,2}$, are statistically independent of ω_1 and ω_2, i.e.,

$$p(\mathbf{a}_{\substack{1\\2}}, \omega_{\substack{1\\2}}, \boldsymbol{\psi}_{\substack{1\\2}}) = p(\mathbf{a}_{\substack{1\\2}})p(\omega_{\substack{1\\2}})p(\boldsymbol{\psi}_{\substack{1\\2}}) \tag{7.57}$$

from which it follows that (7.56) takes the form

$$\Lambda = \frac{\int \cdots \int d\mathbf{a}_1 \, e^{-\frac{1}{2\sigma_n^2}\int_{-\infty}^{\infty} dt' a_1^2(t')} p(\mathbf{a}_1)\, d\omega_1\, p(\omega_1)\, d\boldsymbol{\psi}_1\, p(\boldsymbol{\psi}_1)\, e^{\frac{1}{\sigma_n^2}\int_0^T dt' v(t') a_1(t') \cos(\omega_1 t' + \psi_1(t'))}}{\int \cdots \int d\mathbf{a}_2 \, e^{-\frac{1}{2\sigma_n^2}\int_{-\infty}^{\infty} dt' a_2^2(t')} p(\mathbf{a}_2)\, d\omega_2\, p(\omega_2)\, d\boldsymbol{\psi}_2\, p(\boldsymbol{\psi}_2)\, e^{\frac{1}{\sigma_n^2}\int_0^T dt' v(t') a_2(t') \cos(\omega_2 t' + \psi_2(t'))}} \tag{7.58}$$

7.2.1 Coherent Binary Detection

A special case of great interest is that wherein all parameters of both signals are a priori known, i.e., all PDF's are proportional to impulse functions. The assignment of PDF's follows from the generalization of Eq. (7.10) to include *both* signals $s_1(t)$ and $s_2(t)$. In this case,

$$\ln \Lambda = \frac{1}{4}\left(\int_0^T dt' \, [a_1^2(t') - a_2^2(t')]\right) + \frac{1}{\sigma_n^2}\left(\int_0^T dt' \, v(t') a_2(t') \cos(\omega_2 t' + \psi_2(t')) - \int_0^T dt' \, v(t') a_1(t') \cos(\omega_1 t' + \psi_1(t'))\right) \tag{7.59}$$

Recalling the discussions of Section 4.1 (concentrating especially on extension of the discussion in Section 4.1.3.1 to the case at hand), we can deduce that the indicated optimum processing operation is to drive the incoming waveform simultaneously through two coherent matched filter channels and compare the outputs. If minimization of the over-all error probability is the optimization criterion, and if the two signals are a priori equally likely, then the discussions of Section 4.1 imply that the optimal decision rule is:
Call $s_i(t)$ if

$$\int_0^T dt' \, v(t') s_2(t') - \int_0^T dt' \, v(t') s_1(t') > \frac{1}{2}\left[\int_0^T dt' \, s_2^2(t') - \int_0^T dt' \, s_1^2(t')\right] \tag{7.60.a}$$

Call $s_2(t)$ if

$$\int_0^T dt' \, v(t') s_2(t') - \int_0^T dt' \, v(t') s_1(t') < \frac{1}{2}\left[\int_0^T dt' \, s_2^2(t') - \int_0^T dt' \, s_1^2(t')\right] \tag{7.60.b}$$

The signal energies (see (6.32)) are given by

$$\begin{aligned} E_{s1} &= \int_0^T dt' \, s_1^2(t') \simeq \frac{1}{2}\int_0^T dt' \, a_1^2(t') \\ E_{s2} &= \int_0^T dt' \, s_2^2(t') \simeq \frac{1}{2}\int_0^T dt' \, a_2^2(t') \end{aligned} \tag{7.61}$$

The threshold that divides the $s_1(t)$ decision and the $s_2(t)$ decision is at the point

$$\int_0^T dt'\, v(t')s_1(t') - \int_0^T dt'\, v(t')s_2(t') = \frac{1}{2}(E_{s2} - E_{s1}) \tag{7.62}$$

If the two signals are equally energetic, then

$$E_{s1} = E_{s2} = E_s \tag{7.63}$$

and the threshold condition is

$$\int_0^T dt'\, v(t')s_1(t') = \int_0^T dt'\, v(t')s_2(t') \tag{7.64}$$

The decision rule in this case involves a simple differencing of two matched filter outputs, deciding $s_2(t)$ if the difference is positive, $s_1(t)$ if it is negative.

In some of the analysis of binary digital communication systems appearing in the literature, the assumption (7.63) is made at the outset. This excludes the case of on-off keying (see Chapter 12, Section 12.1.2). Otherwise, it covers many of the systems in common use. However, for the sake of completeness, we will not make this assumption, which is really not required for simplification of analysis in the coherent detection case.

The error probabilities for coherent detection are merely the probabilities that the difference between the two matched filter outputs falls into a region where the wrong signal would be indicated. They are given by

$$P_{12} = P(1) \int_{-\infty}^{\Delta E_s/2} p(\Delta V_o^{(1)})\, d(\Delta V_o^{(1)}) \tag{7.65.a}$$

$$P_{21} = P(2) \int_{\Delta E_s/2}^{-\infty} p(\Delta V_o^{(2)})\, d(\Delta V_o^{(2)}) \tag{7.65.b}$$

where

$P(1)$ = probability that $s_1(t)$ is the correct signal

$P(2)$ = probability that $s_2(t)$ is the correct signal

$V_{o1}^{(1)}$ = #1 filter output in the case where $s_1(t)$ is the correct signal

$V_{o1}^{(2)}$ = #1 filter output in the case where $s_2(t)$ is the correct signal

$V_{o2}^{(1)}$ = #2 filter output in the case where $s_1(t)$ is the correct signal

$V_{o2}^{(2)}$ = #2 filter output in the case where $s_2(t)$ is the correct signal

$\Delta V_o^{(1)} = V_{o1}^{(1)} - V_{o2}^{(1)}$

$\Delta V_o^{(2)} = V_{o1}^{(2)} - V_{o2}^{(2)}$

$p(\Delta V_o^{\binom{(1)}{(2)}})$ = PDF for $V_o^{\binom{(1)}{(2)}}$

$\Delta E_s = E_{s1} - E_{s2}$

$V_{o1}^{(1)} = S_{11} + N_{11}$

$V_{o2}^{(1)} = S_{12} + N_{12}$

$V_{o1}^{(2)} = S_{21} + N_{21}$

$V_{o2}^{(2)} = S_{22} + N_{22}$

$$S_{11} = \int_0^T dt'\, s_1(t')s_1(t') = E_{s1}$$

$$S_{12} = \int_0^T dt'\, s_1(t')s_2(t') = \sqrt{E_{s1}E_{s2}}\,\chi_{12}$$

$$\chi_{12} = \text{normalized CCF of } s_1(t) \text{ and } s_2(t)$$

$$= \frac{\int_0^T dt'\, s_1(t')s_2(t')}{\sqrt{\int_0^T dt'\, s_1^2(t') \int_{-\infty}^{\infty} dt''\, s_2^2(t'')}}$$

$$S_{21} = \int_0^T dt'\, s_2(t')s_1(t') = \sqrt{E_{s1}E_{s2}}\,\chi_{12} = S_{12}$$

$$S_{22} = \int_0^T dt'\, s_2(t')s_2(t') = E_{s2}$$

$$N_{11} = N_{21} = \int_0^T dt'\, n(t')s_1(t')$$

$$N_{12} = N_{22} = \int_0^T dt'\, n(t')s_2(t')$$

Since the filters are linear and the input noises are Gaussian, the outputs are Gaussian and so is the difference between the outputs (see Section 3.1.3). Consequently, for either superscript (1) or (2) we have

$$p(\Delta V_o) = \frac{e^{-\frac{[\Delta V_o - \langle \Delta V_o \rangle]^2}{2\langle [\Delta V_o - \langle \Delta V_o \rangle]^2 \rangle}}}{\sqrt{2\pi \langle [\Delta V_o - \langle \Delta V_o \rangle]^2 \rangle}} \tag{7.66}$$

where (because the noise inputs and therefore the noise outputs have zero mean)

$$\langle \Delta V_o^{(1)} \rangle = (S_{11} - S_{12}) + \langle N_{11} \rangle - \langle N_{12} \rangle = E_{s1} - \sqrt{E_{s1}E_{s2}}\,\chi_{12} \tag{7.67.a}$$

$$\langle \Delta V_o^{(2)} \rangle = (S_{21} - S_{22}) + \langle N_{21} \rangle - \langle N_{22} \rangle = \sqrt{E_{s1}E_{s2}}\,\chi_{12} - E_{s2} \tag{7.67.b}$$

$$\langle [V_o^{(1)} - V_o^{(1)}]^2 \rangle = \langle N_{11}^2 \rangle + \langle N_{12}^2 \rangle - 2\langle N_{11}N_{12} \rangle \tag{7.67.c}$$

$$\langle [V_o^{(2)} - V_o^{(2)}]^2 \rangle = \langle N_{21}^2 \rangle + \langle N_{22}^2 \rangle - 2\langle N_{21}N_{22} \rangle \tag{7.67.d}$$

where

$$\langle N_{11}^2 \rangle = \langle N_{21}^2 \rangle = \int_{-\infty}^{\infty} dt'\, dt''\, \langle n(t')n(t'') \rangle s_1(t')s_1(t'')$$

$$\langle N_{21}^2 \rangle = \langle N_{22}^2 \rangle = \int_{-\infty}^{\infty} dt'\, dt''\, \langle n(t')n(t'') \rangle s_2(t')s_2(t'')$$

$$\langle N_{11}N_{12} \rangle = \langle N_{21}N_{22} \rangle = \iint_{-\infty}^{\infty} dt'\, dt'' \langle n(t')n(t'') \rangle s_1(t')s_2(t'')$$

Assuming the condition (6.21),

$$\langle n(t')n(t'') \rangle \simeq \frac{\sigma_n^2}{2B_n}\, \delta(t' - t'') \tag{6.21}$$

$$\langle N_{11}^2 \rangle = \langle N_{21}^2 \rangle \simeq \frac{\sigma_n^2}{2B_n} E_{s1} \tag{7.68.a}$$

$$\langle N_{21}^2 \rangle = \langle N_{22}^2 \rangle \simeq \frac{\sigma_n^2}{2B_n} E_{s2} \tag{7.68.b}$$

$$\langle N_{11} N_{12} \rangle = \langle N_{21} N_{22} \rangle \simeq \frac{\sigma_n^2}{2B_n} \sqrt{E_{s1} E_{s2}}\, \chi_{12} \tag{7.68.c}$$

The mean-square deviations of $V_o^{(1)}$ and $V_o^{(2)}$ are

$$\begin{aligned} \langle [\Delta V_o^{(1)} - \Delta V_o^{(1)}]^2 \rangle &= \langle [\Delta V_o^{(2)} - \Delta V_o^{(2)}]^2 \rangle \\ &= \frac{\sigma_n^2}{2B_n} \{E_{s1} + E_{s2} - 2\sqrt{E_{s1} E_{s2}}\, \chi_{12}\} \end{aligned} \tag{7.69.a}$$

In the case of equal energies,

$$\langle [\Delta V_o^{\substack{(2)\\(1)}} - \Delta V_o^{\substack{(2)\\(1)}}]^2 \rangle = \frac{\sigma_n^2}{B_n} E_s(1 - \chi_{12}) \tag{7.69.b}$$

The error probabilities, easily determined from (7.65), (7.66), and (7.69), are

$$P_{12} = \frac{P(1)}{2} \left[1 - \operatorname{erf} \left(\frac{\sqrt{B_n}}{2\sigma_n} \sqrt{E_{s1} + E_{s2} - 2\chi_{12} \sqrt{E_{s1} E_{s2}}} \right) \right] \tag{7.70.a}$$

$$P_{21} = \frac{P(2)}{2} \left[1 - \operatorname{erf} \left(\frac{\sqrt{B_n}}{2\sigma_n} \sqrt{E_{s1} + E_{s2} - 2\chi_{12} \sqrt{E_{s1} E_{s2}}} \right) \right] \tag{7.70.b}$$

We return to the radio signal model (7.54) and observe that

$$E_{\substack{s1\\2}} = \int_0^T dt'\, s_{\substack{1\\2}}^2(t') = \overline{a_{\substack{1\\2}}^2}\, \frac{T_{\substack{1\\2}}}{2} \tag{7.71}$$

where $\overline{a_1^2}$ and $\overline{a_2^2}$ are the time averages of the squared amplitudes of $s_1(t)$ and $s_2(t)$ over the durations of the two signals T_1 and T_2. We define the input SNR's for the two signals as in (6.53),

$$\rho_{\substack{i1\\2}} = \frac{\overline{a_{\substack{1\\2}}^2}}{2\sigma_n^2} \tag{7.71$'$}$$

Now recalling from (6.53) that the output SNR's of the two matched filters are

$$\rho_{\substack{o1\\2}} = \rho_{\substack{i1\\2}}(2B_n T_{\substack{1\\2}}) \tag{7.71$''$}$$

we finally obtain, for Eqs. (7.70.a) and (7.70.b)

$$P_{12} = \frac{P(1)}{2} \left[1 - \operatorname{erf} \left(\frac{\sqrt{B_n}}{2} \sqrt{\rho_{i1} T_1 + \rho_{i2} T_2 - 2\sqrt{\rho_{i1} \rho_{i2} T_1 T_2}\, \chi_{12}} \right) \right] \tag{7.70.a$'$}$$

$$P_{21} = P_{12} \frac{P(2)}{P(1)} \tag{7.70.b$'$}$$

Two cases of great interest are:

1. Signal on or off: $s_2(t) = s(t)$, $s_1(t) = 0$; $\rho_{i1} = 0$; $\rho_{i2} = \rho_i$; $T_2 = T$; $E_{s1} = 0$, $E_{s2} = E_s$.
2. Signals of equal energy: $E_{s1} = \rho_{i1}T_1\sigma_n^2 = \rho_{i2}T_2\sigma_n^2 = E_{s2} = E_s = \rho_i T$.
   (a) Signals uncorrelated: $\chi_{12} = 0$.
   (b) Signals anticorrelated: $\chi_{12} = -1$.

For the first of these cases;

$$\frac{P_{12}}{P_{21}} = \frac{P(1)}{P(2)} \frac{1}{2}\left[1 - \operatorname{erf}\left(\frac{\sqrt{\rho_i B_n T}}{2}\right)\right] \tag{7.72}$$

Note that (7.72) is equivalent to (7.19) and (7.20) specialized to the case $K_T = \frac{1}{2}$. For Case (2)

$$\frac{P_{12}}{P_{21}} = \frac{P(1)}{P(2)} \frac{1}{2}\left[1 - \operatorname{erf}\left(\sqrt{\frac{\rho_i B_n T}{2}}\sqrt{1-\chi_{12}}\right)\right] \tag{7.73}$$

For subcase 2(a)

$$\frac{P_{12}}{P_{21}} = \frac{P(1)}{P(2)} \frac{1}{2}\left[1 - \operatorname{erf}\left(\sqrt{\frac{\rho_i B_n T}{2}}\right)\right] \tag{7.74}$$

and for subcase 2(b)

$$\frac{P_{12}}{P_{21}} = \frac{P(1)}{P(2)} \frac{1}{2}\left[1 - \operatorname{erf}\left(\sqrt{\rho_i B_n T}\right)\right] \tag{7.75}$$

Since by definition, χ_{12} must range between -1 and $+1$, the case of anticorrelated signals gives the highest possible value of the argument of the error function in (7.70). The error function erf (x) is a monotonic increasing function of its argument; therefore, if its argument is positive, the larger the argument the larger erf (x) is. The larger erf (x), the smaller is the quantity $1 - \operatorname{erf}(x)$. Thus the lowest possible error probabilities are attained with anticorrelated signals.

This is important in the design of binary digital communication systems. The two signals transmitted should be designed to be as nearly anticorrelated as consistent with other design constraints. More will be said of this in Chapter 12.

7.2.2 Noncoherent Binary Detection

If the signals are of the form (7.54), with the phases ψ_1 and ψ_2 completely unknown, then coherent detection of binary signals is no longer optimal. Noncoherent or envelope detection becomes optimal, just as in the restricted case of signal presence detection discussed in Section 7.1.

We consider the amplitude and phase as constants here, implying a time scale that is short compared to the time during which $a(t)$ and $\psi(t)$ change appreciably. The discussion in Section 7.1.3, applied to signal presence detection, can be generalized to the binary case. The phase PDF's are given by

$$p(\psi_1) = \frac{1}{2\pi}, \qquad 0 \le \psi_1 \le 2\pi \tag{7.76.a}$$

$$p(\psi_2) = \frac{1}{2\pi}, \qquad 0 \le \psi_2 \le 2\pi \tag{7.76.b}$$

The PDF's of amplitude and frequency as in Eq. (7.21), are impulse functions, indicating complete a priori knowledge of these parameters.

We can now write (7.56) in the form

$$\Lambda = e^{\frac{1}{2\sigma_n^2}\left(\frac{a_1^2}{2}T_1 - \frac{a_2^2}{2}T_2\right)} \frac{\int_0^{2\pi} d\psi_1\, e^{\frac{a_1}{\sigma_n^2}\sqrt{v^2_{c1}+v^2_{s1}}\cos\left(\psi_1 + \tan^{-1}\left[\frac{v_{s1}}{v_{c1}}\right]\right)}}{\int_0^{2\pi} d\psi_2\, e^{\frac{a_2}{\sigma_n^2}\sqrt{v^2_{c2}+v^2_{s2}}\cos\left(\psi_2 + \tan^{-1}\left[\frac{v_{s2}}{v_{c2}}\right]\right)}} \tag{7.77}$$

where

$$V_{c1} \equiv \int_0^{T_1} v(t') \cos \omega_1 t'\, dt'$$

$$V_{s1} \equiv \int_0^{T_1} v(t') \sin \omega_1 t'\, dt'$$

$$V_{c2} \equiv \int_0^{T_2} v(t') \cos \omega_2 t'\, dt'$$

$$V_{s2} \equiv \int_0^{T_2} v(t') \sin \omega_2 t'\, dt'$$

From (7.77) and (3.13)$'$, we have

$$U = \ln \Lambda = -\frac{\Delta E_s}{2\sigma_n^2} + \ln\left[I_0\left(\frac{a_1}{\sigma_n^2}\sqrt{V_{c1}^2 + V_{s1}^2}\right)\right] - \ln\left[I_0\left(\frac{a_2}{\sigma_n^2}\sqrt{V_{c1}^2 + V_{s1}^2}\right)\right] \tag{7.78}$$

where ΔE_s is defined, as in Eq. (7.65), as the difference between the signal energies E_{s1} and E_{s2}.

Again, as in Section 7.1.3, we consider the limiting cases of large and small SNR. For large SNR, we apply (7.36) to the present case and obtain

$$U \simeq -\frac{\Delta E_s}{2\sigma_n^2} + \left(\frac{a_1}{\sigma_n^2}\sqrt{V_{c1}^2 + V_{s1}^2} - \frac{a_2}{\sigma_n^2}\sqrt{V_{c2}^2 + V_{s2}^2}\right) \tag{7.79}$$

and for small SNR, through application of (7.37), we have

$$U \simeq \frac{1}{4}\left\{-\frac{2\Delta E_s}{\sigma_n^2} + \left(\frac{a_1}{\sigma_n^2}\sqrt{V_{c1}^2 + V_{s1}^2}\right)^2 - \left(\frac{a_2}{\sigma_n^2}\sqrt{V_{s2}^2 + V_{s2}^2}\right)^2\right\} \tag{7.80}$$

From (7.79) and (7.80) we conclude that the optimal detection techniques for (1) large and (2) small SNR, respectively, are: noncoherent matched filtering of both signals followed by (1) linear and (2) square-law envelope rectification, respectively. If the a priori probabilities of $s_1(t)$ and $s_2(t)$ are equal, if the ideal observer criterion is used (Section 4.1.2.1), and if the SNR is large, the optimum threshold that divides $s_1(t)$ and $s_2(t)$ decision spaces is that wherein the difference between the binary rectified outputs is $\Delta E_s/2$. Under the same conditions with small SNR, the optimum threshold value of the difference between quadratically rectified outputs is $2\,\Delta E_s$. Otherwise the two cases are, in principle, the same. They are essentially similar in practice as well, as was remarked in Section 7.1.3 in connection with signal presence detection. The optimal processing in either case involves computation of the absolute values of the two signal envelopes at the matched filter output. If the decision threshold is zero, as it would be if the two signals were equally energetic, it would make no theoretical difference in the decision process whether or not we squared the absolute values of the filter outputs. Note that $I_0(x)$ is a monotonic increasing function of x.† It is evident that all that would be required in general, whether SNR's were low, intermediate, or high, is to calculate some monotonic increasing function of $\sqrt{V_{c1}^2 + V_{s1}^2}$ and the same function of $\sqrt{V_{c2}^2 + V_{s2}^2}$ and take their difference. If the difference exceeds zero, then call $s_1(t)$; if not, call $s_2(t)$. This process would give the same error probabilities regardless of the precise form of the functions used, as long as they are monotonic increasing.

The error probability calculations are somewhat more complicated than for coherent detection. To simplify matters, we will specialize immediately to the case of equally energetic signals. This excludes only one important case in digital communications, that of on-off keying, which has already been treated in Section 7.1.3. A further simplification results from the assumption of orthogonal signals.

In general, without either of the above simplifying assumptions, the error probabilities are given by

$$P_{12} = P(1) \int_0^\infty d\,|V_{o1}^{(1)}| \int_{|V_{o1}^{(1)}|}^\infty d\,|V_{o2}^{(1)}|\, p(|V_{o1}^{(1)}|, |V_{o2}^{(1)}|) \tag{7.81.a}$$

$$P_{21} = P(2) \int_0^\infty d\,|V_{o2}^{(2)}| \int_{|V_{o2}^{(2)}|}^\infty d\,|V_{o1}^{(2)}|\, p(|V_{o1}^{(2)}|, |V_{o2}^{(2)}|) \tag{7.81.b}$$

where $p(|V_{o1}^{(j)}|, |V_{o2}^{(j)}|)$ is the joint PDF of the #1 and #2 filter output envelopes

$$|V_{o1}^{(j)}| = \sqrt{[V_{o1c}^{(j)}]^2 + [V_{o1s}^{(j)}]^2} \quad \text{and} \quad |V_{o2}^{(j)}| = \sqrt{[V_{o2c}^{(j)}]^2 + [V_{o2s}^{(j)}]^2}$$

the superscript (j) being 1 if $s_1(t)$ is the correct signal, 2 if $s_2(t)$ is the correct signal.

---

† Again, as cited in the footnote below Eq. 7.24, see Abramowitz and Stegun, 1965, Figure 9.7, p. 374.

If the assumptions of equal energies ($E_{s1} = E_{s2} = E_s$) and orthogonality of signals ($\chi_{12} = 0$) are invoked, the outputs $|V_{o1}|$ and $|V_{o2}|$ become statistically independent and Eqs. (7.8.a) and (7.81.b) assume the forms

$$P_{12} = P(1) \int_0^{\infty} d|V_{o1}^{(1)}|\, p(|V_{o1}^{(1)}|) \int_{|V_{o1}^{(1)}|}^{\infty} d|V_{o2}^{(1)}|\, p(|V_{o2}^{(1)}|) \tag{7.82.a}$$

$$P_{21} = P(2) \int_0^{\infty} d|V_{o2}^{(2)}|\, p(|V_{o2}^{(2)}|) \int_{|V_{o2}^{(2)}|}^{\infty} d|V_{o1}^{(2)}|\, p(|V_{o1}^{(2)}|) \tag{7.82.b}$$

where $p(|V_{o1}^{(j)}|)$ and $p(|V_{o2}^{(j)}|)$ are the PDF's of $|V_{o1}^{(j)}|$ and $|V_{o2}^{(j)}|$, respectively, given by (use (3.46) or (3.48) with appropriate changes in parameter nomenclature)

$$p(|V_{o1}^{(1)}|) = \frac{2B_n}{\sigma_n^2} \frac{|V_{o1}^{(1)}|}{E_s} I_0\left(\frac{2B_n}{\sigma_n^2} |V_{o1}^{(1)}|\right) e^{-\frac{B_n}{\sigma_n^2 E_s}(|V_{o2}^{(2)}|^2 + E_s^2)} \tag{7.83.a}$$

$$p(|V_{o2}^{(2)}|) = \frac{2B_n}{\sigma_n^2} \frac{|V_{o2}^{(2)}|}{E_s} I_0\left(\frac{2B_n}{\sigma_n^2} |V_{o2}^{(2)}|\right) e^{-\frac{B_n}{\sigma_n^2 E_s}(|V_{o1}^{(1)}|^2 + E_s^2)} \tag{7.83.b}$$

$$p(|V_{o1}^{(2)}|) = \frac{2B_n}{\sigma_n^2 E_s} |V_{o2}^{(2)}|\, e^{-\frac{B_n}{\sigma_n^2 E_s}|V_{o1}^{(2)}|^2} \tag{7.83.c}$$

$$p(|V_{o2}^{(1)}|) = \frac{2B_n}{\sigma_n^2 E_s} |V_{o1}^{(1)}|\, e^{-\frac{B_n}{\sigma_n^2 E_s}|V_{o2}^{(1)}|^2} \tag{7.83.d}$$

Equations (7.83.c) and (7.83.d) reflect the assumption of orthogonality between the two filter outputs, i.e., if $s_1(t)$ is the correct input signal, the #2 filter output is pure noise, and if $s_2(t)$ is the correct input signal then the output of #1 is pure noise.

Substituting Eqs. (7.83.a) through (7.83.d) into Eqs. (7.82.a) and (7.82.b) and using the definition of the Marcum Q-function given in (7.34), we obtain†

$$\begin{aligned}
\frac{P_{12}}{P(1)} = \frac{P_{21}}{P(2)} &= \int_0^{\infty} dx\, x I_0(\sqrt{2\rho_o}x)\, e^{-(x^2+2\rho_o)/2} \int_x^{\infty} dy\, y\, e^{-y^2/2} \\
&= \int_0^{\infty} dx\, x I_0(\sqrt{2\rho_o}x)\, e^{-(x^2+\rho_o)} = \frac{1}{2} e^{-\rho_o/2} \int_0^{\infty} dx'\, x' I_0(\sqrt{\rho_o}x')\, e^{-(x'^2+\rho_o)/2} \\
&= \frac{1}{2} e^{-\rho_o/2}\, Q\left(\sqrt{\frac{\rho_o}{2}}, 0\right) = \frac{1}{2} e^{-\rho_o/2}
\end{aligned} \tag{7.84}$$

where

$$\rho_o = \frac{B_n E_s}{\sigma_n^2} = \rho_i B_n T = \text{output SNR of either noncoherent filter}$$

The conditional error probabilities $P_{12}/P(1)$ or $P_{21}/P(2)$ as obtained from (7.84) are shown plotted as a function of $2\rho_i B_n T$ in Figure 7-7. In using the result (7.84) in later discussions of digital communication systems, it must

---

† See footnote following Eqs. (7.35).

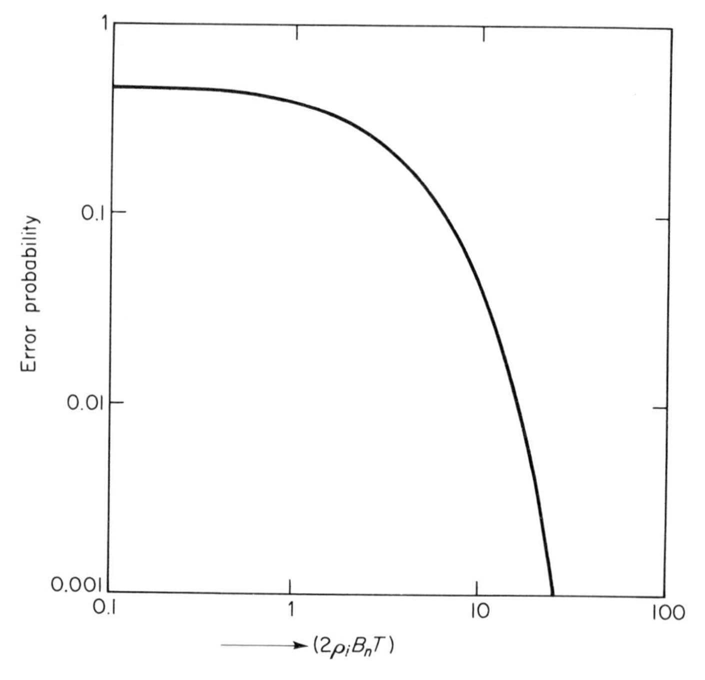

Figure 7-7 Conditional error probability versus output SNR —noncoherent binary detection—equal energy signals.

be remembered that it applies only to the case of orthogonal signals with equal energies.

7.3 MULTIPLE ALTERNATIVE (M'ARY) DETECTION

In some cases it is required to decide, in a background of noise, which of M possible signals is present at the receiver, where M exceeds 2. This is a task of inherently greater difficulty than binary detection. In determining which of two equally likely signals is present, we are asking for a single bit of information. In the M'ary case, where each of the M signals are equally likely, we request $\log_2 M$ bits of information in determining which one is present (see Section 3.6).

An approach† to M'ary detection theory, which will be invoked here, is to bypass the usual optimization via decision theory because it is mathematically cumbersome and to "generalize" intuitively from the decision-

† Kotelnikov, 1959, Chapter 3, pp. 20–24.

theoretic results for binary detection. According to these results, with Gaussian noise as the interfering agent, the optimum processor is the likelihood ratio computer. Extending this idea to the M alternative case, we would use Bayes' theorem (Eq. (2.3)) and compute the conditional probabilities that, *if* a given signal-plus-noise waveform is observed at the receiver, *then* the signal $s_m(t)$ is present, where the index m runs from 1 through M. Bayes' theorem tells us that

$$P(s_m/v_1, \ldots, v_N) = \frac{p(v_1, \ldots, v_N/s_m)P(s_m)}{p(v_1, \ldots, v_N)} \tag{7.85}$$

where

$p(v_1, \ldots, v_N/s_m)$ = conditional PDF for received signal-plus-noise waveform $v_1, \ldots, v_N$

$P(s_m)$ = probability that $s_m(t)$ is the correct signal

$P(s_m/v_1, \ldots, v_N)$ = conditional probability that $s_m(t)$ is the correct signal, given that $(v_1, \ldots, v_N)$ is observed

By the same reasoning as was used in the analogous problem in the binary case (see Section 7.2.1) we determine through (7.85) that the processor which computes $P(s_m/v_1, \ldots, V_N)$ is a set of coherent matched filters whose reference signals have the forms of the signals $s_m(t)$, $m = 1, \ldots, M$. Further generalization of the theory to include an a priori unknown phase of $s_m(t)$, would lead to noncoherent matched filters (see Section 7.2.2). If the noise were Gaussian, the same conclusions would follow from the application of maximum likelihood estimator theory, where the M possible signals are regarded as if they were M regions of a parameter to be estimated.

Let us consider first the case where signal phases are a priori unknown so that envelope detection follows matched filtering in the receiver. The error probability for M'ary detection is defined as the probability that the wrong signal is chosen at the receiver as the one actually present. Assuming that $s_m(t)$ is the true signal, the probability of successful detection is the probability that the output of the filter matched to $s_m(t)$ and the envelope detector that follows will exceed the envelope-detected outputs of the filters matched to the other signals $s_1(t), s_2(t), \ldots, s_{m-1}(t), s_{m+1}(t), \ldots, s_M(t)$. The error probability, by definition, is unity minus the success probability. If $(|V_{o1}|, \ldots, |V_{oM}|)$ are the amplitudes of the outputs of the filters matched to $(s_1, \ldots, s_M)$, then the error probability is

$$\begin{aligned} P_\varepsilon = 1 - \int_0^\infty d|V_{om}^{(m)}|\, p(|V_{om}^{(m)}|) \int_0^{|V_{om}{}^{(m)}|} d|V_{o1}^{(m)}|\, p(|V_{o1}^{(m)}|) \\ \int_0^{|V_{om}{}^{(m)}|} d|V_{o2}^{(m)}|\, p(|V_{o2}^{(m)}|) \cdots \int_0^{|V_{om}{}^{(m)}|} d|V_{o,m-1}^{(m)}|\, p(|V_{o,m-1}^{(m)}|) \\ \int_0^{|V_{om}|} d|V_{o,m+1}^{(m)}| \cdots \int_0^{|V_{om}|} d|V_{oM}^{(m)}|\, p(|V_{oM}^{(m)}|) \end{aligned} \tag{7.86}$$

In words, Eq. (7.86) tells us that (1) the probability of error is unity minus the probability of success, and (2) the probability of success is the product of the probabilities that all of the matched filter output amplitudes except the mth are smaller than the mth, integrated over all possible values of the mth output from 0 to ∞.

All of the PDF's in (7.86) except $p(|V_{om}|)$ are PDF's for the output of a filter matched to a signal different than the pure input signal. Hence they are all PDF's of the amplitude of pure Gaussian noise with rms value (see (6.53)) $\sigma_{no} = \frac{\sigma_n}{2}\frac{T}{B_n}$, σ_n being the rms noise input to the matched filter, T the integration time, and $2B_n$ the bandpass bandwidth of the input noise. The rms output noise values are the same for all the matched filters, since the integration times are the same. Any one of the inner integrals in (7.86), say the first, has the form (see Eqs. (7.83.c,d))

$$\int_0^{|V_{om}|} d|V_{01}|p(|V_{01}|) = \int_0^{|V_{om}|} d|V_{01}|\frac{|V_{01}|}{\sigma_{no}^2} e^{-|V_{01}|^2/2\sigma_{no}^2} = \{1 - e^{-|V_{om}|^2/2\sigma_{no}^2}\} \tag{7.87}$$

We note that the mth filter output, which contains both signal and noise, has a Rice distribution (Eqs. (7.83.a,b)). Using this fact and substituting (7.87) into (7.86), we obtain

$$P_\varepsilon = 1 - \int_0^\infty dx\, xI_0(\sqrt{2\rho_o}\, x)\, e^{-(x^2+2\rho_o)/2}[1 - e^{-x^2/2}]^{M-1} \tag{7.88}$$

(where ρ_o has the same meaning as in (7.84)). Applying a binomial expansion to the bracketed quantity, we have (using the special property of the Q-function given in the footnote below Eqs. (7.35) and defining ρ_o as in (7.84))

$$\begin{aligned} P_\varepsilon &= 1 - \sum_{\ell=0}^{M-1} \frac{(M-1)!(-1)^\ell}{\ell!(M-1-\ell)!}\left[\frac{1}{(\ell+1)} e^{-\rho_o\left[\frac{1}{\ell+1}\right]} Q\left(\sqrt{\frac{2\rho_o}{(\ell+1)}}, 0\right)\right] \\ &= 1 - \sum_{\ell=0}^{M-1} \frac{(M-1)!(-1)^\ell}{(\ell+1)!(M-1-\ell)!} e^{-\rho_o\left(\frac{\ell}{\ell+1}\right)} \end{aligned} \tag{7.89}$$

Note that (7.89) is equivalent to (7.84) in the case $M = 2$.

The noncoherent detection error probability as given by (7.89) is shown plotted against $(2\rho_i B_n T)$ for a few values of M in Figure 7-8. The curves show how error probability increases with M at fixed SNR for small values of M. Most of these results were obtained from Marcum's Q-function tables.† Curves showing error probability versus SNR for both coherent and non-coherent M'ary detection can be found in various other references.‡

---

† Marcum, 1950.

‡ For example, Jones, Pierce, and Stein, 1964, Figures 2.6-3, 2.6-4, p. 188, Figures 2.6-5 and 2.6-6, p. 194; Schwartz, Bennett, and Stein, 1966, Figure 2.7-4, p. 95 (from Kotelnikov, 1960, Figure 5-1, p. 49). These curves are in the context of M'ary digital signalling.

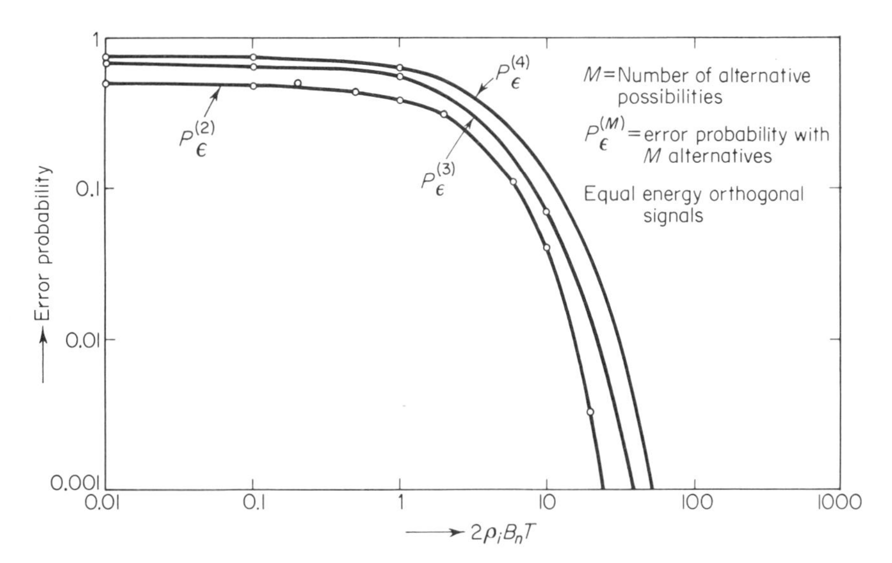

Figure 7-8 Error probability versus output SNR for multiple alternative detection (noncoherent).

If all signal phases are a priori known, so that coherent detection is feasible, then the filter outputs are not amplitudes of the matched filter outputs $|V_{om}|$, but the matched filter outputs V_{om} themselves.

In treating the coherent case, we can simplify matters by noting, as we did with coherent binary detection, that the *differences* between outputs of different filters are Gaussian. The probability of success is now the probability that none of the differences $(V_{oj} - V_{om})$ $(j = 1, \ldots, m-1, m+1, \ldots, M)$ will exceed zero in the case where the mth filter contains the signal. Therefore the analog of (7.86) is (where $\Delta V_{oj}^{(m)}$ now represents $(V_{oj}^{(m)} - V_{om}^{(m)})$, the difference between jth and mth filter outputs in the case where the mth signal is the correct one)

$$P_\varepsilon = 1 - \left[\int_{-\infty}^{0} d(\Delta V_{o1}^{(m)})p(\Delta V_{o1}^{(m)}) \ldots \int_{-\infty}^{0} d(\Delta V_{o,m-1}^{(m)})p(\Delta V_{o,m-1}^{(m)}) \int_{\infty}^{0} d(\Delta V_{o,m+1}^{(m)})p(\Delta V_{o,m+1}^{(m)}) \ldots \int_{\infty}^{0} d(\Delta V_{oM}^{(m)})p(\Delta V_{oM}^{(m)})\right] \tag{7.90}$$

The integrals within the square bracket (see (7.66)) are all of the form

$$\int_{-\infty}^{0} d(\Delta V_{oj})p(\Delta V_{oj}) = \frac{1}{\sqrt{2\pi\langle[\Delta V_{oj} - \langle\Delta V_{oj}\rangle]^2\rangle}} \int_{-\infty}^{0} d(\Delta V_{oj})\, e^{-\frac{[\Delta V_{oj} - \langle\Delta V_{oj}\rangle]^2}{2\langle[\Delta V_{oj} - \langle\Delta V_{oj}\rangle]^2\rangle}}$$
$$\frac{1}{2}\left\{1 + \text{erf}\left[\frac{\langle\Delta V_{oj}\rangle}{\sqrt{2\langle[\Delta V_{oj} - \langle\Delta V_{oj}\rangle]^2\rangle}}\right]\right\} \tag{7.91}$$

The analog of (7.88) and (7.89) in the coherent case, using (7.90) and (7.91) and the assumption that each of the quantities ΔV_{oj} and $\langle[\Delta V_{oj} - \Delta V_{oj}]^2\rangle$ is the same for all j, is

$$P_\varepsilon = 1 - \frac{1}{2^{M-1}}\left[1 + \operatorname{erf}\left(\frac{\langle \Delta V_{01}\rangle}{\sqrt{2\langle[\Delta V_{01} - \langle \Delta V_{01}\rangle]^2\rangle}}\right)\right]^{M-1} \tag{7.92}$$

Using the same definitions of the mean and variance of ΔV_{01} as were used in Section 7.2.1 in the analysis of coherent binary detection, it can easily be shown that (7.92) degenerates into the equivalent of Eqs. (7.70.a)′ and (7.70.b)′ in the case $M = 2$, $\rho_{i1} = \rho_{i2} = \rho_i$, $T_1 = T_2 = T$, $\chi_{12} = 0$. In the general M-alternative case, where $\rho_{ik} = \rho_i$, $T_k = T$ for all k and $\chi_{jk} = 0$ for all j,k, (7.92) degenerates into

$$P_{\underset{J}{\varepsilon}} = 1 - \frac{1}{2^{M-1}}\left[1 + \operatorname{erf}\left(\sqrt{\frac{\rho_i B_n T}{2}}\right)\right]^{M-1} \tag{7.92$'$}$$

For $M = 2$, (7.92)′ degenerates into the equivalent of (7.73).

The error probability P_ε obtained from (7.92)′ is shown plotted against $2\rho_i B_n T$ for a few small values of M in Figure 7-9. The increase in error probability with increasing M for a fixed value of SNR is observed to be significantly different in the coherent and noncoherent cases.

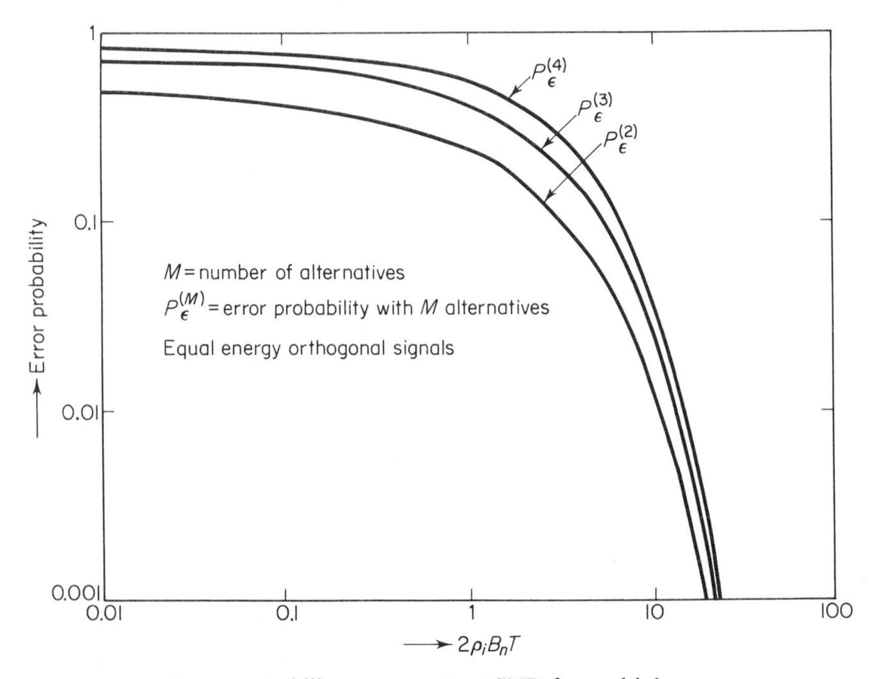

Figure 7-9 Error probability versus output SNR for multiple alternative detection (coherent).

Comparisons between binary detection and multiple alternative detection are important in digital communication systems analysis. Such comparisons will be discussed in Chapter 12 (Section 12.2.1) and will be based on the material covered in the present section and in Sections 7.2.1 and 7.2.2.

7.4 DETECTION OF FADING RADIO SIGNALS

Analysis of the detection of an incoming radio signal that is subject to fading requires an extension of the theory used in Sections 7.1 through 7.3.

For background on what follows, the reader is referred to the discussion of fading in Section 5.5.4.7 and the discussion of processing against fading in Section 6.10. In Section 6.10, the assumption of "slow fading" was made, i.e., the fading bandwidth was assumed small compared to the signal information bandwidth. The same assumption implies that, while the signal may still be regarded as a constant amplitude sinusoid in formulating and solving detection problems, the amplitude must now be regarded as a random variable at the receiver. To calculate error probabilities with fading signals, we must know the amplitude PDF associated with the fading. The error probabilities calculated in Sections 7.1.2, 7.1.3, 7.2.1, 7.2.2, and 7.3, which appear as functions of SNR, must now be averaged over the possible range of SNR's using the fading amplitude PDF in the averaging process.

Because of the "slow fading" assumption, this change in the error probabilities is the only modification that will be introduced in extending the theory to the fading case. Strictly speaking, modifications are also required in the optimization theory itself, i.e., the analysis that results in specification of a processing operation that minimizes the error probability (see Eqs. (7.11), (7.23), (7.53), and (7.77)). This can be seen in the case of signal-presence detection by returning to Eqs. (7.4) or (7.56) and observing that the amplitude a_i is a random variable in the fading case and hence we must include it in the components of the vector $\boldsymbol{\beta}$; thus, given a fading amplitude PDF, such as the Rayleigh PDF (Eq. (3.40)), it would be inserted into the integrands of (7.4) or (7.56) and the integration would then be carried out. This results, in general, in a different likelihood ratio computation. However, matched filtering or crosscorrelation will still be a part of the optimal processing dictated by the theory, provided the amplitude fluctuations due to fading are not so rapid that they affect the integrations in the exponentials of (7.4) or (7.56). Thus as long as the fading is sufficiently slow, we need not consider the change in the optimal processing filters, only in the calculation of error probabilities, assuming the same processing filters as would have been used in the nonfading case.

As a practical matter, it should be noted that the signal's phase as well as its amplitude is influenced by fading. Hence a knowledge of phase at the receiver cannot usually be assumed in a serious fading situation, and coherent detection is impractical. Therefore realistic calculations of error probabilities in the fading case usually presuppose a noncoherent detection technique.

The averaging out of the error probabilities over the distribution of SNR's leads to the following generic form for the error probability in the fading case:

$$P_\varepsilon = \int_0^\infty P_\varepsilon(\rho)p(\rho)\,d\rho \tag{7.93}$$

where $P_\varepsilon(\rho)$ represents the error probability as a function of the SNR, denoted by ρ, and $p(\rho)$ is the PDF of the SNR as determined by the nature of the fading.

Specializing immediately to the case of Rayleigh fading (see Section 5.5.4.7) to which our discussion will be restricted, we have, from (3.40),

$$p(\rho) = \frac{1}{\langle\rho\rangle} e^{-\rho/\langle\rho\rangle} \tag{7.94}$$

where $\langle\rho\rangle \equiv \int_0^\infty d\rho\, \rho p(\rho) =$ average SNR.

Substituting (7.94) into (7.93), we obtain

$$P_\varepsilon = \frac{1}{\langle\rho\rangle} \int_0^\infty d\rho\, P_\varepsilon(\rho)\, e^{-\rho/\langle\rho\rangle} \tag{7.95}$$

The simplest case to analyze is that of noncoherent binary detection as discussed in Section 7.2.2. Consider the case where the two alternatives are equally probable, i.e.,

$$P(1) = P(2) = \tfrac{1}{2} \tag{7.96}$$

Then from (7.84) and (7.96), the total error probability is

$$P_\varepsilon(\rho) = P(1)P_{21} + P(2)P_{12} = \tfrac{1}{2}e^{-\rho/2} \tag{7.97}$$

where the subscript o has been dropped from ρ_o.

From (7.95) and (7.97)

$$P = \frac{1}{2\langle\rho\rangle} \int_0^\infty d\rho\, e^{-\left[\frac{1}{\langle\rho\rangle} + \frac{1}{2}\right]\rho} = \frac{1}{[\langle\rho\rangle + 2]} \tag{7.98}$$

In Figure 7-10, error probability is shown plotted against $\langle\rho\rangle$ in the fading case and against ρ in the nonfading case, using (7.97) and (7.98) in the nonfading and fading cases, respectively. Unlike most of our error probability curves, those of Figure 7-10 show the SNR in decibels. The comparison of these two curves is illustrative of the way in which fading increases error probabilities for a given SNR. For example, noting that $\rho = \langle\rho\rangle$ in the nonfading case, we see from the curves that the error probability is $\simeq .5$ in the fading case, $\simeq .45$ in the nonfading case if $\rho = .1$; $\simeq .33$ and $.18$ in fading and nonfading cases, respectively, if $\rho = 10$, and $\simeq .14$ and $\simeq .003$ for fading and nonfading respectively if $\rho = 5$. For small average SNR, where $\rho \ll 1$, it makes little difference whether or not fading is present, because the error probability is high in either case and fading does not change its value substantially. For high SNR, however, the error probability is many times greater

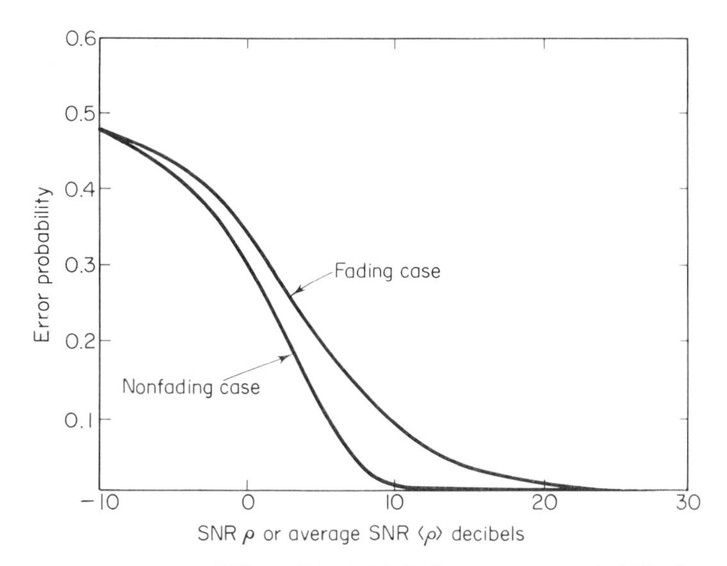

Figure 7-10 Effect of Rayleigh fading on error probability for noncoherent binary detection.

with fading than without it. This is due to the finite probability that a fade will drop the signal level into a low SNR region, although its average SNR is high.

Another way to compare fading and nonfading detection performance through Figure 7-10 is to consider the relative values of SNR required to attain a given error probability. For example, an error probability of .000023 or less requires that ρ be at least 10 dB in the absence of fading. To attain the same error probability with fading, it is required that $\langle \rho \rangle$ be at least 46 dB. Thus in this case, the presence of fading with $\langle \rho \rangle$ equal to 10 dB imposes a requirement that at least 36 dB of additional SNR be furnished in order that an error probability as low as .000023 be maintained.

There are cases other than that discussed above that are analytically tractable in general and still others that are tractable only in the case of very high SNR. For further discussion of error probabilities in the presence of fading, the reader is referred to the bibliography at the end of this chapter.

REFERENCES

See the reference list in Chapter 4 on detection and estimation theory. Also see the list in Chapter 3 on optimization of SNR and matched filters. Some of the following are somewhat pertinent to certain topics in Chapter 7.

Helstrom PGCS, 1960

Helstrom, 1960; Chapters IV, V, VI, IX

IRE, PGIT, June, 1960

Middleton, 1960; Part 4

Peterson, Birdsall, and Fox, 1954

Reich and Swerling, 1953

Binary Detection of Radio Signals

The following are in addition to references previously cited, especially in Chapter 4, that bear on this topic.

Helstrom, PGCS, 1960

Jones, Pierce, and Stein, 1964; Sections 2.1 through 2.5

Middleton, 1960; Chapters 19, 20

Reiger, 1953

Stein, 1964

Stein and Jones, 1967; Chapters 9, 10, 11, 12

Multiple Alternative Detection of Radio Signals

The following are in addition to references previously cited, especially in Chapter 4, that bear on this topic.

Helstrom, PGCS, 1960

Jones, Pierce, and Stein, 1964; Section 2.6, pp. 181–196

Kotelnikov, 1960; various parts of book, especially Chapter 5

Lucky and Hancock, 1962

Middleton, 1960; Section 23.1, pp. 1024–1045

Nuttall, 1962

Schwartz, Bennett, and Stein, 1966; Section 2.7, pp. 85–99

Stein and Jones, 1967; Chapter 14

Fading Signal Detection

Barrow, 1962

Bello and Nelin, June, 1962

Jones, Pierce, and Stein; Chapter II, Sections 2.9, 2.10

Kailath, 1960

Schwartz, Bennett, and Stein, 1966; Chapters 10, 11

Stein and Jones, 1967; Chapter 16

8

EXTRACTION OF INFORMATION FROM RADIO SIGNALS

The problem of extraction of information from a radio signal, or equivalently, measurement of parameters contained in the signal waveshape, is a different and generally more difficult one than that of mere detection of the signal. One rather important difference is that accurate information extraction usually requires a high input SNR while detection can often be accomplished with a low or moderate SNR. (These remarks do not apply in the absolute sense, because theoretically the SNR can always be increased indefinitely if enough integration time is available. However, in most practical situations there are strong limitations on available time, hence the SNR is also limited.)

The approach to the extraction problem to be presented here will be based on maximum likelihood estimator theory.† It is a specialization to radio signals in Gaussian noise of the theory presented in Section 4.2.

The model of a radio signal and the associated notation to be used in the discussions in this chapter is the same as that of Chapters 6 and 7, as given by Eqs. (6.2) or (7.1). The symbols $v(t)$, $s(t)$, and $n(t)$ represent signal-plus-noise, signal, and noise, respectively. The noise is assumed to be Gaussian. Subscript i represents input and subscript o represents output.

† The use of maximum likelihood estimator theory (MLE) here is very elementary and heuristic. There is a great deal more to MLE theory that would be of interest to communication theorists and is thoroughly covered in more sophisticated texts. The use of the theory here is to provide very rough guidelines in showing how this particular communication theory concept enters into the design philosophy behind techniques used to extract information from radio signals.

The discussions to follow will concern the application of some of the elementary ideas of maximum likelihood estimation theory† to general radio signal parameter measurement (Section 8.1), the measurement of amplitude (Section 8.2), phase (Section 8.3) and joint amplitude and phase (Section 8.4) of purely sinusoidal radio signals of duration T, the extraction of large amounts of information contained in the waveshape of an incoming radio signal (Section 8.5 and 8.6) and measurement of occurrence time and frequency composition of RF pulses (Section 8.7 and 8.8).

Fading effects will not be discussed in this chapter. The deleterious effects of fading on radio transmission and the way this problem is dealt with through diversity techniques were discussed in Sections 5.5.4.7 and 6.10. It will become apparent later in the present chapter that information extraction, like detection, is basically a question of SNR where Gaussian noise is the principal interfering agent. Thus many of the effects of fading on the quality of measurement of radio signal parameters can be deduced from a knowledge of the SNR degradation. The SNR improvement attainable through diversity can usually be translated directly into an improvement in the quality of parameter measurement.

8.1 MAXIMUM LIKELIHOOD ESTIMATION APPLIED TO RADIO SIGNALS IN GAUSSIAN NOISE

To apply maximum likelihood estimation theory to radio signals in Gaussian noise we return to Section 4.2.1, wherein the optimum measurement of a signal parameter β in a Gaussian noise background is discussed. To specialize the discussion to the problem at hand, we focus attention on Eq. (4.41) and the text that follows it. If the noise at the receiver input can be treated as a bandlimited white noise with bandpass bandwidth $2B_n$, then, as pointed out in Sections 2.4.4.2 and 3.6.2, its samples at intervals of $1/2B_n$ are uncorrelated and therefore statistically independent. This follows from the fact that the noise ACF is a sinc function whose zeroes occur at intervals of $1/2B_n$. The joint PDF for the samples of the incoming signal-plus-noise $v(t)$ at instants t_1, $t_1 + 1/2B_n$, $t_1 + 1/B_n$, . . . , $t_1 + N/2B_n$ is given by

$$p(v_i, \ldots, v_{Ni}\beta) = \frac{1}{[2\pi\sigma_n^2]^{N/2}} e^{-\frac{1}{2\sigma_n^2} \sum_{k=1}^{N} [v_k - s_i(t_k;\beta)]^2} \tag{8.1}$$

† There are other estimation criteria, e.g., minimum variance estimation, which will not be discussed in this book. In the practical sense, much of the processing dictated by other estimation theory becomes roughly equivalent to that of MLE theory whenever the interfering agent is Gaussian noise. The reader is referred to the bibliographies at the ends of Chapters 4 and 8 for further material on estimation theory.

where

$$v_k = v(t_k) = v\left(t_i + \frac{k}{2B_n}\right)$$

and

$$s_i(t_k\,;\boldsymbol{\beta}) = a_i(t_k\,;\boldsymbol{\beta})\cos\left(\omega_c(\boldsymbol{\beta})t_k + \psi_i(t_k\,;\boldsymbol{\beta})\right)$$

$a_i(t_k;\boldsymbol{\beta})$ and $\psi_i(t_k;\boldsymbol{\beta})$ being the slowly varying signal amplitude and phase, respectively, at time t_k, ω_c is the radio frequency, and $\boldsymbol{\beta}$ represents the signal parameters, at least some of which are a priori unknown and are to be measured.

According to Eq. (4.41), the maximum likelihood estimator (MLE) of one of the components of the vector $\boldsymbol{\beta}$, which we shall call β_k, is

$$\frac{\partial}{\partial\beta_k}p(v_1, \ldots, v_N\,;\boldsymbol{\beta}) = 0 \tag{8.2}$$

The specialization of Eq. (4.42) to this case is

$$\frac{\partial}{\partial\beta_k}\left\{e^{-\frac{1}{2\sigma_n^2}\sum_{k=1}^{N}[v_k - s_i(t_k\,;\boldsymbol{\beta})]^2}\right\}$$
$$= \left\{-\frac{1}{2\sigma_n^2}\sum_{k=1}^{N}2[v_k - s_i(t_k\,;\boldsymbol{\beta})]\frac{\partial s_i}{\partial\beta_k}(t_k\,;\boldsymbol{\beta})\right\}e^{-\frac{1}{2\sigma_n^2}\sum_{k=1}^{N}[v_k - s_i(t_k\,;\boldsymbol{\beta})]^2} \doteq 0 \tag{8.3}$$

or equivalently (since the exponential function is never zero)

$$\sum_{k=1}^{N}[v_k - s_i(t_k\,;\boldsymbol{\beta})]\frac{\partial s_i}{\partial\beta_k}(t_k\,;\boldsymbol{\beta}) = 0 \tag{8.4}$$

In our radio signal-plus-noise model, it has been assumed (see Eq. (6.20)) that the noise bandwidth is much greater than the signal bandwidth. The samples have been taken at intervals of the reciprocal of the bandpass bandwidth of the noise, which must be considerably smaller than the reciprocal signal bandwidth. But the reciprocal signal bandwidth is the Nyquist sampling rate (see Section 1.5) for the function $s_i(t, \boldsymbol{\beta})$ while the reciprocal noise bandwidth is the approximate Nyquist sampling rate for the function $v(t)$.† The bandwidth of the waveform $\partial s_i(t\,;\boldsymbol{\beta})/\partial\beta$ should be much closer to that of $s_i(t\,;\boldsymbol{\beta})$ than to that of the noise, and since the noise bandwidth is very much greater than the bandwidth of $s_i(t\,;\boldsymbol{\beta})$, it is also much greater than that of $\partial s_i/\partial\beta$. Thus the time function $[v(t)\partial s_i(t\,;\boldsymbol{\beta})/\partial\beta]$ has a bandwidth near that of the noise, while $[s_i(t\,;\boldsymbol{\beta})\partial s_i(t\,;\boldsymbol{\beta})/\partial\beta]$ has a bandwidth of an order of magnitude not much in excess of the signal bandwidth itself, or at least not as large as the noise bandwidth. By these qualitative arguments, we conclude that the summand in Eq. (8.4) when considered as a continuous

† Since $v(t) = s(t) + n(t)$, and the bandwidth of $n(t)$ far exceeds that of $s(t)$, the bandwidth of $v(t)$ is approximately that of $n(t)$.

function of time, has a bandwidth not far removed from that of the noise. Its Nyquist sampling rate is therefore not far from $1/2B_n$, and the sum can be approximated by an integral.† Equation (8.4), then, assumes the approximate form

$$\int_0^T dt' \, [v(t') - s_i(t\,;\boldsymbol{\beta})] \frac{\partial s_i}{\partial \beta_k}(t'\,;\boldsymbol{\beta}) \simeq 0 \tag{8.5}$$

or, accounting notationally for the fact that the unknown parameter β_k could be contained in either the amplitude a_i, the phase ψ_i, or the frequency ω_c, we can write

$$\begin{aligned}\int_0^T dt' \, v(t') \frac{\partial}{\partial \beta_k} [a_i(t'\,;\boldsymbol{\beta}) \cos(\omega_c(\boldsymbol{\beta})t' + \psi_i(t'\,;\boldsymbol{\beta}))] \\ = \int_0^T dt' \, a_i(t'\,;\boldsymbol{\beta}) \cos(\omega_c(\boldsymbol{\beta})t' + \psi_i(t'\,;\boldsymbol{\beta})) \\ \frac{\partial}{\partial \beta_k} [a_i(t'\,;\boldsymbol{\beta}) \cos(\omega_c(\boldsymbol{\beta})t' + \psi_i(t'\,;\beta))]\end{aligned} \tag{8.6}$$

Since $v(t)$ and T are not functions of β_k, Eq. (8.6) can be written in the form

$$\begin{aligned}\frac{\partial}{\partial \beta_k} \int_0^T dt' \, v(t') a_i(t'\,;\boldsymbol{\beta}) \cos(\omega_c(\boldsymbol{\beta})t' + \psi_i(t'\,;\boldsymbol{\beta})) \\ = \frac{1}{2} \frac{\partial}{\partial \beta_k} \left(\int_0^T dt' \, [a_i(t'\,;\boldsymbol{\beta})]^2 \cos^2(\omega_c(\boldsymbol{\beta})t' + \psi_i(t'\,;\boldsymbol{\beta})) \right)\end{aligned} \tag{8.6$'$}$$

Most of the remainder of this chapter will be devoted to various important specializations of Eq. (8.6) or (8.6)′.

8.2 OPTIMAL MEASUREMENT OF AMPLITUDE OF A SINUSOIDAL RADIO SIGNAL

8.2.1 Case of Known Phase

Consider the special case of (8.6) where the signal is a perfect sinusoid over the interval 0–T, i.e.,

$$\begin{aligned} a_i(t\,;\boldsymbol{\beta}) &= a = \text{constant}, \qquad 0 \le t \le T \\ \omega_c(\boldsymbol{\beta}) &= \omega_c = \text{constant}, \qquad 0 \le t \le T \\ \psi_i(t\,;\boldsymbol{\beta}) &= \psi = \text{constant}, \qquad 0 \le t \le T \end{aligned} \tag{8.7}$$

The unknown variable β_k is assumed to be the amplitude itself, i.e.,

$$\beta_k = a \tag{8.8}$$

while the other variables ψ and ω_c are assumed to be a priori known.

† See the footnote at the end of Section 4.1.3.1 for the justification of this step.

Then, substituting (8.7) and (8.8) into (8.6), we have

$$\begin{aligned}\int_0^T dt'\, v(t') \frac{\partial}{\partial a}[a\cos(\omega_c t' + \psi)] &= \int_0^T dt'\, v(t') \cos(\omega_c t' + \psi) \\ &= \int_0^T dt'\, a\cos(\omega_c t' + \psi)\frac{\partial}{\partial a}[a\cos(\omega_c t' + \psi)] \\ &= \frac{a}{2}\left[T + \int_0^T dt' \cos(2\omega_c t' + 2\psi)\right]\end{aligned} \tag{8.9}$$

As in Chapters 6 and 7, the assumption

$$T \gg \frac{1}{\omega_c} \tag{8.10}$$

assures us that the integral involving $\cos(2\omega_c t' + 2\psi)$ is negligible, resulting in,

$$\begin{aligned} a_{\text{MLE}} &= \text{maximum likelihood estimator of } a \\ &= \frac{2}{T}\int_0^T dt'\, v(t') \cos(\omega_c t' + \psi)\end{aligned} \tag{8.11}$$

i.e., the MLE of the amplitude is simply twice the time average of the product of the incoming signal and a sinusoidal reference waveform designed to match the incoming signal in amplitude and phase.

Thus, as in the case of SNR maximization and optimum detection, as long as additive Gaussian noise is the unwanted disturbance, the optimum processing of a signal whose frequency and phase are a priori known involves a crosscorrelator operation. (Optimum processing in this case means optimum measurement of amplitude.)

To illustrate the meaning of this method of measuring amplitude, consider the noise-free case, in which

$$v(t) = s_i(t) = a\cos(\omega_c t + \psi) \tag{8.12}$$

Substituting (8.12) into (8.11) and invoking (8.10) results in

$$a_{\text{MLE}} = \frac{2a}{T}\int_0^T dt' \cos^2(\omega_c t' + \psi) \simeq a \tag{8.13}$$

The addition of a noise term $n_i(t)$ to the right-hand side of (8.12) and substitution into (8.13) yields

$$a_{\text{MLE}} = a + \frac{2}{T}\int_0^T dt'\, n_i(t') \cos(\omega_c t' + \psi) \tag{8.14}$$

The rms fractional error in measurement of amplitude, from (8.14), is

$$\begin{aligned}\frac{\mathcal{E}_a}{a} &\equiv \frac{\sqrt{\langle (a_{\text{MLE}} - a)^2 \rangle}}{a} \\ &= \frac{2}{aT}\int_0^T \int_0^T dt'\, dt''\, \langle n_i(t') n_i(t'') \rangle \cos(\omega_c t' + \psi)\cos(\omega_c t'' + \psi)\end{aligned} \tag{8.15}$$

Since the noise has been assumed bandlimited white with a bandwidth greatly in excess of $1/T$, we can make use of the approximation (6.21) in (8.15), resulting in

$$\frac{\mathcal{E}_a}{a} \simeq \frac{2}{aT}\sqrt{\frac{\sigma_n^2}{2B_n}\int_0^T dt' \cos^2(\omega_c t' + \psi)} \simeq \frac{\sigma_n}{a\sqrt{B_n T}} = \frac{1}{\sqrt{2\rho_i B_n T}} \tag{8.16}$$

where ρ_i is the input SNR, equal to $(a_i^2/2\sigma_n^2)$.

From (8.16), we see that the fractional error is the reciprocal of the square root of the quantity $(2\rho_i B_n T)$, which, as we will recall from Eq. (6.53), is the output SNR of a coherent matched filter. This would suggest that the fractional error might be inversely proportional to the square root of the SNR of any linear bandpass filter we might have used on the signal. It can be shown that this is the case, by repeating the calculation (8.16) for other filters. This is left as an exercise for the reader.

The degradation in accuracy resulting from the use of a bandpass filter that is not quite a perfect matched filter, e.g., an *RLC* filter of the same bandwidth, will in some cases be very small.†

Another criterion of the quality of the optimum amplitude measurement characterized by (8.11) is the probability that the indicated amplitude at the output of the measurement is within some specified region centered at the true amplitude. If the true amplitude is a, the probability that a_{MLE} is between $a - (\Delta a/2)$ and $a + (\Delta a/2)$ is given by

$$p(\Delta a; a) = \int_{a-(\Delta a/2)}^{a+(\Delta a/2)} p(v)\, dv \tag{8.17}$$

where $p(v)$ is the PDF for the filter output.

With Gaussian noise at the input, the filter output noise is also Gaussian (see Section 3.1.4).

The mean voltage at the filter output is

$$v_0(t) = s_0(t) + n_0(t) \tag{8.18}$$

where

$$\langle s_0(t)\rangle = \frac{2}{T}\int_0^T a\cos^2(\omega_c t' + \psi)\, dt' \simeq a$$

$$\langle n_0(t)\rangle = \frac{2}{T}\int_0^T dt' \langle n_i(t')\rangle \cos(\omega_c t' + \psi) \simeq 0$$

The mean-square output is

$$\begin{aligned}\langle [v_0(t)]^2 &= [s_0(t)]^2 + \langle [n_0(t)]^2\rangle \\ &= a^2 + \frac{4}{T^2}\int_0^T\int_0^T dt'\, dt'' \langle n_i(t')n_i(t'')\rangle \cos(\omega_c t' + \psi)\cos(\omega_c t'' + \psi) \\ &\simeq a^2\left[1 + \frac{1}{2\rho_i B_n T}\right]\end{aligned} \tag{8.19}$$

† A particularly good discussion of this point in the context of communication systems can be found in Jones, Pierce, and Stein, 1964; Section 2.5.5, pp. 175–178.

The mean-square deviation of the output is

$$\begin{aligned}\langle [v_0(t) - v_0(t)]^2 \rangle &= \langle [v_0(t)]^2 \rangle - [\langle v_0(t) \rangle^2] \\ &= a^2 \left[1 + \frac{1}{2\rho_i B_n T} \right] - a^2 \\ &= \frac{a^2}{2\rho_i B_n T} = \frac{\sigma_n^2}{B_n T} \equiv \sigma_{no}^2\end{aligned} \tag{8.20}$$

The output PDF is

$$p(v) = \frac{1}{2\pi\sigma_{no}^2} e^{-(v-a)^2/2\sigma^2_{no}} \tag{8.21}$$

Hence, from (8.17) and (8.21),

$$p(\Delta a\,; a) = \frac{1}{\sqrt{2\pi\sigma_{no}^2}} \int_{a-(\Delta a/2)}^{a+(\Delta a/2)} dv\, e^{-(v-a)^2/2\sigma^2_{no}} = \operatorname{erf}\left(\frac{1}{2} \frac{\Delta a}{a} \sqrt{\rho_i B_n T} \right) \tag{8.22}$$

Plots of $P(\Delta a_i a)$ versus $x = (\Delta a/a)(10^2)$ for various values of $\rho_i B_n T$ are shown in Figure 8-1. The probabilities increase significantly with increases in $\rho_i B_n T$, as would be expected. For high SNR (e.g., $\rho_i B_n T = 100$), the probability that the amplitude falls within $\Delta a/2$ of the true amplitude increases very rapidly with $\Delta a/a$ until a certain value is reached (in the

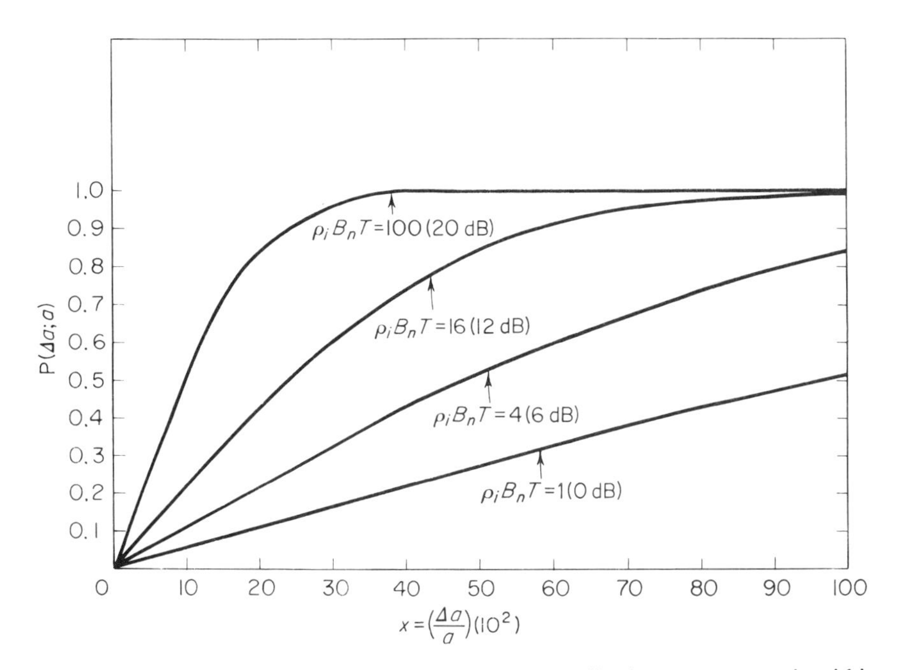

Figure 8-1 Probability that amplitude measurement is within x percent of true value versus percentage error.

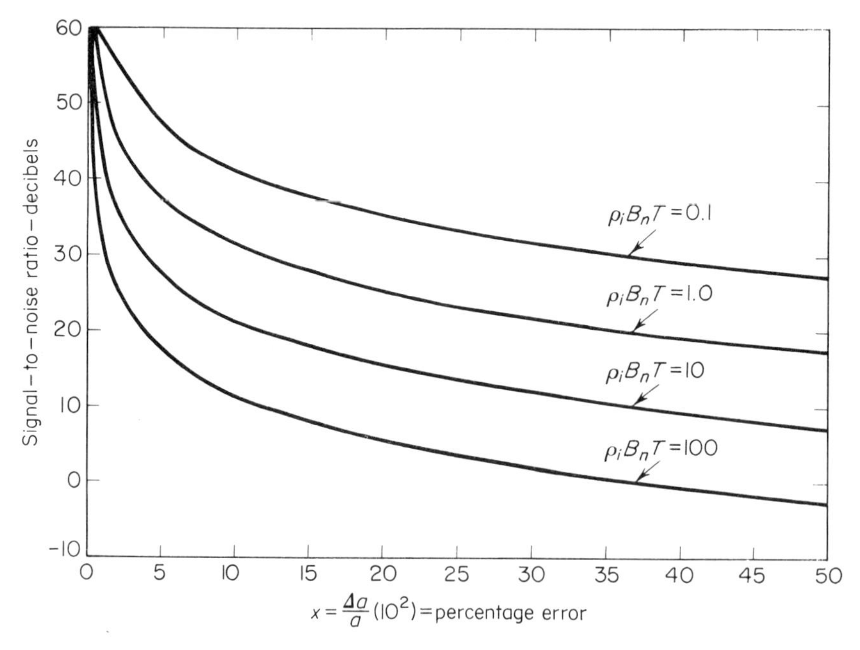

Figure 8-2 SNR consistent with a given percentage error.

specific case shown, the value is about 40 percent), then the curve saturates. Beyond this point, the probability is so nearly unity that there is no significantly greater chance of attaining a measurement within, say, 60 percent of the true amplitude than that of attaining 40 percent of the true amplitude.

The input SNR required for a given value of $(\Delta a/a)$ is shown plotted as a function of $(\Delta a/a)$ in Figure 8-2. The abscissa on the figure is designated as the error. Error is here defined as the percentage of the true amplitude within which the indicated amplitude is 99 percent certain to fall. In different terms, one specifies that the amplitude indicated at the output of the measurement has a 99 percent probability of being within a certain percentage of the true amplitude. That percentage is defined as the error. The figure shows how the input SNR corresponding to a given error decreases with the error, and how the required input SNR for a given error decreases with an increase in the time-bandwidth product.

8.2.2 Case of Unknown Phase

In many applications it is quite unrealistic to assume an a priori knowledge of phase in the measurement of the amplitude of a sinusoidal radio signal. To account for this phase ignorance, we can reformulate the MLE principle, writing (8.3) in the form

$$\frac{\partial}{\partial a}\left\{\int_0^{2\pi} e^{-\frac{1}{2\sigma_n^2}\sum_{k=1}^{N}[v_k - a\cos(\omega_c t_k+\psi)]^2}\,d\psi\right\} = 0 \tag{8.23}$$

The quantity we try to maximize is now the *average* of the joint PDF of amplitude and phase, where the averaging is over all possible phase angles between zero and 2π. This average is simply the PDF of amplitude.

Approximating the sum with an integral† and using the assumption (8.10), we obtain from (8.23)

$$\begin{aligned}&\frac{\partial}{\partial a}\int_0^{2\pi} d\psi\, e^{\frac{a}{\sigma_n^2}\int_0^T dt'\, v(t')\cos(\omega_c t'+\psi) - \frac{a^2}{2\sigma_n^2}\int_0^T dt'\cos^2(\omega_c t'+\psi)}\\ &\simeq \frac{\partial}{\partial a}\left\{\int_0^{2\pi} d\psi\, e^{\frac{a}{\sigma_n^2}\sqrt{v_c^2+v_s^2}\cos\left(\psi+\tan^{-1}\left[-\frac{v_s}{v_c}\right]\right)-\frac{a^2T}{4\sigma_n^2}}\right\} = 0\end{aligned} \tag{8.24}$$

where

$$v_c \equiv \int_0^T dt'\, v(t')\cos\omega_c t'$$

$$v_s \equiv \int_0^T dt'\, v(t')\sin\omega_c t'$$

Integrating with respect to phase from 0 to 2π in (8.24), we obtain

$$\frac{\partial}{\partial a}\left\{e^{-(a^2T/4\sigma_n^2)}\, I_0\!\left(\frac{a}{\sigma_n^2}\sqrt{v_c^2+v_s^2}\right)\right\} = 0 \tag{8.25}$$

where $I_0(x)$ is the modified Bessel function of zero order (see Eq. (3.45)′).

A reasonably high SNR is required for accurate measurement. We will assume a high SNR in the discussion to follow. Hence, we will use the asymptotic form of $I_0(x)$ for large argument,‡ i.e.,

$$I_0(x) \simeq \frac{e^x}{\sqrt{2\pi x}} \tag{8.26}$$

and therefore

$$\begin{aligned}\frac{\partial}{\partial a}\left\{e^{-(a^2T/4\sigma_n^2)}\, I_0\!\left(\frac{a}{\sigma_n^2}\sqrt{v_c^2+v_s^2}\right)\right\} &\simeq \frac{\sigma_n^2}{\sqrt{2\pi(v_c^2+v_s^2)}}\frac{\partial}{\partial a}\left\{\frac{e^{-\frac{a^2T}{4\sigma_n^2}+\frac{a}{\sigma_n^2}\sqrt{v_c^2+v_s^2}}}{\sqrt{a}}\right\}\\ &= \frac{\sigma_n^2 e^{-\frac{a^2T}{4\sigma_n^2}+\frac{a}{\sigma_n^2}\sqrt{v_c^2+v_s^2}}}{\sqrt{2\pi(v_c^2+v_s^2)}}\left\{a^{-1/2}\left[-\frac{aT}{2\sigma_n^2}+\frac{\sqrt{v_c^2+v_s^2}}{\sigma_n^2}\right]-\frac{1}{2}a^{-3/2}\right\} = 0\end{aligned} \tag{8.27}$$

For a large SNR, it follows that

$$a \simeq \frac{2}{T}\sqrt{v_c^2+v_s^2} = a_{\text{MLE}} \tag{8.28}$$

Equation (8.28) tells us that the MLE of amplitude, in the case where phase is a priori unknown and the input SNR is sufficiently high, is the

† Again supported by the argument in the footnote at the end of Section 4.1.3.1.

‡ See Goldman, 1948, Appendix E, p. 420.

output of a noncoherent matched filter applied to the incoming signal (see Section 6.4). This is perfectly reasonable in the light of the discussions in Chapter 6. If phase is unknown and we want to apply matched-filtering to a sinusoidal radio signal, we simply envelope-detect the matched-filter output. This result appeared in Section 7.1.3 in connection with signal detection. It is no surprise that it should also appear in the context of signal amplitude measurement, since the optimal filtering technique is the same in both cases.

The noise-free output is easily evaluated here. In the absence of noise, the outputs v_c and v_s are

$$v_{cs} = \frac{aT}{2} \cos \psi \tag{8.29.a}$$

$$v_{ss} = \frac{aT}{2} \sin \psi \tag{8.29.b}$$

$$\sqrt{v_{cs}^2 + v_{ss}^2} = \frac{aT}{2} \tag{8.29.c}$$

$$a_{\text{MLE}} = \frac{2}{T} \sqrt{v_{cs}^2 + v_{ss}^2} = a \tag{8.29.d}$$

Again, as in Section 8.2.1, in the absence of noise, a_{MLE} turns out to be the true amplitude.

The rms fractional error in amplitude measurement is

$$\begin{aligned}\frac{\mathcal{E}_a}{a} &= \frac{\sqrt{\langle (a_{\text{MLE}} - a)^2 \rangle}}{a} \\ &= \frac{1}{a}\sqrt{\left\langle \left[\frac{2}{T}\sqrt{\frac{a^2T^2}{4} + 2(v_{cs}v_{cn} + v_{ss}v_{sn}) + (v_{cn}^2 + v_{sn}^2)} - a\right]^2 \right\rangle} \\ &= \sqrt{\left\langle \left[\sqrt{1 + \frac{4}{a^2T^2}[aT(v_{cn}\cos\psi + v_{sn}\sin\psi) + (v_{cn}^2 + v_{sn}^2)]} - 1\right]^2 \right\rangle}\end{aligned} \tag{8.30}$$

where

$$v_{cn} \equiv \int_0^T dt'\, n(t') \cos \omega_c t'$$

$$v_{sn} \equiv \int_0^T dt'\, n(t') \sin \omega_c t'$$

For high SNR, we can use the power series expansion for $\sqrt{1 + x}$ and neglect all but the sum of the first two terms $(1 + x/2)$, resulting in

$$\begin{aligned}\frac{\mathcal{E}_a}{a} &\simeq \sqrt{\left\langle \left\{1 + \frac{2}{a^2T^2}[aT(v_{cn}\cos\psi + v_{sn}\sin\psi) + (v_{cn}^2 + v_{sn}^2)] - 1\right\}^2 \right\rangle} \\ &\simeq \frac{2}{a^2T^2}\sqrt{\begin{aligned}&a^2T^2(\langle v_{cn}^2\rangle \cos^2\psi + \langle v_{sn}^2\rangle \sin^2\psi + 2\langle v_{cn}v_{sn}\rangle \cos\psi \sin\psi) \\ &+ (\langle v_{cn}^4\rangle + \langle v_{sn}^4\rangle + 2\langle v_{cn}^2 v_{sn}^2\rangle) + 2aT(\cos\psi\,[\langle v_{cn}^3\rangle + \langle v_{cn}v_{sn}^2\rangle] \\ &+ \sin\psi\,[\langle v_{sn}v_{cn}^2\rangle + \langle v_{sn}^3\rangle])\end{aligned}}\end{aligned} \tag{8.31}$$

From properties of Gaussian noise given in Appendix III,

$$\langle v_{cn}\rangle = \langle v_{sn}\rangle = \langle v_{cn}v_{sn}\rangle = \langle v_{cn}v_{sn}^2\rangle = \langle v_{sn}v_{cn}^2\rangle = \langle v_{cn}^3\rangle = \langle v_{sn}^3\rangle = 0 \tag{8.32.a}$$

$$\langle v_{cn}^2\rangle = \langle v_{sn}^2\rangle = \frac{\sigma_n^2 T}{4B_n} \tag{8.32.b}$$

$$\langle v_{cn}^4\rangle = \langle v_{sn}^4\rangle = \frac{3\sigma_n^4 T^2}{16B_n^2} \tag{8.32.c}$$

$$\langle v_{cn}^2 v_{sn}^2\rangle = \frac{\sigma_n^4 T^2}{16B_n^2} \tag{8.32.d}$$

Substitution of Eqs. (8.32) into (8.31) results in

$$\frac{\mathcal{E}_a}{a} \simeq \frac{2}{a^2T^2}\sqrt{\frac{a^2T^3\sigma_n^2}{4B_n} + \frac{\sigma_n^4T^2}{2B_n^2}} = \frac{1}{\sqrt{2\rho_i B_n T}}\sqrt{1 + \frac{1}{\rho_i B_n T}} \tag{8.33}$$

For high SNR, where the second term inside the radical in (8.33) can be neglected, we obtain

$$\frac{\mathcal{E}_a}{a} \simeq \frac{1}{\sqrt{2\rho_i B_n T}} \tag{8.34}$$

From comparison of (8.34) with (8.16) we see that with very high SNR, the rms fractional error is equivalent to that attained in the case where phase is a priori known. However, we should note that (8.16) was obtained without the restrictive assumption of high SNR, whereas (8.34) requires this assumption. If we assume a moderate SNR (e.g., $\rho_i B_n T \simeq 1$) and return to the form (8.33), we note that there is a significantly larger error in the present case than in the case where phase is a priori known. For example, if $\rho_i B_n T \simeq 1$, then the error given by (8.33) exceeds that given by (8.16) by a factor $\sqrt{2}$, or 3 dB.

The probabilistic criterion of performance can also be used in the present case. The output PDF to be used in the counterpart of (8.17) for the case under discussion is simply the PDF of the *amplitude* of a signal plus a Gaussian noise with zero mean and rms value σ_{no}^2, i.e., the Rice distribution (see Eq. (3.46)).

$$p(v) = v\, e^{-\frac{1}{2}(v^2 + [2\rho_i B_n T])}\, I_0(v\sqrt{2\rho_i B_n T}) \tag{8.35}$$

Thus from (8.17) and (8.35), we have

$$\begin{aligned} p(\Delta a\,;a) &= \int_{\sqrt{2\rho_i B_n T}\,(1-\Delta a/2a)}^{\sqrt{2\rho_i B_n T}\,(1+\Delta a/2a)} dv\, v\, e^{-\frac{1}{2}(v^2+[2\rho_i B_n T])}\, I_0(v\sqrt{2\rho_i B_n T}) \\ &= Q\left(\sqrt{2\rho_i B_n T}, \sqrt{2\rho_i B_n T}\left[1 - \frac{\Delta a}{2a}\right]\right) \\ &\quad - Q\left(\sqrt{2\rho_i B_n T}, \sqrt{2\rho_i B_n T}\left[1 + \frac{a}{2a}\right]\right) \end{aligned} \tag{8.36}$$

A family of curves showing $p(\Delta a\,;a)$ versus $|\Delta a/a|$ for various values of $\sqrt{2\rho_i B_n T}$ is presented in Figure 8-3; another family showing $P(\Delta a, a)$ as

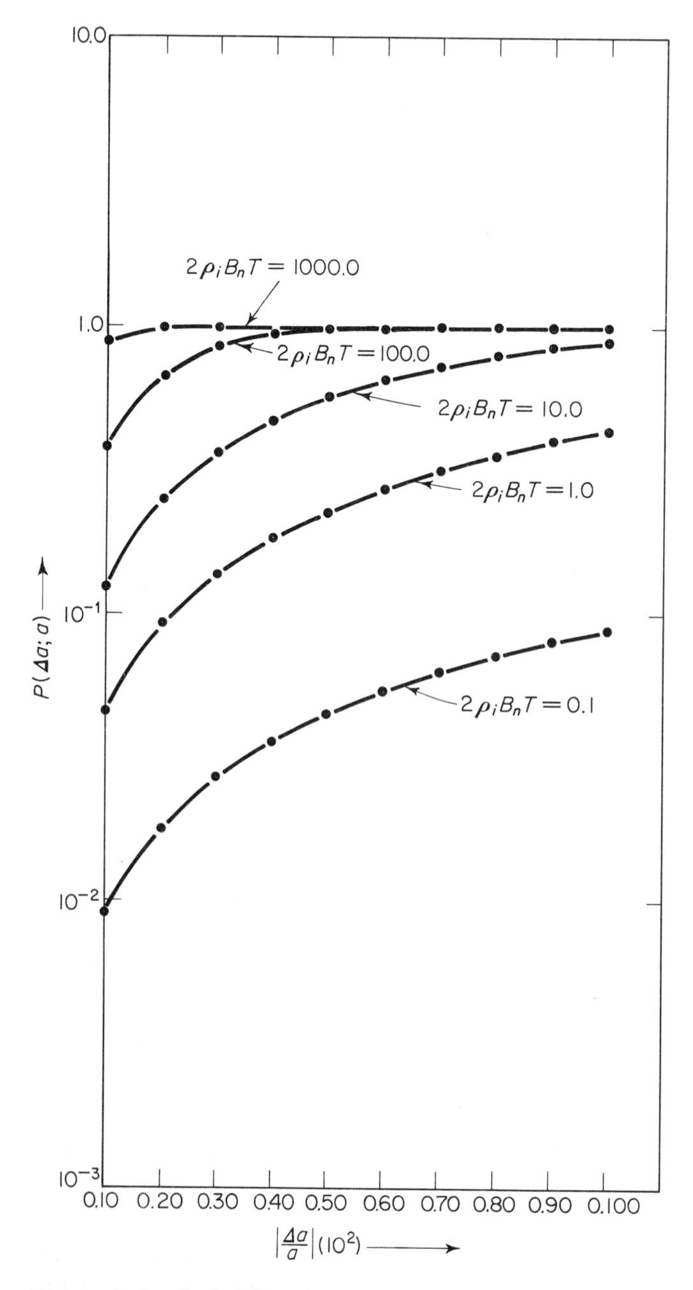

Figure 8-3 Probability that amplitude measurement falls between $\left(a - \frac{\Delta a}{2}\right)$ and $\left(a + \frac{\Delta a}{2}\right)$ versus $\left|\frac{\Delta a}{2}\right|(10^2)$.

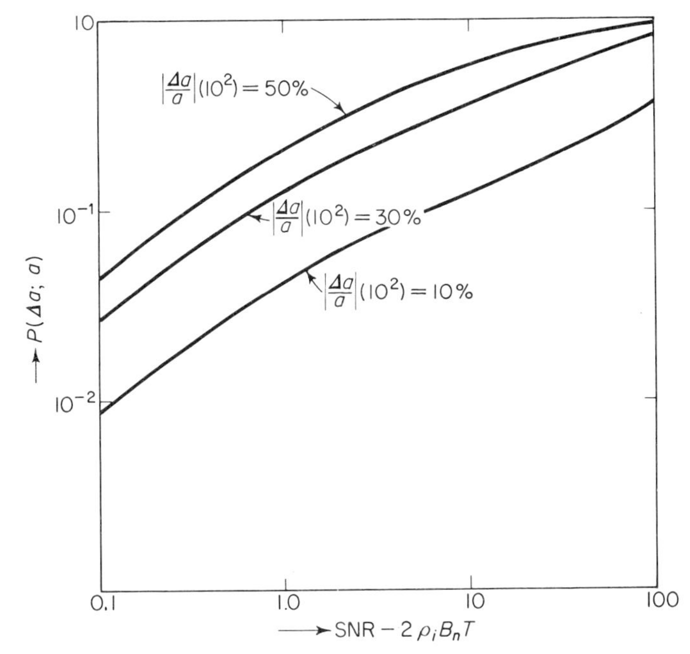

Figure 8-4 Probability that amplitude measurement falls between $\left(a - \frac{\Delta a}{2}\right)$ and $\left(a + \frac{\Delta a}{2}\right)$ versus output SNR.

a function of $\rho_i B_n T$ for various values of $|\Delta a/a|$ is presented in Figure 8-4. It is evident from (8.36) that the difference between the two Q-functions becomes very small as $|\Delta a/a|$ approaches zero, i.e., as the allowed error in measurement of amplitude becomes very small. The curves show that the probability of an amplitude measurement between $a - \Delta a/2$ and $a + \Delta a/2$ is extremely small if $|\Delta a|$ is 10 percent of a or less, unless the SNR is quite high (at least 10 dB). Also, the probability that the measurement will be within 50 percent of the correct value is very high if the SNR is at least 10 dB. For small SNR this probability is extremely low.

8.3 OPTIMAL MEASUREMENT OF PHASE OF A SINUSOIDAL RADIO SIGNAL

Let us now apply (8.6) in the case where the amplitude, phase, and frequency are constant (i.e., Eq. (8.7) holds) and the unknown parameter is the phase, the amplitude and frequency being a priori known. In this case (8.6) assumes the form

$$\int_0^T dt'\, v(t') \frac{\partial}{\partial \psi} [a \cos(\omega_c t' + \psi)] = -a \int_0^T dt'\, v(t') \sin(\omega_c t' + \psi)$$
$$= -a^2 \int_0^T dt' \cos(\omega_c t' + \psi) \sin(\omega_c t' + \psi) \tag{8.37}$$

By virtue of the assumption (8.10),

$$\int_0^T dt' \sin(\omega_c t' + \psi) \cos(\omega_c t' + \psi) \simeq 0 \tag{8.38}$$

From (8.37) and (8.38), it follows that

$$\int_0^T dt'\, v(t') \sin(\omega_c t' + \psi_{\text{MLE}}) = v_s \cos \psi_{\text{MLE}} + v_c \sin \psi_{\text{MLE}} \simeq 0 \tag{8.39}$$

where, as in Section 8.2.2,

$$v_c \equiv \int_0^T dt'\, v(t') \cos \omega_c t'$$
$$v_s \equiv \int_0^T dt'\, v(t') \sin \omega_c t'$$

Equation (8.36) can be written in the form

$$\psi_{\text{MLE}} = \tan^{-1}\left[\frac{-v_s}{v_c}\right] \tag{8.40}$$

The interpretation of (8.40) is perfectly straightforward. It tells us that the optimal method of measuring signal phase in a Gaussian noise background, from the MLE viewpoint, is to first feed the incoming signal into two sinusoidal matched filters at the same frequency and in phase quadrature,† then measure the arctangent of the negative ratio of the sine filter output to the cosine filter output. The matched filtering is the key idea here, its purpose being to build up the SNR to as high a value as possible before measuring the phase. This is the same sort of message as was delivered to the reader in Section 8.2; first matched filtering is applied in order to maximize the SNR, then the desired parameter is measured. This sort of result appears repeatedly in MLE optimization problems involving Gaussian noise.

In the noise-free case, with the aid of (8.10), the cosine and sine terms v_{ss} and v_{cs}, respectively, are

$$v_{ss} = a \int_0^T dt' \cos(\omega_c t' + \psi) \sin \omega_c t' \simeq -\frac{aT}{2} \sin \psi \tag{8.41.a}$$

$$v_{cs} = a \int_0^T dt' \cos(\omega_c t' + \psi) \cos \omega_c t' \simeq \frac{aT}{2} \cos \psi \tag{8.41.b}$$

and therefore, from (8.40)

† That is, having a 90° phase difference.

$$\psi_{\text{MLE}} = \tan^{-1}\left(\frac{\frac{aT}{2}\sin\psi}{\frac{aT}{a}\cos\psi}\right) = \psi \tag{8.42}$$

indicating that the maximum likelihood estimator of phase is the true phase in the absence of noise, as must be the case if the MLE computation is to be used directly as a measurement of phase.

The details of calculation of the rms error in phase measurement are presented in Appendix IV. In the case of very high SNR, the phase error is approximated by (see Eq. (IV–9)).

$$\varepsilon_\psi = \frac{1}{2}\frac{1}{\sqrt{\rho_i B_n T}} \tag{8.43}$$

Equation (8.43) tells us that for large input SNR, the rms phase error is independent of the true value of the phase angle and is inversely proportional to the square root of the product $\rho_i B_n T$. This is the same functional dependence on this product as shown by the fractional error in amplitude measurement (see Eq. (8.16)) except for a factor $\sqrt{2}$ by which the fractional amplitude error exceeds the phase error.

If a probabilistic criterion of phase measurement accuracy is desired, we must calculate the PDF for the maximum likelihood estimator of phase.

We begin the calculation by recognizing that the quantity

$$v_o = v_c \cos\omega_c t + v_s \sin\omega_c t = |v_o|\cos(\omega_c t + \psi_o) \tag{8.44}$$

where

$$|v_o| \equiv \sqrt{v_c^2 + v_s^2}$$

is a sample function of a narrowband Gaussian random process, its sine and cosine components both being outputs of linear filters whose inputs are random functions with Gaussian statistics.† The PDF of the output phase ψ_o can be conceived as the average (over all possible amplitudes) of the joint PDF of amplitude and phase. The joint PDF of amplitude a and phase ψ of a narrowband Gaussian process is given by (see Section 3.3, Eq. (3.44))

$$\begin{aligned} p(a;\psi) &= \frac{1}{2\pi\sigma^2} e^{-\frac{1}{2\sigma^2}[(x-x_s)^2+(y-y_s)^2]} \\ &= \frac{1}{2\pi\sigma^2} e^{-\frac{1}{2\sigma^2}[a^2-2a\,a_s\cos(\psi-\psi_s)+a_s^2]} \end{aligned} \tag{8.45}$$

where

$$x \equiv a\cos\psi$$
$$y \equiv a\sin\psi$$

† See Section 3.1.4.

$$x_s \equiv a_s \cos \psi_s$$
$$y_s \equiv a_s \sin \psi_s$$
$$\sigma^2 = \langle (x - x_s)^2 \rangle = \langle (y - y_s)^2 \rangle$$

Then

$$p(\psi) = \int_0^\infty da\, ap(a\,; \psi) = \frac{e^{-\frac{a_s^2}{2\sigma^2}}}{2\pi\sigma^2} \int_0^\infty da\, a\, e^{-\frac{1}{2\sigma^2}[a^2 - 2a\, a_s \cos(\psi - \psi_s)]}$$
$$= \frac{e^{-\frac{a_s^2}{2\sigma}}}{2\pi} + \frac{a_s\, e^{-\frac{a_s^2 \sin^2(\psi - \psi_s)}{2\sigma^2}}}{2\sqrt{2\pi\sigma^2}} \cos(\psi - \psi_s)\left[1 + \operatorname{erf}\left(\frac{a_s}{\sqrt{2\sigma^2}} \cos(\psi - \psi_s)\right)\right] \tag{8.46}$$

where the method of integration in (8.46) involves the completion of the square in the exponential appearing in the integrand. As expected, $p(\psi) = 1/2\pi$, (random phase) in the zero-signal case.

In the particular problem of interest here,

$$a_s = \frac{aT}{2} \tag{8.47.a}$$

$$\sigma^2 = \frac{\sigma_n^2 T}{4B_n} \tag{8.47.b}$$

$$\frac{a_s}{\sqrt{2\sigma^2}} = \sqrt{\rho_i B_n T} \tag{8.47.c}$$

$$\psi_s = \tan^{-1}\left(\frac{-v_s}{v_c}\right) \tag{8.47.d}$$

Substituting Eqs. (8.47) into (8.46) and invoking a simple trigonometric identity, we obtain

$$p(\psi_0) = \frac{e^{-\rho_i B_n T}}{2\pi} + \left\{\frac{1}{2\sqrt{\pi}} \sqrt{\rho_i B_n T}\, e^{-\frac{\rho_i B_n T}{2} + \frac{\rho_i B_n T}{2} \cos 2(\psi_0 - \psi_s)} \cos(\psi_o - \psi_s)\right.$$
$$\left.[1 + \operatorname{erf}(\sqrt{\rho_i B_n T} \cos(\psi_o - \psi_s))]\right\} \tag{8.48}$$

If the filter output SNR ($\rho_i B_n T$) is very high, then ψ_o is near ψ_s, so that $(\psi_o - \psi_s)$ is sufficiently close to zero to assure us that $\cos(\psi_o - \psi_s)$ is near unity. Then $\operatorname{erf}\left[\frac{\sqrt{\rho_i B_n T}}{2} \cos(\psi_o - \psi_s)\right] \simeq \operatorname{erf}(\sqrt{\rho_i B_n T}) \simeq 1$ if $\rho_i B_n T \gg 1$ and $|\psi_o - \psi_s| \ll \pi/2$. Hence,

$$p(\psi_o) \simeq \frac{e^{-\rho_i B_n T}}{2\pi} [1 + 2\sqrt{\pi} \sqrt{\rho_i B_n T}\, e^{\frac{\rho_i B_n T}{2}[1 + \cos 2(\psi_o - \psi_s)]} \cos(\psi_o - \psi_s)] \tag{8.49}$$

The second term of (8.49) is much greater than the first under the large SNR assumption; consequently, if

$$\rho_i B_n T \gg 1 \tag{8.50}$$

then

$$p(\psi_o) \simeq \frac{1}{\sqrt{\pi}} e^{-\rho_i B_n T} \sqrt{\rho_i B_n T}\, e^{\frac{\rho_i B_n T}{2}[1+\cos 2(\psi_o - \psi_s)]} \cos(\psi_o - \psi_s) \tag{8.51}$$

We can use (8.51) to calculate the probability that ψ_o is within some specified region around ψ, provided $|\psi_o - \psi_s| < \pi/2$.

This calculation is simple if ψ_{MLE} is confined to extremely small regions around ψ, i.e., if

$$2(\psi_o - \psi_s) \ll \frac{\pi}{4}; \qquad \cos^2(\psi_o - \psi_s) \simeq 1 - \frac{[2(\psi_o - \psi_s)]^2}{2} \tag{8.52}$$

in which case,

$$p(\psi_o) \simeq \frac{1}{\sqrt{\pi}} \sqrt{\rho_i B_n T}\, e^{-\rho_i B_n T(\psi_o - \psi_s)^2} \tag{8.53}$$

and

$$p(\Delta\psi;\psi) = p\left\{\psi - \frac{\Delta\psi}{2} \leq \psi_{\text{MLE}} \leq \psi + \frac{\Delta\psi}{2}\right\} = \int_{\psi_s - (\Delta\psi/2)}^{\psi_s + (\Delta\psi/2)} d\psi_o p(\psi_o)$$

$$\simeq \frac{\sqrt{\rho_i B_n T}}{\pi} \int_{-(\Delta\psi/2)}^{(\Delta\psi/2)} d\theta\, e^{-\rho_i B_n T \theta^2} = \operatorname{erf}\left(\frac{\Delta\psi\sqrt{\rho_i B_n T}}{2}\right) \tag{8.54}$$

Plots of $p(\Delta\psi;\psi)$ versus $\Delta\psi$ for various values of $(\rho_i B_n T)$ and $p(\Delta\psi;\psi)$ versus $(2\rho_i B_n T)$ for various values of $\Delta\psi$ are presented in Figures 8-5 and

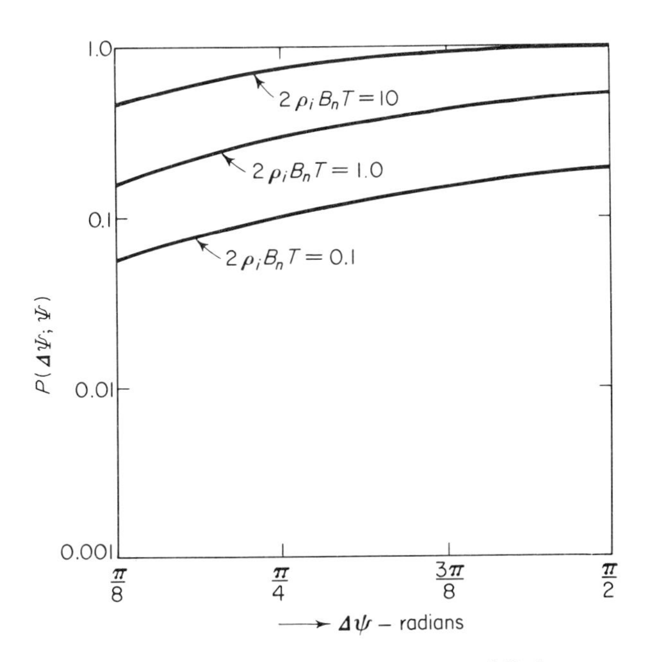

Figure 8-5 Probability that phase measurement falls between $\psi_o - \frac{\Delta\psi}{2}$ and $\psi_o + \frac{\Delta\psi}{2}$ versus $\Delta\psi$.

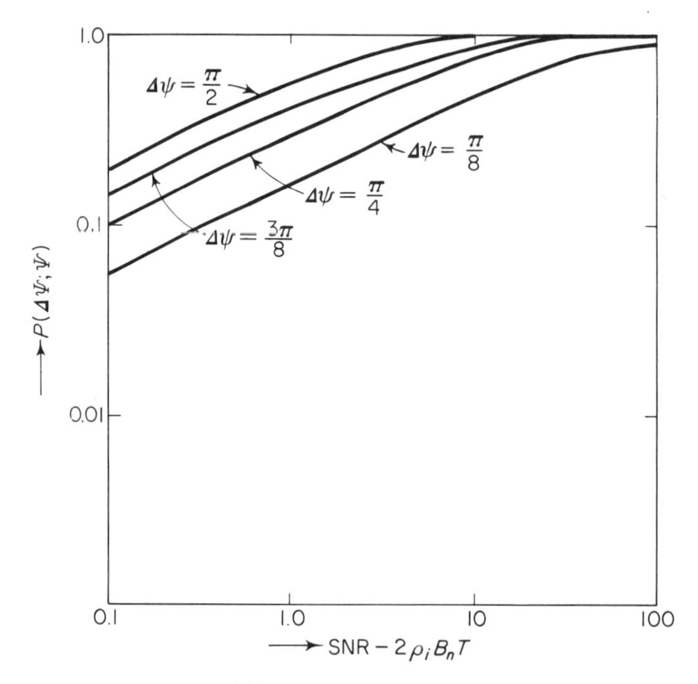

Figure 8-6 Probability that phase measurement falls between $\psi_o - \frac{\Delta\psi}{2}$ and $\psi_o + \frac{\Delta\psi}{2}$ *versus* output SNR.

8-6, respectively. It is possible to extract from these curves the *threshold SNR*, defined as that input SNR required in order that $p(\Delta\psi\ ; \psi)$ be (say) 90 percent or higher. For example, from Figure 8-6, we note that a 90 percent probability that the phase measurement falls within 22.5° of the true phase (i.e., $\Delta\psi/2 = \pi/8$) requires an SNR of at least 20, or about 13 dB, and a 90 percent probability that the measurement is within 45° of the true phase (i.e., $\Delta\psi/2 = \pi/4$) requires an SNR of 5, or about 7 dB.

8.4 JOINT MEASUREMENT OF AMPLITUDE AND PHASE OF A SINUSOIDAL RADIO SIGNAL

Suppose we should wish to jointly measure the amplitude and phase of a sinusoidal radio signal of known frequency. We note that a knowledge of both the amplitude and phase is completely equivalent to a knowledge of the x and y-component of a signal with representation

$$s_i(t) = x \cos \omega_c t + y \sin \omega_c t \tag{8.55}$$

In this case, we can express (8.6) in the form

$$\int_0^T dt'\, v(t') \frac{\partial}{\partial x}(x \cos \omega_c t' + y \sin \omega_c t') = \int_0^T dt'\,(x \cos \omega_c t' + y \sin \omega_c t') \frac{\partial}{\partial x}(x \cos \omega_c t' + y \sin \omega_c t') \tag{8.56.a}$$

$$\int_0^T dt'\, v(t') \frac{\partial}{\partial y}(x \cos \omega_c t' + y \sin \omega_c t') = \int_0^T dt'\,(x \cos \omega_c t' + y \sin \omega_c t') \frac{\partial}{\partial y}(x \cos \omega_c t' + y \sin \omega_c t') \tag{8.56.b}$$

or equivalently

$$\int_0^T dt'\, v(t') \cos \omega_c t' = x \int_0^T dt' \cos^2 \omega_c t' + y \int_0^T dt' \cos \omega_c t' \sin \omega_c t' \simeq \frac{xT}{2} \tag{8.57.a}$$

$$\int_0^T dt'\, v(t') \sin \omega_c t' = x \int_0^T dt' \cos \omega_c t' \sin \omega_c t + y \int_0^T dt' \sin^2 \omega_c t' \simeq \frac{yT}{2} \tag{8.57.b}$$

or

$$x_{\text{MLE}} = \frac{2}{T} \int_0^T dt'\, v(t') \cos \omega_c t' \tag{8.58.a}$$

$$y_{\text{MLE}} = \frac{2}{T} \int_0^T dt'\, v(t') \sin \omega_c t' \tag{8.58.b}$$

In the noise-free case

$$x_{\text{MLE}} = \frac{2x}{T} \int_0^T dt' \cos^2 \omega_c t' + \frac{2y}{T} \int_0^T dt' \cos \omega_c t' \sin \omega_c t' \simeq x \tag{8.59.a}$$

$$y_{\text{MLE}} = \frac{2x}{T} \int_0^T dt' \cos \omega_c t' \sin \omega t' + \frac{2y}{T} \int_0^T dt' \sin^2 \omega_c t' \simeq y \tag{8.59.b}$$

If noise is present, then

$$x_{\text{MLE}} = x + \frac{2}{T} \int_0^T dt'\, n_i(t') \cos \omega_c t' \tag{8.60.a}$$

$$y_{\text{MLE}} = y + \frac{2}{T} \int_0^T dt'\, n_i(t') \sin \omega_c t' \tag{8.60.b}$$

and the measurement errors in x and y, with the aid of (6.21) and Equations (8.60.a) and (8.60.b), are given by

$$\mathcal{E}_x = \sqrt{\langle (x_{\text{MLE}} - x)^2 \rangle}$$

$$\mathcal{E}_y = \sqrt{\langle (y_{\text{MLE}} - y)^2 \rangle}$$

$$\simeq \frac{2}{T} \sqrt{\int_0^T \int_0^T dt'\, dt''\, \langle n_i(t') n_i(t'') \rangle \begin{matrix} (\cos \omega_c t' \cos \omega_c t'') \\ \\ (\sin \omega_c t' \sin \omega_c t'') \end{matrix}} \simeq \frac{\sigma_n}{\sqrt{B_n T}} \qquad \begin{matrix} (8.61.\text{a}) \\ \\ (8.61.\text{b}) \end{matrix}$$

If we should wish to use the probabilistic criterion of measurement quality, we can merely observe that the noise components of x_{MLE} and y_{MLE} are both outputs of linear filters whose noise inputs are Gaussian with zero mean and hence are themselves Gaussian. Thus, the probability that x_{MLE} and y_{MLE} fall within the regions $(x - \Delta x/2)$ to $(x + \Delta x/2)$ and $(y - \Delta y/2)$ to $(y + \Delta y/2)$, respectively, are

$$P(\Delta x\,; x) = P\left\{x - \frac{\Delta x}{2} \leq x_{\text{MLE}} \leq x + \frac{\Delta x}{2}\right\}$$

$$= \operatorname{erf}\left(\frac{x + \frac{\Delta x}{2}}{\sqrt{4B_n T\sigma_n^2}}\right) - \operatorname{erf}\left(\frac{x - \frac{\Delta x}{2}}{\sqrt{4B_n T\sigma_n^2}}\right) \tag{8.62.a}$$

$$P(\Delta y\,; y) = P\left\{y - \frac{\Delta y}{2} \leq y_{\text{MLE}} \leq y + \frac{\Delta y}{2}\right\}$$

$$= \operatorname{erf}\left(\frac{y + \frac{\Delta y}{2}}{\sqrt{4B_n T\sigma_n^2}}\right) - \operatorname{erf}\left(\frac{y - \frac{\Delta y}{2}}{\sqrt{4B_n T\sigma_n^2}}\right) \tag{8.62.b}$$

The rms values of x_{MLE} and y_{MLE} are simply the rms values of the output noise $\sqrt{2B_n T\sigma_n^2}$.

$P(\Delta x\,; x)$ and $P(\Delta y\,; y)$ are of the same form and $\mathcal{E}_x$ and $\mathcal{E}_y$ are equal. The probability can be expressed in the form

$$P(\Delta x\,; x) = \operatorname{erf}\left(\rho_i \frac{\left(x + \frac{\Delta x}{2}\right)}{a} \frac{\mathcal{E}_x}{a}\right) - \operatorname{erf}\left(\rho_i \frac{\left(x - \frac{\Delta x}{2}\right)}{a} \frac{\mathcal{E}_x}{a}\right) \tag{8.63}$$

with an equivalent expression for $P(\Delta y\,; y)$ where x's are replaced by y's. Equation (8.63) relates the rms error to the probability criterion.

8.5 EXTRACTION OF COMPLETE WAVESHAPE INFORMATION FROM AN ARBITRARY RADIO SIGNAL

In the discussions of Sections 8.1 through 8.4 the incoming radio signal has been assumed sinusoidal over the interval 0–T. In the present section, we will consider the case where the signal is an information-bearing carrier of arbitrary duration, i.e., a perfectly general type of RF signal, and see whether we can carry over to this realistic case any of the conclusions of the simplified sinusoidal model.

8.5.1 Sampling Approach

Suppose the highest frequency in the signal waveform (the lowpass bandwidth of the signal) is B_s and the observation interval is 0–T. Then if we use the signal representation (6.2) of Section 6.1, we have

$$s_i(t) = x_i(t) \cos \omega_c t - y_i(t) \sin \omega_c t \tag{6.2}$$

the two time waveforms $x_i(t)$ and $y_i(t)$ can be assumed limited to frequencies from 0 to B_s. This implies (see Section 3.2, especially Eq. (1.70)) that we can describe both $x_i(t)$ and $y_i(t)$ through the sampling theorem (where $N/T = 2B_s$) as follows:

$$x_i(t) = \sum_{n=0}^{2B_sT} x_i\left(\frac{n}{2B_s}\right) \operatorname{sinc}\left[2\pi B_s\left(t - \frac{n}{2B_s}\right)\right] \tag{8.64.a}$$

$$y_i(t) = \sum_{n=0}^{2B_sT} y_i\left(\frac{n}{2B_s}\right) \operatorname{sinc}\left[2\pi B_s\left(t - \frac{n}{2B_s}\right)\right] \tag{8.64.b}$$

Our signal information waveform is now characterized by $2B_sT$ parameters (see Section 1.5), which could be $2B_sT$ Fourier coefficients or $2B_sT$ sample values, depending on whether one is interested in measuring parameters associated with certain frequencies or those associated with certain time regions.

Concentrating for the moment on the sample value approach, we might want to know any one of the following items:

1. $x_i\left(\frac{n}{2B_s}\right)$ and $y_i\left(\frac{n}{2B_s}\right)$—complete amplitude and phase information about the signal at every sample point.
2. $a_i\left(\frac{n}{2B_s}\right) = \sqrt{\left[x_i\left(\frac{n}{2B_s}\right)\right]^2 + \left[y_i\left(\frac{n}{2B_n}\right)\right]^2}$—the signal amplitude at every sample point.
3. $\psi_i\left(\frac{n}{2B_s}\right) = \tan^{-1}\left(\frac{y_i\left(\frac{n}{2B_s}\right)}{x_i\left(\frac{n}{2B_s}\right)}\right)$—the signal phase at every sample point.

We will consider in turn each of the information requirements (1), (2) and (3).

First, we recognize that the MLE theory (Eq. (8.3)) gives us the following generic expression for (1), (2), or (3), where β represents the parameter to be measured.

$$\frac{\partial}{\partial\beta}\left\{e^{-\frac{1}{2\sigma_n^2}\sum_{k=1}^{N}\left(v_k - \sum_{p=0}^{2B_sT} \operatorname{sinc}\left(2\pi B_s\left[t_k - \frac{p}{2B_s}\right]\right)\left[x_i\left(\frac{p}{2B_s}\right)\cos\omega_c t_k - y_i\left(\frac{p}{2B_s}\right)\sin\omega_c t_k\right]\right)^2}\right\} = 0 \tag{8.65.a}$$

or equivalently

$$\frac{\partial}{\partial\beta}\left\{e^{-\frac{1}{2\sigma_n^2}\sum_{k=1}^{N}\left(v_k - \sum_{p=0}^{2B_sT} \operatorname{sinc}\left(2\pi B_s\left[t_k - \frac{p}{2B_s}\right)\right] a_i\left(\frac{p}{2B_s}\right)\cos\left(\omega_c t_k + \psi_i\left(\frac{p}{2B_s}\right)\right)^2}\right\} = 0 \tag{8.65.b}$$

In case (1), the form (8.65.a) is the most convenient, while in cases (2) and (3), the form (8.65.b) is the most convenient.

Using the form (8.65.a), we have

$$\frac{\partial}{\partial\beta}\left\{\sum_{k=1}^{N}\sum_{\ell=0}^{2B_sT} v_k \operatorname{sinc}\left(2\pi B_s\left[t_k-\frac{\ell}{2B_s}\right]\right)\left[x_i\left(\frac{\ell}{2B_s}\right)\cos\omega_c t_k - y_i\left(\frac{\ell}{2B_s}\right)\sin\omega_c t_k\right]\right.$$
$$-\frac{1}{2}\sum_{k=1}^{N}\sum_{\ell=1}^{2B_sT}\sum_{m=0}^{2B_sT}\operatorname{sinc}\left(2\pi B_s\left[t_k-\frac{\ell}{2B_s}\right]\right)\operatorname{sinc}\left(2\pi B_s\left[t_k-\frac{m}{2B_s}\right]\right)\cdots$$
$$\left[x_i\left(\frac{\ell}{2B_s}\right)x_i\left(\frac{m}{2B_s}\right)\cos^2\omega_c t_k + y_i\left(\frac{\ell}{2B_s}\right)y_i\left(\frac{m}{2B_s}\right)\sin^2\omega_c t_k\right.$$
$$\left.\left.-\left(x_i\left(\frac{\ell}{2B_s}\right)y_i\left(\frac{m}{2B_s}\right)+x_i\left(\frac{m}{2B_s}\right)y_i\left(\frac{\ell}{2B_s}\right)\right)\cos\omega_c t_k\sin\omega_c t_k\right]\right\}=0 \quad (8.66)$$

Guided by arguments analogous to that advanced in the footnote below (4.26) in Section 4.1.3.1, we can approximate the k-indexed sums by integrals. We can then cancel the integrals involving frequencies $2\omega_c$. The latter step is based on the slowly varying nature of $x_i(t)$ and $y_i(t)$ compared with cos $\omega_c t$ or sin $\omega_c t$.

The final result of these operations is

$$\frac{\partial}{\partial\beta}\left\{\sum_{\ell=0}^{2B_sT}\left[x_i\left(\frac{\ell}{2B_s}\right)\int_0^T dt'\, v(t')\operatorname{sinc}\left(2\pi B_s\left[t'-\frac{\ell}{2B_s}\right]\right)\cos\omega_c t'\right.\right.$$
$$\left.-y_i\left(\frac{\ell}{2B_s}\right)\int_0^T dt'\, v(t')\operatorname{sinc}\left(2\pi B_s\left[t'-\frac{\ell}{2B_s}\right]\right)\sin\omega_c t'\right]$$
$$-\frac{1}{4}\sum_{\ell,m=0}^{2B_sT}\left[x_i\left(\frac{\ell}{2B_s}\right)x_i\left(\frac{m}{2B_s}\right)+y_i\left(\frac{\ell}{2B_s}\right)y_i\left(\frac{m}{2B_s}\right)\right]\cdots$$
$$\left.\int_0^T dt'\operatorname{sinc}\left(2\pi B_s\left[t'-\frac{\ell}{2B_s}\right]\right)\operatorname{sinc}\left(2\pi B_s\left[t'-\frac{m}{2B_s}\right]\right)\right\}=0 \quad (8.67.a)$$

or equivalently (either from (8.67.a) or directly from the form (8.65.b))

$$\frac{\partial}{\partial\beta}\left\{\sum_{\ell=0}^{2B_sT}\left[a_i\left(\frac{\ell}{2B_s}\right)\int_0^T dt'\, v(t')\operatorname{sinc}\left(2\pi B_s\left[t'-\frac{\ell}{2B_s}\right]\right)\cos\left(\omega_c t'+\psi_i\left(\frac{\ell}{2B_s}\right)\right)\right]\right.$$
$$-\frac{1}{4}\sum_{\ell,m=0}^{2B_sT}a_i\left(\frac{\ell}{2B_s}\right)a_i\left(\frac{m}{2B_s}\right)\left[\cos\left(\psi_i\left(\frac{\ell}{2B_s}\right)-\psi_i\left(\frac{m}{2B_s}\right)\right)\right]\cdots$$
$$\left.\int_0^T dt'\operatorname{sinc}\left(2\pi B_s\left[t'-\frac{\ell}{2B_s}\right]\right)\operatorname{sinc}\left(2\pi B_s\left[t'-\frac{m}{2B_s}\right]\right)\right\}=0 \quad (8.67.b)$$

Let us first consider case (1), wherein β represents the set of parameters $x_i(n/2B_s)$, $y_i(n/2B_s)$, which we will designate as x_{in}, y_{in} for convenience (not to be confused with x_n and y_n as used to represent quadrature components of noise). In this case we will use the form (8.67.a), leading to

$$\int_0^T dt'\, v(t')\cos\omega_c t'\operatorname{sinc}\left(2\pi B_s\left[t'-\frac{p}{2B_s}\right]\right)$$
$$=\frac{1}{2}\sum_{m=0}^{2B_sT}x_i\left(\frac{m}{2B_s}\right)\int_0^T dt'\operatorname{sinc}\left(2B_s\left[t'-\frac{m}{2B_s}\right]\right)\operatorname{sinc}\left(2B_s\left[t'-\frac{p}{2B_s}\right]\right)$$
$$(8.68.a)$$

$$\int_0^T dt'\, v(t') \sin \omega_c t' \operatorname{sinc}\left(2\pi B_s\left[t' - \frac{p}{2B_s}\right]\right) =$$
$$-\frac{1}{2}\sum_{m=0}^{2B_sT} y_i\left(\frac{m}{2B_s}\right)\int_0^T dt' \operatorname{sinc}\left(2\pi B_s\left[t' - \frac{m}{2B_s}\right]\right)\operatorname{sinc}\left(2B_s\left[t' - \frac{p}{2B_s}\right]\right) \tag{8.68.b}$$

We assume that

$$T \gg \frac{1}{2B_s} \tag{8.69}$$

which implies that the upper integration limit can be approximated by $+\infty$ and, consequently,†

$$\lim_{T\to\infty}\int_0^T dt' \operatorname{sinc}\left(2\pi B_s\left[t' - \frac{m}{2B_s}\right]\right)\operatorname{sinc}\left(2\pi B_s\left[t' - \frac{p}{2B_s}\right]\right)$$
$$\simeq \delta_{mn}\int_0^\infty dt'\left[\operatorname{sinc}\left(2\pi B_s\left[t' - \frac{m}{2B_s}\right]\right)\right]^2 \simeq \frac{\delta_{mp}}{2\pi B_s}\int_0^\infty dx\,[\operatorname{sinc}(x)]^2 = \frac{\delta_{mp}}{4B_s} \tag{8.70}$$

Using (8.70) in Eqs. (8.68) and changing the index p to k, we obtain

$$\left[x_i\left(\frac{k}{2B_s}\right)\right]_{\text{MLE}} = 8B_s\int_0^T dt'\, v(t')\cos\omega_c t' \operatorname{sinc}\left(2\pi B_s\left[t' - \frac{k}{2B_s}\right]\right) \tag{8.71.a}$$

$$\left[y_i\left(\frac{k}{2B_s}\right)\right]_{\text{MLE}} = -8B_s\int_0^T dt'\, v(t')\sin\omega_c t' \operatorname{sinc}\left(2\pi B_s\left[t' - \frac{k}{2B_s}\right]\right) \tag{8.71.b}$$

By again going through the procedures outlined in Sections 8.2 and 8.3, we see immediately that the MLE of the kth amplitude sample point is

$$\left[a_i\left(\frac{k}{2B_s}\right)\right]_{\text{MLE}} = \sqrt{\left[x_i\left(\frac{k}{2B_s}\right)\right]^2_{\text{MLE}} + \left[y_i\left(\frac{k}{2B_s}\right)\right]^2_{\text{MLE}}} = (v_c^{(k)})^2_{\text{MLE}} + (v_s^{(k)})^2_{\text{MLE}} \tag{8.72}$$

where

$$(v_c^{(k)})_{\text{MLE}} = 8B_s\int_0^T dt'\, v(t')\cos\omega_c t' \operatorname{sinc}\left(2\pi B_s\left[t' - \frac{k}{2B_s}\right]\right)$$

$$(v_s^{(k)})_{\text{MLE}} = 8B_s\int_0^T dt'\, v(t')\sin\omega_c t' \operatorname{sinc}\left(2\pi B_s\left[t' - \frac{k}{2B_s}\right]\right)$$

and the MLE of the kth sample phase is

$$\left[\psi\left(\frac{k}{2B_s}\right)\right]_{\text{MLE}} = \tan^{-1}\left(\frac{(v_s)_{\text{MLE}}}{(v_c)_{\text{MLE}}}\right) \tag{8.73}$$

The results (8.71), (8.72), and (8.73) are analogous to Eqs. (8.58) of Section 8.4, (8.28) of Section 8.2.2, and (8.40) of Section 8.3, respectively, all obtained for the sinusoidal case by direct application of MLE theory to the cosine and sine components x and y, the amplitude a, and the phase ψ, respectively.

† When B_s is extremely large, the sinc functions in the integrand will overlap only when $m = n$.

A comparison of (8.71), (8.72), and (8.73) with the results for sinusoidal signals (8.58), (8.28), and (8.40) indicates that the optimum MLE processing of a general radio signal with lowpass bandwidth B_s differs from that of a perfectly sinusoidal signal in one significant way; the latter involves crosscorrelation of the signal with sinusoids or cosinusoids of the same frequency, while the former involves crosscorrelation of the signal with sinusoids or cosinusoids of the same frequency *and* a sinc function that plays the role of a time-bounding agent, i.e., it limits the integration to a time domain region within the immediate vicinity of the sample point about which information is to be acquired. This is certainly not surprising if it is recognized that in the absence of a priori knowledge of the waveform, information about a given sample point could not be extracted from a time domain region that does not contain that sample point, and that information about other sample points would be superfluous.

It is instructive to develop (8.71.a) and (8.71.b) in terms of frequency domain functions. We can do this by taking Fourier transforms of $v(t)$, $\cos\omega_c t$, and $\text{sinc}\,(2\pi B_s[t' - k/2B_s])$, and the unit step functions $u(t)$ and $u(T - t)$, resulting in

$$\left[x_i\left(\frac{k}{2B_s}\right)\right]_{\text{MLE}} = \frac{1}{2\pi}\int_{-\infty}^{\infty} d\omega\, V(\omega)\, F_x(\omega)\, e^{-(jk\omega/2B_s)} \tag{8.74.a}$$

$$\left[y_i\left(\frac{k}{2B_s}\right)\right]_{\text{MLE}} = \frac{1}{2\pi}\int_{-\infty}^{\infty} d\omega\, V(\omega)\, F_y(\omega)\, e^{-(jk\omega/2B_s)} \tag{8.74.b}$$

where

$$F_x(\omega) \propto \cos\left(\frac{k\omega_c}{2B_s}\right) \quad \text{if} \quad \pm\,\omega_c - 2\pi B_s \leq \omega \leq \pm\omega_c + 2\pi B_s$$

$$0 \quad \text{otherwise}$$

$$F_y(\omega) \propto \sin\left(\frac{k\omega_c}{2B_s}\right) \quad \text{if} \quad \pm\omega_c - 2\pi B_s \leq \omega \leq \pm\omega_c + 2\pi B_s$$

$$0 \quad \text{otherwise}$$

provided that T is large enough compared to $1/\omega_c$ and $4\pi B_s$ to justify approximating $\text{sinc}\,(\omega T/2)$ by $\pi\delta(\omega T/2)$ within the integrals.

Equations (8.74.a) and (8.74.b) tell us, in effect, that the approximate optimum filters for estimation of $x_i(k/2B_s)$ and $y_i(k/2B_s)$ are (for very large T) bandpass filters centered on the carrier frequency, approximately flat over the signal band, and containing factors $\cos\,(k\omega_c/2B_s)$ and $\sin\,(k\omega_c/2B_s)$, respectively. The cosine factor in the estimation filter for $x_i(k/2B_s)$ passes the cosine component of the signal in a region near the kth sample point, and the sine factor in the filter for estimation of $y_i(k/2B_s)$ passes the corresponding sine component.

To assess the performance of the processing scheme dictated by MLE theory, we will first calculate the rms errors in measurement of $x_i(k/2B_s)$ and $y_i(k/2B_s)$.

Before doing this, we will first assure ourselves that in the noise-free case, the optimal processing yields the true values of $x_i(k/2B_s)$ and $y_i(k/2B_s)$. The indicated values of these quantities (denoted by $x_i^{(k)}$ and $y_i^{(k)}$ in the noise-free case are (from Eqs. (8.64))

$$x_i^{(k)} = +8B_s \sum_{\ell=0}^{2B_sT} \left\{ x_i^{(k)} \int_0^T dt' \operatorname{sinc}\left(2\pi B_s\left[t' - \frac{\ell}{2B_s}\right]\right) \operatorname{sinc}\left(2\pi B_s\left[t' - \frac{k}{2B_s}\right]\right) \right.$$
$$\cos \omega_c t' \cos \omega_c t' - y_i^{(\ell)} \int_0^T dt' \operatorname{sinc}\left(2\pi B_s\left[t' - \frac{\ell}{2B_s}\right]\right)$$
$$\left. \operatorname{sinc}\left(2\pi B_s\left[t' - \frac{k}{2B_s}\right]\right) \sin \omega_c t' \cos \omega_c t' \right\} \tag{8.75.a}$$

$$y_i^{(k)} = -8B_s \sum_{\ell=0}^{2B_sT} \left\{ x_i^{(k)} \int_0^T dt' \operatorname{sinc}\left(2\pi B_s\left[t' - \frac{\ell}{2B_s}\right]\right) \operatorname{sinc}\left(2\pi B_2\left[t' - \frac{k}{2B_s}\right]\right) \right.$$
$$\cos \omega_c t' \sin \omega_c t' - y_i^{(\ell)} \int_0^T dt' \operatorname{sinc}\left(2\pi B_s\left[t' - \frac{\ell}{2B_s}\right]\right)$$
$$\left. \operatorname{sinc}\left(2\pi B_s\left[t' - \frac{k}{2B_s}\right]\right) \sin \omega_c t' \sin \omega_c t' \right\} \tag{8.75.b}$$

Again because of the assumption (8.10) the sine-cosine terms and all other terms involving the frequency $2\omega_c$ can be neglected, and by virtue of (8.69) the upper limit can be approximated by $+\infty$, resulting in

$$(x_i^{(k)})_{\text{MLE}} \simeq (8B_s)\frac{1}{2} \sum_{\ell=0}^{2B_sT} \left\{ x_i^{(\ell)} \int_0^\infty dt' \operatorname{sinc}\left(2\pi B_s\left[t' - \frac{\ell}{2B_s}\right]\right) \right.$$
$$\operatorname{sinc}\left(2\pi B_s\left[t' - \frac{k}{2B_s}\right]\right) \simeq \frac{4B_s}{2\pi B_s} \sum_{\ell=0}^{2B_sT} x_i^{(\ell)} \int_0^\infty dx\,[\operatorname{sinc}(x)]^2\, \delta_{\ell k} = x_i^{(k)} \tag{8.76.a}$$

$$(y_i^{(k)})_{\text{MLE}} \simeq -(8B_s)\frac{1}{2} \sum_{\ell=0}^{2B_sT} \left\{ y_i^{(\ell)} \int_0^\infty dt' \operatorname{sinc}\left(2\pi B_s\left[t' - \frac{\ell}{2B_s}\right]\right) \right.$$
$$\left.\operatorname{sinc}\left(2\pi B_s\left[t' - \frac{k}{2B_s}\right]\right)\right\} \simeq \frac{4B_s}{2\pi B_s} \sum_{\ell=0}^{2B_sT} y_i^{(\ell)} \left(\int_0^\infty dx\,[\operatorname{sinc}(x)]^2\right) \delta_{\ell k} = y_i^{(k)} \tag{8.76.b}$$

The rms fractional errors in measurement of $x_i^{(k)}$ and $y_i^{(k)}$ with the aid of (6.21), (8.10), (8.76.a, b), and (8.69) are:†

$$\mathcal{E}_x = \frac{1}{x_i^{(k)}} \sqrt{\langle [(x_i^{(k)})_{\text{MLE}} - x_i^{(k)}]^2 \rangle}$$

$$\mathcal{E}_y = \frac{1}{y_i^{(k)}} \sqrt{\langle [(y_i^{(k)})_{\text{MLE}} - y_i^{(k)}]^2 \rangle}$$

$$\simeq \begin{matrix} \dfrac{1}{x_i^{(k)}} \\ \dfrac{1}{y_i^{(k)}} \end{matrix} \sqrt{(8B_s)^2 \frac{\sigma_n^2}{2B_n} \sum_{\ell,m=0}^{2B_sT} \int_0^T dt' \begin{matrix}\cos^2 \\ \sin^2\end{matrix} (\omega_c t') \operatorname{sinc}\left(2\pi B_s\left[t' - \frac{\ell}{2B_s}\right]\right) \times \operatorname{sinc}\left(2\pi B_s\left[t' - \frac{m}{2B_s}\right]\right)}$$

† Since $2B_sT \gg 1$, we approximate $(2B_sT + 1)$ by $2B_sT$.

$$\simeq \frac{\frac{1}{x_i^{(k)}}}{\frac{1}{y_i^{(k)}}} \sqrt{\frac{16B_s^2\sigma_n^2}{(2\pi B_s)B_n} \sum_{\ell=0}^{2B_sT} \int_0^\infty dx\,[\mathrm{sinc}\,(x)]^2} \simeq \frac{\frac{1}{x_i^{(k)}}}{\frac{1}{y_i^{(k)}}} \sqrt{\frac{4B_s\sigma_n^2}{2\pi B_n}\frac{\pi}{2}(2B_sT)}$$

$$= \frac{\left(\frac{1}{x_i^{(k)}}\right)}{\left(\frac{1}{y_i^{(k)}}\right)} B_sT\sigma_n \frac{2}{B_nT} = \frac{(2B_sT)}{\sqrt{2\rho_{ik}^{\binom{x}{y}} B_nT}} \tag{8.77.a.b}$$

where

$$\rho_{ik}^{(x)} \equiv \frac{[x_i^{(k)}]^2}{\sigma_n^2} \equiv \text{input SNR for } x_i^{(k)}$$

$$\rho_{ik}^{(y)} \equiv \frac{[y_i^{(k)}]^2}{\sigma_n^2} \equiv \text{input SNR for } y_i^{(k)}$$

From Eqs. (8.77.a, b), we see that the fractional errors are inversely proportional to a quantity that resembles the familiar parameter $\sqrt{2\rho_i B_n T}$ (the coherent matched filter output SNR) provided we define the input SNR ρ_i in the proper way. We also see that the errors are proportional to the product of signal bandwidth and integration time, which is a large number according to (8.69). Thus, if (8.69) holds, a required condition for the validity of our analysis, then rather large values of $\rho_{ik}^{\binom{x}{y}} B_nT$ are required to make the fractional errors small. The reason for this lies in the filter noise outputs, which unlike the signal outputs include significantly large contributions from time regions other than that near the sampling point.

8.5.2 Fourier Series Approach

In the present section, we have heretofore used the time domain sampling theorem (1.70) to represent the signal. The problem can be treated from the frequency domain point of view, either through Fourier series or Fourier integral methods or through the frequency domain sampling theorem (1.69), all of which turn out to be essentially equivalent. Suppose we use the Fourier series approach, which will be sufficient to illustrate the essential ideas.

Again we use (6.2) to represent the signal and expand the quadrature components $x_i(t)$ and $y_i(t)$ in exponential Fourier series as follows:

$$x_i(t) = \sum_{n=-2B_sT}^{2B_sT} c_n^{(x)} e^{\frac{2\pi njt}{T}} \tag{8.78.a}$$

$$y_i(t) = \sum_{n=-2B_sT}^{2B_sT} c_n^{(y)} e^{\frac{2\pi njt}{T}} \tag{8.78.b}$$

where

$$c_n^{\binom{x}{y}} = a_n^{\binom{x}{y}} - jb_n^{\binom{x}{y}}, \qquad c_n^{\binom{x}{y}} = a_n^{\binom{x}{y}} + jb_n^{\binom{x}{y}}$$

$a_n^{\binom{x}{y}}$ and $b_n^{\binom{x}{y}}$ are cosine and sine Fourier components, respectively.

MLE theory, together with Eqs. (8.78.a) and (8.78.b), gives for the MLE of the Fourier coefficients $c_n^{(x)}$, $c_n^{(y)}$, $c_{-n}^{(x)}$, $c_{-n}^{(y)}$ (from which MLE's for $a_n^{(x)}$, $b_n^{(x)}$, $a_n^{(y)}$, $b_n^{(y)}$ can be derived) the following expressions:

$$\frac{\partial}{\partial c_{\substack{n\\-n}}^{\substack{(x)\\(y)}}} \left\{ e^{-\frac{1}{2\sigma_n^2}\sum_{k=1}^{N}[v_k - \sum_{n=-2B_sT}^{2B_sT}(c_n{}^{(x)}\, e^{(2\pi n j t_k/T)} \cos \omega_c t_k + c_n{}^{(y)}\, e^{(2\pi n j T_k/T)} \sin \omega_c t_k)]^2} \right\} = 0 \qquad (8.79)$$

Following exactly the line of reasoning used to proceed from Eqs. (8.65) through Eqs. (8.74) and invoking the assumption (8.69) along the way, we arrive at an interim result analogous to Eqs. (8.67),

$$\begin{aligned} \frac{\partial}{\partial c_{\substack{n\\-n}}^{\substack{(x)\\(y)}}} \Bigg\{ \sum_{n=-2B_sT}^{2B_sT} \Bigg[& c_n^{(x)} \int_0^T dt'\, v(t') \cos \omega_c t'\, e^{(2\pi n j t'/T)} \\ & + c_n^{(y)} \int_0^T dt'\, v(t') \sin \omega_c t'\, e^{(2\pi n j t'/T)} \Bigg] \\ & - \frac{T}{4} \sum_{m,\,n=-2B_sT}^{2B_sT} [c_m^{(x)} c_n^{(x)} c_m^{(y)} c_n^{(y)}] \int_0^T dt'\, e^{(2\pi j(m+n)t'/T)} \Bigg\} = 0 \end{aligned} \qquad (8.80)$$

or equivalently (since the second integral equals $T\,\delta_{mn}$)

$$\begin{aligned} \frac{\partial}{\partial c_{\substack{n\\-n}}^{\substack{(x)\\(y)}}} \Bigg\{ \sum_{n=-2B_sT}^{2B_sT} \Bigg[& c_n^{(x)} \int_0^T dt'\, v(t') \cos \omega_c t'\, e^{(2\pi n j t'/T)} + c_n^{(y)} \int_0^T dt'\, v(t') \sin \omega_c t'\, e^{(2\pi n j t'/T)} \\ & - \frac{T}{4} (|c_n^{(x)}|^2 + |c_n^{(y)}|^2) \Bigg] \Bigg\} = 0 \end{aligned} \qquad (8.80)'$$

Returning to the real Fourier coefficients a_n and b_n, we arrive at the analog of Eqs. (8.71),

$$\begin{matrix} (a_n^{(x)})_{\text{MLE}} \\ (b_n^{(x)})_{\text{MLE}} \\ (a_n^{(y)})_{\text{MLE}} \\ (b_n^{(y)})_{\text{MLE}} \end{matrix} = \frac{2}{T} \int_0^T dt'\, v(t') \begin{matrix} \cos \omega_c t' \\ \sin \omega_c t' \\ \cos \omega_c t' \\ \sin \omega_c t' \end{matrix} \quad \begin{matrix} \cos\left(\dfrac{2\pi n t'}{T}\right) \\ \sin\left(\dfrac{2\pi n t'}{T}\right) \end{matrix} \qquad \begin{matrix} (8.81.a) \\ (8.81.b) \\ (8.81.c) \\ (8.81.d) \end{matrix}$$

The results (8.81) tell us in effect that in order to optimally measure the Fourier coefficients $a_n^{(x)}$, $b_n^{(x)}$, we must crosscorrelate them with products of RF cosine and sine waveforms and waveforms containing sine and cosine functions with the desired harmonic frequency. This is a result that can be guessed intuitively by observing that the detection of a particular frequency component to the exclusion of others would be best accomplished by cross-correlating the incoming signal with that particular frequency component. This is essentially equivalent to the mathematical computation of the nth Fourier coefficient.

The analogs of (8.28) and (8.40), applicable to measurement of amplitude and phase, respectively, at a particular frequency, are

MLE of amplitude of nth Fourier component $= (A_n)_{\text{MLE}}$

$$= \sqrt{\left[\int_0^T dt'\, v(t') \cos \omega_c t' \cos\left(\frac{2\pi n t'}{T}\right)\right]^2 + \left[\int_0^T dt'\, v(t') \sin \omega_c t' \sin\left(\frac{2\pi n t'}{T}\right)\right]^2} \tag{8.82}$$

MLE of phase of nth Fourier component $= (\psi_n)_{\text{MLE}}$

$$= \tan^{-1}\left[\frac{-\int_0^T dt'\, v(t') \sin \omega_c t' \sin\left(\frac{2\pi n t'}{T}\right)}{\int_0^T dt'\, v(t') \cos \omega_c t' \cos\left(\frac{2\pi n t'}{T}\right)}\right] \tag{8.83}$$

Evaluation of the accuracy of measurement of Fourier coefficients proceeds along the same lines as in the case of sample values. A development analogous to that leading from Eqs. (8.69) through Eqs. (8.71) can be used to show that the rms error in measurement of a Fourier coefficient a_k or b_k is inversely proportional to the square root of the ratio of the power contained in the nth harmonic to the noise. This ratio is

$$\rho_{ik} = \frac{a_k^2}{2\sigma_n^2} \tag{8.84}$$

where the noise is assumed to be spectrally uniform over all signal frequencies. Note that ρ_{ik} is only a fraction of the input SNR ρ_i, the latter being the sum of the ratios ρ_{ik} over all harmonics, i.e.,

$$\frac{\rho_{ik}}{\rho_i} = \frac{(a_k^2/2\sigma_n^2)}{\sum_{k=1}^{\infty} \frac{a_k^2}{2\sigma_n^2}} = \frac{a_k^2}{\sum_{k=1}^{\infty} a_k^2} \tag{8.85}$$

8.6 MEASUREMENT OF TIME OF OCCURRENCE OF RF PULSES

We will now develop an approach to the measurement of the time of occurrence of a pulse of radio frequency energy. Consider the incoming signal to be an RF pulse of the form

$$s_i(t) = a \cos(\omega_c t + \psi) f_p(t - t_0) \tag{8.86}$$

where $f_p(t - t_0)$ is the pulse shaping function, t_0 being an instant of time characterizing the pulse position, e.g., the leading edge, trailing edge, or center.

The pulse duration τ_p is assumed to be large compared to the RF period, i.e.,

$$\tau_p \gg \frac{1}{\omega_c} \tag{8.87}$$

From (8.87) it will follow that integrals whose integrands are sinusoids of frequency $2\omega_c$, when integrated over the pulse period, are negligibly small. Such integrals will arise in what follows and will be assumed negligible.

To treat the maximum likelihood estimation of the pulse position, we express (8.6)′ in the form

$$\frac{\partial}{\partial t_0}\int_{-\infty}^{\infty} dt'\, v(t')\, a\cos(\omega_c t' + \psi_r) f_p(t' - t_0) = \frac{\partial}{\partial t_0}\int_{-\infty}^{\infty} dt\, a^2 \cos^2(\omega_c t' + \psi_r)[f_p(t' - t_0)]^2 \tag{8.88}$$

By virtue of the assumption (8.87), the right-hand side of (8.88) is approximately

$$\frac{a^2}{2}\frac{\partial}{\partial t_0}\left(\int_{-\infty}^{\infty} dt'\,[f_p(t' - t_0)]^2\right) = \frac{a^2}{2}\frac{\partial}{\partial t_0}\int_{-\infty}^{\infty} dx\,[f_p(x)]^2 = 0 \tag{8.89}$$

The vanishing of the derivative in (8.89) is brought about by the fact that the integral is independent of t_0.

From (8.88) and (8.89),

$$\frac{\partial}{\partial t_0}\int_{-\infty}^{\infty} dt'\, v(t')\cos(\omega_c t' + \psi_r) f_p(t' - t_0) = 0 \tag{8.90}$$

Equation (8.90) suggests† a procedure for estimating t_0; that is, (1) to crosscorrelate the incoming pulse with a reference pulse whose radio frequency and RF phase match those of the incoming signal and whose pulse shaping function $f_p(t - t_1)$ also matches that of the incoming signal, except for a time shift $(t_1 - t_0)$ and then (2) to vary t_1 in the reference pulse until a value of t_1 is found that maximizes the crosscorrelator output. Intuition and previous discussion (see Appendix II) tell us that the value of t_1 that will do this is t_0. To show that this is indeed true, consider the noise-free case, for which

$$v(t) = s_i(t) = a\cos(\omega_c t + \psi_r) f_p(t - t_0) \tag{8.91}$$

and (with the aid of the assumption (8.87))

$$\int_{-\infty}^{\infty} dt'\, v(t')\cos(\omega_c t' + \psi_r) f_p(t' - t_1) \simeq \frac{1}{2}\int_{-\infty}^{\infty} dt'\, f_p(t' - t_0) f_p(t' - t_1) \tag{8.92}$$

The expression on the right-hand side of (8.92) is the crosscorrelation function between the pulse shaping waveform of the incoming and reference

---

† This suggested procedure does not necessarily follow from (8.90) as a unique optimal estimation technique. In this case, as in others in this chapter, we are using MLE theory only to provide rough guidelines for a suggested implementation. See Helstrom, 1960, Chapter VIII, especially pp. 203–223 for a more mathematically sophisticated discussion of estimation of signal arrival time.

signals, where the occurrence times of these two pulses differ by $t_1 - t_0$. This crosscorrelation function for the case of a square pulse has already been discussed in Section 6.6 (also see Appendix II). It was demonstrated for a square pulse that the maximum crosscorrelation occurs when the two pulses coincide in time. The precise functional variation of the CCF with $t_1 - t_0$ depends on the shape of the function $f_p(t)$, but it always reaches a maximum when $t_1 - t_0$ is zero.

Evaluation of the noise-induced error in measurement of pulse occurrence time is not simple if one attempts to accomplish it by the standard procedure, i.e., solving (8.90) for t_0 and measuring the rms difference between indicated and true values of t_0. The evaluation procedure to be used here will be a different one, based on probabilistic considerations and geared to digital measurement techniques. The details will be deferred to Chapter 9, but the general idea will be summarized verbally below.

Consider a measurement procedure in which the time domain is divided into slots, each a pulse length in duration. The pulse is known to be in one of those. It may be in two of them, depending on whether or not its placement coincides exactly with one of the slots. The strip of signal corresponding to the kth time slot is crosscorrelated with a reference waveform shaped like the incoming pulse and coinciding with the kth time-slot. The crosscorrelator outputs are then compared and a decision is made as to which slot contains the pulse, based on the relative outputs. Our evaluation of measurement error will be based on this technique. It will be discussed in Section 9.3.1 in a connection with range measurement by pulsed radar.

8.7 OPTIMAL MEASUREMENT OF FREQUENCY

The optimization of frequency measurement through the MLE theory in the highly idealized case where amplitude and phase are a priori known begins with the application of Eq. (8.6) in the form

$$\begin{aligned}
\int_0^T dt'\, v(t') \frac{\partial}{\partial \omega_c} [a \cos(\omega_c t' + \psi_r)] &= -a \int_0^T dt'\, t' \sin(\omega_c t' + \psi_r) v(t') \\
= \int_0^T dt'\, a \cos(\omega_c t' + \psi_r) \frac{\partial}{\partial \omega_c} (a \cos(\omega_c t' + \psi_r)) & \\
= -a^2 \int_0^T dt'\, t' \cos(\omega_c t' + \psi_r) \sin(\omega_c t' + \psi_r) & \qquad (8.93) \\
= -\frac{a^2}{2} \int_0^T dt'\, t' \sin(2\omega_c t' + 2\psi_r) - \int_0^T dt'\, t' \sin(0) & \\
= \frac{a^2}{2} \frac{\partial}{\partial(2\omega_c)} \int_0^T dt' \cos(2\omega_c t' + 2\psi_r) &
\end{aligned}$$

By virtue of assumption (8.10), the integral on the right-hand side of Eq. (8.93) is effectively zero; then

$$\int_0^T dt'\, t'\, v(t') \sin((\omega_c)_{\text{MLE}} t' + \psi) = 0 \tag{8.94}$$

or

$$\left[\frac{\partial}{\partial \omega_c} \int_0^T dt'\, v(t') \cos(\omega_c t' + \psi)\right]_{\omega_c = (\omega_c)_{\text{MLE}}} = 0 \tag{8.94$'$}$$

An analogous argument for the case where phase is not a priori known leads to

$$\left[\frac{\partial}{\partial \omega_c} [v_c^2 + v_s^2]\right]_{\omega_c = (\omega_c)_{\text{MLE}}} = 0 \tag{8.95}$$

According to (8.94)′ and (8.95) the maximum likelihood estimator of the angular frequency is the frequency of a cosine signal which when crosscorrelated with the incoming signal, results in a larger correlator output than any other frequency. If phase is known, the crosscorrelation is done coherently; if not, it is done noncoherently. Actually, this statement is not exactly equivalent to the mathematical statements (8.94)′ and (8.95). The vanishing of the derivative with respect to ω_c implies only a stationary point, which could be a maximum, minimum, or inflection point. The necessity that it be a maximum is the fundamental requirement of maximum likelihood estimator theory.

To implement Eq. (8.94)′, the incoming signal could be crosscorrelated with sine waves of various frequencies and the outputs compared. Alternatively, the frequency of the reference signal could be slowly varied and the correlator output observed as a function of time. These are the same two alternatives presented in Section 7.1.4 in the discussion of detection of a radio signal with unknown frequency. In effect, the procedure involves the use of a spectrum analyzer, which can be implemented with a bank of matched filters or a single matched filter with slowly varying central frequency. The only difference between detection with unknown frequency and measurement of frequency is that in the former case the frequency information made available by the optimum processing operation can be discarded, whereas in the latter case, it is the essence of the processing operation.

The crosscorrelator output in the noise-free case has already been determined in Section 6.5. It is given by (6.58), which in the context of our present discussion has the form

$$v_o(T; \omega_c) = \frac{aT}{2} \operatorname{sinc}(\Delta\omega T) \tag{8.96}$$

where $\Delta\omega$ is $\omega_c - \omega_i$, the difference between the actual frequency of the incoming signal and the frequency of the reference signal. The sinc function reaches its maximum absolute value when $\Delta\omega = 0$, i.e., when the incoming signal and reference signal frequencies are equal. If no noise were present, the processing operation for optimal frequency measurement would yield

$$\left\{\frac{\partial}{\partial\omega}[v_o(T;\omega_c]\right\}_{\omega_c=(\omega_c)_{\text{MLE}}} = \left\{\frac{aT}{2}\frac{\partial}{\partial\omega_c}[\text{sinc}((\omega_c - \omega_i)T)]\right\}_{\omega_c=(\omega_c)\text{MLE}} \tag{8.97}$$

If

$$|\omega_c - \omega_i| \ll \frac{\pi}{T} \tag{8.98}$$

then we can expand the sinc function in a power series and neglect all but the first few terms; i.e., (8.97) has the form

$$\begin{aligned}\frac{\partial}{\partial x}\{\text{sinc}(x)\} &\simeq \frac{\partial}{\partial x}\left\{1 - \frac{x^2}{3!} + \frac{x^4}{5!} - \frac{x^6}{7!} + \dots\right\} \\ &= -\frac{x}{3!}\left\{1 - 2\left(\frac{3!}{5!}\right)x^2 + 3\left(\frac{3!}{7!}\right)x^4 - \dots\right\} = 0\end{aligned} \tag{8.99}$$

where

$$x \equiv (\omega_c - \omega_i)T$$

In the noise-free case, the MLE of frequency is the solution of Eq. (8.99). If we restrict the reference signal frequencies to those close enough to ω_i to allow assurance that (8.98) is fulfilled, then all terms beyond the first could be neglected without serious error; thus we would have, approximately,

$$\frac{x}{T} \equiv (\omega_c - \omega_i) \simeq 0 \tag{8.100}$$

Equation (8.100) says, in effect, that the computation of the maximum likelihood estimator of frequency will yield the true incoming signal frequency in the absence of noise, provided the trial frequencies used in generating the crosscorrelator reference signal are sufficiently close to the true frequency. In practice, this means that the incoming signal frequency is a priori known to be within a given range of frequencies, and the measurement is for the purpose of "pinpointing" the frequency.

Evaluation of the effect of noise on accuracy of frequency measurement is somewhat more complicated than in the case of amplitude measurement or even that of phase measurement. The degree of difficulty is about comparable to that in pulse occurrence-time measurement (see Section 8.6).

As in the latter case, we will involve a digital measurement scheme as a basis for discussion of accuracy. The frequency domain will be divided into slots of equal bandwidth and the decision as to which slot contains the signal will be based on the relative outputs of the slots. The evaluation of accuracy is in terms of the probability of correctly determining in which slot the signal lies. This has already been discussed in Section 7.3 in the general context of multiple alternative detection. It will be discussed further in Section 9.3.2 in connection with measurement of radar range rate through Doppler shift and in Section 12.2 in connection with M'ary frequency shift keying as a digital modulation technique.

REFERENCES

See the reference list in Chapter 4 and the comments in the reference list in Chapter 7. The same comments apply to the references on the material in Chapter 8. In addition to these, the following are somewhat pertinent to Chapter 8.

Bello, 1960

Hancock and Wintz, 1966

Helstrom, 1960; Chapter VII, VIII

Middleton, 1960; Chapter 21

Slepian, 1954

Swerling, 1959

9

APPLICATIONS TO RADAR SYSTEMS—I

Many of the earliest practical applications of statistical communication theory in the United States were to radar technology. This was largely because of the existence of the M.I.T. Radiation Laboratory during World War II. The scientific staff of that laboratory was directly responsible for much American progress in radar technology during the 1940's and indirectly responsible, through its writings,† for much progress after the 1940's. The application of statistical concepts to problems of radar detection, tracking, and fire control was one aspect of such progress. This work has continued to have an effect on radar research and development during the years since 1946.

Before discussing the types of noise phenomena affecting radar systems, the techniques of analysis used to explain them, and the various methods of combating them, we will first briefly describe the principles of radar systems and define the classes of such systems that will be of concern to us later in this chapter.

A radar system environment will be defined as consisting of (1) a transmitter and antenna emitting radio frequency electromagnetic waves (usually microwaves, i.e., 1000 to 30,000 MHz) but possibly extending into the mil-

† That is, the M.I.T. Radiation Laboratory Series, of which a few volumes (e.g., Vol. 1, Ridenour, 1947; Vol. 24, Lawson and Uhlenbeck, 1950; Vol. 13, Kerr, 1951; Vol. 25, James, Nichols, and Phillips, 1947) are referenced several times in this book, are now classical. It is no exaggeration to say that they nourished an entire generation of radar engineers and, though now superseded by other texts, are still useful as reference books on many of the basic ideas in radar theory and practice.

limeter wave region (i.e., 30,000 to 300,000 MHz), sometimes known as *hyper-microwave*, (2) a target (e.g., aircraft, ship, etc.) from which the waves are reflected, and (3) a receiver located at the same position as the transmitter. The system's purpose is detection of the target or extraction of information about the target. The *bistatic* radar, in which transmitter and receiver are not spatially coincident, and the *passive* radar, which is simply a microwave or millimeter receiver acting on energy emitted by the target, could also be included in the general definition of radar. However, the discussions in this chapter will be directed toward the conventional *monostatic radar transceiver* system where transmitter and receiver are at the same location and operate with the same antenna. Some discussion of bistatic and passive radar will be presented in Chapter 10.

9.1 RADAR SYSTEM ANALYSIS

We postulate an idealized radar transceiver,† isolated in free space (i.e., not influenced by the presence of the earth or irregularities in the atmosphere) and an idealized target which subtends a negligibly small angle at the radar. The transmitter sends an electromagnetic wave into the surrounding space, the

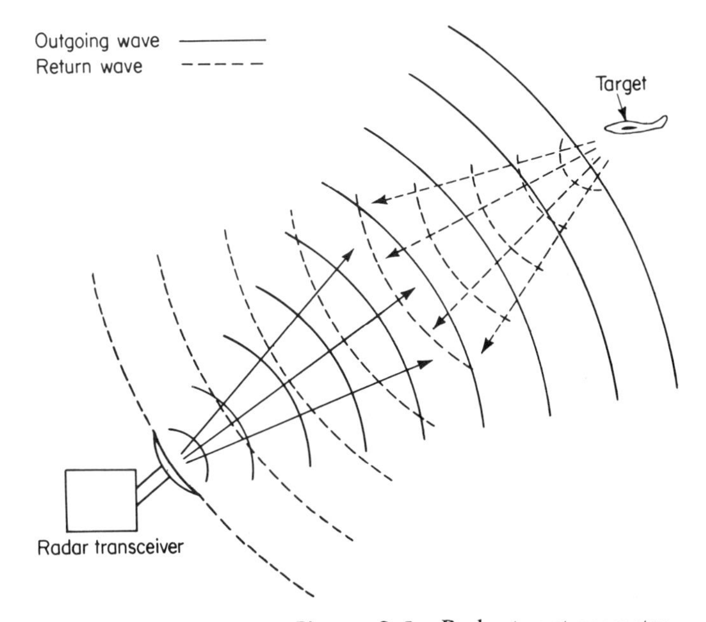

Figure 9-1 Radar-target geometry.

† That is, a transmitter and receiver at the same location.

energy in the wave being angularly distributed in a manner determined by the design of the antenna. The energy traveling in the direction of the target is intercepted by the target. Some of it is absorbed by the target and some is scattered out in various directions. That fraction of the incident energy scattered back toward the radar is picked up by the receiver, indicating the presence of the target, and usually containing information about the target's trajectory. The radar geometry is shown in Figure 9-1.

In the discussion to follow it will be useful for the reader to know the standard frequency bands used in radar. These are as follows: P, L, S, X, K, centered roughly at 2 m, 30 cm, 10 cm, 3 cm and 1.25 cm, respectively. The P and L bands are in the UHF region (see Table 5-1). The S, X and K bands are largely in SHF.

9.1.1 The Radar Equation

If the total peak power radiated into space is P_T and the transmitting antenna has a radiation pattern that is *isotropic* or uniform over all spherical polar angles θ and azimuthal angles ϕ, the power per unit area incident on a target at a distance r (called *range*) from the radar and at angles θ and ϕ is

$$\left(\frac{dP_i}{dA}\right)_{\text{target}} = \frac{P_T G_T(\theta, \phi)}{4\pi r^2} \tag{9.1}$$

where $G_T(\theta, \phi)$ is the one-way power gain of the transmitting antenna at angles θ and ϕ (see Section 5.3).

The target intercepts an amount of incident power proportional to its area A_T (normal to the direction of the incident wave) but only scatters back a fraction b_T of this power, depending on its geometry. The target may be thought of as a transmitting antenna whose excitation is provided by the incident wave. In response to this excitation, it returns to the receiving antenna an amount of power per unit area

$$\left(\frac{dP_R}{dA}\right)_{\text{receiver}} = \left(\frac{P_T G_T(\theta, \phi) b_T A_T}{4\pi r^2}\right) \tag{9.2}$$

The latter has an effective aperture area $A_R(\theta, \phi)$ (see Sections 1.3.2 and 5.3) which defines the fraction of the reflected power from direction (θ, ϕ) that is intercepted by the antenna and effectively delivered into the receiver. Multiplying dP_R/dA by $A_R(\theta, \phi)$ and defining $b_T A_T$ as the *radar cross section* of the target (denoted by σ), we obtain for the power delivered to the radar receiver

$$P_R = \frac{P_T G_T(\theta, \phi)\, \sigma A_R(\theta, \phi)}{(4\pi)^2 r^2} \tag{9.3}$$

A relationship between the gain and effective aperture area of an antenna is (see Eq. 5.13)

$$\frac{4\pi A_R(\theta, \phi)}{\lambda^2} = G_T(\theta, \phi) \tag{9.4}$$

where λ is the radar wavelength. Note that both the gain and the aperture area are the same for reception as for transmission, by virtue of the reciprocity theorem (see Section 5.5).

A_R and G_T may be written in the form

$$G_T(\theta, \phi) = G_0 |g(\theta, \phi)|^2 \tag{9.5}$$

$$A_R(\theta, \phi) = A_0 |g(\theta, \phi)|^2 \tag{9.6}$$

where G_0 and A_0 refer to peak values of gain and effective area, respectively, and $g(\theta, \phi)$ represents an antenna field pattern shaping function, normalized to unity at the peak, whose absolute square gives the angular distribution of relative power.

Substitution of (9.4), (9.5), and (9.6) into (9.3) results in

$$P_R = \frac{P_T G_0^2 \lambda^2}{(4\pi)^3 r^4} \sigma\{|g(\theta, \phi)|^2\}^2 \tag{9.7.a}$$

or alternatively

$$P_R = \frac{P_T A_0^2}{4\pi\lambda^2 r^4} \sigma\{|g(\theta, \phi)|^2\}^2 \tag{9.7.b}$$

The *radar equation* (9.7.a) can be used to determine the variation of received power with certain design parameters, provided one knows the target's radar cross section. In spite of the fact that the latter is usually known only in an order-of-magnitude sense, the radar equation is a very useful analytical and design tool. Its principal utility is in showing functional dependences of various parameters. Extreme accuracy in prediction of received power is rarely required, rough estimates to within a few decibels being quite sufficient for most engineering purposes. Because of the target's linearity in backscattering the waves (i.e., the fraction of the power backscattered is independent of the amplitude of the incident wave), the cross section can be considered as an intrinsic property of the target that does not change with the radar design parameters. Consequently, analyses of the effects of design changes on system performance can often be performed without knowledge of the absolute value of the radar cross section.

From the analytical results in Chapter 7, it is apparent that the parameter of interest in detection of a signal in additive bandlimited white Gaussian noise whose bandwidth is much wider than that of the signal is (E_s/N_o), where E_s is the total energy in the received signal, independent of its waveform, and N_o is the mean noise power per unit bandwidth. We note that (E_s/N_o) is identical to the parameter $\rho_o^{(o)} = 2\rho_i B_n T$, the coherent matched filter output SNR.

Since this parameter will be so important in discussions, it is desirable to write the radar equation in terms of it. To this end we assume that all radar transmission takes place within a time interval between $t = 0$ and $t = t_a$. The time period t_a is looked upon as the *total available time* for the radar to detect or locate the target. During this period, the peak transmitter power is either P_T or zero. In a CW radar, for example, it is always P_T. In a pulsed radar it is P_T for the duration of each pulse interval and zero within the interval between pulses.

The *duty cycle d* is defined as the fraction of the time interval 0–t_a during which the transmitter is activated. The total transmitted energy is

$$E_T = P_T t_a d = P_T t_a f_p \tau_p \tag{9.8}$$

where it is recognized that the duty cycle is actually the ratio of pulse duration τ_p to the interpulse period, and that the latter is the reciprocal of the pulse repetition frequency f_p, often abbreviated as PRF.

The signal energy received during the time 0–t_a, from (9.7.b) and (9.8), is

$$E_s = \frac{P_T t_a f_p \tau_p A_o^2 |g(\theta, \phi)|^4 f_c^2 \sigma}{4\pi c^2 r^4} \tag{9.9}$$

where the carrier frequency in hertz is denoted by f_c. The noise per unit bandwidth is (see Eq. (3.62))

$$N_o = \frac{\sigma_n^2}{2B_n} = kT_{\text{abs}} F_n \tag{9.10}$$

where k is Boltzmann's constant, T_{abs} is the absolute temperature, and F_n is the receiver noise figure.

In radar system engineering, there are certain controllable parameters and certain parameters fixed by nature or, in military applications, by noncooperative enemies. It is useful from a design point of view to separate these classes of parameters. The radar cross section is an example of a noncooperative parameter, which an enemy who does not want his aircraft or missiles to be detected will make as small as possible (e.g., by black-coating his vehicles to make them poor radio wave reflectors). The available time t_a and the minimum range at which we require radar to do its work are usually fixed by strategic considerations. The receiver noise figure F_n and the absolute temperature T_{abs} will be assumed uncontrollable in the discussions to follow, although F_n can be minimized by proper design of receiver components and T_{abs} would be minimized by cooling, as is done in low noise receivers using maser amplifiers.†

The problem upon which we will focus attention is that of optimizing the controllable parameters for particular radar functions (e.g., detection, ranging, angular measurement, etc.).

† L. S. Nergaard, Chapter 4 of Berkowitz, 1965, "Modern Low Noise Devices," pp. 432–467, especially pp. 442–445.

These controllable parameters, from our viewpoint, are P_T, f_p, τ_p, A_o and f_c. The other parameters, except for t_a and r, will be lumped into a single constant called κ. We take the ratio of (9.9) to (9.10),

$$\frac{E_s}{N_o} \equiv \rho_o^{(o)} \frac{\kappa}{r^4} P_T t_a f_p \tau_p A_o^2 f_c^2 |g(\theta, \phi)|^4 \tag{9.11}$$

where $\kappa \equiv \sigma/4\pi T_{abs} F_n c^2$

The minimum value of $\rho_o^{(o)}$ required for a given performance level will be determined by the radar function and the receiver processing technique. In any given application, the quantity usually of interest is the peak transmitter power required to achieve the minimum acceptable value of $\rho_o^{(o)}$ (to be denoted by $(\rho_o^{(o)})_{\min}$).

This value of P_T will be denoted by $(P_T)_{\min}$. It is the parameter $(P_T)_{\min}$ that we would like to minimize in many applications, i.e., we would like to achieve the specified performance level with as little expenditure of transmitter power as possible. Thus a form of the radar equation that will sometimes be used in the discussions to follow is

$$\begin{aligned}(P_T)_{\min} \text{ (in decibels)} &= -10 \log_{10} k + (\rho_o^{(o)})_{\min} \text{ (in decibels)} \\ &\quad - 10 \log_{10} (t_a f_p \tau_p A_o^2 f_c^2) - 40 \log_{10} |g(\theta, \phi)| + 40 \log_{10} r\end{aligned} \tag{9.12.a}$$

Another form, based on the idea of maximization of the range at which a given performance level can be attained with a fixed transmitter power, is

$$(r)_{\max} = \sqrt[4]{\frac{\kappa P_T t_a f_p \tau_p A_o^2 f_c^2 |g(\theta, \phi)|^2}{\rho_o^{(o)}}} \tag{9.12.b}$$

where $(r)_{\max}$ is the maximum attainable range.

In forms similar to (9.12.b), our equation is usually called the *radar range equation.* Its theoretical significance is enhanced by its demonstration that apparently large increases in transmitter power, antenna aperture area, target cross section, etc., cause very small and sometimes insignificant increases in maximum range. For example, a 16-fold increase in transmitter power is required for a doubling of maximum range. A doubling of transmitter power increases the range by only about 20 percent.

9.1.2 Pulsed and CW Radars

Most radars can be classified as either (1) pulsed radars, (2) CW doppler radars, or (3) pulsed doppler radars, In a pulsed radar, a pulse of RF energy is transmitted and range information is extracted from the return by measurement of the time required for the pulse to travel to the target and back. Since it travels a distance $2r$ at the velocity of light (3×10^8 m/s), the range r in meters is determined by computing $\frac{1}{2}(3 \times 10^8)t$, where t is the pulse travel time in seconds. The latter is typically of the order of tens or hundreds of microseconds. A target at a range of 15 km will correspond to a pulse travel time of 100 μs or 0.1 ms. In a CW doppler radar, a continuous monochromatic wave is transmitted and the doppler shift of the return is measured, providing an indication of the target's velocity component parallel to the line of sight (range rate). No range information is obtained directly from a CW doppler radar, but range can be continuously determined through integration of the doppler range rate measurements, provided the range is known at one point on the trajectory.

The pulsed doppler radar measures both pulse travel time and doppler shift and thereby acquires both range and range rate information.

9.1.3 The Radar Receiver

The radar receiver is usually of the superheterodyne type (see Section 5.4 and Section 6.8) as shown in Figure 9-2. The RF *head* consists of the antenna (usually a parabolic or paraboloidal dish with an aperture area of several square wavelengths for directivity and gain), a waveguide to carry the microwave signal from the antenna through the RF section of the receiver, and

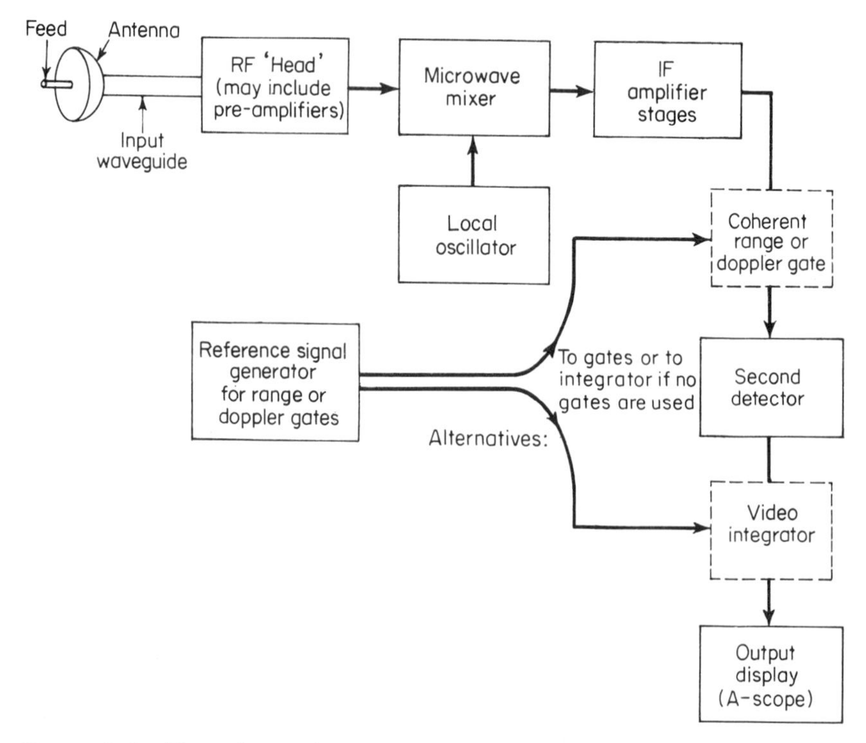

Figure 9-2 The radar receiver.

possibly a microwave amplifier.† The microwave signal is mixed with a local oscillator signal designed to produce an IF signal frequency of the order of megacycles per second. The IF signal travels from the mixer through one or more stages of IF amplification to the second detector, where it is rectified (e.g., square law, linear half wave, linear full wave: see Section 3.3) and then through a video amplifier which strips off the remaining IF "wiggles" from the signal, possibly amplifies it, and produces a clean video pulse (or dc signal in the case of CW radar) for display. A doppler radar must have an additional feature, namely a frequency detector in the IF stage to detect the doppler shift.

9.2 RADAR DETECTION THEORY

The problem of detecting the presence of targets by radar has been studied extensively.‡ The same is true of problems of range, angle, and velocity measurement by radar. At first glance these appear to be different problems, but in fact they are specialized aspects of the same problem.

The general question asked by the radar observer is as follows: "Is there a target within a specified region of space?" There is a difference between the way in which this question is "asked" by a detection radar and by a radar whose major purpose is range or velocity measurement. The former chooses a very large volume of space, perhaps the entire radar field of view, and sends out signals to interrogate the target. Figuratively speaking, these signals ask the target "Are you somewhere out there?" The range or velocity measuring radar refines the detection process by successively examining small regions of space and inquiring of the target "Are you within this region?" (The word "space" here refers to both position space, i.e., range and angle, and the space of all possible line-of-sight velocities or "range rates.") An affirmative answer provides accurate position and/or velocity information, as well as existence information.

In practice, detection radars successively examine small regions of space and interrogate the target only as to its presence within the given region. This means that information about target trajectory is inherently present in the process of detection. Therefore the process of detection of a target and that of measurement of its trajectory cannot be completely separated.

One important difference between detection and measurement does exist, however. It is often desired to accomplish the former at low SNR, while the

† Sometimes it is desirable to provide gain in the RF stage, in which case a preamplifier is used. The need for this is usually circumvented by the heterodyning process (see Section 6.8).

‡ See Lawson and Uhlenbeck, 1950, Chapter 7, especially Sections 7.3, 7.4, pp. 165–172; Skolnick, 1962, Chapter 9, pp. 408–452; Berkowitz, 1965, Part III, pp. 105–193.

latter is usually attempted only at high SNR. For this reason, high SNR approximations, which often simplify calculation of statistical quantities such as error probability, can usually be invoked safely in analysis of measurement but not necessarily that of detection.

The radar signal at RF has parameters not precisely known to the observer, for example, range r, cross section σ, phase ψ_r, angular positions θ and ϕ, and range rate $\dot{r}$, as given by the doppler shift f_D. The noise in the RF and IF stages of a radar is predominantly thermal and shot noise and external noise (atmosphere and extraterrestrial) of Gaussian character. For all practical purposes, as in other kinds of radio receivers (see Section 5.5.3), much of the RF noise can usually be regarded as Gaussian and as spectrally flat over the receiver bandwidth. Thus the theory of detection of signals with random parameters in bandlimited white Gaussian noise (as discussed in Section 4.1) and its application to radio signals (as discussed in Chapter 7) is applicable to detection of radar signals if we omit cases where impulse noise and tone interference are important.

9.2.1 Idealized Radar Detection Theory

Before discussing specific radar detection problems, let us consider a highly idealized and usually fictitious situation in which the form of the incoming radar signal is a priori known in complete detail at the receiver, such that a perfect matched filter can be used. This case represents the best possible detection performance. We can invoke the theory discussed in Section 7.1.2. The error probabilities are given by (7.19) and (7.20):

$$P_{fr} = \tfrac{1}{2}\{1 - \operatorname{erf}((1 - K_T)\sqrt{\rho_i B_n T})\} \tag{7.19}$$

$$P_{fa} = \tfrac{1}{2}\{1 - \operatorname{erf}(K_T\sqrt{\rho_i B_n T})\} \tag{7.20}$$

Using the ideal observer detection criterion (Section 4.1.2.1), assuming equal a priori probabilities for "signal present" and "noise alone," and choosing the threshold level to minimize the over-all error probability P_ϵ(i.e., $K_T = \frac{1}{2}$, see Section 4.1.2.1), we have

$$P_\epsilon = \frac{1}{2}\left\{1 - \operatorname{erf}\left(\frac{1}{2\sqrt{2}}\rho_o^{(o)}\right)\right\} \tag{9.13}$$

We can now combine (9.13) with the radar equation in the form (9.12.a) and determine the peak transmitter power required for detection with a radar using idealized matched filtering, where some predetermined value of P_ϵ defines the threshould value of $\rho_o^{(o)}$. This is not a realistic value of $(P_T)_{\min}$ but it can be used as a standard against which calculations made with more realistic assumptions can be compared.

To accomplish this idealized calculation, we should first plot a curve of P_ϵ versus $\rho_o^{(o)}$. But false rest and false alarm probability curves have already been plotted for the case of an arbitrary signal in Gaussian noise, a case that contains the situation of interest here (Chapter 7, Figure 7-2). The over-all error probability with false alarms and false rests assumed equally likely can be determined from Figure 7-2. Examining the curves of Figure 7-2 we observe that for $K_T = 0.5$ (the case for which false rest and false alarm probabilities coincide) an over-all error probability of 10^{-1} corresponds to $\rho_o^{(o)} \cong 7$, and an error probability of 10^{-2} corresponds to $\rho_o^{(o)} \cong 20$, and $P_\epsilon = 10^{-3}$ corresponds to $\rho_o^{(o)} \cong 33$, etc.† In general, we can always specify a unique value of $\rho_o^{(o)}$ given a maximum acceptable value of P_ϵ (denoted by $(P_\epsilon)_{\max}$). To incorporate this into (9.12), we express $(\rho_o^{(o)})_{\min}$ as a function of P_ϵ in (9.12) and simply choose a value of $(\rho_o^{(o)})_{\min}$ from the curves of Figure 7-2 to correspond to any desired value of P_ϵ.

9.2.2 Effects of A Priori Uncertainty in Target Parameters

We have not yet accounted for the factor of a priori ignorance of target signal parameters in radar detection. The idealized theory discussed above presupposes a complete knowledge of r, $\dot{r}$, ψ_r, θ, and ϕ. This is extremely unrealistic.

We can immediately dismiss the problem of uncertainty in ψ_r by remarking that filtering should be noncoherent to account for phase ignorance (see Section 7.1.3). To account mathematically for ignorance in the parameters r, $\dot{r}$, θ, and ϕ, we will first assume the antenna pattern function $|g(\theta, \phi)|$ to be flat over a solid angle Ω_b, and zero elsewhere. This is not a pattern that is realizable in practice, but this oversimplified model is useful for bringing out the important features of the problem of angular uncertainty. The results developed from it can be extrapolated approximately to realistic antenna pattern functions.

Suppose the initial uncertainty in angular target position is characterized by a solid angle Ω_u, the solid angle within which the target is known to be located before the beam begins to search for it. We assume that $\Omega_b \leq \Omega_u$, i.e., the beam solid angle is smaller than the solid angle of the uncertainty region. The beam, then, must be swept through the uncertainty region in a search for the target. Before the search begins, the target is as likely to be at one point within the uncertainty region as any other; thus, on the average, the beam will spend "on target" a fraction (Ω_b/Ω_u) of the total available search time t_a, i.e., a period $t_a(\Omega_b/\Omega_u)$. But since $A_0 = \lambda^2/\Omega_b$ (see Eq. (5.9)) and $\lambda = c/f_c$,

$$\Omega_b = \frac{c^2}{f_c^2 A_0} \tag{9.14}$$

† Note that $\rho_o^{(o)} = 2\rho_i B_n T$ and that the over-all error probability in the case $K_T = 0.5$ is given by $P_\epsilon = \frac{1}{2}(P_{frc} + P_{fac}) = P_{frc} = P_{fac}$. Thus we can read the over-all error probability directly from the curves on Figure 7-2.

Also,

$$|g(\theta, \phi)| = 1 \text{ when the beam is "on target"} \qquad (9.15)$$
$$= 0 \text{ when the beam is "off target"}$$

Now (9.11) takes the form

$$\rho_v^{(o)} = \frac{\kappa c^2 P_T f_p \tau_p A_0}{r^4 \Omega_b} t_a \frac{\Omega_b}{\Omega_u} = \frac{\kappa c^2 P_T f_p \tau_p A_0 t_a}{r^4 \Omega_u} \qquad (9.16)$$

with corresponding modifications in (9.12.a) and (9.12.b).

Equation (9.16) would apply to a case where range, range rate, and incoming RF signal phase and pulse shape were precisely known.

The problems of uncertainty in range and range rate can be treated in much the same way. If the radar is pulsed, then the length of the reference pulses in the receiver will determine the interval of target ranges that will be allowed into the receiver, i.e., the matched filter acts as a range gate. If the reference pulses are coincident with the target pulses, then large signals will appear at the matched filter output; if not, only noise will be present in the output. It is a consequence of matched filtering arguments that optimum detection will be achieved if the reference and signal pulses have precisely the same shape, i.e., if we assume the incoming pulses to be square, the reference pulses should also be square and of the same duration. The reference pulse duration, then, determines the proper range gate width.

Suppose the reference pulses are swept slowly along the time base until the proper range has been located. If the sweep rate is uniform, the gate spends a fraction $w_{rg}/\Delta r$ of the available time "on target," where w_{rg} is the range gate width and Δr is the a priori range uncertainty. But according to the above arguments, w_{rg} should optimally be the equivalent in range units of τ_p, i.e.,

$$\text{fractional "on range" time} = \frac{w_{rg}}{\Delta r} = \frac{c\tau_p}{2\,\Delta r} \qquad (9.17)$$

Analogous considerations apply to the range rate uncertainty problem. The receiver responds to a range of doppler shifts determined by its bandwidth. If one wished to design a receiver to extract very accurate range rate information from the signal, it would be desirable to construct a *doppler gate* or *velocity gate*. This would be a flatband filter, receptive to a small band of frequencies around a given central frequency f_D (equivalent to f_c) believed to correspond to the target doppler shift. This central frequency would be varied slowly across the doppler uncertainty region. We denote the doppler frequency uncertainty by Δf. The doppler gate width in velocity units is denoted by w_{vg} and the corresponding bandpass filter bandwidth is $2B_R$. If the range rate uncertainty is $\Delta v = (c/2)(\Delta f/f_D) = (c/2)(\Delta f/f_c)$, the fractional time spent within the correct doppler region is $w_{vg}/\Delta v$. Moreover, the doppler gate width is equivalent to the receiver's bandwidth $2B_R$ which should be equal to about twice reciprocal pulse duration for optimum detection,† i.e.,

$$\text{fractional “on-doppler” time} = \frac{w_{vg}}{\Delta v} = \frac{c\Delta f}{2f_D\,\Delta v} = \frac{cB_R}{f_D\,\Delta v} = \frac{c}{f_c\tau_p\,\Delta v} \tag{9.18}$$

If available time is severely limited, it might be necessary to reduce the time required to search for the correct range and range rate. This can be done (see the discussion in Section 7.1.4.2) by providing multiple range gates and multiple doppler gates, each gate covering a fraction of the total uncertainty region in range and range rate. Suppose there are N_{rg} equal-width range gates, each being swept over a region of width $\Delta r/N_{rg}$. This would increase the fraction of the total available time spent on target by a factor $N_{rg}N_{vg}$. To amplify this point, we note that if one range gate and one doppler gate is used, each gate covering the entire uncertainty region Δr or Δv, then the radar will be on target as long as the target is within the beam, i.e., the actual on-target time will be $t_a(\Omega_b/\Omega_u)$ and (9.16) will then hold as given. If single range and doppler gates are used and the range gate is swept over the entire uncertainty region Δr in the time $t_a(\Omega_b/\Omega_u)$ while the doppler gate, of width Δv, remains stationary, then the actual on-target time will be $t_a(\Omega_b/\Omega_u)(w_{rg}/\Delta r)$ (as obtained from (9.17)) for t_a. By similar reasoning, if the doppler gate sweeps over its uncertainty region Δv in the time $t_a(\Omega_b/\Omega_u)$ while the range gate, of width Δr, remains stationary, then the effective on-target time will be $t_a(\Omega_b/\Omega_u)(w_{rg}/\Delta v)$. If both gates are swept simultaneously over their uncertainty regions, then the effective on-target time is that in which the beam, range gate, and doppler gate are simultaneously on target, which is $t_a(\Omega_b/\Omega_u)(w_{rg}/\Delta r)(w_{vg}/\Delta v)$. If multiple range or doppler gates are used, then each of the range and doppler gates are assigned fractions $\Delta r/N_{rg}$ and $\Delta v/N_{vg}$, respectively, of their uncertainty regions to be swept within the available time; hence the fractions $w_{rg}/\Delta r$ and $w_{vg}/\Delta v$ that appear above are changed to $w_{rg}/(\Delta r/N_{rg})$ and $w_{vg}/(\Delta v/N_{vg})$, respectively. The original available time t_a now becomes

$$\left(t_a\,\frac{b}{\Omega_u}\,\frac{w_{rg}}{\Delta r}\,\frac{w_{vg}}{\Delta v}\,N_{rg}\,N_{vg}\right)$$

Equation (9.16), then, must be multiplied by the factor $w_{rg}w_{vg}N_{rg}N_{vg}/\Delta r\Delta v$ to account for the use of multiple searching range and doppler gates.

The net result of these arguments and Eqs. (9.1) and (9.18) is the following expression for the minimum transmitter power required for detection:

$$\underset{\text{in dB}}{(P_T)_{\min}} = -10\log_{10}K + 40\log_{10}r + \underset{\text{in dB}}{(\rho_o^{(o)})_{\min}} + 10\log_{10}\left(\frac{V_uB_r}{t_af_pA_0N_{rg}N_{vg}}\right) \tag{9.19}$$

---

† See Lawson and Uhlenbeck, 1950, Section 8.6, pp. 199–210, especially Figure 8-7, or Berkowitz, 1965, Figure 2-7, p. 24.

where $K \equiv \kappa c^4/2f_c$ and V_u is equal to the product $(\Omega_u \, \Delta r \Delta v)$, and is designated as the *volume of uncertainty*. Strictly speaking, it is not a volume at all, but it is representative of a kind of four-dimensional volume in a rectangular coordinate space whose coordinates are (r, θ, ϕ, and $\dot{r}$).

The interpretation of (9.19) may be summarized in the following statements:

1. The power required to detect a target at a given range and with a given degree of a priori parameter uncertainty is inversely proportional (in absolute power units) to the antenna aperture area.† This is a weaker dependence on aperture area than would be indicated by the usual form of the radar equation (i.e., $\propto$ (area)^2), and the weakening is due to the fact that the decrease in fractional time spent on target during the search operation when the aperture area is increased offsets the increased gain accompanying a larger aperture area.
2. Required power is inversely proportional to available time. This is simply due to the increase in signal energy with fixed peak transmitter power.
3. Required power is proportional to the product of a priori uncertainties in range, range rate, and solid angle. The larger these uncertainties, for a given available time, the shorter the fractional time spent on target.
4. To overcome the effects of parameter uncertainties, we can sweep the antenna beam, the range gate, and the velocity gate through the uncertainty regions. The signal is received only while the beam and gates are in the regions corresponding to the correct target parameters. To increase the time spent on target, we can employ multiple gates, each occupying a fraction of the total uncertainty region. The increase in received signal energy is then proportional to the number of range gates and to the number of doppler gates.
5. The range gate width in time units is simply the pulse duration, and the doppler gate width in frequency units is the receiver bandwidth. But optimum pulse duration is the reciprocal bandwidth; hence the fractional on-target time, being proportional to the range and velocity gate-width product, is independent of receiver bandwidth. But the received signal energy with a fixed peak transmitted power is proportional to pulse width, or (optimally) inversely proportional to receiver bandwidth; thus the required power is proportional to receiver bandwidth.

It is common in radar engineering practice to ask for the maximum range of a radar for a given transmitted power and a given level of detection or location performance rather than the required transmitted power for a given range and performance level. This is perhaps the most familiar application of the radar equation, and in this form it is usually called the *radar range equa-*

---

† As noted in Section 5.3, the effective area is proportional to the true physical aperture area A_0. The constant of proportionality is determined by the type of antenna employed.

tion (as noted below Eq. (9.12.b)) Solving (9.19) for r, we obtain

$$r_{\max} = \sqrt[4]{\frac{KP_T t_a f_p A_0 N_{rg} N_{vg}}{V_u B_R (\rho_o^{(o)})_{\min}}} \tag{9.20}$$

where $r_{\max}$ is the maximum attainable range and $(\rho_o^{(o)})_{\min}$ is the threshold value of E_s/N_o, i.e., the minimum value required to attain a range $r_{\max}$.

9.3 TARGET TRAJECTORY MEASUREMENT

The problem of continuously locating a target, as previously stated, (Section 9.2), can be regarded as an extension of the detection problem. In locating targets, the a priori probabilities are somewhat different than in the pure target detection problem. It is assumed a priori certain that the target is present within the radar field of view, having previously been detected. The rough trajectory information contained in the detection process has been used to reduce the a priori uncertainty in target parameters. The parameter V_u is, in general, smaller than its values for a typical detection situation. Also, some sort of range, range rate, and angle tracking is provided, thus circumventing the necessity for scanning the beam over a large angular region and searching for the correct range and velocity.

Strictly speaking, the tracking radar cannot be analyzed without going into the theory of automatic control, or *closed loop* analysis. This will not be done here, and the treatment to follow will be an *open loop* analysis, confining attention to optimization of the process of feeding trajectory information into the tracking system.

The basis of optimization of the measurement of target parameters is maximum likelihood estimator theory applied to a sinusoidal signal (usually a pulsed sinusoid) superposed on a bandlimited white Gaussian noise. The theory was discussed in Chapter 8 in connection with extraction of information from radio signals. The results of the analysis are applicable to the present discussion: The optimum processing for estimation of range rate (equivalently, doppler shift on the carrier frequency) is given by (8.94), (8.94)′, or (8.95), i.e., determination of the reference signal frequency which maximizes the output of a crosscorrelator, coherent if phase is a priori known, noncoherent otherwise.

The same theory applied to the problem of estimating range through measurement of pulse travel time leads to the result (8.90), where time is converted into range through a proportionality constant equal to half the velocity of light. The processing of a radar signal dictated by the theory for optimum estimation of range is noncoherent crosscorrelation, using as a reference pulse that which maximizes the correlator output. This tells us that the processing of a pulsed radar signal for optimum estimation of range is equivalent to the processing for optimum detection. The only difference is that the range information inherent in the processing is useful in the case of range estimation and is superfluous in the case of pure detection. The same can be said of range rate estimation. The doppler shift of the signal in a pure radar detection problem is simply an additional unknown parameter to reckon with; in range rate measurement it is the item to be measured. In both cases, however, the processing method is basically the same, information about the doppler shift becoming available as a result of the processing. This information can be discarded if one is only interested in detecting the target's presence, but it is available and can be used if so desired.

Having established that crosscorrelation or matched filtering is the optimum processing scheme for both detection and trajectory measurement in a Gaussian noise environment, we can now consider trajectory measurement as a specialized form of detection.

The theory discussed in Sections 8.6 and 8.7 carries over into our present problem, and the results presented in those sections are applicable thereto with appropriate proportionality factors. The process of optimal measurement of target parameters is equivalent to that of optimally answering the question, "Is the target within a particular region of solid angle Ω_0, within the range interval $r_0 - (\Delta r/2)$ to $r_0 + (\Delta r/2)$, and within the range rate interval $v_0 - (\Delta v/2)$ to $v_0 + (\Delta v/2)$?" It will be recalled that this is precisely the question asked in a detection process, the only modification being a reduction in the size of the region of a priori uncertainty in parameters.

In the detection case, it is important only that the unknown target parameters be averaged out in the processing, but detailed information about them acquired in the detection process can be discarded. In target location, however, such is not the case. Each separate interrogation of the target must provide a highly reliable answer to the question of whether it is within the region covered by the interrogation. The SNR should be very high, from which it follows that both false rest and false alarm probabilities are very low, in order that a single interrogated region and none other will clearly indicate target presence. If the SNR is sufficiently high, there will be only negligible ambiguity in deciding which of the regions contains the target. By making these regions small enough and the SNR large enough the accuracy of the measurement can be made arbitrarily high. Thus the accuracy would be limited only by available transmitter power if the propagation medium were perfectly "clean." In practice, accuracy is further limited by refraction effects and other propagation vagaries (see Section 5.5.4).

9.3.1 Range Measurement

Consider the range gating process applied to target detection, as discussed in Section 9.2.2. With the quasi-static approach to the range searching operation, the crosscorrelator output at any one position of the range gate is a superposition of pulses belonging to a time interval equal to the gate duration. Comparison of outputs corresponding to different gate positions should yield one output that is overwhelmingly larger than all others. It is then decided that the target is within the range interval corresponding to the high-output gate position. If a bank of range gates covering the unknown range region is used, the high output is that corresponding to one particular gate. In order to be explicit in the discussions to follow, we will assume a bank of nonoverlapping gates, as shown in Figure 9-3.

The range measurement problem can be regarded as a decision problem. Strictly speaking, it is not a binary decision problem, but involves a decision between N_{rg} alternatives, where N_{rg} is the number of gates. The decision is an answer to the question "Which gate contains the target?" It can be analyzed through decision theory if it is assumed that the noises corresponding to different gates are statistically independent.

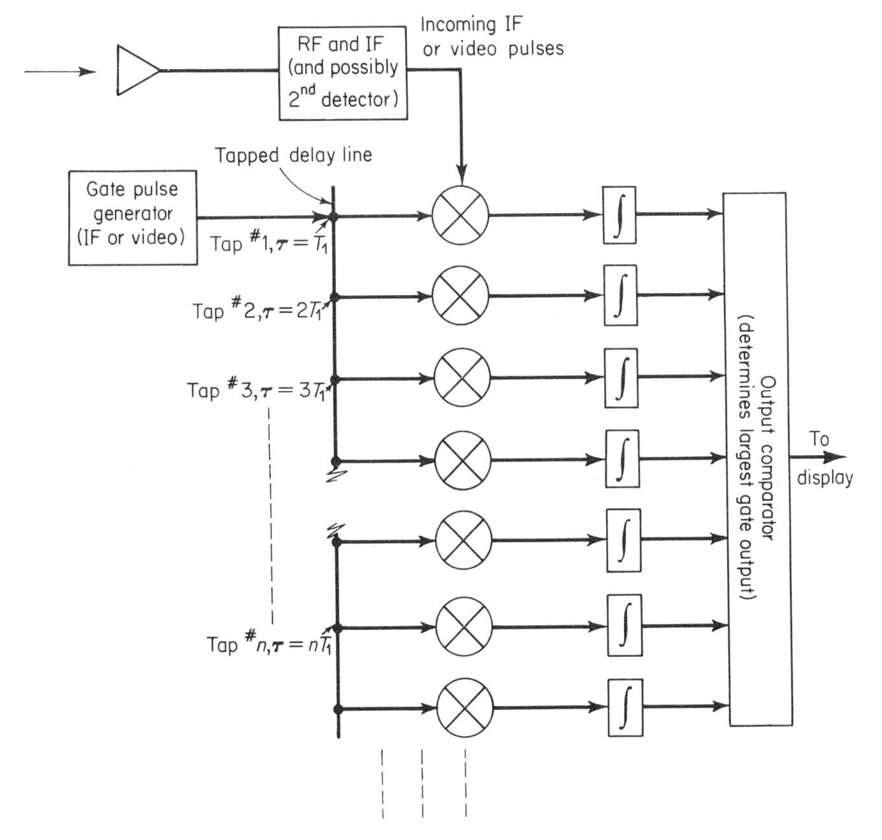

Figure 9-3 Range measurement with multiple gates (IF coherent or video gating).

Note that only one of the gates actually contains the target. Therefore that gate, say the kth in the gate bank, contains signal-plus-noise and the others contain noise alone. The gate with the largest output will be interpreted as that containing the largest signal. The probability of correctly deciding which gate contains the target, then, is merely the probability that the signal-plus-noise output of the kth gate exceeds the noise-alone output of any other gate. This is nothing more than the probability that $V_{s+n} - V_n$, the difference between output amplitudes for the gate containing the signal and all the gates with noise alone, exceeds zero.

This sort of probability was calculated in general in Section 7.3 in connection with multiple alternative detection. The general problem discussed in that section was that of correctly deciding which of M possible signals is present in a background of additive Gaussian noise, where the M signals and noises are mutually statistically independent and the signals have equal energies. In the case now under discussion, the M possible signals are the radar returns corresponding to M different range or range rate regions. Detection may be coherent or noncoherent, depending on whether phase is a priori known.

Note that target signals from two different ranges r_1 and r_2 do not actually have equal energies. Their energies differ by a factor $(r_1/r_2)^4$. The theory becomes somewhat more complicated if account is taken of these energy differences. To apply the equal energy theory leading to (7.89) or (7.92) to the case at hand, we have to (1) consider the a priori uncertainty region Δr as sufficiently small so that energies of signals from different ranges within that region do not differ greatly, or (2) if condition (1) is not met, think of the parameter ρ_o that appears in the error probabilities as an *average* SNR, i.e., averaged over the a priori range uncertainty region. This latter procedure is not rigorous, because the error probabilities would differ from (7.89) or (7.92) if energy differences were accounted for accurately. However, the theory will still provide us with meaningful answers if we are willing to accept a certain amount of ambiguity in the precise numerical value of ρ_o. In view of the difficulty of precisely specifying radar cross section and other parameters, this is not a serious limitation in rough radar system analysis.

If it is decided that the range or doppler gate with the largest output will be regarded as that containing the target, then the probability of determination of the correct gate is given by unity minus the error probability. From (7.92)′, this is given for the coherent case by

$$P_s = 1 - P_\epsilon = \frac{1}{2^{M-1}}\left\{1 + \operatorname{erf}\left(\sqrt{\frac{\rho_o}{2}}\right)\right\}^{M-1} \tag{9.21}$$

and for the noncoherent case (from (7.89)) by

$$P_s = 1 - P_\epsilon = \sum_{\ell=0}^{M-1} \frac{(M-1)!(-1)^\ell}{(\ell+1)!(M-1-\ell)!}\, e^{-\rho_o \frac{\ell}{\ell+1}} \tag{9.22}$$

where ρ_o is the noncoherent matched filter output SNR, equal to $\rho_i B_n T$. Note that the expressions (7.67), (7.69), (7.71), (7.71)', and (7.71)'' have been used to obtain (9.21) from (7.92)

The expression (9.22) is applicable to both (1) the quasi-static range search model and (2) the multiple range gate model, provided the width of the range gate is at least as large as the noise correlation time.† In the case of (1) this condition implies that the noises at times corresponding to different gate positions are mutually uncorrelated, and in the case of (2), that the noises in different gates are mutually uncorrelated. Because the cosine and sine components of the gate outputs are Gaussian, the absence of correlation implies statistical independence (see Section 3.1). Note that the noise correlation time is the reciprocal of the noise bandwidth, which is, in turn, equal to the receiver bandwidth (see Section 2.4.4.4). If the gate duration is comparable to a pulse width, and the receiver is optimally designed, the gate duration, being equal to the reciprocal bandwidth, is equal to the noise correlation time. The condition of statistical independence between gate outputs required for validity of (9.22) will be met.

The success probability P_s for desired values of $\rho_o^{(o)}$ and M can be inferred from multiple alternative error probabilities like those plotted in Figures 7-8 and 7-9. The curves of Figure 7-8 correspond to (9.22) and those of Figure 7-9 correspond to (9.21). By subtracting the error probabilities plotted in these figures from unity, one obtains the success probabilities of (9.21) and (9.22).

It must now be decided what constitutes a successful measurement. It is desirable that the success probability (9.22) should be extremely high e.g., 99.99 percent, before we can regard it as virtually certain that the gate we have decided upon actually contains the target. We then set a threshold on the input SNR and the integration time, i.e., minimum values of these quantities which will produce a 99.99 percent probability of correctly deciding which gate contains the target.

It is implied by discussions in Sections 9.2.1 and 9.2.2 that the optimum range gate width is about one received pulse length. To allow for pulse stretching, this is best increased, perhaps to two pulse lengths. In any case, the receiver (bandpass) bandwidth is designed to be about twice the reciprocal of the expected duration of an incoming pulse. The smaller the range gate width the more precisely can the target's range be determined. Hence, the potential accuracy of range measurement is inversely proportional to pulse duration, or directly proportional to bandwidth. In fact, if we take the viewpoint of the above discussion with respect to range measurement, the range

† Equation (9.21) will be excluded from the discussions to follow, on the grounds that coherent filtering is usually impractical in the cases to be discussed. If range and range rate are a priori uncertain, then a priori phase ignorance can be assumed.

to which the target can be confined with near certainty after the measurement is made is roughly equal to $c/2B_R$.

The measure of quality of the measurement can be defined as the fractional reduction in range uncertainty. We will call this quality measure L_r. If the a priori uncertainty is Δr, and the a posteriori uncertainty is δr (equal to the range gate width w_{r_g}), then from the above arguments

$$L_r = \frac{\Delta r}{\delta r} = \frac{\Delta r}{w_{r_g}} = \frac{2\,\Delta r}{c}B_R = \frac{2\,\Delta r}{c\tau_p} \tag{9.23}$$

According to (9.23), the quality of the range measurement is directly proportional to receiver bandwidth. But an increase in receiver bandwidth must be purchased at a price of increased noise. This increased noise, if signal power remains constant, has the effect of reducing the success probability as given by (9.21) or (9.22). Therefore, if a given success probability is required before one is willing to regard the measurement as sufficiently reliable, there is a limit imposed on L_r with fixed signal power. To determine this limit, we note that the parameter which determines success probability is $\rho_o^{(o)}$, as given by (9.16) for a flat antenna beam with solid angle of coverage Ω_b searching for the target and a priori uncertainty Ω_u in solid angle. Incorporating the arguments leading to (9.17), (9.18), and (9.23) into (9.16), and setting $(w_{v_g}/\Delta v)N_{v_g}$ equal to unity in (9.18) to eliminate doppler gating, we have

$$\rho_o^{(o)} = \left[\frac{\kappa c^3 P_T f_p A_0 t_a \tau_p N_{r_g}}{\Omega_u r^4}\right]\frac{1}{L_r} \tag{9.24}$$

The value of $\rho_o^{(o)}$ required for a given success probability according to (9.22) can be read from a curve like those in Figure 7-8. This value is denoted by $[\rho_o^{(o)}]_{L_r}$. Note that $\rho_o^{(o)}$ depends on the number M of possible regions in which the target may lie before the measurement is made. In the case at hand, as indicated through the subscript L_r on $\rho_o^{(o)}$, the number M is $L_r = \Delta r/w_{r_g}$. Prior to the measurement, the radar observer knows only that the target is somewhere in the range uncertainty region Δr. After the measurement, he has confined it to a region w_{r_g}. Hence the decision he is making in the measurement process is: "In which of $\Delta r/w_{r_g}$ possible range regions does the target reside?"

Substitution of the required value of $[\rho_o^{(o)}]_{L_r}$ into (9.24) results in an expression for the ratio of the transmitter power required to reduce range uncertainty by a factor L_r to that required to merely detect the target, if it is within the region $\Omega_u\,\Delta v\,\Delta r$, the parameters $t_a, f_p, \Omega_u, \Delta v, r, \sigma$ and N_{r_g} being equal in the two cases;

$$\left[\frac{(P_T)_{L_r}}{(P_T)_{\text{detection}}}\right]_{\text{dB}} = 10\log_{10}[L_r] + 10\log_{10}\frac{[\rho_o^{(o)}]_{L_r}}{[\rho_o^{(o)}]_{L_r=1}} \tag{9.25}$$

The parameters in (9.24) can all be specified in any given radar application. The parameters L_r and $\rho_o^{(o)}$ can be specified if we are willing to specify a minimum success probability for acceptance of the measurement, say 99.99 percent, and a minimum acceptable measure of quality, say $L_r = 10$. If these parameters are given, the maximum range $r_{\max}$ at which the specified performance levels can be attained is

$$r_{\max} = \frac{C_0 \sqrt[4]{P_T}}{\sqrt[4]{L_r [\rho_o^{(o)}]_{L_r}}} \tag{9.26}$$

where

$$C_0 = \sqrt[4]{\frac{\kappa c^3 f_p A_0 t_a N_{r_g}}{\Omega_u f_0 \Delta v}}$$

In Figure 9-4, the dimensionless ratio $(r_{\max}/r_0)$ is shown plotted against another dimensionless ratio (P_T/P_{T_0}) for various values of L_r and the success probability P_s. Here, P_{T_0} is an arbitrary reference power level, where $r_0 = C_0 \sqrt[4]{P_{T_0}}$ (the range corresponding to the case where L_r and $\rho_o^{(o)}$ are unity, P_{T_0} is the transmitted power and other radar parameters are the same as those used in computing $r_{\max}$). The success probability enters (9.26) indirectly through $\rho_o^{(o)}$, whose value for a partucular success probability can be determined from curves like those in Figure 7-8 or 7-9, depending on whether the detection process is coherent or noncoherent.

A table appears on Figure 9-4 indicating the success probability, (i.e., the probability of correctly deciding which range region contains the target) corresponding to a given value of $\rho_o^{(o)}$, a given form of detection (coherent or incoherent) and a given number of possible range regions. The table shows, for a few illustrative cases, how the success probability for a given value of $\rho_o^{(o)}$ decreases with an increasing number of range regions. It also shows that the success probability is higher for coherent than for noncoherent detection, and indicates the values of L_r corresponding to the given values of $\rho_o^{(o)}$.

An important point to be extracted from the discussion above is that pure detection of the presence of a target within the range interval Δr could take place with $L_r = 1$, at least in principle, while range measurement specifically requires that L_r exceed unity, otherwise nothing has been accomplished by the measurement. The range at which a target within Δr could be detected with a given set of radar parameters exceeds that at which a measurement with quality L_r can take place (with the same radar parameters and the same required error probability or, equivalently, the same value of $\rho_o^{(o)}$) by a factor $\sqrt[4]{L_r}$. For example, to obtain a quality L_r of 16 in a parameter regime where $[\rho_o^{(o)}]_{L_r}$ is not very sensitive to L_r, the maximum range will be reduced by a factor of approximately 2 from its value in the case where detection of the presence of the target is all that is required.†

† There is a subtle point in this argument. The error probability for *detection* is really a conditional probability, i.e., the probability that the target will not be detected *if* it is within Δr. If one equates this probability to the error probability for range measurement and equates all radar parameters in the two cases, then L_r relates to the ratio of *detection* range to the maximum range for range *measurement*.

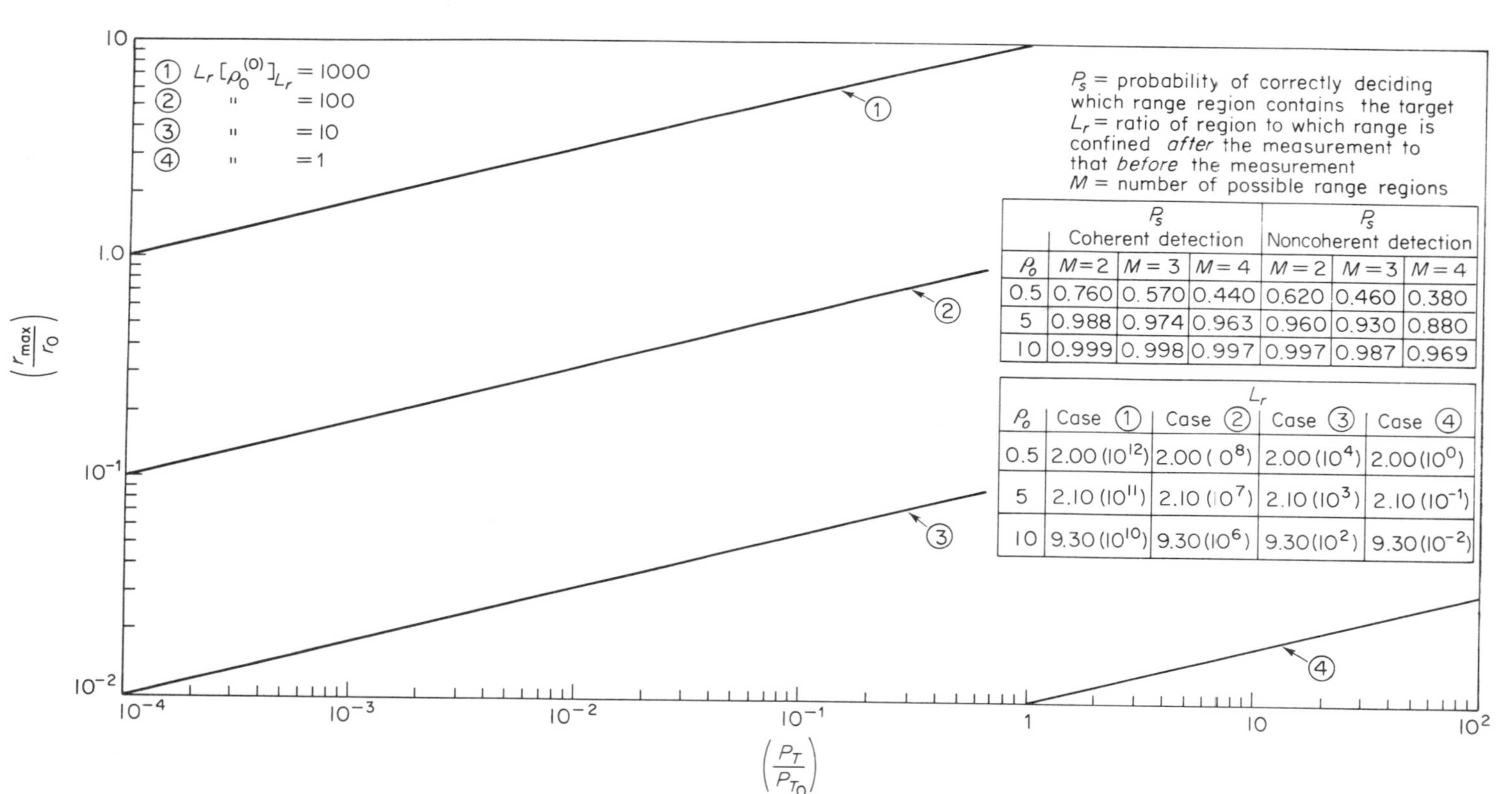

	P_s Coherent detection			P_s Noncoherent detection		
ρ_0	$M=2$	$M=3$	$M=4$	$M=2$	$M=3$	$M=4$
0.5	0.760	0.570	0.440	0.620	0.460	0.380
5	0.988	0.974	0.963	0.960	0.930	0.880
10	0.999	0.998	0.997	0.997	0.987	0.969

	L_r			
ρ_0	Case ①	Case ②	Case ③	Case ④
0.5	2.00 (10^{12})	2.00 (10^{8})	2.00 (10^{4})	2.00 (10^{0})
5	2.10 (10^{11})	2.10 (10^{7})	2.10 (10^{3})	2.10 (10^{-1})
10	9.30 (10^{10})	9.30 (10^{6})	9.30 (10^{2})	9.30 (10^{-2})

Figure 9-4 Maximum range (normalized) versus transmitted power (normalized) in radar range measurement.

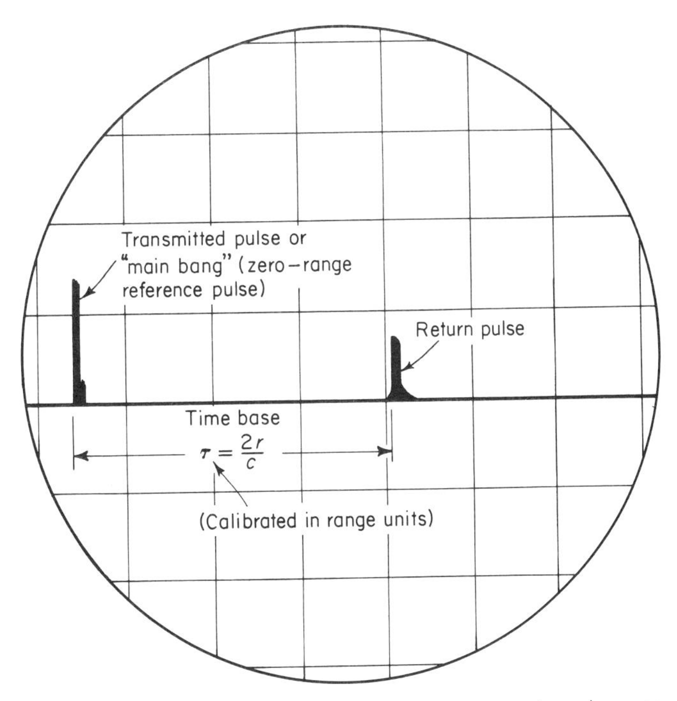

Figure 9-5 Range measurement on A-scope.

There are many ways to look at the range measurement process other than that discussed above.† For example, if SNR is sufficiently large, range gating is not required for an accurate measurement of target range. The range can be estimated visually on the time base of an oscilloscope after straightforward bandpass filtering (see Figure 9-5). Let us consider this process, which is suboptimum provided there is a priori pulse information available on which to base a crosscorrelation process. Otherwise, it may be all that is feasible. It is, in fact, a very common method of radar range measurement.

Suppose the pulse enters the receiver and undergoes bandpass filtering and rectification. Bandpass filtering is the nearest thing to crosscorrelation one can do if a reasonably accurate range estimate can be made before the precise measurement of range begins.

The range measurement is based on a determination of the instant of time at which one of the pulse edges, say the leading edge, crosses a certain threshold level. If noise is present, an error in reading this instant of time on the oscilloscope will result. The analysis of the noise-induced error incurred

† See Lawson and Uhlenbeck, 1950, pp. 20–21; Skolnick, 1962, Section 105, pp. 462–474.

in this measurement technique will not be discussed here. An excellent account of it has been given by Skolnick.†

9.3.2 Range Rate Measurement

Determination of the target trajectory can be partially accomplished indirectly by measuring range rate (doppler shift) and continuously integrating it to find the range. This is feasible if an initial value of range is available as a reference. In some cases, range rate is of direct interest and must be measured by the radar.

The CW doppler radar is ordinarily used for this type of measurement. The considerations of Section 9.2 regarding detection radars can be applied here and those of Section 9.3.1 regarding range measurement with pulsed radars have an analog for range rate measurement with CW doppler radars. In the case of a CW doppler radar, the usual type of range gating is not used, because one cannot isolate range regions through the measurement of pulse travel time.

The analog of (9.23) obtained through (9.18),‡ is

$$L_v = \frac{\Delta v}{\delta v} = \frac{\Delta v}{w_{v_g}} = \frac{fc\Delta v}{cB_R} \tag{9.27}$$

where L_v the quality measure for range rate measurement, is defined as the ratio of Δv, the a priori region of uncertainty in range rate, to δv, the a posteriori region of uncertainty in range rate. The latter is by definition the width of the doppler gate.

Again, as in the case of range measurement, we adopt the philosophy that the range rate region to which the target is confined after the measurement is that corresponding to the doppler gate with largest output, if multiple stationary gates covering the a priori uncertainty region are used. If a single gate sweeps through the uncertainty region, we can use the quasi-static analysis approach, dividing the region into equal intervals of width δv and assuming that the gate switches discretely from one interval to the next, remaining in each interval for a period T_v. In this case, it can also be said that the a posteriori uncertainty region is that corresponding to the gate with the largest output, where the kth gate now refers to the position of the gate during the kth time interval.

The probability of a correct decision as to which gate contains the target is again given by (9.21) or (9.22).

A minimum acceptable value can again be set for this probability (e.g., 99.99 percent), and the corresponding required value of $\rho_o^{(o)}$ can be deter-

† Skolnick, pp. 464–465.

‡ The last expression in (9.18) cannot be used here, because pulse width τ_p has no meaning in a CW radar. Thus we must think in terms of the receiver bandwidth B_R.

mined from (9.22) or Figure 9-4. The value of $\rho_o^{(o)}$ for a CW radar with N_{v_g} doppler gates each of width δv and each searching over a region $\Delta v/N_{v_g}$ units wide is (from (9.16), (9.17), (9.18), and (9.27) and with $f_p\tau_p$ and $(w_{r_g}/\Delta r)N_{r_g}$ set equal to unity in (9.16) and (9.17) to eliminate pulsing and range gating)

$$\rho_o^{(o)} = \left[\frac{\kappa c^2 P_T A_0 t_a N_{v_g}}{\Omega_u r^4}\right]\frac{1}{L_v} \tag{9.28}$$

Equation (9.28) is the analog of (9.24) for range rate measurement. It tells us the value of $\rho_o^{(o)}$ attained in range rate measurement with a given set of values of the parameters inside the brackets and with a range rate uncertainty reduction L_v.

The analogs of (9.25) and (9.26), respectively, are

$$\left[\frac{(P_T)_{L_v}}{(P_T)_{\text{detection}}}\right]_{\text{dB}} = 10\log_{10}[L_v] + 10\log_{10}\frac{[\rho_o^{(o)}]_{L_v}}{[\rho_o^{(o)}]_{L_v=1}} \tag{9.29}$$

and

$$r_{\max} = \frac{C_0'\sqrt[4]{P_T}}{\sqrt[4]{L_v[\rho_o^{(o)}]_{L_v}}} \tag{9.30}$$

where

$$C_0' = \sqrt[4]{\frac{\kappa c^2 A_0 t_a N_{v_g}}{\Omega_u}}\;\frac{1}{\sqrt[4]{L_v[\rho_o^{(o)}]_{L_v}}}$$

The value of $[\rho_o^{(o)}]_{L_v}$ to be used in (9.30) is obtained from (9.22) where the value of M to be used is in this case the range rate uncertainty reduction L_v. As with range measurement, the maximum range does not increase very rapidly with L_v because of the fourth root dependence in (9.30). The variation of the dimensionless ratio $(r_{\max}/r_o)$ with another dimensionless ratio (P_T/P'_{T_0}) where $r'_0 = C'_0\sqrt[4]{P'_{T_0}}$ for various values of L_v and P_s, is shown in Figure 9-6, the range rate analog of Figure 9-4 (which applies to range measurement). The curves and the table that accompanies them are equivalent to those of Figure 9-6 except that L_v is substituted for L_r.

A common way of looking at radar range rate measurement (or any kind of frequency measurement), more often used than that discussed in this chapter, is through axis-crossing theory. This viewpoint will not be treated in this book. Excellent discussions of axis-crossing theory appear elsewhere in the literature.†

9.3.3 Joint Range and Range Rate Measurement—Pulsed Doppler Radars

A pulsed doppler radar can be used to continuously measure both range and range rate. Since knowledge of range throughout the trajectory is

† See, for example, Rice, 1945, in Wax, 1954, pp. 189–209; Bendat, 1958, pp. 125–130 and Chapter 10; Lee; 1960, pp. 177–186; McFadden, 1956, 1958.

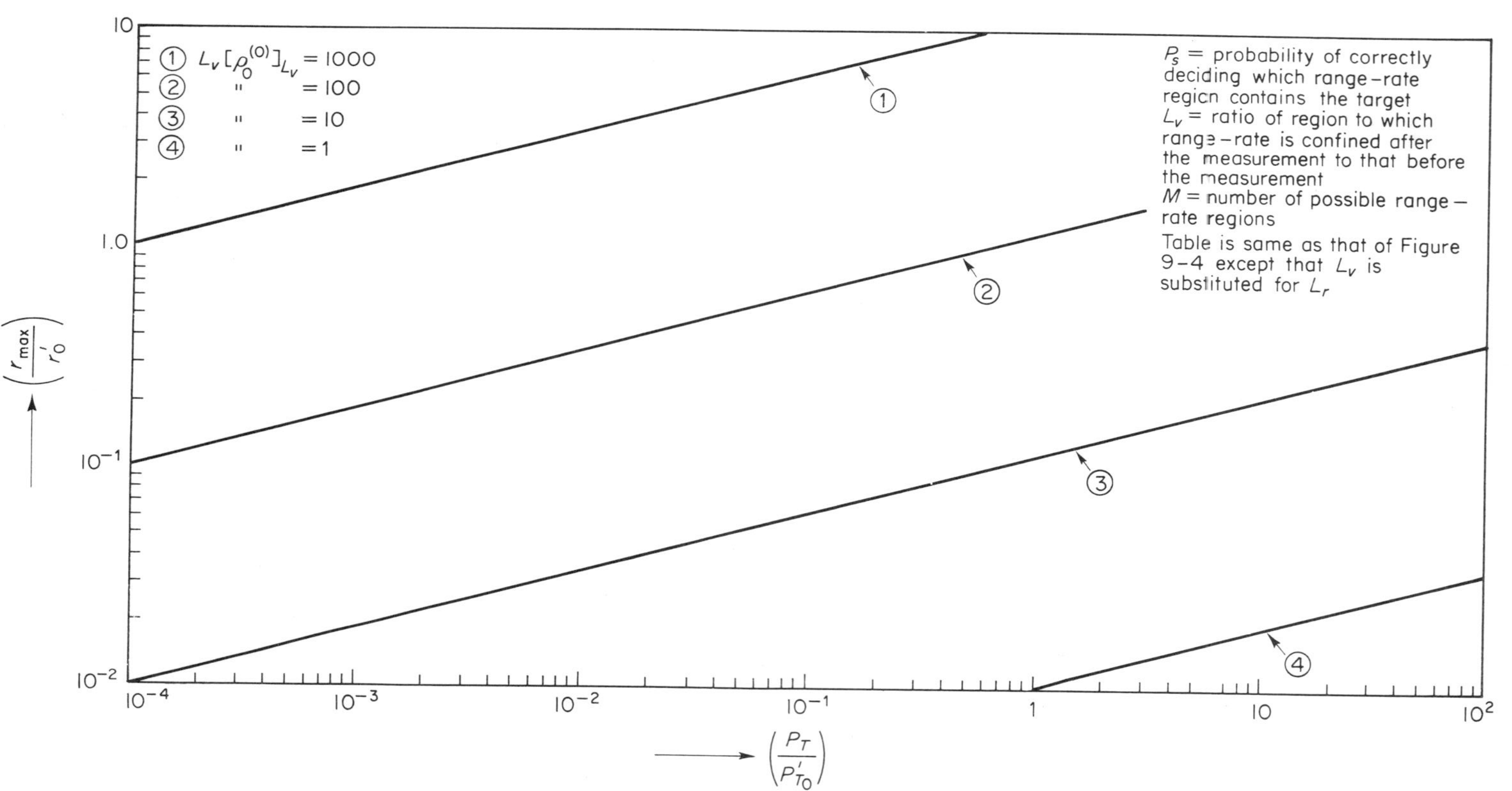

Figure 9-6 Maximum range (normalized) versus transmitted power (normalized) in radar range-rate measurement.

tantamount to that of range rate, it would seem that such a system would provide redundant information. This is, in fact, exactly what it does, but such redundancy may serve a useful purpose. It can be used in a computing scheme to enhance the accuracy of the trajectory measurement.

It was assumed in deriving (9.28) that no range gating is used, as is the case with a CW radar. In a pulsed doppler radar, both range and doppler gating may be used. In applying (9.21) or (9.22) to this case, we must think of the "gate" as a two-dimensional region, the two dimensions being range and range rate. The two-dimensional uncertainty region ($\Delta r\, \Delta v$) is now divided into range-range rate regions by the gating operations. The measurement process now involves a decision as to which of these two-dimensional gating regions contains the target (see Figure 9-7).

The analog of (9.24) and (9.28) for the pulsed doppler case, obtained from

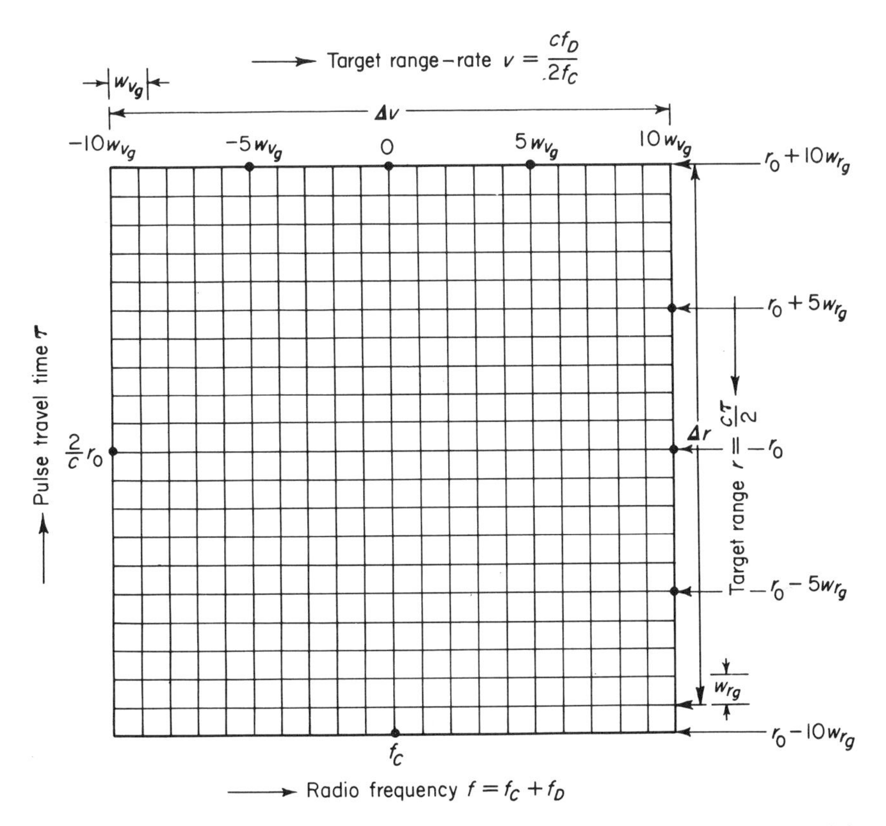

Figure 9-7 Range and range-rate uncertainty regions. Divisions are those corresponding to widths of range and doppler gates.

(9.16), (9.17,) (9.18), (9.23), (9.27), and the arguments preceding and following these equations, is

$$\rho_o^{(o)} = \left[\frac{\kappa c^2 P_T f_p A_0 t_a N_{rg} N_{vg}}{r^4 \Omega_u B_R}\right]\left[\frac{1}{L_r L_v}\right] \tag{9.31}$$

If maximum range is desired, an analog of (9.26) and (9.30) can easily be found from (9.31). An expression for the maximum attainable value of $(L_r L_v)$ for fixed values of the bracketed parameters in (9.31) is

$$(L_r L_v)_{\max} = \left[\frac{\kappa c^2 P_T f_p A_0 t_a N_{rg} N_{vg}}{r^4 \Omega_u B_R}\right] \frac{1}{([\rho_o^{(o)}]_{L_r L_v})_{\min}} \tag{9.32}$$

where $([\rho_o^{(o)}]_{L_r L_v})_{\min}$ is the minimum value of (E_s/N_o) required to achieve a given success probability, obtained from (9.22). In the present case, the number of alternatives M to be used in (9.22) is equal to the product $L_r L_v$. Equation (9.32) tells us that, with fixed values of radar parameters, enhancement of direct range information must be bought at the price of reduction in direct range rate information and vice versa, if the indicated methods of range and range rate measurements are used.

However, this need not be a hindrance in acquiring reliable trajectory information if one is willing, for example, to (1) integrate the range rate continuously to determine range, and (2) average together the result of the integrated range rate readings and the direct range readings, or alternatively to (1) differentiate range continuously to determine range rate and (2) take the average of direct range rate measurements and differentiated range measurements. Obviously, the first set of procedures applies to the case where a continuous indication of range is desired and the second set to that where continuous indication of range rate is desired. In either case, the averaging process should help to reduce inaccuracy due to noise fluctuations.

It is also possible to use range rate measurement or to use range indications to continuously reset the range gate for optimization of range rate measurement. In fact, if sufficiently elaborate instrumentation is allowed, the range and range rate measurements can continuously feed back on each other and "close in" on the correct trajectory parameters. This kind of arrangement is of necessity a tracking system and is beyond the scope of the present discussions.

9.3.4 Angle Measurement

There are two basically different ways in which target angle is measured by radar, as follows:

1. The radar interferometer technique, where the phase difference between signals arriving at two different antennas acts as a measure of target angle, and antenna pattern shape has no important effect on the measurement.

2. Techniques based on measurement of amplitude or differential amplitude, wherein target angle is directly or indirectly indicated through the use of antenna pattern shape.

9.3.4.1 Radar Interferometry

Let us first consider the standard radar interferometry technique. The basic idea is illustrated in Figure 9-8. Two antennas A_1 and A_2 are separated by a distance d. Radar waves emitted simultaneously by these antennas both strike a target T at an angle θ relative to the normal to $A_1 - A_2$. The wave originating at A_1, striking T and returning to A_1 travels a distance $2r$ in its round trip to target and back. The wave from A_2 travels a greater distance $2r + 2d \sin\theta$ in its round trip to T and back. The electrical path length difference between the target returns at A_1 and A_2 is $(4\pi\, d/\lambda) \sin\theta$. The two

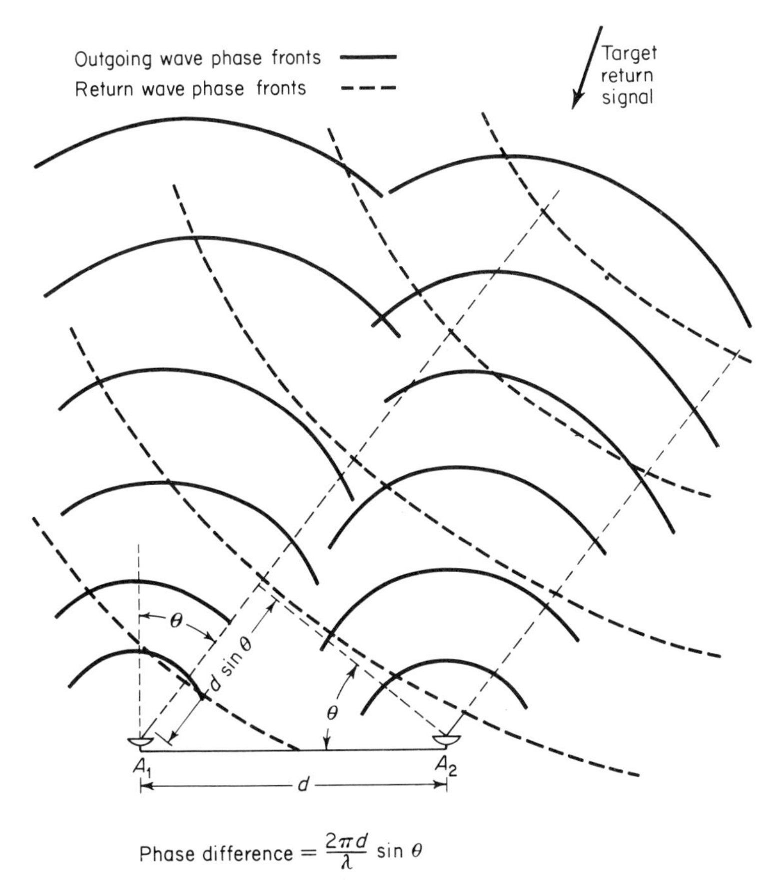

Figure 9-8 Basic idea of radar interferometry.

signals are fed into a *phase comparator*, a device whose output is proportional to the sine of the phase difference, i.e., to sin $[(4\pi d/\lambda), \sin\theta)]$.

There are certain practical limitations in radar interferometry that should be mentioned before we embark on a discussion of its statistical aspects.

If the absolute value of the angle θ is very small, e.g., less than 15°, it can be assumed that

$$\sin\theta \simeq \theta \tag{9.33}$$

and the approximate phase comparator output is

$$u_o(\theta) \simeq \sin\left(\frac{4\pi d\theta}{\lambda}\right) \tag{9.34}$$

In general with a calibration constant of unity, the comparator output is

$$u_o(\theta) \simeq \sin\left(\frac{4\pi d}{\lambda}\sin\theta\right) \tag{9.35}$$

To illustrate some of these limitations, we present in Figure 9-9 the plots of $u_o(\theta)$ versus θ for two cases; (a) spacing small compared to wavelength ($d = 0.1\lambda$) and (b) spacing comparable to wavelength ($d = \lambda$). In the case $d = 0.1\,\lambda$, the variation with θ is smooth and monotonic increasing, and it is nearly linear for $\theta \lesssim 40°$.

The striking point about the curve corresponding to the case $d = \lambda$, obvious from Eq. (9.33), is the oscillatory variation of $u_o(\theta)$ with θ: An ambiguity in angle measurement appears as soon as $(4\pi d/\lambda)\sin\theta$ exceeds $\pi/2$. A practical measurement technique requires that the variable being measured be a single-valued function of the output of the measuring instrument. This is not the case if $(4\pi d/\lambda)\sin\theta > \pi/2$. Thus, to preclude ambiguity in the measurement of angle, we must have the conditions

$$|\theta| < \frac{\pi}{2} \tag{9.36}$$

$$|\sin\theta| < \frac{\lambda}{8d} \tag{9.37}$$

For a given wavelength and a given antenna separation the conditions (9.36) and (9.37) place limits on the size of the angle that can be measured.

Now let us study noise as a factor in limiting measurement accuracy. We will circumvent the precise application of maximum likelihood estimator theory to the problem of phase difference measurement, which was not covered in Chapter 8. In lieu of processing for optimum phase *difference* measurement, we will base our discussion on measurement of the phase difference between two signals, each of which is processed for (approximately) optimum phase measurement, according to the discussions in Section 8.3. This implies that we would first do something as close as possible to matched filtering on the signals entering the two antenna channels (assuming that the

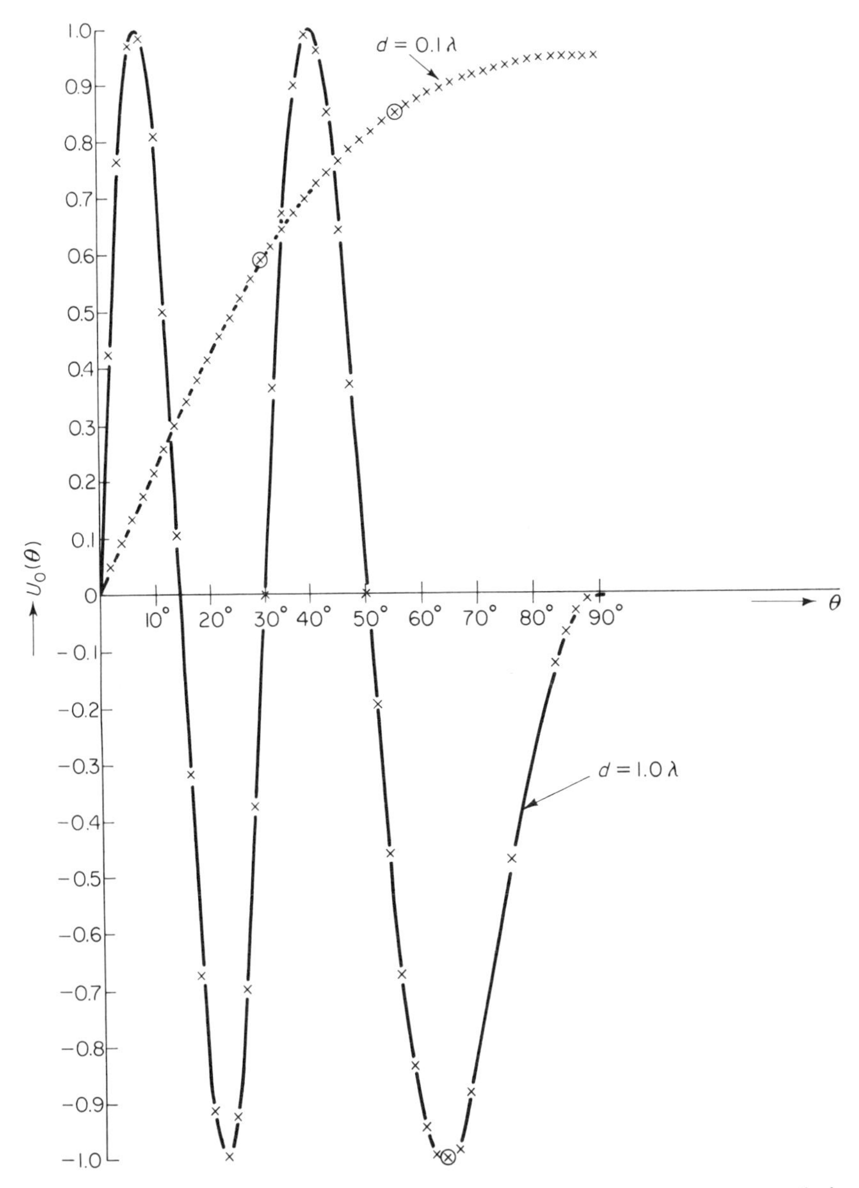

Figure 9-9 Phase comparator output versus target angle in radar interferometry.

two antennas are connected to the same RF head, so that the noise waveforms in the two channels are identical) and then take the phase difference between the two matched filter outputs. If the signal is a pure sinusoidal pulse of length T_p, the signal-plus-noise matched filter output complex envelopes of the receiver channels corresponding to A_1 and A_2 are, respectively,

$$[\hat{v}_o(t)]_{A_1} \simeq [\hat{s}_o + \hat{n}_o(t)]\, e^{j\omega_c t} \tag{9.38}$$

$$[\hat{v}_o(t)]_{A_2} = \left[\hat{s}_o\, e^{j\left(-\frac{\pi}{2}+\frac{4\pi d}{\lambda}\sin\theta\right)} + \hat{n}_o(t)\right] e^{j\omega_c t} \tag{9.39}$$

where $\hat{s}_o = a_o e_{j\psi_o}$ (see Sections 6.1 and 6.2, Eqs. (6.4) and (6.9)).

The output of an ideal phase comparator whose input complex envelopes are $[\hat{v}_o(t)]_{A_1}$ and $[\hat{v}_o(t)]_{A_2}$ is

$$\begin{aligned} u_o(\theta) &= \frac{\lambda}{4\pi d a_o^2 T_p} \operatorname{Re}\left\{\int_0^{T_p} dt'\,[\hat{s}_o + \hat{n}_o(t')][\hat{s}_o^*\, e^{-j\left(-\frac{\pi}{2}+\frac{4\pi d}{\lambda}\sin\theta\right)} + \hat{n}_o^*(t')]\right\} \\ &= \operatorname{Re}\left\{\frac{\lambda j}{4\pi d}\, e^{-j\frac{4\pi d}{\lambda}\sin\theta}\right. \\ &\quad + \frac{\lambda a_o e^{j\psi_o}}{4\pi d a_o^2 T_p}\left[\int_0^{T_p} dt'\,\hat{n}_o^*(t') + j\, e^{-j\left(2\psi_o + \frac{4\pi d}{\lambda}\sin\theta\right)} \int_0^{T_p} dt'\,\hat{n}_o(t')\right] \\ &\quad \left. + \frac{\lambda}{4\pi d a_o^2 T_p}\int_0^{T_p} dt'\,|\hat{n}_o(t')|^2\right\} \end{aligned} \tag{9.40}$$

If $|(4\pi d/\lambda)\sin\theta| \ll \pi/2$ and $\theta \ll \pi/2$, then (9.40) takes the approximate form

$$\begin{aligned} u_o(\theta) \simeq \theta + \frac{\lambda}{4\pi d a_o^2 T_p} \operatorname{Re}\left\{e^{j\psi_o}\left[\int_0^{T_p} dt'\,\hat{n}_o^*(t') + j e^{-2\left(2\psi_o + \frac{4\pi d}{\lambda}\theta\right)}\right.\right. \\ \left.\left. \int_0^{T_p} dt'\,\hat{n}_o(t')\right]\right\} + \frac{\lambda}{4\pi d a_o^2 T_p}\int_0^{T_p} dt'\,|\hat{n}_o(t')|^2 \end{aligned} \tag{9.41}$$

where the nonlinear terms in θ can be neglected if the above assumptions hold. If the SNR is sufficiently high to justify neglect of the noise-noise term in (9.41), then the rms error in the measurement is approximated by a quantity proportional to the square root of the ensemble average of the square of the quantity in braces. This calculation, which is somewhat cumbersome, is left as an exercise for the reader. For wideband noise and high SNR, the rms error is roughly proportional to the rms noise, as can be shown by detailed evaluation of (9.41) with the noise-noise term neglected.

9.3.4.2 Use of Antenna Patterns in Angle Measurement

In search and detection radars, the azimuthal angle of the target can often be determined to substantial accuracy by scanning azimuthally past the target with a narrow beam and observing the time interval over which the return is maximized (see Figure 9-10). To illustrate the kind of accuracy

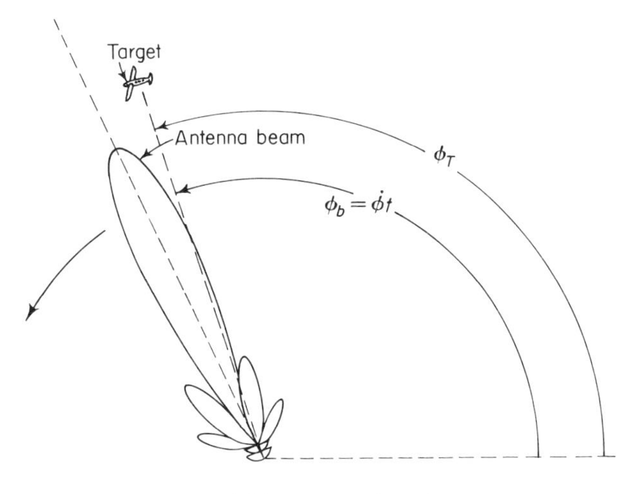

(a) Scanning antenna beam

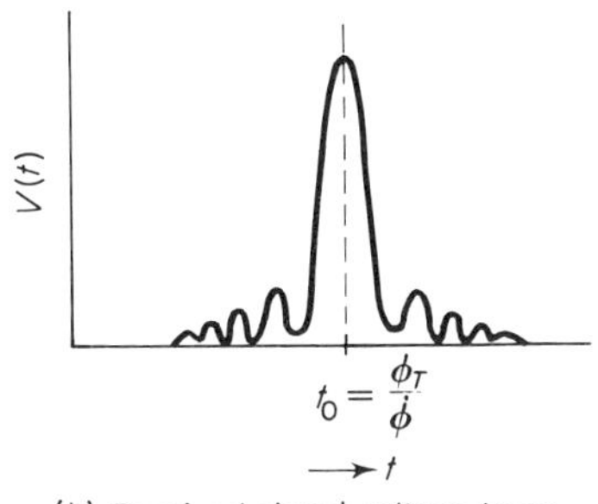

(b) Received signal voltage trace

Figure 9-10 Amplitude measurement of target angle with scanning antenna.

attainable by this method, we will assume a beam extremely narrow in the azimuthal angle ϕ, e.g., a beam width of less than 10°. If the antenna aperture is rectangular, as is often the case in such radars, the beam shape (one-way power pattern) in the direction of the azimuthal angle (see Section 1.3.2) is approximately,

$$f(\phi) = \left[\operatorname{sinc}\left(\frac{\pi A}{\lambda}\left(\phi - \phi_b\right)\right)\right]^2 \tag{9.42}$$

where ϕ_b is the angle at the peak of the beam, and A is the horizontal aperture dimension. The small beam width assumption has been used in replacing the sine of the angle $(\phi - \phi_b)$ by the angle itself, the return signal power being assumed negligible at angles where this approximation doesn't hold. The beam is swept at a constant rate $\dot{\phi}$ degrees per second, and the target stands at angle ϕ_T. The return signal traced out as the beam sweeps past the target according to (9.7.a) and (9.42) has the approximate amplitude

$$a_s(t) = \sqrt{\frac{P_T G_0^2 \lambda^2 \sigma}{(4\pi)^3 r^4}} \left[\operatorname{sinc}\left(\frac{\pi A}{\lambda}(\phi_T - \dot{\phi} t)\right)\right]^2 \tag{9.43}$$

where the quantities used were defined in Section 9.1.1, $t = 0$ is the instant at which the scan begins, and the coordinates are chosen such that $\phi_b = 0$ at $t = 0$. (This choice has no effect on the generality of the arguments.)

The radar observer will decide that the target angle is ϕ_T if he observes that the return signal amplitude reaches its peak at $t_0 = \phi_T/\dot{\phi}$ which it will in the absence of noise. If the additive receiver noise is taken into account, we see that the instant at which the amplitude reaches its peak will in general differ from t_0; thus an error will occur in reading the target angle.

To analyze this effect, let us add to the radar return noise with complex envelope $\hat{n}(t)$. The signal-plus-noise complex envelope is

$$\hat{v}(t) = C\left[\operatorname{sinc}\frac{\pi A}{\lambda}(\phi_T - \dot{\phi} t)\right]^2 e^{j\psi_s} + \hat{n}(t) \tag{9.44}$$

where

$$C \equiv \sqrt{\frac{P_T G_0^2 \lambda^2 \sigma}{(4\pi)^3 r^4}}$$

and where ψ_s is the signal phase, which depends on the path and target characteristics in such an unpredictable way that we can regard it as randomly distributed from 0 to 2π.

If ϕ_T and $(\dot{\phi} t)$ differ by only a few degrees (as they must, by assumption, in angular regions where there is significant target return), we can expand the sinc function in powers of its argument and neglect all but the first two terms. Doing so, and calculating the amplitude of $v(t)$, we have

$$a_v(t) = \sqrt{\begin{aligned}&\left\{C^2\left[1 - \frac{1}{6}\left(\frac{\pi A}{\lambda}\right)^2(\phi_T - \dot{\phi} t)^2\right]^4\right. \\ &\left.+ 2\,\mathrm{Re}\left(C\, e^{j\psi_s}\left[1 - \frac{1}{6}\left(\frac{\pi A}{\lambda}\right)^2(\phi_T - \dot{\phi} t)^2\right]^2 \hat{n}^*(t)\right) + |\hat{n}(t)|^2\right\}\end{aligned}} \tag{9.45}$$

In the noise-free case, we would have

$$a_v(t) = a_s(t) \simeq C\left[1 - \frac{1}{6}\left(\frac{\pi A}{\lambda}\right)^2(\phi_T - \dot{\phi} t)^2\right]^2 \tag{9.46}$$

We process the signal as if there were no noise present, i.e., we compute ϕ_T from (9.46) assuming all other quantities in (9.46) to be known. The indicated value of ϕ_T in the noise-free case, which is the true value within the approximations that have been made, is

$$\phi_T = \dot{\phi} t + \frac{\lambda}{\pi A}\sqrt{6\left(1 - \sqrt{\frac{a_s(t)}{C}}\right)} \tag{9.47}$$

If the SNR is sufficiently high to justify neglect of the noise-noise term $|\hat{n}(t)|^2$ inside the radical in (9.45) and the assumption that the signal-noise term is small compared to the signal-signal term, then with the aid of a Taylor series expansion, we can approximate $a_v(t)$ (through (9.45) and (9.46)) by

$$a_v(t) \simeq a_s(t) + \text{Re}\,\{e^{j\psi_s}\hat{n}^*(t)\} \tag{9.48}$$

The computation of ϕ_T to be made in the receiver is given by (9.47) with $a_v(t)$ replacing $a_s(t)$. The indicated target angle is

$$(\phi_T)_{\text{ind}} = \dot{\phi}t + \frac{\lambda}{\pi A}\sqrt{6\left(1 - \sqrt{\frac{a_v(t)}{C}} - \frac{1}{C}\,\text{Re}\,\{e^{j\psi_s}\hat{n}^*(t)\}\right)} \tag{9.49}$$

Again invoking the large SNR assumption and a Taylor series expansion in (9.49), and subtracting (9.47) from (9.49), we obtain (very roughly)

$$(\phi_T)_{\text{ind}} - \phi_T \propto \text{Re}\,[e^{j\psi_s}\hat{n}^*(t)] \tag{9.50}$$

The rms error in measurement of ϕ_T, from (9.47), (9.49), and (9.50), is

$$\begin{aligned}\epsilon &= \sqrt{\langle[(\phi_T)_{\text{ind}} - \phi_T]^2\rangle} \\ &\propto \sqrt{\langle e^{2j\psi_s}\rangle\langle(\hat{n}^*(t))^2\rangle + \langle e^{-2j\psi_r}\rangle\langle(\hat{n}(t))^2\rangle + 2\langle|\hat{n}(t)|^2\rangle} \\ &= \sqrt{\langle|\hat{n}(t)|^2\rangle} = \sqrt{2}\,\sigma_n\end{aligned} \tag{9.51}$$

where we have used the fact that $\langle|\hat{n}(t)|^2\rangle = 2\langle[n(t)]^2\rangle = 2\sigma_n^2$ and where the randomness of the phase ψ_s is reflected in the assumption that $\langle e^{\pm 2j\psi_s}\rangle = 0$.

Thus for very high SNR, the rms error in measurement of the angle by amplitude is proportional to the rms noise level.

The amplitude comparison technique of target angle measurement makes direct use of the antenna pattern and measures differences in pattern amplitude to determine the angle. The diagram of Figure 9-11 illustrates the point. Various techniques are employed to measure angle, e.g., conical scan, monopulse, sequential lobing, etc., but they all have one thing in common, namely that they take the difference between amplitudes of the target signal on two sides of a *crossover axis*, which is the line AA' on the diagram. The two lobes L_1 and L_2 shown to the left and right of the antenna pattern, returns from which are measured simultaneously and compared (as in monopulse) or the same lobe is measured at two different, closely spaced instants of time† (as in conical scan or sequential lobing). In conical scan, the beam is swept conically and differential readings are taken between positions 180° apart. In sequential lobing, the beam is switched rapidly from one quadrant to the next, and differential readings are taken between returns from opposite quadrants.

---

† So closely spaced that the target has no chance to move during the time the antenna beam is moving from L_1 to L_2.

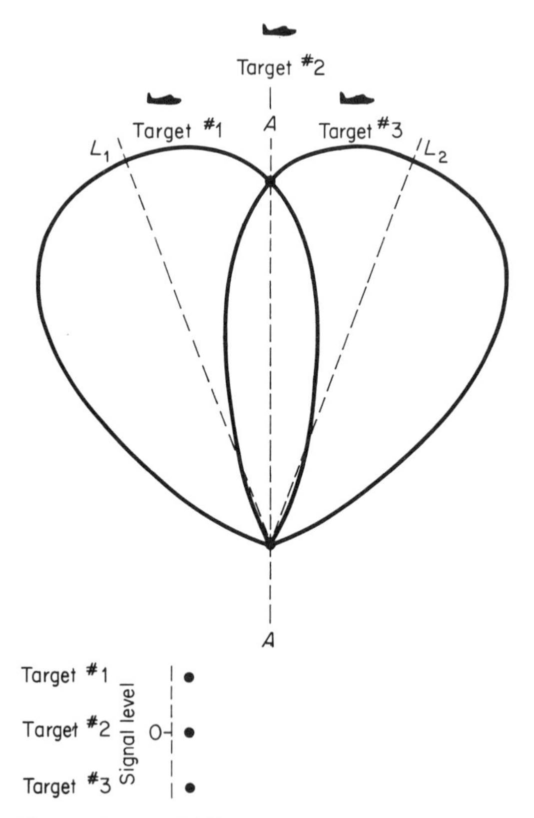

Figure 9-11 Differential angle measurement with beam pattern.

To study the effects of noise on angular measurement accomplished through differential amplitude reading, we will simplify the mathematical model of the antenna pattern to remove the effect of side lobes, which greatly complicates the differential amplitude measurement process. We will treat side lobes as we did phase ambiguity in interferometry, that is, as something that limits the size of angles that can be measured. Any situation in which the target is in the side lobes of the left- or right-hand beam introduces an amplitude ambiguity. In fact, the target must be between the peaks of L_1 and L_2 to avoid ambiguity.

A convenient beam pattern model that brings out the important features of the process is one in which the pattern shape is assumed Gaussian, and in which it is assumed that the beam is so narrow and the target so far from the radar, that we can think of the angular arc about the radar as a plane, as shown in the diagram. This is a *small angle approximation*. From the target antenna geometry shown in the diagram, we can write

$$[\hat{v}_o(t)]_{L_2} = \hat{s}_o\, e^{-\frac{(\theta_0-\theta_T)^2}{\theta_b^2}} + \hat{n}_{o2}(t) = \hat{s}_o\, e^{-\frac{(\theta_0^2+\theta_T^2}{\theta_b^2}}\, e^{-\frac{2\theta_0\theta_T}{\theta_b^2}} + \hat{n}_{o2}(t) \qquad (9.52.a)$$

$$[\hat{v}_o(t)]_{L_1} = \hat{s}_o\, e^{-\frac{(\theta_0+\theta_T)^2}{\theta_b^2}} + \hat{n}_{o1}(t) = \hat{s}_o\, e^{-\frac{(\theta_0^2+\theta_{T_b})}{\theta_b^2}}\, e^{-\frac{2\theta_0\theta_T}{\theta_b^2}} + \hat{n}_{o1}(t) \qquad (9.52.b)$$

where the assumed pattern shaping function is $e^{-(\theta/\theta_b)^2}$, θ_b is the beam width of either lobe (i.e., the spacing between the peak and a point down from the peak by a factor $1/e$), θ_T is the target angle relative to the offset axis, and $\hat{s}_o$ and $\hat{n}_o$ are peak signal and noise complex envelopes, respectively.

Note that the peak signal complex envelop $\hat{s}_o$ is assumed to be the same for both lobes while the noise complex envelopes $\hat{n}_{o1}$ and $\hat{n}_{o2}$ are different.

The amplitude comparator is calibrated to produce the output θ_T in the noise-free case. Its output is:

$$\begin{aligned} u_o(\theta_T) &= \operatorname{Re}\left\{\frac{[\hat{v}_o(t)]_{L_2} - [\hat{v}_o(t)]_{L_1}}{\frac{4\theta_0}{\theta_b^2}\hat{s}_o\, e^{-(\theta_0^2+\theta_T^2)/\theta_b^2}}\right\} \\ &= \operatorname{Re}\left\{[e^{2\theta_0\theta_T/\theta_b^2} - e^{-2\theta_0\theta_T/\theta_b^2}] + \frac{[\hat{n}_{o2}(t) - \hat{n}_{o1}(t)]}{\frac{4\theta_0}{\theta_b^2}\hat{s}_o\, e^{-(\theta_0^2+\theta_T^2)/\theta_b^2}}\right\} \\ &\simeq \operatorname{Re}\left\{\theta_T + \frac{[\hat{n}_{o2}(t) - \hat{n}_{o1}(t)]}{\left(\frac{4\theta_0}{\theta_b^2}\hat{s}_o\, e^{-(\theta_0^2+\theta_T^2)/\theta_b^2}\right)}\right\} + 0(\theta_T^2) \end{aligned} \qquad (9.53)$$

The terms that are nonlinear in θ_T are negligible if the beam is sufficiently narrow. If they are present, the measurement could still be carried out, but in practice they generally complicate matters and it is best to confine this measurement technique to target angles that are small compared to beam width.

If the noise waveforms associated with left and right beam positions are completely uncorrelated, then noise has the greatest effect on the measurement, since the two noises add noncoherently. If completely correlated, noise has the least effect since the noises cancel. In conical scan radar, if the scanning rate is f_s scans per second, the left and right beam positions are separated in time by $1/2f_s$ s, and the important question is the ratio of the noise correlation time to $1/2f_s$, or, equivalently, the ratio of noise bandwidth to $2f_s$. To put this on a mathematical basis let us calculate the rms accuracy of the angle measurement in the case where θ_T/θ_b is small enough to neglect the nonlinear terms in θ_T, i.e.,

$$\begin{aligned} \epsilon = \sqrt{\langle|u_o(\theta_T) - \theta_T|^2\rangle} &\simeq \sqrt{\frac{|\hat{n}_{o2}(t)|^2 + |\hat{n}_{o1}(t)|^2 - 2\operatorname{Re}\langle\hat{n}_{o1}(t)\hat{n}_{o2}^*(t)\rangle}{\frac{4\theta_0}{\theta_b^{\,2}}a_o\, e^{-(\theta_0^2+\theta_T^2)/\theta_b^2}}} \\ &= \frac{\theta_b\sigma_n}{\sqrt{\theta_0 a_o}}\, e^{-(\theta_0^2+\theta_T^2)/\theta_b^2}\sqrt{1 - \operatorname{Re}[\chi_{12}]} \end{aligned} \qquad (9.54)$$

where $\chi_{12} = (1/2\sigma_n^2)\langle\hat{n}_{o1}(t)\hat{n}_{o2}^*(t)\rangle$, a_o is the (real) signal amplitude (i.e.,

$\hat{s}_o = a_o e^{j\psi_o}$, but $\psi_o = 0$), and where it has been assumed that $|\hat{n}_{o1}(t)|^2 = |\hat{n}_{o1}(t)|^2 = \sigma_n^2/2$.

The crosscorrelation function between $n_{o1}(t)$ and $n_{o2}(t)$ depends critically on the lobing technique, i.e., conical, sequential, monopulse, etc. One can see, in general, that the case in which the noises are completely uncorrelated, ($\chi_{12} = 0$) would correspond to a large error, while that in which there is complete correlation ($\chi_{12} = 1$), would correspond to zero error. It is easy to see why this is so. Two completely correlated noises with the same rms values are two identical noises; consequently, their difference will vanish. Two completely uncorrelated noises will add incoherently; hence the two noise powers will combine additively rather than subtractively when the returns from the two lobes are subtracted. The result of this is twice the noise power that would accompany the returns from either of the two lobes individually irradiating the target.

REFERENCES

Radar Systems in General and Microwave Propagation as it Affects Radar Systems

Barton, 1964

Berkowitz, 1965

Kerr, 1951

Merrill, 1961

Middleton and Van Vleck, 1946

Reintjes and Coate, 1952

Ridenour, 1947

Skolnick, 1962

Application of Statistical Communication Theory Ideas to Radar Detection, Measurement of Range and Angle

The following are in addition to references already cited on theoretical background for radar detection and estimation.

Barton, 1964 (many sections apply to this category)

Bello, 1960

Berkowitz, 1965 (many subsections apply to this category)

Bussgang and Middleton, 1955

Davenport and Root, 1958; Sections 14-1, 14-7

Helstrom, 1960 (much of this book has a radar "flavor")

James, Nichols, and Phillips, 1947; Chapter 6 (S. Phillips), Chapters 7, 8 (Dowker and Phillips) (primarily applicable to radar tracking in noise)

Lawson and Uhlenbeck, 1950 (most of the book is written in the context of radar systems)

Marcum and Swerling, 1960
Peterson, Birdsall, and Fox, 1954
Schwartz, 1959; Section 7-10
Selin, PGIT, 1965
Siebert, 1956
Skolnick, 1962; especially Chapters 9, 10
Swerling, 1956
Swerling, 1957
Swerling, 1966
Urkowitz, 1953
Woodward, 1953; Chapters 5, 6, 7

10

APPLICATIONS TO RADAR SYSTEMS—II

This chapter extends the discussion of applications of statistical communication theory to radio detection and location systems. Chapter 9 covered some of the elementary aspects of this subject. A few topics somewhat more specialized than those treated in Chapter 9 have been reserved for the present chapter.

Section 10.1 is a direct extension of Chapter 9. It covers further applications of statistical communication theory to conventional *monostatic* radar systems, in which transmitter and receiver are at the same location. Section 10.2 treats two interesting and important topics that have not been covered elsewhere in this book, bistatic radar systems (those in which transmitter and receiver are at different locations) and passive detection and location systems (essentially receivers that detect and/or locate a target through the target's emission of radio energy). Strictly speaking, the latter are not radars by the usual definition of that term. We are enlarging our definition of radar to include them. This is not without precedent; many engineers informally refer to a passive detection system as a *passive radar*.

10.1 FURTHER TOPICS IN MONOSTATIC RADAR

Below are discussions of topics in monostatic radar systems that were not treated in Chapter 9.

10.1.1 External Radar Noise Sources

There are certain sources of radio noise that are of special interest in connection with radar, either because they occur within radar frequency bands, or because they are peculiar to the propagation geometry associated with radar. Three of these will be discussed below. They are:

1. Atmospherical and extraterrestrial radar noise.
2. Clutter.
3. Target noise.

The first of these is of interest because it lies within radar bands, while the last two are special types of noise peculiar to radar.

10.1.1.1 Atmospheric and Extraterrestrial Radar Noise

Although lightning-induced atmospheric noise of the type that degrades VLF, LF, and MF radio systems is usually not much of a problem at radar frequencies, there is a great deal of *atmospheric absorption noise* and *extraterrestrial noise* at these frequencies (see Sections 5.5.3.2 and 5.5.3.3).

The ionosphere is effectively transparent to transmissions above 1 GHz. Consequently, noise of extraterrestrial origin easily penetrates the ionosphere in the higher radar frequency bands. Even with a highly directive antenna pointing horizontally, enough of this noise energy enters the beam side lobes to present a problem, especially in highly sensitive radars. In particular, for radars using modern low noise amplifiers, extraterrestrial noise is far more intense than internal receiver noise; hence, it is the ultimate limiting factor in those radars.

The curve of Figure 5-4 shows the spectral distribution of extraterrestrial noise at radio frequencies including those characteristic of radar. The noise is given in terms of incident noise power per unit area in decibels relative to ideal receiver noise. To calculate the actual noise received by the radar in absolute units of power, we must first convert the decibel figures on the curve to power units, then multiply by the ideal receiver noise power kTB and by the effective receiving area of the antenna.

In addition to the extraterrestrial noise, there is noise due to absorption of radiation from various natural sources by the earth and the atmosphere. Assuming the earth and atmosphere are in thermal equilibrium with their surroundings, they must radiate as much power as they absorb. Hence a noise background due to this mechanism is always present in the radar environment.

The Planck radiation law† gives the wavelength spectrum of radiation from an ideal black body as a function of the body's absolute temperature.

---

† See any one of many classical physics texts, e.g., Joos, 1934, pp. 585–587 or Sears, Vol. III, 1948, p. 312.

Most bodies are not ideal black bodies, but the Planck law can nevertheless be used to estimate the relative intensities of radiation at different wavelengths from bodies at a given temperature. According to the Planck law and experimental observation, there is considerable radiation within the microwave spectrum from bodies at the temperature encountered in the environment in which radars operate.

10.1.1.2 Clutter

A noise phenomenon peculiar to radar systems is that of clutter.† The noise known by this name is that arising through backscattering of energy from the transmitted radar beam. Another way of characterizing clutter is to consider it as return from targets in the radar beam other than the one to be detected or located.

There are several varieties of backscatter phenomena giving rise to clutter. One of these, especially troublesome in some radars, is known as ground clutter. If a radar beam is pointed at the ground, for example, or even if its side lobes are so directed, energy will be backscattered from various objects, e.g., trees, buildings, or any bits of matter with a reasonably high reflection coefficient at microwave frequencies. Some of this energy will come from specular reflection, e.g., a return from a smooth slab of ground whose area is many square wavelengths. Some of it arises from discrete scatterers on the ground that can be considered mutually statistically independent. The same general idea applies to other types of clutter, such as return from precipitation particles.

The clutter signal, then, can usually be represented quite generally as a sum of returns from independent scatterers with different delays, where some of these returns may be overwhelmingly larger than the others because they have their origin in specular reflection. It is important to note that some of the discrete scatterers may be moving relative to the radar and hence each may have a doppler shift. In addition to the fixed doppler shift that will sometimes be present on all scatterers due to relative motion between radar and ground (e.g., in airborne radar situations), there may be a fluctuating doppler shift on each scatterer due to motions that are essentially random. These fluctuating dopplers are statistically independent.

A mathematical model of clutter return embodying these ideas gives for the clutter signal voltage

$$v_c(t) = \operatorname{Re} \sum_{k=1}^{N} \frac{A_k}{r_k^2} e^{-j\omega_{c0}\left(t - \frac{2r_k}{c}\right) - j\omega_{c0}\frac{v_k}{c}\left(t - \frac{2r_k}{c}\right)} \tag{10.1}$$

where r_k is the range of the kth scatterer, ω_{c0} is RF angular frequency, A_k

† Now classical references on radar clutter are Lawson and Uhlenbeck, 1950, Chapters 6 and 11 and Kerr, 1951, Chapter 6, "The Fluctuations of Clutter Echoes" by D. Kerr and S. Goldstein, pp. 550–571.

is a complex number containing such parameters as backscatter cross section of the kth target and its angular position in the beam, $v_k \omega_{c0}/\pi c$ is the doppler frequency shift, v_k being target line-of-sight velocity and c the velocity of light, and the quantities $2r_k/c$ are the delays in the target returns due to differences in range.

The arguments of Section 3.5 on the central limit theorem can be applied to many clutter problems. If one signal in the sum (11.1) is a specular reflection and much larger than the others, it can be isolated from them. If the remaining returns are of comparable amplitude, that is, if they arise from an assembly of identical or similar scatters at nearly the same range, then the central limit theorem arguments (culminating in (3.77)) show that the statistics of the sum are Gaussian, possibly with a superposed large signal from the specular reflection. The amplitude PDF of the clutter is in general that given by (3.46) (the Rice distribution).

Because the scatterers are statistically independent,† the mean clutter power P_c is the sum of the means of absolute squares of the complex amplitudes, i.e.,

$$P_c = \sum_{k=1}^{N} \frac{|A_k|^2}{r_k^4} \tag{10.2}$$

The spectrum of the clutter signal can be calculated from a consideration of relative doppler shifts. The easiest way to get it is to determine the ACF and take its Fourier transform. Many years ago, Siegert‡ calculated the ACF of the clutter signal under certain conditions. We can obtain the spectrum without explicitly invoking Siegert's result. To do so we first calculate $\langle \overline{v_c(t)\; v_c(t + (\tau)} \rangle$ from (10.1), obtaining

$$\begin{aligned} R_{v_c}(\tau) &= \langle \overline{v_c(t) v_c(t+\tau)} \rangle \\ &= \frac{1}{2} \operatorname{Re} \left\{ \sum_{k=1}^{N} \int dA_k\, p(A_k) |A_k|^2 \int dr_k \frac{p(r_k)}{r_k^4} \int dv_k\, p(v_k) e^{j\omega_{co}\left(1+\frac{2v_k}{c}\right)\tau} \right\} \end{aligned} \tag{10.3}$$

where $p(A_k)$, $p(r_k)$,and $p(v_k)$ are PDF's for amplitude, range, and range rate, respectively. These variables are assumed statistically independent, and the independence between scatterers has been accounted for by removing cross terms in (10.3).

If we assume a sufficiently narrow range gate, so that $1/r_k$ can be regarded as equal to $1/r$ for all scatterers, and further assume that all scatterers have the same statistics, Eq. (10.4) takes the form

$$R_{v_c}(\tau) = \frac{N|A|^2}{2r^4} \operatorname{Re}\left[e^{j\omega_{co}\tau} \int_{-\infty}^{\infty} dv\, p(v)\, e^{2j\omega_{co} v\tau/c} \right] \tag{10.4}$$

where the subscript k has been dropped on all statistical parameters.

---

† Because of this statistical independence, the cross terms between contributions from different scatterers vanish in the process of calculating the mean-square voltage.

‡ See Kerr, 1951, pp. 556–557.

If we assume a distribution of the doppler velocity v that is symmetrical about a mean zero,† carry out the integration in (10.4) and Fourier transform the result to get the spectrum (see Eq. (2.41)), we obtain

$$G_{v_c}(\omega) = \pi c \frac{N\langle |A|^2\rangle}{4\omega_{co}r^4}\left\{p\left(\left(\frac{c}{2\omega_{co}}\right)(\omega - \omega_{co})\right) + p\left(\left(\frac{c}{2\omega_{co}}(\omega + \omega_{co})\right)\right\} \tag{10.5}$$

The result (10.5) confirms that (1) the mean clutter power is the sum of powers from individual scatterers (which we already know from (10.2)), and (2) it tells us that if all scatterers have the same statistics, the clutter spectrum is centered at a frequency equal to the transmitted RF plus the average doppler shift and is shaped like the probability density function of the doppler velocities, with appropriate proportionality constants to convert from velocity to frequency. Thus the clutter spectrum is very directly related to the distribution of line-of-sight velocities of the contributing scatterers.

We began this discussion with explicit mention of ground clutter. The same theory applies to the many other types of radar clutter that can arise in applications. Clutter from precipitation particles in rainstorms can occur significantly at frequencies above X-band.‡ This is a nuisance in most radars, but is the phenomenon that renders some types of radars feasible, e.g., storm warning radars on aircraft. In military applications, *chaff* particles or *window*, can be thrown out by an enemy to present a false target to the radar. These particles are tiny metallic dipoles with high radar cross sections. There are many of them and they are independent; thus the clutter model discussed above applies to them.

From the point of view of radar performance, we can simply regard clutter as an additive noise which may have a mean value due to a specular component and which is normally distributed about this mean, The basic differences between clutter and internal noise as disturbing agents are:

1. Internal noise is generally white over the receiver passband and has zero mean, while clutter may have a colored noise spectrum within the passband and a finite mean.
2. Increasing transmitter power can overcome internal noise but enhances clutter in the same proportion as the desired target signal; thus increases in transmitted power do not increase the signal-to-clutter ratio.

Because of the spread in doppler velocities of the moving scatterers giving rise to clutter, a way to process against it is to determine the target range rate precisely enough to construct a very narrow doppler gate containing the target doppler and excluding much of the clutter energy. The narrower the doppler gate, the more clutter exclusion, i.e., only energy from scatterers with line-of-sight velocity components nearly equal to target range rate will be received.

---

† The mean doppler velocity is assumed to correspond to the frequency ω_{co}.

‡ See Kerr, 1951, Chapter 7, pp. 588–640.

10.1.1.3 Target Noise

There are two basic types of noise emanating from the target itself. One of these, called *target scintillation*, is a fluctuation in the amplitude of the return signal. The other, known as *glint noise*, is a fluctuation in the apparent angular position of the target. Target scintillation is due to the fact that a typical radar target, such as an aircraft or missile, has modes of vibration that give rise to a time variation in its reflection characteristics. A way of expressing this is to say that the radar cross section σ is a function of time. To explain this, one need only examine the reflectivity pattern of a typical aircraft at a representative radar frequency, roughly sketched in Figure 10-1.†

Because of complex lobe structure, as the target vibrates during its natural flight motion, the aspect angle with respect to the radar fluctuates in a more or less random fashion. The spectrum of this fluctuation is not white, but favors certain frequencies related to the vibration frequencies of the target.

In its effects, target scintillation has the characteristics of fading, and can be analyzed as such. Statistically, its amplitude will usually be Rice distributed, because it arises from a combination of specular reflection and fluctuating returns from large numbers of random scatterers.

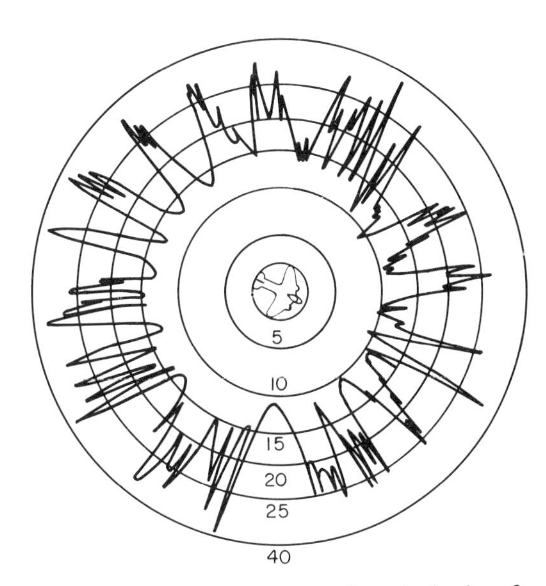

Figure 10-1 Radar reflectivity pattern of typical aircraft. Based on diagrams in Figures 3-8 through 3-11 in Ridenour, 1947.

† Figure 10-1 is based on diagrams like those shown for World War II aircraft in Figures 3-8, 3-9, 3-10, and 3-11 on pages 76, 77, 78, and 79, respectively of Ridenour, 1947.

Glint noise only occurs at very short ranges, i.e., when the target is so close to the radar that it subtends a large angle with respect to the radar. To show in a very rough manner how it affects the process of determining target angle by differential amplitude measurement, we use the diagram of Figure 10-2, in which an extended target is pictured in the left and right antenna lobes. (See Section 9.3.4.2 for background on this.) Suppose this target had a finite number N of important reflecting points, or glint points, as is characteristic of actual physical targets. Suppose the glint points are at angles $\theta_{T1}, \theta_{T2}, \ldots, \theta_{TN}$, relative to the crossover axis and that the comparator is calibrated to produce a noise-free output θ_T. Neglecting additive noise in the receiver, the comparator output (see Eq. (9.53)) will be†

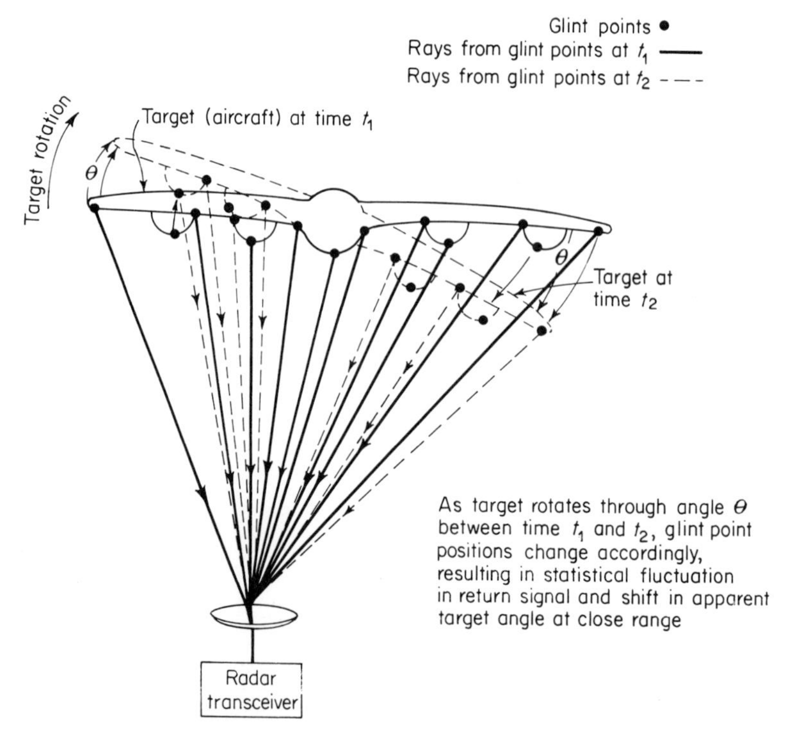

Figure 10-2 Glint noise.

† To derive (10.6), we assume that the comparator is calibrated to produce an output proportional to the target angle in the absence of glint noise and for sufficiently small target angles. One way to do this is divide the *difference* between the amplitudes of the radar return signals from the left and right lobes by the *sum* of the left and right lobe signals with the proportionality factor $(\theta_b^2/2\theta_0)$ to produce the output θ_T in the absence of glint.

$$u_0(\theta_{T1}, \ldots, \theta_{TN}) = \frac{1\{(e^{-(\theta_0-\theta_{T1})^2/\theta_b{}^2} - e^{-(\theta_0+\theta_{T1})^2/\theta_b{}^2}) + \ldots + (e^{-(\theta_0-\theta_{TN})^2/\theta_b{}^2} - e^{-(\theta_0+\theta_{TN})^2/\theta_b{}^2})\}}{(2\theta_0/\theta_b^2)\{(e^{-(\theta_0-\theta_{T1})^2/\theta_b{}^2} + e^{-(\theta_0+\theta_{T1})^2/\theta_b{}^2}) + \ldots + (e^{-(\theta_0-\theta_{TN})^2/\theta_b{}^2} + e^{-(\theta_0+\theta_{TN})^2/\theta_b{}^2})\}} \tag{10.6}$$

If the glint point angles θ_{T1}, θ_{TN}, are slowly varying due to vibration or other motion of the target, the effect is a fluctuation in the apparent target angle. To show this, we expand the exponentials in power series and retain only linear terms (on the assumption that the angles θ_{Tn} are small compared to θ_b). Then, if we are willing to assume that

$$\frac{2\theta_0^2\theta_{Tn}^2}{\theta_b^4} \ll 1, \qquad n = 1, \ldots, N \tag{10.7}$$

(consistent with the assumption of small target angle)

$$\frac{e^{-\theta_{Tn}{}^2}}{e^{-\theta_T{}^2}} \simeq 1, \qquad n = 1, \ldots, N \tag{10.8}$$

(roughly valid if the targets are not too widely separated), then it follows from (10.6), (10.7), and (10.8) that

$$u_0(\theta_{T1}, \ldots, \theta_{TN}) \simeq \frac{\theta_{T1} + \theta_{T2} + \ldots + \theta_{TN}}{N} \tag{10.9}$$

The approximate rms error due to fluctuations in the positions of the glint points (from 10.9) is

$$\epsilon = \langle (u_o - \theta_T)^2 \rangle \simeq \sqrt{\left\langle \left[\frac{(\theta_{T1} + \theta_{T2} + \ldots + \theta_{TN})}{N} - \theta_T \right]^2 \right\rangle} \tag{10.10}$$

From (10.9) it follows that the comparator signal will indicate an *average* glint point position or *centroid* of radar reflection. As the glint point positions fluctuate in random fashion, as they will with a moving target, the angle indicated by the comparator will also fluctuate randomly. This fluctuation appears as a noise in the indicated target angle. The rms error due to this noise, as given by (10.10), is the rms difference between the centroid and the true target angle.

If there are a very large number of glint points distributed densely along the target and they fluctuate independently, then the mean comparator output will indicate an angle very close to a central point on the target, If there are only a few, the mean of the indicated target angle may be far from the true target angle. It is in the latter case that glint constitutes a serious problem in certain radar applications. Otherwise, its only effect is a fluctuation in the indicated target angle about the mean, which may be of little consequence unless very precise angle tracking information is required.

The model used here to describe glint is a highly simplified one. Actually, glint is a complex phenomenon that defies attempts at precise analysis and whose characteristics are sensitive to the details of the radar-target geometry.

For further information, the reader is advised to see references cited in the bibliography at the end of this chapter.

10.1.2 Multiple Target Resolution

In many radar applications, it is important that a set of signal returns from multiple targets be recognized as such and not erroneously interpreted as a signal from a single target. A radar that rarely makes this type of error is a high resolution radar, resolution here referring to the ability to pick out a single target from an aggregate of targets. A "target" here need not be defined as a single aircraft or missile, but may be, for example, an engine nacelle or a missile exhaust.

Certain obvious qualitative statements can be made about the variation of resolution capabilities of a radar with certain parameters. If we are concerned with the ability to resolve two targets having *nearly* equal line-of-sight velocities, then we are speaking of *doppler resolution*, i.e., the ability to separate two radar returns having nearly the same doppler frequency shift. If we want to separate two targets at nearly the same range then our radar receiver must be able to resolve two pulses closely spaced on the time base. Resolution of two targets at nearly the same angle from the radar† requires either an antenna pattern with sharp angular definition or, if interferometry is used for measurement, a high phase resolution receiver.

In general one can state, guided by the discussions in Section 1.4, that: high doppler resolution, like high range rate measurement accuracy, requires long signal observation time (Section 1.4.2); high range resolution, like high range measurement accuracy, requires large bandwidth (Section 1.4.1); and high angular resolution requires a large antenna aperture if the angular measurement is based on the antenna pattern shape (Section 1.4.4).

In the discussions below, an attempt will be made to express some of these ideas quantitatively within the context of radar systems.

10.1.2.1 Range and Range Rate Resolution

In Sections 1.4.1 and 1.4.2, we discussed time and frequency resolution of waveforms. The considerations discussed there can be applied to radar. To apply them, we must first recall that range and range rate measurements in radar are actually time and frequency measurements, respectively. Consequently, the problem of resolving returns from two targets with nearly equal range is the problem of resolving two pulse trains whose pulses are so close together that they overlap. Similarly, resolution of returns from two

† Processing techniques for improving resolution without increasing antenna aperture size also exist (see Pedinoff and Ksienski, 1962), but will not be discussed here.

targets with nearly equal range rates requires the ability to resolve two sinusoids that differ slightly in frequency. These problems were discussed in Sections 1.4.1 and 1.4.2, in a general context.

To discuss them in the context of radar, we simply transform *pulse travel time* to *range* and *doppler frequency* to *range rate*. These transformations are, respectively:

1. Range $= r = c\tau/2$, where c is the velocity of light and τ is the pulse round-trip travel time.
2. Range rate $= v = (c/2)(\omega_D/\omega_c)$, where ω_D is the doppler frequency shift and ω_c is the carrier frequency.

The general conclusions drawn from the examples discussed in Sections 1.4.1 and 1.4.2 are:

1. The *limit of time resolution*, i.e., minimum separation required for the resolution of two adjacent pulses, can be taken to be the reciprocal of the bandwidth of a filter through which both pulses have been passed.
2. The *limit of frequency resolution*, i.e., the minimum frequency separation required for resolution of two pulsed sinusoids of the same duration, can be taken to be the reciprocal of their common pulse duration.

We can translate these statements into the context of radar as follows:

1. Two pulses returned from two targets that are very closely spaced in range are "just resolved" when the targets are separated by the quantity (c/B_R), where c is the velocity of light and B_R is the receiver bandwidth.
2. Two pulses of duration τ_p returned from targets closely spaced in range rate can be considered as "just resolved" when range rate separation of the targets is roughly equal to the quantity $(c/2\omega_c\tau_p)$. If the pulse width is designed to equal the reciprocal bandwidth for maximum SNR,† then the limit of range rate resolution can be taken to be $cB_R/2\omega_c$. This implies (see (1) above) that range and range rate resolution may be incompatible. This point will be elaborated upon in Section 10.1.2.2.

10.1.2.2 The Ambiguity Function Applied to Radar‡

In Section 1.45, the ambiguity function is discussed in a general context as an analytical device to measure the time and frequency resolution capabilities of a system. The reader is referred to that discussion, which culminated in the definitions and expressions (1.58) through (1.63). The discussion can

† See references cited in the footnote below Eq. (9.18) in Section 9.2.2.

‡ Woodward, to whom engineering application of the concept of ambiguity function is sometimes attributed, discussed its application to radar in his well-known monograph (1953), Chapters 6 and 7.

be applied to radar by simply recognizing the relationship between time and range and that between frequency and range rate, as has already been done in the discussions of range and range rate measurement in Sections 9.3.1 and 9.3.2 and those of range and range rate resolution in Section 10.1.2.1.

Before specializing to radar the ambiguity functions defined by (1.58), (1.59), and (1.60), it is instructive to consider their origins more carefully. The task at hand is to resolve two *real* waveforms of finite duration which can be denoted by $v(t, \omega_c) = g(t) \cos \omega_c t$ and $v(t, \omega_c; \tau, \omega_1) = g(t + \tau) \cos((\omega_c + \omega_1)(t + \tau))$, where $g(t)$ represents a pulse shaping function. The time-frequency ambiguity function $\Lambda(\tau, f)$ defined by (1.60) arises from study of the quantity

$$\overline{D^2} = \int_{-\infty}^{\infty} dt\, [v(t, \omega_c) - v(t, \omega_c; \tau, \omega_1)]^2 \tag{10.11}$$

which represents the integral over all time of the squared difference between the value of $v(t, \omega_c)$ and its value at a shifted time $(t + \tau)$ and a shifted frequency $(\omega_c + \omega_1)$.

Straightforward manipulation of (10.11), using complex envelope ideas† and noting that "high-frequency integrals" (i.e., those involving $e^{\pm j\omega_c t}$) vanish, results in the expression

$$\frac{\overline{D^2}}{\int_{-\infty}^{\infty} dt\, g^2(t)} = \left\{1 - \mathrm{Re}\left(e^{-j(\omega_c + \omega_1)\tau} \frac{\int_{-\infty}^{\infty} dt\, g(t) g(t + \tau) e^{-j\omega_1 t}}{\int_{-\infty}^{\infty} dt\, g^2(t)}\right)\right\} \tag{10.12}$$

If we now think of our signal as $g(t)$ and note that ω_1 as used above is equivalent to $2\pi f$ as used in (1.60), we can write

$$\frac{\overline{D^2}}{\int_{-\infty}^{\infty} dt\, g^2(t)} = \{1 - \mathrm{Re}(e^{-j(\omega_c + \omega_1)\tau} \Lambda(\tau, \omega_1))\} \tag{10.13}$$

where the time-frequency ambiguity function to be used in the discussion to follow is defined by

$$\Lambda(\tau, \omega_1) = \frac{\int_{-\infty}^{\infty} dt\, g(t) g(t + \tau) e^{-j\omega_1 t}}{\int_{-\infty}^{\infty} dt\, g^2(t)} \tag{10.14}$$

Specializing (10.14) to the case of zero frequency shift, we arrive at the counterpart of (1.58)

$$\rho(\tau) = \Lambda(\tau, 0) = \frac{\int_{-\infty}^{\infty} dt\, g(t) g(t + \tau)}{\int_{-\infty}^{\infty} dt\, g^2(t)} \tag{10.15}$$

which represents the ambiguity function for time-resolution.

† Note that $v(t, \omega_c) = \mathrm{Re}\{g(t) e^{j\omega_c t}\}$ and $v(t, \omega_c; \tau, \omega_1) = \mathrm{Re}\{[g(t + \tau)\, e^{j\omega_1 t}, e^{j(\omega_c + \omega_1)\tau}]\, e^{j\omega_c t}\}$. The complex envelope of the shifted signal is the square-bracketed quantity.

The counterpart of (1.59), to be used in discussing radar frequency resolution, is obtained by setting the time-shift τ to zero in (10.14), resulting in

$$\Psi(\omega_1) = \Lambda(0, \omega_1) = \frac{\int_{-\infty}^{\infty} dt\, g^2(t) e^{-j\omega_1 t}}{\int_{-\infty}^{\infty} dt\, g^2(t)}, \tag{10.16}$$

which represents the frequency ambiguity function. It is easy to show through straightforward Fourier transformation or equivalently by Parseval's theorem that the form (10.16) is equivalent to the frequency-domain form (1.59).

We will now consider two specific pulse shapes that are particularly convenient for illustrative analysis. These are: (1) the square pulse of width T_p, and (2) the Gaussian pulse $g(t) = e^{-(\pi t^2)/(2T_p^2)}$ (where the form of the exponent is chosen such that $\int_{-\infty}^{\infty} dt\, [g^2(t)/g^2(0)] = T_p$ for both square and Gaussian pulses).

By straightforward integration of (10.14), we obtain

$$|\Lambda(\tau, \omega_1)| = \begin{cases} \left(1 - \frac{|\tau|}{T_p}\right) \operatorname{sinc}\left(\frac{\omega_1 T_p}{2}\left[1 - \frac{|\tau|}{T_p}\right]\right); & |\tau| \leq T_p \\ 0; & |\tau| > T_p \end{cases} \tag{10.17.a}$$

with a rectangular pulse of width T_p,

$$|\Lambda(\tau, \omega_1)| = e^{-\frac{\pi}{4}\left[\left(\frac{\tau}{T_p}\right)^2 + \frac{(\omega_1 T_p)^2}{\pi^2}\right]} \tag{10.17.b}$$

with a Gaussian pulse of width T_p (where "width" is defined as $\int_{-\infty}^{\infty} dt\, [g^2(t)/g^2(0)]$). From (10.15) and (10.16) or (10.17a) and (10.17.b), we have

$$|\rho(\tau)| = \begin{cases} 1 - \frac{|\tau|}{T_p}; & |\tau| \leq T_p \\ 0; & |\tau| > T_p \end{cases} \tag{10.15)$'$}$$

with rectangular pulse,

$$|\rho(\tau)| = e^{-\frac{\pi \tau^2}{4T_p^2}} \tag{10.15)$''$}$$

with Gaussian pulse,

$$|\Psi(\omega_1)| = \operatorname{sinc}\left(\frac{\omega_1 T_p}{2}\right) \tag{10.16)$'$}$$

with rectangular pulse, and

$$|\Psi(\omega_1)| = e^{-\frac{(\omega_1 T_p)^2}{4\pi}} \tag{10.16)$''$}$$

with Gaussian pulse.

To apply the above results meaningfully to multiple target resolution in radar, we should think of the delay time τ as the difference between arrival times of the centers of two received pulses. These pulses result from reflection of a single pulse transmitted at time $t = 0$ from two targets, one at range r, the other at range $(r + \delta r)$. Both targets are assumed to have the same doppler velocity. In this context, the parameter τ is $(2/c)\delta r$. If we express the pulse duration in radar range units rather than time units, and denote it by $R_p = (c/2)T_p$, we can express (10.15)′ and (10.15)″ in the following forms:

$$\left|\rho\left(\frac{2}{c}\delta r\right)\right| = \begin{cases} 1 - \dfrac{|\delta r|}{R_p}; & |\delta r| \leq R_p \\ 0; & |\delta r| > R_p \end{cases} \tag{10.15$'''$}$$

with rectangular pulse,

$$\left|\rho\left(\frac{2}{c}\delta r\right)\right| = e^{-\frac{\pi}{4}\left(\frac{\delta r}{R_p}\right)^2} \tag{10.15$''''$}$$

with Gaussian pulse.

The frequency ω_1 in (10.16)′ and (10.16)″ is interpreted as the difference between the frequencies of two received radar signals arising from reflection of the same transmitted signal from two targets with different range rates. One target has range rate v and the other has range rate $v + \delta v$. The doppler frequency difference corresponding to this range rate difference is $2\omega_c \delta v/c = 4\pi\delta v/\lambda$, where λ is radar wavelength. Expressed in terms of the range rate difference δv, (10.16)′ and (10.16)″ take the forms:

$$\left|\Psi\left(\frac{4\pi\delta v}{\lambda}\right)\right| = \operatorname{sinc}\left(\frac{4\pi R_p}{\lambda}\frac{\delta v}{c}\right) \tag{10.16$'''$}$$

with rectangular pulse

$$\left|\Psi\left(\frac{4\pi\delta v}{\lambda}\right)\right| = e^{-\frac{1}{\pi}\left(\frac{4\pi R_p}{\lambda}\right)^2\left(\frac{\delta v}{c}\right)^2} \tag{10.16$''''$}$$

with Gaussian pulse.

Expressing Eqs. (10.17.a) and (10.17.b) in terms of the range and doppler velocities, we have

$$\left|\Lambda\left(\frac{2}{c}\delta r, \frac{4\pi\delta v}{\lambda}\right)\right| = \begin{cases} \left(1 - \dfrac{|\delta r|}{R_p}\right)\operatorname{sinc}\left(\left[\dfrac{4\pi R_p}{\lambda}\right]\left[\dfrac{\delta v}{c}\right]\left[1 - \dfrac{|\delta r|}{R_p}\right]\right); & |\delta r| \leq R_p \\ 0; & |\delta r| > R_p \end{cases} \tag{10.18}$$

with a rectangular pulse whose duration in range units is R_p, and

$$\left|\Lambda\left(\frac{2}{c}\delta r, \frac{4\pi\delta v}{\lambda}\right)\right| = e^{-\frac{\pi}{4}\left[\left(\frac{\delta r}{R_p}\right)^2 + \frac{4}{\pi^2}\left(\frac{4\pi R_p}{\lambda}\right)^2\left(\frac{\delta v}{c}\right)^2\right]} \tag{10.19}$$

with a Gaussian pulse of duration R_p in range units.

The range resolution of a radar pulse has already been defined in Section 10.1.2.1 as the minimum value of δr that can be distinguished when the pulse returns to the radar, i.e., the smallest range separation that two targets could have and still be perceived as two targets. The doppler resolution is defined similarly in terms of δv. Using this definition and applying ambiguity functions, we say that high range resolution requires that $|\rho|$ should be as "sharp" as possible, i.e., should decay to zero rapidly as $|\delta r|$ increases. High doppler resolution requires that $|\Psi|$ should decay rapidly as $|\delta v|$ increases. A way of using ambiguity functions that brings out the tradeoff between range and doppler resolution is to require that, for high range resolution $|\Lambda|$ decays to zero rapidly as $|\delta r|$ increases and for high doppler resolution $|\Lambda|$ decays to zero rapidly as $|\delta v|$ increases. Plots of $|\rho|$ versus δr and $|\Psi|$ versus δv for both rectangular and Gaussian pulses are shown in Figures 10-3 and 10-4, respectively. A contour plot in the δr-, δv-plane, showing contours of constant $|\Lambda|$ for the Gaussian pulse is presented in Figure 10-5.

Suppose we define the minimum resolvable range difference as the value of $|\delta r|$ corresponding to a 50 percent decay in the ambiguity function from its peak value at $|\delta r| = 0$. Figure 10-3 shows that this value of $|\delta r|$ is about 0.5 R_p for the rectangular pulse and about 0.93 R_p for the Gaussian pulse. Thus, the range resolution capability of a rectangular pulse seems to be about twice as good as that of a pulse with Gaussian shape. This is partially due to the way we have defined R_p for the Gaussian pulse, and the discrepancy could be reduced through a different definition of R_p. However, the superiority in range resolution capability is in part a real one and is due to the

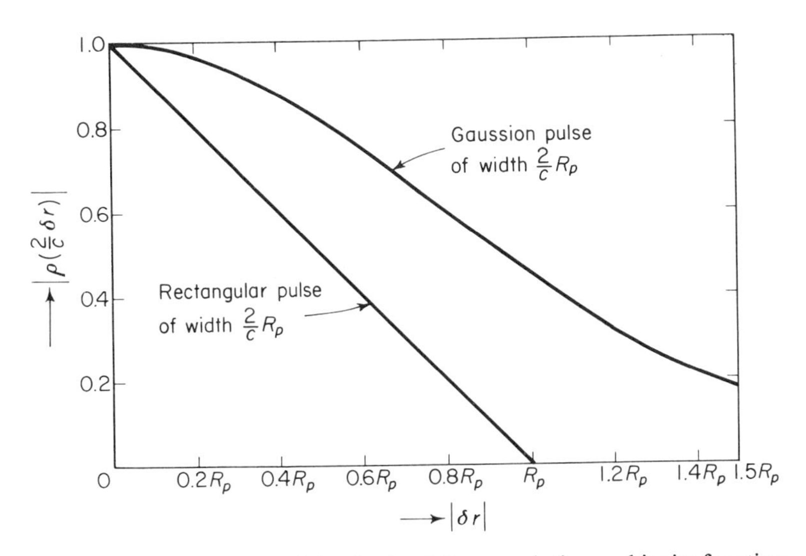

Figure 10-3 Magnitude of time-resolution ambiguity function versus range-difference magnitude.

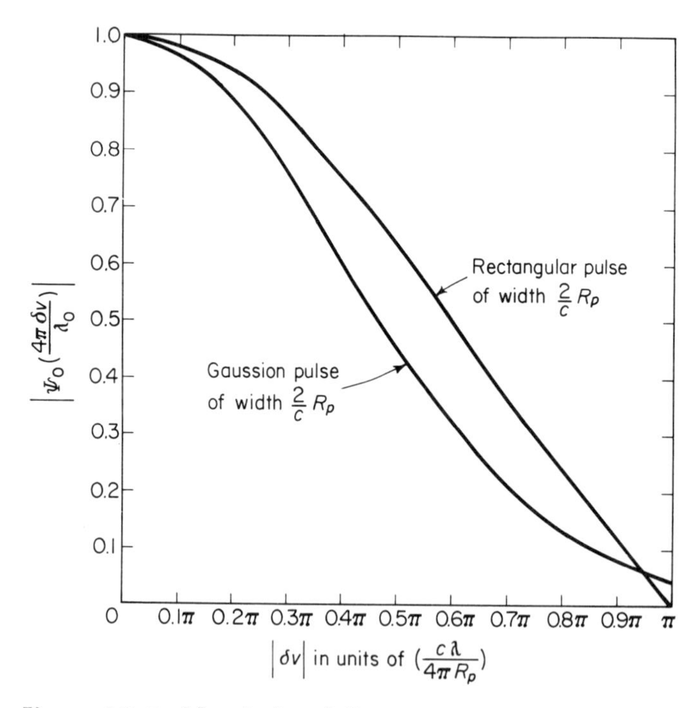

Figure 10-4 Magnitude of frequency-resolution ambiguity function versus range-rate difference magnitude.

sharp edges of the rectangular pulse as contrasted with the gradual decay of the Gaussian pulse. These sharp edges require large bandwidth and, hence, this result corroborates our previous observation that large bandwidth implies high range resolution.

We now define the minimum resolvable range rate as that value of $|\delta v|$ corresponding to a 50 percent decay in the frequency-resolution ambiguity function from its peak value. On this basis, Figure 10-4 shows the Gaussian pulse to be a slightly better range rate resolver than the rectangular pulse. In this case, we can attribute the superiority of the Gaussian pulse largely to its greater duration, i.e., to the fact that at least some pulse energy exists far beyond the point on the time-scale that is defined as the pulse width. This corroborates previous observations that frequency resolution increases with signal duration.

Figure 10-5 applies only to the Gaussian pulse. The curves shown are plots of the minimum resolvable range rate in meter/s against the minimum resolvable range in meters (where those values of $|\delta r|$ and $|\delta v|$ correspond to a 50 percent decay in $|\Lambda|$ from its peak value), for a 1 cm wavelength

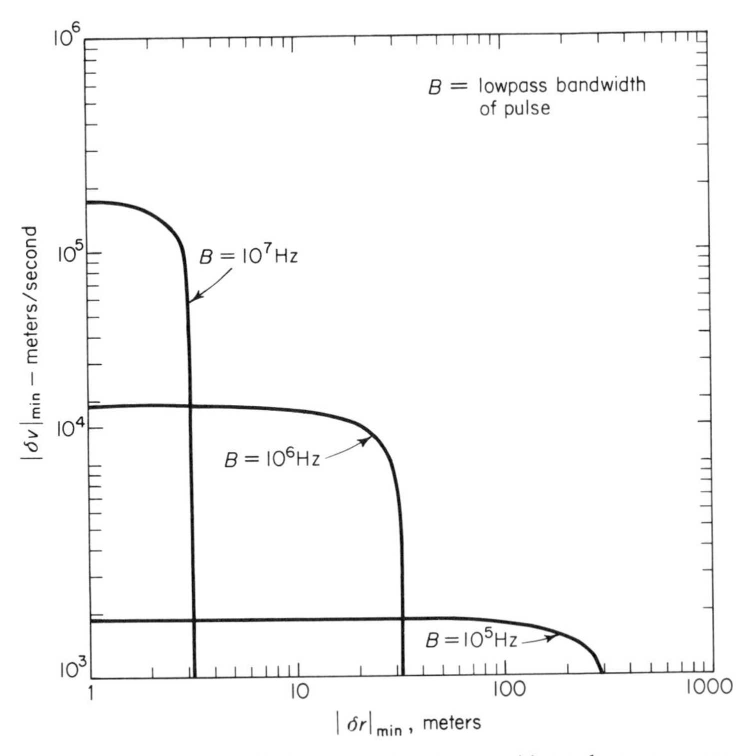

Figure 10-5 Minimum resolvable range (through range-range-rate ambiguity function) Gaussian pulse, 1 cm wavelength.

and for three different values of a parameter B, defined as the "lowpass bandwidth" of the pulse and equal to $(c/8R_p)$.† It is evident from these curves that a tradeoff exists between good range resolution and good range rate resolution and that one is attained only at the expense of the other through adjustment in bandwidth. Figure 10-5 shows that an increase in bandwidth tends to increase range resolution while reducing range rate resolution. For example, with a bandwidth of 100 KHz, the minimum resolvable range $|\delta r|_{min}$ is about 300 meters for a minimum resolvable range rate $|\delta v|_{min}$ of 1 km/s. If a much lower value of $|\delta r|$ is desired, say 10 meters, it is necessary to accept a $|\delta v|_{min}$ as high as 1.8 km/s. With a bandwidth of 10 MHz, values of $|\delta r|_{min}$ as low as three meters can be attained without great sacrifices in range rate resolution. However, if a minimum resolvable range as small

† The bandwidth is defined through the expression $2B = (1/2\pi) \int_{-\infty}^{\infty} d\omega \, |G(\omega)|^2 / |G(0)|^2$, where $G(\omega)$ is the Fourier transform of $g(t)$.

as 1 meter were desired, if would be necessary to accept a minimum resolvable range rate slightly higher than 100 km/s to attain it.

It is often asserted that the time-frequency resolution tradeoff brought out by the use of noise-free ambiguity functions is not strictly valid because there is no real limitation on resolution except that imposed by noise. The basis of this assertion is that *any* difference in the signal due to difference in the value of time or frequency, *however small*, can be perceived if the noise level is sufficiently low. Strictly speaking, this assertion is true. To account for it in our analysis, we should study the effect on the noise level of attempts to enhance resolution. This more accurate approach to the resolution problem is discussed with great clarity by Helstrom in his text on detection theory.† Helstrom's approach focusses on the task of resolving two signals $Af(t)$ and $Bg(t)$ in a noise background, where $f(t)$ and $g(t)$ are different waveforms or time-displaced versions of the same waveform and where A and B are constants which may or may not be different. The task of resolving $Af(t)$ and $Bg(t)$ is that of deciding between four possibilities on the basis of a given received signal-plus-noise. These possibilities are:

1. Neither of the two signals are present.
2. Only the signal $Af(t)$ is present.
3. Only the signal $Bg(t)$ is present.
4. Both $Af(t)$ and $Bg(t)$ are present.

To determine a strategy for optimally deciding between these possibilities, Helstrom first considers the use of minimum average risk criteria and observes that, as in simple detection, the difficulty in using it is in specification of relative risks associated with different classes of error. He then invokes maximum likelihood estimator theory and after considerable analytical work arrives at prescribed processing techniques to optimize the resolution of the two signals in the presence of additive white Gaussian noise. The optimal resolution techniques dictated by the theory with additive white Gaussian noise involve matched filtering of the incoming signal with reference signals made up from $f(t)$ and $g(t)$ and comparison of the matched filter outputs. The crosscorrelation function between $f(t)$ and $g(t)$ turns out to be a critical quantity in the analysis. As might be expected, the less correlated are the two signals, the higher the probability of successfully resolving them.

Helstrom pursues this topic in far greater generality and analytical detail than indicated here, and the reader is referred to his book for the details, which make interesting reading. As a general philosophical point, it may be said here that the theory of signal *resolution* in additive white Gaussian noise bears some resemblance in its arguments and results to the theories

† Helstrom, 1960, Chapter X, pp. 267–296.

of signal *detection* and parameter measurement in additive white Gaussian noise. Regardless of the crust, when the pie is sliced, the filling can always be seen to contain some variation of matched filtering.

10.1.3 Special Aspects of Pulsed Radar Detection

Consider a pulsed radar illuminating a target. Suppose the target signal phase is a priori known at the receiver, so that matched filtering or crosscorrelation can be accomplished coherently. In this case, the reference signal is simply a train of RF or IF pulses of the same frequency as that of the target signal, corrected for the doppler shift if it is known. Both the PRF and the RF or IF must be matched to the target signal. In this case, the considerations of Section 9.2 and 9.3 are applicable and the analysis can be based on them. The fact that we are dealing with pulsed, (rather than CW) waveforms is not important in principle. Only the total signal *energy* is important, and the way in which this energy is distributed in time is irrelevent, as long as the reference pulse train matches the target pulse train.

In the case where a priori ignorance of phase and/or target doppler shift precludes the use of coherent crosscorrelation, this is no longer true. The correlation process must be applied to pulse envelopes, i.e., at the video stage. Special considerations apply in that case, and it is our purpose here to discuss these considerations. In the process of doing this, we will also discuss certain aspects of coherent crosscorrelation of radar pulse trains.

The pulses used in the crosscorrelation process are assumed to be rectangular† and of width τ_g. To allow for a priori uncertainty in range, multiplication of these pulses by the target pulses can be accomplished on a multiple channel or sweep basis, in analogy to the two alternative procedures available for doppler velocity gating. In either case, the operation is known as range gating. If processing is done at RF or IF, it is sometimes necessary to simultaneously range gate and doppler gate, since in the usual case, neither range nor range rate are known precisely before radar measurements begin.‡

In a pulsed radar, the RF passband should be designed in such a manner as to maximize the SNR at the IF input. According to matched filter theory this objective would be attained by matching the receiver passband to the spectrum of the incoming pulse. Since the detailed shape of the incoming pulse is a priori unknown to the receiver designer, he can approximate this optimization condition in a practical way by setting the receiver (bandpass)

---

† Optimally designed reference pulses would be shaped like the return pulses, i.e., distorted by medium and target. But since the distortion is not generally known in detail this is not feasible.

‡ In principle, this could be circumvented by using range rate information from the range gate to set the frequency reference in the doppler velocity gate.

bandwidth $2B_R$ (equivalent to the internal noise bandwidth) equal to the bandwidth of the pulse as was assumed in previous discussions in Chapter 9.

To specialize our thinking about crosscorrelation to the case of a radar pulse train, it is necessary only to recognize that:

1. The noise correlation time, being roughly the reciprocal of the receiver bandwidth, is small compared to the interpulse interval, assuming that the latter is designed to be large compared to pulse duration. Therefore the noise bursts associated with different reference pulses are uncorrelated. Since the noise is Gaussian, this implies that they are statistically independent.

2. Because of (1), the outputs resulting from integration of products of noise reference pulses can be added incoherently, i.e., their power adds algebraically in computing the average power in their sum.† Thus if N reference pulses appear during a given time interval, the average integrated noise power is $N\sigma_n^2$, where σ_n^2 is the mean noise power associated with a single reference pulse.

3. Also because of (1), the products of target signal pulses and reference pulses add coherently, i.e., the total integrated signal power is $(N\sigma_s)^2$, where σ_s^2 is the mean signal power resulting from multiplication of a single reference pulse by a single target pulse followed by integration of the product.

4. Because of (2) and (3), the increase in power SNR over the single pulse case brought about by integration is $N(\sigma_s/\sigma_n)^2$.

5. The points indicated in (2) and (3) are modified by integrating with an exponential weighing factor e^{-t/T_c} with a time constant T_c that is large compared to the interpulse interval t_i. An optimum value of time constant exists, such that signal inputs are "remembered" as long as possible consistent with the condition that noise inputs are quickly "forgotten." To show this, note that if all reference pulses have the same amplitude, then the power SNR at the integrator output is,‡

$$\mathrm{SNR} = \frac{\sigma s^2}{\sigma n^2} \frac{\left(\sum_{k=1}^{N} e^{(-kt/T_c)i}\right)^2}{\sum_{k=1}^{N} e^{(-2kt/T_c)i}} \simeq \left(\frac{\sigma_s}{\sigma_n}\right)^2 \frac{2T_c}{t_i} \frac{[1 - e^{(-Nt/T_c)i}]^2}{[1 - e^{(-2Nt/T_c)i}]} \qquad (10.20)$$

where t_i is the interpulse interval, assumed small compared to T_c. It is easily shown by a power series expansion of $e^{(-Nt/T_c)i}$ in (10.20) that the SNR degenerates into $[(\sigma_s/\sigma_n)^2 N]$ in the case where $Nt_i/T_c \ll 1$.

---

† See Section 3.4.2, text below Eq. (3.66), and Section 5.5.3.5, footnote below Eq. (5.19).

‡ Equation (10.20) is based on the geometric series $1/(1-x) = \sum_{k=1}^{\infty} x^k$ and its consequence,

$$\sum_{k=1}^{N} x^k = \sum_{k=0}^{\infty} x^k - 1 - \sum_{k=N+1}^{\infty} x^k = \left(\frac{1}{1-x}\right) - 1 - x^{N+1}\left(\frac{1}{1-x}\right) = \frac{x(1-x^N)}{(1-x)}$$

As indicated above, there are two basically different methods of applying crosscorrelation techniques to the detection of pulsed radar signals. The first is IF *crosscorrelation*, i.e., to operate on the signals at RF or IF, constructing the reference signal as a train of RF or IF pulses whose frequency matches that of the incoming signal at RF or IF. This requires either a knowledge of the target doppler shift or possibly the use of doppler gates. It does not necessarily require a knowledge of phase, since it can be (and usually is) done noncoherently, i.e., each pulse is envelope detected and the envelopes are then summed.

The second method is called *video crosscorrelation*. The IF pulses are first rectified, either linearly or quadratically, multiplied by a reference train of video pulses whose PRF matches that of the target return, and the product is then integrated. The advantage of this method over IF crosscorrelation is that it circumvents the requirement that the doppler shift be a priori known.† The only target signal parameter that must be specified in the video crosscorrelator is the PRF,‡ which is always known at the radar. The disadvantage is a loss in output SNR that is the inevitable price to be paid for lack of a priori knowledge of signal parameters,

The reader is referred to the discussions of Sections 6.5 and 7.1.4 concerning the use of the square-law detector with integration and subtraction of the mean output noise to circumvent the necessity for a priori frequency knowledge. The use of the video crosscorrelator in lieu of the RF or IF crosscorrelator is dictated by the same considerations, and the degradation in performance resulting from its use is comparable. Another alternative is the use of an autocorrelator (see Section 6.9) whose delay time is roughly a multiple of the PRF. This can be done in different ways, each of which has its advantages and disadvantages.

Consider first the IF coherent autocorrelator (see Figure 10-6), with a delay introduced at IF, the delay time τ being roughly a multiple of the interpulse interval but precisely a multiple of the IF period. The incoming signal is multiplied by the delayed signal and the product is integrated. The implementation of this type of autocorrelator requires that the delay time be well within a pulse length of the interpulse interval; otherwise no significant overlap between the incoming pulse and its delayed version will occur and zero signal output will result. This is analogous to the requirement in a pulse crosscorrelator (either IF or video) that the interval between reference pulses be well within a pulse length of the interpulse interval of the radar return. Both of these objectives require a rather precise a priori knowledge of PRF.

---

† Doppler shifts in the PRF as well as the RF or IF are also present, but are usually small enough so that the resulting relative displacement of video reference and target pulses is tolerable.

‡ Aside from pulse width, which need not be exactly equal for target and reference signals.

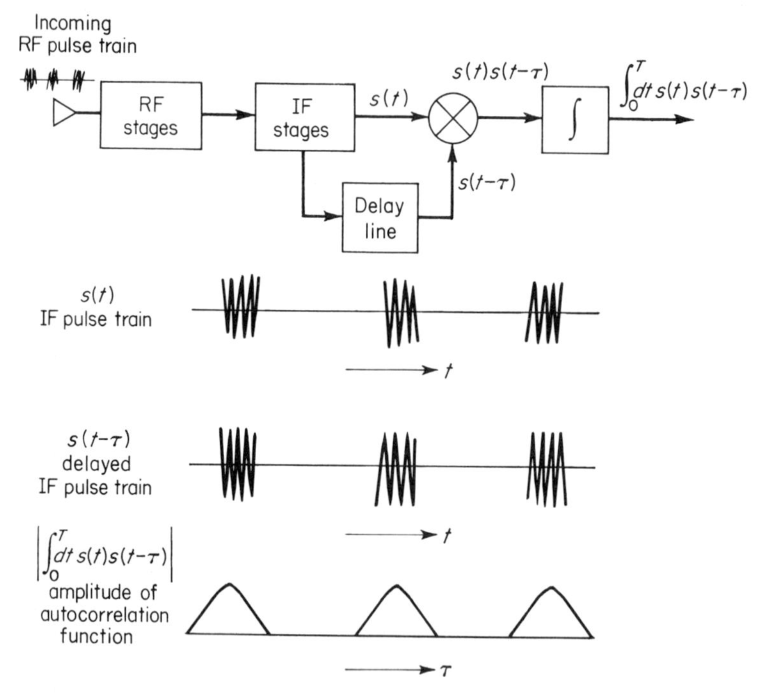

Figure 10-6 IF coherent crosscorrelator.

The requirement that the delay time be precisely a multiple of the IF period is analogous to the requirements in IF coherent crosscorrelators that the frequency of the reference sinusoid be equal to that of the incoming signal and that the phases of the two be matched. In the coherent crosscorrelator case, a precise knowledge of the target doppler shift and the incoming signal phase is required. Since the latter is hardly ever a priori known, the noncoherent correlator scheme, using the envelope of a matched filter output, can be adapted either to the crosscorrelator or the autocorrelator. To adapt it to the autocorrelator, two delayed versions of the incoming signal can be used, both being delayed by an amount within a pulse length of the interpulse period, but the two being in phase quadrature. The outputs of the autocorrelation process with these two delayed signals can then be squared and summed as in the case of the noncoherent crosscorrelator. This technique, in addition to circumventing the requirement for precise phase knowledge, also accounts for the possibility of pulse-to-pulse phase fluctuations. Again, as in the case of the noncoherent crosscorrelator, these advantages are gained at a cost of very roughly 3dB in SNR (see Section 6.4).

Consider the case where a priori knowledge of frequency and phase is very poor. In radar, this would correspond to a case with an extremely large uncertainty in range rate. The operation of square law rectification and integration prescribed for this case, as discussed in Section 6.5, would be ineffective if applied to pulsed radar. The interval between pulses is usually very long compared to the pulse duration. Therefore during a large percentage of the time no target signal is present, and any integration process applied during these intervals would result in integration of pure noise. This is bound to reduce SNR and degrade performance. Therefore, it is always desirable to deactivate the processing during the interpulse interval. The video crosscorrelator serves this purpose without requiring a priori knowledge of either frequency or phase for its implementation. Each pulse is rectified, either linearly or quadratically, and the crosscorrelation process is applied to the resulting pulse envelope. The reference signal is a train of video pulses which, if matched in time to the video pulses in the incoming signal, will result in integration only during those periods when target signal energy is present, just as in the case of the IF crosscorrelator.

The important items of information in applying video crosscorrelation (also IF crosscorrelation) are target range and pulse duration. The former is a priori known only within an interval Δr. In analogy to velocity gating, the alternatives are (1) to use a single gate, setting the gate width τ_g equal to $(2/c)\Delta r$ (where c is the velocity of light) if $(2/c)\Delta r$ is small compared to an interpulse interval; (2) to construct M gates each of width $(1/M)[(2/c)\Delta r]$; (3) to construct a single gate whose width is a fraction of $(2/c)\Delta r$ and sweep it through the interval $(2/c)\Delta r$; (4) to use some combination of procedures (1), (2), and (3).

It is easy to determine the relationship between the ratio of τ_g to the pulse duration τ_p and the SNR at the correlator output, without recourse to elaborate mathematical analysis. If $\tau_g = \tau_p$, then signal-plus-noise and noise alone outputs of the correlator are of equal duration. The analysis discussed in Section 6.6 and detailed in Appendix II then applies precisely, if the correlation is done coherently at IF. The SNR can be obtained with guidance from Figure 6-4. The error probabilities can be obtained from (7.19) and (7.20) or from Figures 7-1 or 7-2. The total *effective* integration time T_{ie} to be used in applying the general results to the specific case at hand is given by

$$T_{ie} = \tau_p f_p \ell T_{ia} \tag{10.21}$$

where T_{ia} is the *actual* time occupied by the pulse train, f_p is the pulse repetition frequency and ℓ is the ratio of effective integration time to true integration time, determined by the extent of overlap between target pulses and gate pulses (see Appendix II).

If $\tau_p < \tau_g$, then during the time interval wherein the reference pulse is on but the target pulse is off, noise alone is being integrated. The signal integration time of each pulse is τ_p while noise integration time is τ_g. The signal power is proportional to τ_g^2 and the noise power is proportional to τ_g. Hence output SNR is proportional to τ_p^2/τ_g. This can be accounted for by setting ℓ equal to (τ_p/τ_g). If τ_p is equal to τ_g, output SNR is proportional to τ_p. In this case we set ℓ equal to unity. If $\tau_p > \tau_g$, then there is a time interval wherein the gate pulse is off but the target pulse is on. This means that some signal energy is being lost. Signal integration time is τ_g, while noise integration time is τ_p. Since signal integration is coherent, the fractional loss in signal power is $(\tau_g/\tau_p)^2$. Noise integration being incoherent, the reduction in noise power relative to the case $\tau_g = \tau_p$ is in the ratio (τ_g/τ_p). The net result is a fractional loss in SNR of (τ_g/τ_p). Hence in this case ℓ is set equal to (τ_g/τ_p).

The error probabilities for the video crosscorrelator cannot be determined from (7.19) and (7.20), since the noise at the output of the rectifier is no longer Gaussian. Hence, the output probability density function for each pulse is given by (3.50) if quadratic envelope rectification is used and by (3.46) if linear rectification is used. Since the successive noise pulses are statistically independent, the false rest and false alarm probabilities with detection threshold v_T are given respectively, by

$$P_{fr} = \int_0^\infty dv_1\, p_{s+n}(v_1) \int_0^\infty dv_2\, p_{s+n}(v_2) \ldots \int_0^\infty dv_{N-1}\, p_{s+n}(v_{N-1}) \int_0^{v_T-(v_1+\ldots+v_{N-1})} dv_N\, p_{s+n}(v_N) \tag{10.22.a}$$

and

$$P_{fa} = \int_0^\infty dv_1\, p_n(v_1) \int_0^\infty dv_2\, p_n(v_2) \ldots \int_0^\infty dv_{N-1}\, p_n(v_{N-1}) \int_{v_T-(v_1+\ldots+v_{N-1})}^\infty dv_N\, p_n(v_N) \tag{10.22.b}$$

where $p_{s+n}(v)$ is the correlator output probability density function for a single pulse of signal-plus-noise and $p_n(v)$ is the corresponding function for noise alone. These functions are given by (3.50) and by (3.33) respectively (with $a_s = 0$), with quadratic envelope detection and by (3.46) and (3.40), respectively, with linear detection.

The error probabilities (10.22.a) and (10.22.b) have been calculated by Marcum and Swerling (1960) for a wide range of pulsed radar cases. The reader interested in further details is referred to their monograph.

It should be noted that noise is not the only agent that limits pulsed radar performance. In pulsed doppler radar, even in the absence of noise, there are limits to the degree to which we can simultaneously confine both range and range rate to narrow regions. To see this, we apply the duration bandwidth uncertainty principle of Section 1.4.3. to radar pulses. Referring to

(1.52), we invoke the transformation between the pulse travel time and range-to-target and that between the doppler frequency shift and the range rate, in the form

$$\sqrt{\overline{(\Delta t)^2}} = \frac{2}{c}\sqrt{\overline{(\Delta r)^2}} \tag{10.23.a}$$

$$\sqrt{\overline{(\Delta \omega)^2}} = \frac{2\omega_c}{c}\sqrt{\overline{(\Delta v)^2}} \tag{10.23.b}$$

where Δt and $\Delta\omega$ are defined in (1.53) and (1.54), c is the free-space velocity of electromagnetic waves and the symbols r and v wherever they occur refer to range and range rate, respectively.

Substituting Eqs. (10.23) into (1.52), we can express the uncertainty principle in the form

$$\left(\sqrt{\frac{\overline{(\Delta r)^2}}{\lambda}}\right)\left(\sqrt{\frac{\overline{(\Delta v)^2}}{c}}\right) \geq \frac{1}{8\pi} \approx .04 \tag{10.24}$$

We interpret $\sqrt{\overline{(\Delta r)^2}}$ as the *spread* in range (as measured through pulse travel time) contained within the pulse and $\sqrt{\overline{(\Delta v)^2}}$ as the spread in range rate (as measured through doppler frequency shift) contained therein. With this interpretation, we can conclude from (10.24) that a range spread as small as a wavelength implies a range rate spread that is a substantial fraction of the free-space velocity of electromagnetic waves. Since the latter number is 300,000 km/s, this would seem to imply that the range spread must be many hundreds or thousands of wavelengths in order that range rate spread will be comparable to typical vehicle velocities.

In designing a radar pulse to confine target range to a very narrow region, we require a large bandwidth. To confine doppler shift to within a narrow region, we require a large pulse duration. The result (10.24) implies that we cannot have both narrow range confinement and narrow range rate confinement to an unlimited degree, but must compromise between these two goals in designing radar pulses.

10.1.4 Fading in Radar Systems

It is pointed out in Section 5.5.4.7 in the general discussion of multipath and fading that microwave transmissions are sometimes subject to fading. This fading is largely due to atmospheric irregularities along the transmission path. Radar signals, usually being at microwave frequencies, sometimes fade significantly due to such irregularities. Fading of radar signals also occurs due to target scintillation, as pointed out in Section 10.1.1.3. Whether the fading is due to the target or the atmosphere, it generally consists of a specular component superposed on a random component with Rayleigh amplitude statistics. Hence the most general amplitude statistics likely to be encountered is the Rice distribution.

We will simplify the discussion by ignoring the specular component and thereby confining attention to Rayleigh-distributed fading. The reader is referred to Section 7.4, which is a discussion of the effects of fading on error probabilities in signal detection. The early part of that discussion (through Eq. 7.95) applies to the detection of radar signals in Rayleigh fading. However, the special case considered, in which the two possible signals in the receiver are orthogonal and have equal energies, is not applicable to the usual radar detection problem. The on-off case is the one to which we should apply Eqs. (7.94), (7.95), and (7.96), under the assumption that "signal present" and "noise alone" are equally probable events. The error probability $P_\epsilon(\rho)$ appearing in the integrand of (7.95) is given by (from (7.19) and (7.20))

$$P_\epsilon(\rho) = P(1)P_{21} + P(2)P_{12} = \frac{1}{2}(P_{21} + P_{12})\frac{1}{2}\left\{1 - \operatorname{erf}\left(\frac{\rho}{4}\right)\right\} \quad (10.25.a)$$

with coherent detection, and (from (7.35.a) and (7.35.b))

$$P_\epsilon(\rho) = \frac{1}{2}\left\{1 - Q\left(\sqrt{2\rho}\,\frac{\sqrt{2\rho}}{4}\right) + e^{-\rho/16}\right\} \quad (10.25.b)$$

with noncoherent detection, where the detection threshold in (7.19), (7.20), (7.35.a) and (7.35.b) have been chosen at one-half the expected signal amplitude (i.e., $K_T = \frac{1}{2}$) and the SNR ρ is set equal to $2\rho_i B_n T$.

The error probability with fading is given by

$$P_\epsilon = \frac{1}{\langle\rho\rangle}\int_0^\infty d\rho\, P_\epsilon(\rho)e^{-\rho/\langle\rho\rangle} = \frac{1}{2} - \frac{1}{2\langle\rho\rangle}\int_0^\infty d\rho \operatorname{erf}\left(\frac{\rho}{4}\right)e^{-\rho/\langle\rho\rangle} \quad (10.26.a)$$

with coherent detection

$$2\frac{1}{\langle\rho\rangle}\int_0^\infty d\rho\left\{1 - Q\left(\sqrt{2\rho}, \frac{\sqrt{2\rho}}{4}\right) = e^{-\rho/16}\right\}e^{-\rho/\langle\rho\rangle} \quad (10.26.b)$$

with noncoherent detection.

Expansion of erf $(\rho/4)$ into power series and subsequent integration of (10.26.a) results in

$$P_\epsilon = \frac{1}{2} + \frac{1}{\sqrt{\pi}}\sum_{n=0}^{\infty}\frac{(-1)^{n+1}(2n)!\langle\rho\rangle^{2n+1}}{4^{2n+1}n!} \quad (10.27.a)$$

with coherent detection.

In the noncoherent detection case, we have, from (10.26.b),

$$P_\epsilon \simeq \frac{1}{2}\left\{\frac{\langle\rho\rangle + 32}{\langle\rho\rangle + 16}\right\} - \frac{1}{2\langle\rho\rangle}\int_0^\infty d\rho\, Q\left(\sqrt{2\rho}, \frac{\sqrt{2\rho}}{4}\right)e^{-\rho/\langle\rho\rangle} \quad (10.27.b)$$

Without going further, one can see from (10.27.a) and (10.27.b) that the functional dependence of the error probabilities on $\langle\rho\rangle$ in the fading case are very different than the dependence of the error probabilities on ρ in the nonfading case. The analysis will not be carried further here. The reader is referred to the bibliography at the end of this chapter for more extensive treatments of the subject in which calculations like those indicated here are carried out and error probabilities associated with fading are compared quantitatively with those in the nonfading case.†

10.2 TOPICS IN DETECTION AND LOCATION

There are two major classes of radio detection and location systems that have not yet been discussed in this book, *bistatic radar* and *passive radio detection and location systems.* The sections below will be devoted to very brief introductory treatments of these topics, with particular emphasis on the ways in which the ground rules for application of statistical communication theory ideas differ from those applicable to conventional or monostatic radar.

10.2.1 Bistatic Versus Monostatic Radar

A *bistatic radar* is a radar in which the transmitter and receiver are placed at different locations. A conventional *backscatter* type of radar may be called *monostatic* to distinguish it from the bistatic radar. The geometry of the bistatic radar system is illustrated in Figure 10-7.

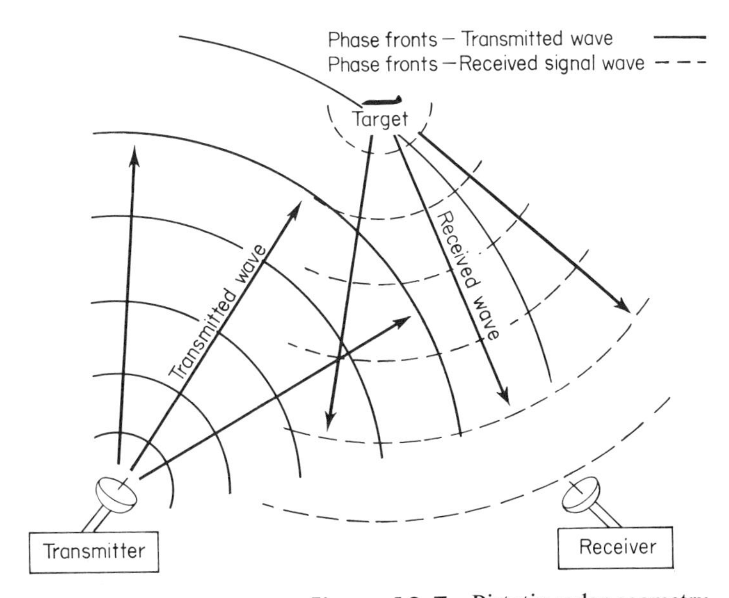

Figure 10-7 Bistatic radar geometry.

† See Barrow, 1962.

There are some rather important differences between bistatic and monostatic radars. Some of these significantly change the way in which statistical communication theory can be applied to problems of detection and location with these two different classes of radar.

First, we note that the radar equations (9.7.a) and (9.7.b) must be generalized to account for the displacement between transmitter and receiver. To accomplish this, we would carry through steps analogous to those leading from Eqs. (9.1) through (9.7) for the geometry depicted in Figure 10-7, leading to the result

$$P_R = \frac{P_T G_{0T} G_{0R} g_T(\theta, \phi) \sigma(\theta', \phi', \theta'', \phi'') g_R(\theta''', \phi''')}{(4\pi)^3 r_1^2 r_2^2} \tag{10.28}$$

where

P_R = received power

P_T = transmitted power

G_{0T} = peak gain of transmitting antenna

$g_T(\theta, \phi)$ = relative transmitting gain as function of target angles (θ, ϕ) in transmitter's frame of reference (see Figure 10-7(a)).

$\sigma(\theta', \phi'; \theta'', \phi'')$ = cross section of target for scattering in direction (θ'', ϕ'') in target's frame of reference (see Figure 10-7(b)).

G_{R0} = peak gain of receiving antenna

$g_R(\theta''', \phi''')$ = relative receiving antenna gain as a function of target angles (θ''', ϕ''') in receiver's frame of reference (see Figure 10-7).

r_1 = distance between transmitter and target

r_2 = distance between receiver and target.

It is evident that (10.28) approaches (9.7.a) as the transmitter and receiver move closer together.

Certain features of (9.7.a) change as we generalize it to (10.28). First, the radar cross section is no longer the backscattering cross section. It now becomes the cross section for oblique scattering at the particular angles determined by the radar-target geometry. Usually the cross section for oblique scattering is smaller than that for backscattering, resulting in a degradation in received power in the bistatic case. In applying detection theory to bistatic radar situations, we would simply account for this degradation in power and thereby reduce the received SNR associated with detection of a given target. This would result in a larger error probability in detection of a given target bistatically than that incurred with monostatic radar detection. If we were interested in the transmitter power required for a given level of detection performance, we could still calculate it from an equation analogous to (9.12.a), where the functional dependence on various parameters would be much the same, except for the obvious change from $40 \log_{10} r$ to $20(\log_{10} r_1 + \log_{10} r_2)$.

An equation analogous to (9.12.b) could be developed from (10.28) for the purpose of calculating the maximum detection range. This calculation would first require a definition of range, which could be, for example, transmitter-to-target range r_1, target-to-receiver range r_2, the geometric mean $\sqrt{r_1 r_2}$, or the distance between the target and the midpoint between transmitter and receiver, In developing a formula analogous to (9.12.b) from the bistatic equation (10.28), the definition of range would be crucial to the form obtained. For example, if r_2 were defined as range, with r_1 held constant, the fourth root dependence in (9.12.b) would change to a square root dependence.

A much more profound difference between bistatic and monostatic radar concerns the way in which certain target signal parameters relate to target position and velocity. For example, in monostatic pulsed radar, a range measurement has a one-to-one correspondence with a measurement of pulse travel time (see Section 9.3.1). A range rate measurement is equivalent to a measurement of doppler frequency shift (see Section 9.3.2). In generalizing the discussions in Sections 9.3.1, 9.3.2, and 9.3.3 to render them applicable to bistatic radar, it would be necessary to recognize that a measurement of the time required for a pulse to travel from transmitter to receiver would not in itself indicate target range. Also a measurement of the doppler shift of the received signal would not indicate range rate without the aid of auxiliary information.

There is an apparent advantage of bistatic radar which can only be fully realized at the high end of the microwave spectrum, where high directivity is attainable without prohibitively large antenna structures. Referring to Figure 10-7, we note that, if both transmitting and receiving antennas are highly directive, a received signal indicates that the target lies within a small volume common to both beams. Targets that lie within the transmitting beam and not the receiving beam (e.g., T_1) will be illuminated but will not scatter into the receiver and hence will not be detected. Likewise targets that lie within the receiving beam and not the transmitting beam (e.g., T_2) will not be illuminated and hence also will not be detected.

Only signals from targets in both beams (e.g., T_3) will appear at the receiver. The region of intersection of the two beams, known as the *common volume*, can be swept through regions of space by swinging both transmitting and receiving antenna beams simultaneously. This is the way a search for the targets could be carried out with bistatic radar. The measured target position could be confined within a region of the size of the common volume as a result of the search procedure. This could be done without the use of information about pulse travel time or doppler shift, except as refinements on accuracy. Very complicated processing of a set of composite measurements of the angle of the common volume, the pulse travel time, and/or the doppler shift would be required in order to pinpoint the target position and velocity. However, if the antenna beams were extremely narrow, sufficient accuracy in target trajectory measurement could be achieved without range or doppler information, thus circumventing the need for elaborate processing of the received signals.

As is evident from this brief discussion, an analysis of a bistatic radar system involves some very interesting problems in statistical communication theory. These problems are sometimes more complicated than those arising in monostatic radar applications, because of the greater complexity of the geometry. However, the same basic theoretical ideas prevail, e.g., tradeoffs between SNR and target resolution, reduction of error probabilities in detection by using multichannel correlator schemes or searching over long periods of time for unknown parameters, and the purchase of increases in trajectory measurement accuracy with the two key media of exchange, time and equipment complexity.

The reader is referred to the bibliography at the end of this chapter for further information on bistatic radar.

10.2.2 Passive Target Detection and Location

A physical object can sometimes be detected and at least partially located through reception of the electromagnetic energy that it radiates into the space surrounding it. Insofar as some of this radiated energy is within the microwave spectrum, a properly placed microwave receiver should be able to detect the object. A system built around such a receiver, often designated as a passive detection system, could also be called a *passive radar*. The term "passive radar" is used here to contrast such a system with the usual radar system which requires a power source and hence is an "active" system.

From the communication theory point of view, the basic difference between passive and active radar detection and location of targets lies in the a priori target information available in these two cases. In operating an active radar, either monostatic or bistatic, we know at the outset that we have transmitted a signal with particular parameters, i.e., radio frequency, pulse duration and repetition frequency, etc. We can tune the receiver to signals with specifically those parameters of the transmitted wave. We can thereby exclude or "gate out" other classes of signals. In the case of passive detection, because we usually lack a priori knowledge of target emission characteristics, we must "open the gates" and allow a very large class of possible signals into the receiver. This is a disadvantage. It implies that we must ask for a great deal more information to accomplish detection and location tasks passively than is required with active systems. To some degree, this disadvantage is offset by the fact that propagation must occur in only one direction with a consequent $1/r^2$ power dependence from target to receiver, whereas a two-way path is required with active systems, resulting in a $1/r^4$ power dependence. This apparent advantage may be illusory unless the target's strength as a source of microwave radiation is large compared to its strength as a reflector of such radiation. Since good reflection and good natural emission are mutually exclusive,† the range advantage is most meaningful in detecting a target passively through its *man-made* transmissions. If the target transmits a given amount of power P_T and has a given radar cross section σ, it may be easier to detect it passively by receiving its transmissions than to do so with one's own active radar transmissions, because of the $1/r^2$ range advantage. To illustrate this, we invoke the one-way beacon equation (5.12) and the radar equation (9.7.a) or (9.7.b) and compare the ranges required to attain a given received signal level passively and through monostatic radar. In making this comparison, we assume the target's transmitted power and gain over an isotropic radiator in the passive case to be equal to the radar's transmitted power and gain in the active case. The effective area of the receiving antenna is the same in both cases. This comparison is illustrated on the curve of Figure 10-8, showing the ratio of range with passive detection to range with monostatic radar plotted against a convenient dimensionless parameter whose constituents are given on the figure. From this curve one could find the parameter regions in which the r^2 advantage of passive detection is offset by (1) the deficiency of radiated power from the target and (2) the fact that the target's radar cross section is large enough to render it a good radar reflector.

Matched filtering is not usually feasible in passive detection systems. In addition to the lack of a priori parameter information, an additional disadvantage is the fact that the target radiates a wide band of frequencies, thus precluding the usual matched filtering schemes.

Such schemes as autocorrelation detection and squaring followed by integration (see Sections 6.9 and 7.1.4.1) are especially useful in passive detection. These schemes require very little a priori target information. Autocorrelation does require a knowledge of the most likely frequency regions of the radiation, but this requirement can be somewhat offset by sweeping over a wide band of frequencies or the use of multiple channel systems (see Section 7.1.4.2).

In general, devices designed for detection of wideband signals about which the observer knows little are called radiometers.‡ The square-law detector followed by integration is a type of radiometer. Two other common

† See any elementary physics text, e.g., Sears, Vol. III, 1948, pp. 307–311.

‡ See, Goldstein, 1955; Tucker, Graham, and Goldstein, 1957; Galejs, Reich, and Raemer, 1957; Raemer and Reich, 1958.

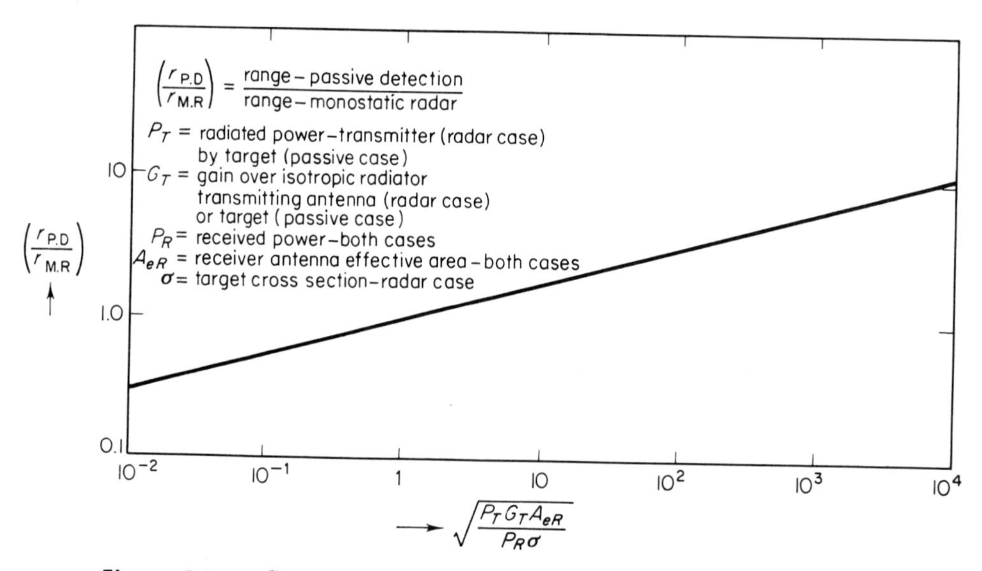

Figure 10-8 Comparison of range-monostatic radar and passive detection.

types are (1) the Dicke radiometer and (2) the two-receiver radiometer. In the Dicke radiometer (see Figure 10-9(a)) the signal is modulated at RF (*before* the internal noise has entered the receiving system) with a low frequency sinusoid and then detected at video by matched filtering with a reference signal of the same low frequency. This might be designated as "tagging" the incoming signal with the low frequency sine wave and then later searching for the "tag." If a signal was present, the tag will be present in the form of energy at the modulation frequency. If no signal was present, the video output is pure noise and will contain no tag.

The two-receiver radiometer (see Figure 10-9(b)) depends for its operation on the existence of two identical receivers, each of which receives the same incoming signal, possibly with a relative time delay (which can be adjusted at RF). The internal noise waveforms in the two receivers are statistically independent. The inputs to the two receivers are multiplied together at RF or IF and the product is then lowpass filtered.

We can calculate the output SNR of the two-receiver radiometer, using methods indicated in Chapter 6. If the complex envelopes of the signal-plus-noise waveforms in receivers 1 and 2 are $[\hat{s}_1(t) + \hat{n}_1(t)]$ and $[\hat{s}_2(t) + \hat{n}_2(t)]$, respectively, the output of the lowpass filter is

$$\hat{s}_o(t) + \hat{n}_o(t) = \frac{1}{2}\,\text{Re}\left\{\int_{-\infty}^{\infty} dt'\, f(t-t')[\hat{s}_1(t') + \hat{n}_1(t')][\hat{s}_2^*(t') + \hat{n}_2^*(t')]\right\} \tag{10.29}$$

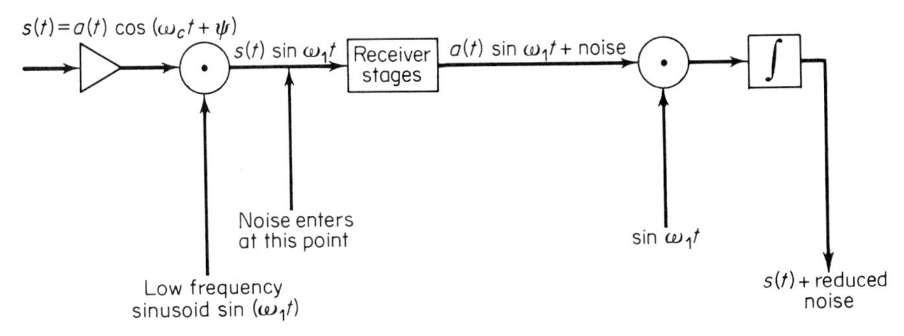

(a) Dicke radiometer

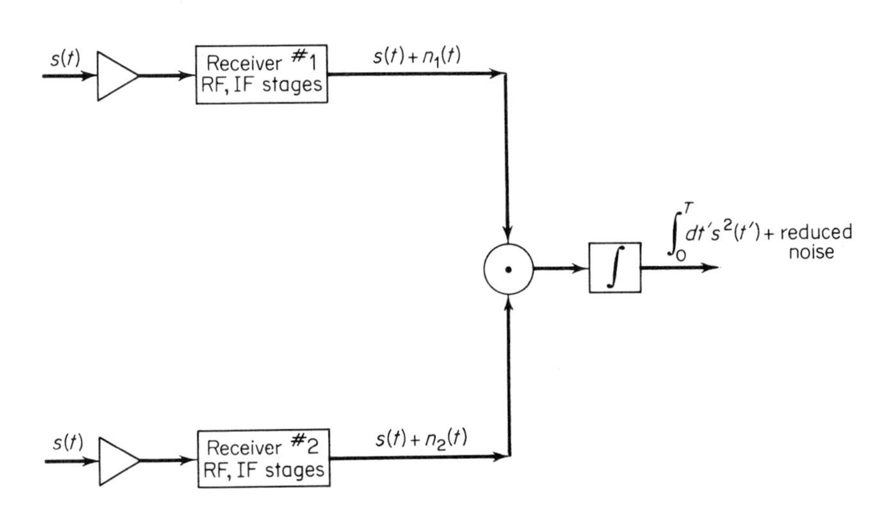

(b) Two-receiver radiometer

Figure 10-9 Examples of radiometers.

where $f(t)$ is the (real) impulse response of the lowpass filter. After adjustment of any relative time delay present, $\hat{s}_1(t)$ and $\hat{s}_2(t)$ are identical and can both be denoted by $\hat{s}(t)$. Thus

$$\hat{s}_o(t) = \frac{1}{2}\int_{-\infty}^{\infty} dt' f(t - t')\,|\hat{s}(t')|^2 \tag{10.30.a}$$

$$\hat{n}_o(t) = \frac{1}{2}\,\text{Re}\left\{\int_{\infty}^{\infty} dt' f(t - t')[\hat{s}^*(t')\hat{n}_1(t') + \hat{s}(t')\hat{n}_2^*(t') + \hat{n}_1(t')\hat{n}_2^*(t')]\right\} \tag{10.30.b}$$

We will now consider both $s(t)$ and $n_{1,2}(t)$ as sample functions of stationary

Gaussian random processes with zero mean (accounting for the fact that a microwave signal radiated from a target has many of the essential characteristics of a random noise) and take the mean square of $[s_o(t) + n_o(t)]$ noting that the two noises $n_1(t)$ and $n_2(t)$ have been assumed mutually statistically independent and also statistically independent of the signal $s(t)$.

The result, after invoking (6.21), justified by the assumption that the internal noise bandwidth is extremely large compared to the bandwidth of the lowpass filter, is†

$$\langle [s_o(t) + n_o(t)]^2 \rangle = \langle [s_o(t)]^2 \rangle + \langle [n_o(t)]^2 \rangle \tag{10.31}$$

where

$$\langle [s_o(t)]^2 \rangle \simeq \frac{1}{4} \int_{-\infty}^{\infty} \int dt'\, dt''\, f(t') f(t'') \,|\langle \hat{s}(t') \hat{s}^*(t'') \rangle|^2 + \sigma_s^4$$

$$\langle [n_o(t)]^2 \rangle \simeq \frac{1}{4} \left(\frac{B_f}{B_n} \right) (\sigma_n^4 + 2\sigma_s^2 \sigma_n^2)$$

where the assumptions and definitions used are

$$\langle s(t') s(t' + \tau) \rangle \equiv \frac{1}{2} \operatorname{Re} \langle \hat{s}(t') \hat{s}^*(t' + \tau) \rangle = \sigma_s^2 \rho_s(\tau),$$

$$(\sigma_s^2 = \text{mean-square signal power},\ \rho_s(\tau) = \text{NACF of signal}) \tag{10.32}$$

$$\langle n_{\substack{1\\2}}(t') n_{\substack{1\\2}}(t' + \tau) \rangle \equiv \frac{1}{2} \operatorname{Re} \langle \hat{n}_{\substack{1\\2}}(t') \hat{n}_{\substack{1\\2}}^*(t' + \tau) \rangle = \sigma_n^2 \rho_n(\tau),$$

$$(\sigma_n^2 = \text{mean-square noise power},\ \rho_n(\tau) = \text{NACF of noise in either receiver}) \tag{10.33}$$

$$\langle n_{\substack{1\\2}}(t') n_{\substack{2\\1}}(t' + \tau) \rangle = 0; \quad \text{(statistical independence of noise waveforms in two receivers)} \tag{10.34}$$

$$\rho_n(\tau) \simeq \frac{1}{2B_n} \delta(\tau); \quad \text{(wideband noise assumption, same as Eq. (6.21))} \tag{10.35}$$

$$\langle \hat{s}(t') \hat{s}^*(t') \hat{s}^*(t'') \hat{s}(t'') \rangle = \langle |\hat{s}(t')|^2 \rangle \langle |\hat{s}(t'')|^2 \rangle + \langle \hat{s}(t') \hat{s}^*(t'') \rangle \langle \hat{s}^*(t') \hat{s}(t'') \rangle,$$

$$\text{(property of zero-mean Gaussian processes)‡} \tag{10.36}$$

$$B_f = \frac{1}{2} \int_{-\infty}^{\infty} dt'\, f^2(t') = \text{lowpass filter bandwidth} \tag{10.37}$$

$$\int_{-\infty}^{\infty} dt'\, f(t') = 1 \tag{10.38}$$

The approximate output SNR of the two-receiver radiometer as calculated

† Note that cross terms between $s_o(t)$ and $n_o(t)$ vanish in the averaging process, because noise and signal and the two noises are uncorrelated and because odd-order moments of Gaussian processes vanish.

‡ See (III.9) in Appendix III and references cited below Eqs. (III.8) through (III.10).

with the aid of Eqs. (10.34) through (10.38) is

$$\rho_o = \frac{B_n \sigma_s^4}{B_f \sigma_n^4} \frac{(1 + L)}{(1 + 2(\sigma_s^2/\sigma_n^2))} = \frac{B_n \rho_i^2}{B_f} \frac{(1 + L)}{(1 + 2\rho_i)} \tag{10.39}$$

where

$$L = \frac{1}{4\sigma_s^2} \int_{-\infty}^{\infty} \int dt'\, dt''\, f(t') f(t'') \, |\langle \hat{s}(t') \hat{s}^*(t'') \rangle|^2$$

$$\rho_i = \frac{\sigma_s^2}{\sigma_n} = \text{input SNR}$$

It would not be quite valid to compare the result (10.39) with the result (6.36) which gives the output SNR of a coherent crosscorrelator, or with (6.93) which gives the output SNR of a square-law device followed by a lowpass filter. These results were obtained under the assumption of a purely deterministic sinusoidal signal. Equation (10.39) applies to a signal waveform that is a sample function of a stationary random process. However, it is possible to make a rough comparison based on examination of the fundamental differences in the noise outputs of (1) a coherent crosscorrelator, (2) a square-law lowpass filter scheme, and (3) a two-receiver radiometer.

First, note that *one* of the benefits of square-law detection is realized with a two-receiver radiometer. The crosscorrelation in the radiometer is performed with a reference signal that is a perfect replica of the incoming signal, without the requirement that the properties of the incoming signal be a priori known. The additional noise-noise term in the quadratic detector output is not present in the two-receiver radiometer output. This is because the internal noises in the two receiver channels are uncorrelated and multiplication followed by lowpass filtering is an approximate computation of the crosscorrelation function between the waveforms in the two channels. The noise-noise part of that output should be very small. These effects enhance the effectiveness of the two-receiver radiometer over the square-law scheme. On the other hand, the two-receiver radiometer should have a lower output SNR than a coherent crosscorrelator. In the crosscorrelator, noise appears only in the incoming signal and not in the reference signals. In the radiometer, there is noise in both channels, thus increasing the mean-square noise output over that of a crosscorrelator.

It should be borne in mind that the noise reduction capabilities of radiometers apply only to *internal* noise. Atmospheric noise entering the radiometer with the desired signal will be treated as if it were part of the desired signal. Discrimination between the target signal and atmospheric background noise entering the receiver is a problem that cannot be solved by radiometers. The basis of its solution depends on the observer's knowledge of the target signal's characteristics and the presence of features in the target signal that distinguish it from background radiation.

For further discussion of radiometers and passive detection systems, the reader is referred to publications listed in the bibliography at the end of this chapter.

REFERENCES

General

See the reference list in Chapter 9 in addition to the list below.

External Radar Noise Sources

Freeman, 1958; Chapter 10 (on target noise)

Kerr, 1951; Chapter 6

Lawson and Uhlenbeck, 1950; Chapter 6 (clutter)

B. D. Steinberg; in Berkowitz, 1966, "Target, Clutter and Noise Spectra," pp. pp. 473–488

Menzel, 1960

Resolution and Ambiguity Functions in Radar

Helstrom, 1955

Helstrom, 1960; Chapter X

Lerner, 1958

Nilsson, 1961

Stewart and Westerfield, 1959

Sussman, 1962

Woodward, 1953; especially Chapters 6 and 7

Urkowitz, 1953

H. Urkowitz; Chapter 1, Part IV, in Berkowitz, 1965, "Ambiguity and Resolution," pp. 197–215

Westerfield, Prager, and Stewart, 1960

Pulsed Radar Detection

A. E. Bailey; in Jackson, 1953, "Integration in Pulse Radar Systems," pp. 216–230

Kelley, Reed, and Root, 1960

Lawson and Uhlenbeck, 1950; Chapters 8, 9, 10, 11

Marcum and Swerling, 1960

Swerling, 1956

Swerling, 1957

Swerling, 1966

References on Bistatic Radar and Passive Microwave Detection, Radiometers

Barton, 1964; Section 4.3, pp. 120–121, also pp. 503–505
Berkowitz, 1965; pp. 11–13
Galejs, Raemer, and Reich, 1957
Goldstein, 1955
Raemer and Reich, 1958
Tucker, Graham, and Goldstein, 1957

11

APPLICATIONS TO RADIO COMMUNICATION SYSTEMS—I†

In the present chapter and in the following one, applications of statistical communication theory to radio communication systems will be discussed. Communication is by far the most common functional area of application of radio technology; hence most of what has been said in Chapters 1 through 8 is applicable to radio communication. This is particularly true of the material in Chapter 5, which concerned radio systems in general regardless of their specific function. Much of the contents of Chapters 6 and 8 is applicable to *analog* communication systems† (Chapter 11) and that of Chapter 7 and some of Chapter 8 is largely applicable to *digital* communication systems.‡

11.1 THE GENERAL COMMUNICATION SYSTEM

Systems for conveying information from a point A to a point B elsewhere in space will be designated here as *communication systems*. This is to differentiate them from other categories of information systems, although in some technical literature the words communication and "information" are used interchangeably. In this book, radar and other radio detection systems are not designated as communication systems, although they are certainly information systems by the definition given in Chapter 1, Section 1.1.

† See the first paragraph in Section 11.2 for a definition of analog communication systems.

‡ The definition of digital communication systems is given at the beginning of Chapter 12.

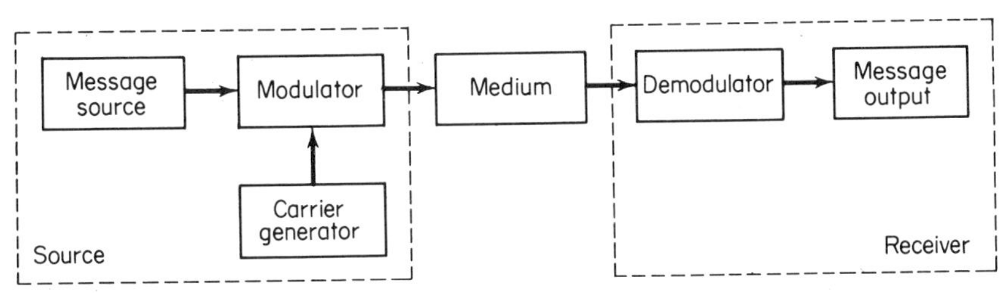

Figure 11-1 The general communication system.

Using the restricted definition given above, we can draw a block diagram illustrating the general constituents of any communication system (Figure 11-1).

The message generated at position A by the *message source* (e.g., a human voice, a Morse code key, etc.) is impressed on a *carrier* (e.g., a radio wave, sound wave, light beam) from the *carrier generator* by means of a *modulation* process instrumented in a *modulator*. The carrier bearing the message is sent through a *medium* (e.g., air, wire lines, sea water) and received at B. Upon reception, it first enters a *demodulator*, wherein the message is stripped off the carrier and the latter is discarded. The message is then presented to the party to whom it was to be communicated at the *message output* (e.g., ear-phones, oscilloscope face, loud speaker).

We will confine our attention here to those systems that can be regarded as linear; that is, if the message is a time waveform $s_i(t)$, then the medium and receiver (prior to the second detector) act on it as linear filters. If the medium and receiver have impulse response $m(t)$ and $r(t)$, respectively, then from (1.5) the message output is

$$s_o(t) = \int_{-\infty}^{\infty} dt'\, r(t - t') \int_{-\infty}^{\infty} dt''\, m(t' - t'')s_i(t'') \tag{11.1}$$

where causality of medium and receiver demand that $m(t)$ and $r(t)$ both vanish for negative t. By (1.4) and (1.5), the frequency response functions of $s_i(t)$, $m(t)$, $r(t)$, and $s_o(t)$, designated by $S_i(\omega)$, $M(\omega)$, $R(\omega)$, and $S_o(\omega)$, respectively, are mutually related through the expression

$$S_o(\omega) = R(\omega)M(\omega)S_i(\omega) \tag{11.2}$$

in terms of angular frequency ω. If cycle frequency is preferred, we can write this in the form

$$S_o(f) = R(f)M(f)S_i(f) \tag{11.2$'$}$$

In general, noise and distortion can be introduced by the source, the medium, and the receiver. The task of communication theory is to determine optimum design techniques against message quality degradation resulting from all three. In order to perform this task, however, it is first necessary to formulate a criterion of message quality to be maximized or alternatively, a criterion of message degradation to be minimized. As with the detection problem, an infinity of such criteria are possible. A few are listed below:

1. Received signal/noise ratio.
2. Channel capacity in bits per second
3. Bit error rate.
4. rms error in the message waveform.
5. Message error probability.

Each of the criteria mentioned has merit in certain applications, but there is no substitute for the subjective judgment of the design engineer (or in many cases, the user) in deciding what criterion to apply. A good system under one criterion may be a poor system under another criterion.

We now focus our attention specifically on *radio* communication systems. A radio communication system, in general, requires (1) a carrier wave, i.e., a continuous wave at a frequency high enough to be propagated efficiently and (2) a modulation, i.e., a time variation impressed on the carrier and containing the information to be transmitted, where the time variations can usually be considered slow compared to those of the carrier.† The information, for example, voice, Morse code, or a set of parameter values to be reported, can always be converted into a voltage or current waveform. This waveform can, in turn, be used to modulate the carrier. The modulated carrier is then transmitted and is demodulated at the receiver, i.e., the carrier is stripped off and only the modulation waveform is retained, the latter containing the desired information.

The only reason for existence of a carrier is the physical difficulty of radiating electromagnetic energy at the frequencies characterizing the information-bearing waveform. In principle it has no effect on the information itself. The same is true of the particulars of the modulation technique, for example, whether it is amplitude, frequency, phase, pulse position or duration, etc. However, there are substantial differences between the noise disturbances characterizing different carriers and modulation methods, and therefore they must be considered separately.

Radio communication systems are of either analog or digital type, or some combination of these. In an entirely analog system, the information is contained in a continuous modulation waveform and displayed as such at the receiver. In an entirely digital system, the information is coded into digits (i.e., pulses or groups of pulses arranged in a special way) to be transmitted and is presented at the receiver in such a manner as to restore the original

† This applies to narrowband systems, the only kind to be considered in this book, but it should be noted that *wideband* radio communication systems exist, in which the time variations associated with the information are comparable to those of the carrier, or in different terms, the information bandwidth is comparable to the carrier frequency.

information. Many communication systems are partially digital and partially analog, e.g., the information may be transmitted digitally and converted to analog form within the receiver.

The common types of modulation are amplitude, frequency, and phase. (The latter two can be lumped into the category of *angle modulation*.) These terms usually apply to analog information, their counterparts in digital systems being amplitude level keying (a special case of which is on-off keying), frequency shift keying (FSK), and phase shift keying (PSK), respectively. There are also modulation schemes using pulses, for example, pulse position modulation (PPM) and pulse duration modulation (PDM). These techniques fall into the general category of pulse modulation, in which some property of a pulse or a set of pulses is used to convey information.

In the discussions of analog systems the modulation and signal processing techniques will be treated from the viewponts of maximization of SNR or maximization of accuracy of measurement of the information-bearing parameters. In discussions of digital systems, the approach will be in terms of error rate. It has already been demonstrated in Chapters 6, 7, and 8 that the same kinds of linear filtering techniques that are contained in the optimum processing from the SNR or measurement accuracy points of view are also optimum when error rate is the criterion of performance. This applies when the noise disturbances are additive and Gaussian. Its implication for communication systems is that the linear filtering that occurs in the early stages of the optimum processing scheme is usually the same for both analog and digital systems.

11.2 ANALOG RADIO COMMUNICATION

Completely analog communication systems employ continuous waves in transmission. The modulation waveform carrying the information is continuous, and the information is not digitized on reception. The information is presented to the human observer at the receiver output in the form of a continuous aural or visual signal. In the present section, the systems dealt with will be of this type.

As indicated in Section 11.1, the common modulation techniques used in analog communication systems are amplitude modulation (AM), phase modulation (PM), and frequency modulation (FM). Each has its special advantages and disadvantages with respect to (1) ease of implementation, (2) resolution of signal detail, and (3) performance in noise. Item (1) will not concern us here. Items (2) and (3) are intimately related and will be integrated in the discussion to follow. Ease of implementation is a sensitive function of the state of the electronic art, which is rapidly changing. The theory that relates resolution and noise performance is fundamental and does not change significantly with the development of new equipment.

11.2.1 Fundamental Ideas of Analog Communication

We will begin our discussion of analog systems at the transmitting end of the system, where we have an information-bearing waveform $s_I(t)$ (a "message") whose highest significant frequency is B_s Hz and whose time duration is T_{I}. We would like to transmit the information contained in this waveform within a time T_{II}, which may be equal to T_{I} or smaller or greater than T_{I}. If T_{I} is equal to T_{II}, we say that the information is being transmitted in *real time*. If time is not critical, we may wish to slow down the transmission and thereby, as we will see later, pick up some increase in fidelity or reception. In this case, $T_{\text{I}} < T_{\text{II}}$. If we have only a limited period of good channel conditions, we may wish to speed up the transmission and release the entire message within a short period of time, such that $T_{\text{I}} > T_{\text{II}}$. In this case, we would sacrifice message fidelity in return for speed.

To approach these ideas more quantitatively, we make use of the sampling theorem (1.70) and a result that depends on it, the Hartley law, given in Eq. (3.89). The reader is referred to the entire discussion in Section 3.6.2, which forms the basis of what follows.

If we were to run the waveform $s_I(t)$ as the output of an instrument in real time, the highest rate R at which we could extract all available information about $s_I(t)$ is given by (3.89), i.e.,

$$R = \frac{dI}{dt} = 2B_s \log_2 M \text{ bits/s} \tag{3.89}$$

where B_s is the highest significant frequency or information bandwidth and M is the number of quantization intervals of $s_I(t)$. Now remember that in applying (3.89), we can allow M to have any value we desire, but with the stipulation that the larger the value of M, the less certain we are in the presence of noise that $s_I(t)$ has actually fallen within the Mth level. Moreover, the larger the SNR at reception, the larger M can be with a given level of fidelity, i.e., the higher the information rate that can be reliably transmitted.

In what follows and in Chapter 12 we will discuss quantitatively the nature of the tradeoff between fidelity (or SNR), information rate, and bandwidth with various modulation techniques.

In digital systems, this tradeoff manifests itself in a straightforward way, as will be shown in Section 11.3. In analog systems, the specification of information rate is arbitrary, being dependent on the desired accuracy at the receiver. As discussed in Section 3.6.2, the peak information signal voltage at the transmitter is V_o, and it is desired to determine at the receiver whether the information signal voltage is within a quantization interval of magnitude ΔV. Then the required number of quantization levels M is $V_o/\Delta V$. The mean noise power level at the receiver is $2B_R N_o$, where N_o is the *noise power density*,† B_R is the lowpass receiver bandwidth (the same as B_n, the low pass noise bandwidth). The highest significant signal information frequency is denoted by B_I and we sample the signal waveform at the Nyquist rate $2B_s$. The signal power loss in transmission between transmitter and receiver is K, and the input impedance of the receiver, then, is

$$\rho_i = \frac{KV_o^2}{Z_o 4 N_o B_R}$$

The number of levels M is

$$\sqrt{\frac{4Z_o N_o B_R \rho_i}{K(\Delta V)^2}}$$

Thus (3.89) takes the form

$$R = 2B_I \log_2 \left[\sqrt{\frac{4Z_o N_o B_R \rho_i}{K(\Delta V)^2}}\right] \qquad (11.3)$$

Equation (11.3) can be used to determine the attainable information rate if the parameters of the transmitter, receiver, and propagation medium are known.

11.2.2 Optimum Design of Analog Communication Systems

The problem of optimizing reception of an analog radio signal in receiver noise is essentially that of optimally estimating values of signal parameters. These parameters may be, e.g., a set of amplitudes or frequencies at different instants of time (in amplitude and frequency modulation systems, respectively). The important point is that they be estimated as accurately as possible in the presence of the noise. Thus the discussions of Chapter 8 are applicable to the optimum design of analog communication systems. In Sections 8.2, 8.3, and 8.7, optimal measurement of amplitude, phase, and frequency, respectively, were discussed from the maximum likelihood estimator viewpoint. In Section 8.5, the theory was extended to the measurement of amplitude and phase of an arbitrary radio signal. In the present section, an attempt will be made to relate some of the conclusions of those discussions to a few of the simplest analog modulation schemes.

The analog modulation techniques to be dealt with in this section will be categorized as amplitude modulation or angle modulation. The latter includes frequency modulation (FM) and phase modulation (PM) as special cases. In AM systems, the amplitude $a(t)$ must be determined as a function of time

† N_o is the quantity we have often denoted by $\sigma_n^2/2B_n$ in previous chapters.

to extract the information, and in PM systems, $\psi(t)$ must be determined as a function of time. In FM systems, it is the time waveform of $d\psi(t)/dt$ that contains the transmitted information.

To adapt maximum likelihood estimator theory to the problem at hand, we must note that the measurement of $a(t)$, as in AM, of $\psi(t)$, as in PM, or of $d\psi/dt$, as in FM, is not quite so simple as the static measurement problems discussed in Sections 8.2, 8.3, and 8.7. The present discussion relates to dynamic measurement problems, whose principles we covered in Section 4.2.1.2 and which was further discussed in Section 8.5. The general conclusion drawn from those discussions is that a dynamic variable can be measured only within a limited accuracy determined by the SNR and the ratio of the noise bandwidth to the bandwidth of the variable as a function of time. The reason for this limitation is that the measurement information must be extracted within the time during which the variable remains essentially constant. This time is roughly the reciprocal of twice the highest frequency of the variable as a function of time, which is equivalent to twice the information bandwidth B_s. Thus the time allowed for the estimation process to take place is roughly $1/2B_s$. We must then begin the process again, taking samples during the next interval $1/2B_s$ time units long and using them to measure the value of our variable during that interval. This is a key limiting factor in designing analog communication systems to operate accurately in the presence of noise.

11.2.3 Amplitude Modulation (AM)

In an amplitude modulation (AM) system, we impress the information-bearing signal $s_I(t)$ on the carrier as its slowly varying amplitude. The operating principle of AM radio is illustrated in Figure 11-2. At the transmitter, the signal $s_I(t)$ is added to the carrier $\cos(\omega_c t)$† with an appropriate weighting factor m and the sum is fed into a nonlinear device followed by a bandpass filter centered at the carrier frequency. The output of the nonlinear device is the AM wave, which because its frequencies are near the RF carrier frequency, can be transmitted into space. At the receiver, the wave enters another nonlinear device, i.e., a *detector*, *rectifier* or *demodulator*, followed by a lowpass filter that recovers the information signal $s_I(t)$.

The sum of carrier and weighted information signal, generated within the transmitter, is

$$e_i(t) = \cos \omega_c t + m s_I(t) \tag{11.4}$$

The signal $e_i(t)$ is the input to the modulator, whose input-output characteristic may be represented with great generality by writing

† We can set the time reference for zero carrier phase without loss of generality.

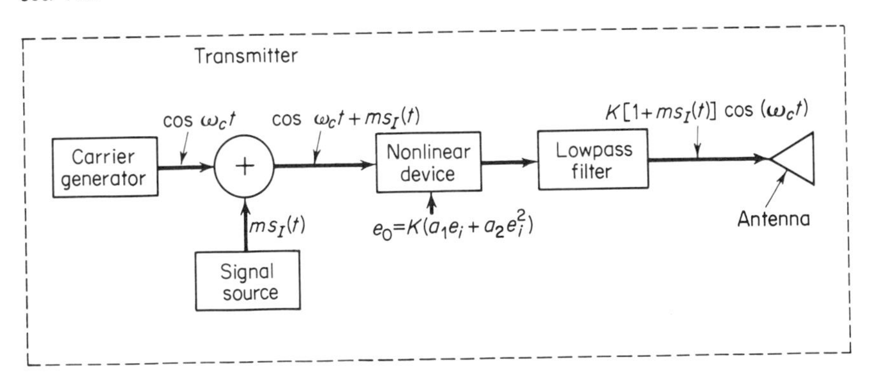

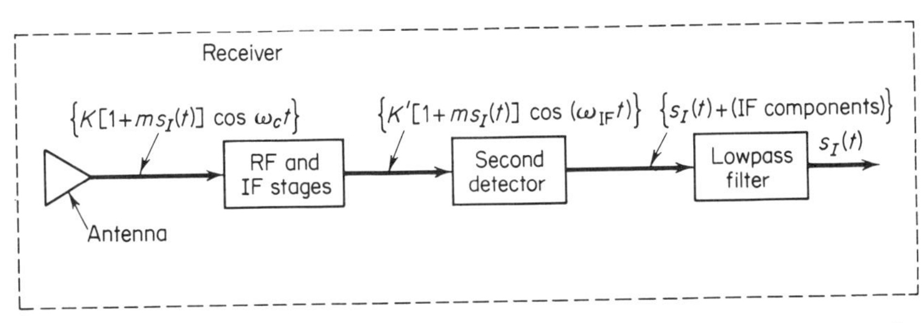

Figure 11-2 Operating principles of AM radio.

$$e_o = K[a_1 e_i + a_2 e_i^2 + a_3 e_i^2 + a_4 e_i^4 + \ldots] \tag{11.5}$$

where e_i and e_o represent voltage or current input and output, respectively. The form of $e_i(t)$ is given by (11.4), K is a gain constant and the a_k's are constants. It is possible to determine the outputs corresponding to the various terms of (11.5) without elaborate mathematics.

We merely recognize that if the bandpass filter following the operation (11.5) is sufficiently narrow, only terms near the carrier frequency will pass the filter. Terms near dc and harmonics of the carrier frequency will be rejected. Denoting the bandpass filtering operation by a wavy line above the output symbol, the outputs corresponding to the terms of the series (11.5) have the following proportionalities

$$\begin{aligned} \tilde{e}_i &\propto \cos \omega_c t \\ \tilde{e}_i^2 &\propto ms_I(t) \cos \omega_c t \\ \tilde{e}_i^3 &\propto m^2[s_I(t)]^2 \cos \omega_c t \text{ and } \cos \omega_c t \end{aligned} \tag{11.6}$$

By adjusting the parameters a_k in (11.5) properly, such that only the cos $\omega_c t$ terms in e_i and e_i^2 come through significantly, it is possible to obtain an output of the form

$$e_o(t) = K[1 + ms_I(t)] \cos \omega_c t \tag{11.7}$$

which is the desired AM waveform.

Figure 11-3 shows a typical AM waveform. The rapid fluctuations are those of the carrier $\cos \omega_c t$. The slow variations are those of the amplitude $[1 + ms_I(t)]$.

The parameter m, called the *modulation index*, must be small enough so that

$$|ms_I(t)| < 1 \quad \text{for all } t \tag{11.8}$$

If the condition (11.8) is not fulfilled, then during periods when $s_I(t)$ becomes very large in the negative direction, the entire amplitude $[1 + ms_I(t)]$ could swing negative. This would be equivalent to a 180° phase reversal, incurred whenever the amplitude crosses the zero axis. The result would be distortion in the pure amplitude signal recovered at the receiver.

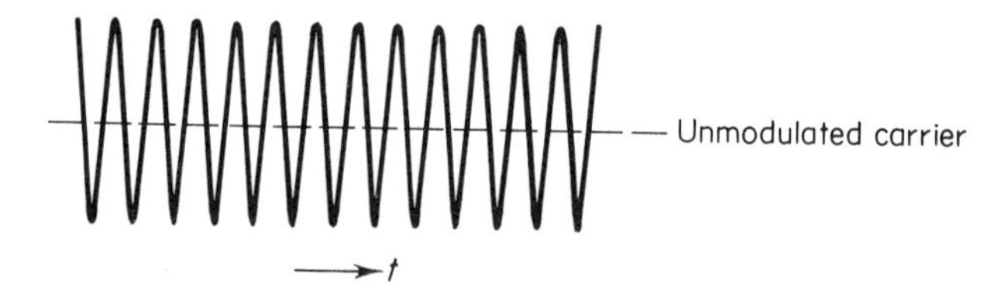

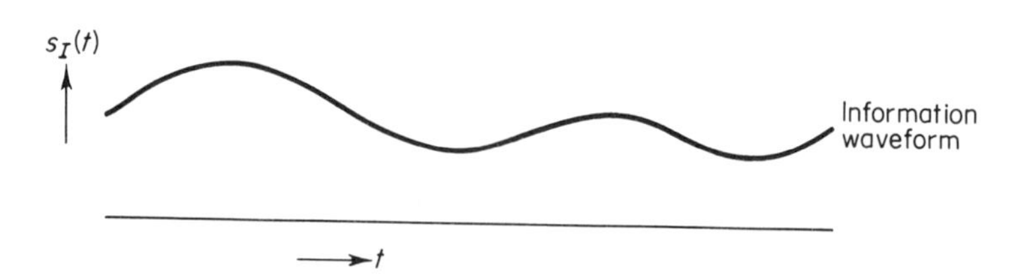

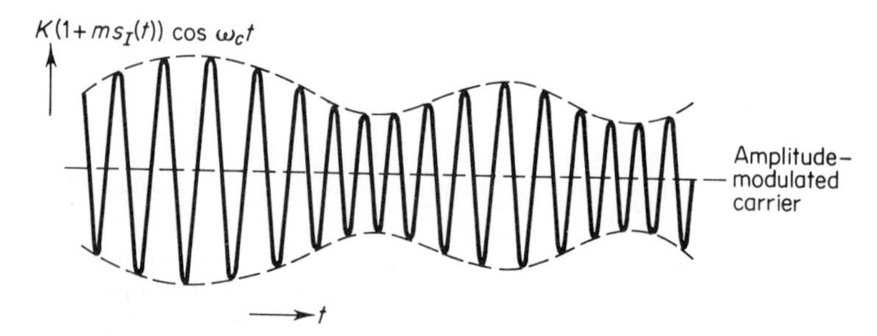

Figure 11-3 Typical AM waveform.

A negative amplitude swing due to a noise burst is highly probable if $|ms_I(t)|$ is near unity and the SNR is not exceptionally high. Therefore, to prevent distortion caused by noise, it is desirable that $|ms_I(t)|$ be somewhat smaller than unity. On the other hand, if it is made too small, the output signal power at the receiver will suffer. This will be demonstrated later in this section.

To illustrate the spectral characteristics of an AM signal, we express $s_I(t)$ in terms of its Fourier cosine transform, i.e.,

$$s_I(t) = \frac{1}{\pi}\int_0^\infty d\omega'\, A_I(\omega') \cos\left[\omega' t + \psi_I(\omega')\right] \tag{11.9}$$

Using (11.9) in (11.7), we have

$$\begin{aligned} e_o(t) = K\cos(\omega_c t) &+ \frac{Km}{2}\left(\frac{1}{\pi}\int_0^\infty d\omega'\, A_I(\omega')\cos\left[(\omega_c+\omega')t + \psi_I(\omega')\right]\right) \\ &+ \frac{Km}{2}\left(\frac{1}{\pi}\int_0^\infty d\omega'\, A_I(\omega')\cos\left[(\omega_c-\omega')t - \psi_I(\omega')\right]\right) \end{aligned} \tag{11.10}$$

In the case where $s_I(t)$ is a pure sinusoidal signal of frequency ω_I, amplitude a_I, and phase ψ_I, i.e.,

$$A_I(\omega) = \delta(\omega - \omega_I)a_I \tag{11.11.a}$$

$$\psi_I(\omega) = \psi_I \tag{11.11.b}$$

then

$$\begin{aligned} e_o(t) = K\cos\omega_c t &+ \frac{Km}{2}\left\{a_I\cos\left[\omega_c+\omega_I)t + \psi_I\right]\right. \\ &+ \left.\frac{Km}{2}\left\{a_I\cos\left[(\omega_c-\omega_I)t - \psi_I\right]\right\}\right. \end{aligned} \tag{11.12}$$

In this degenerate case, the AM waveform consists of a pure carrier term (see Figure 11-4(1) for illustration of the carrier spectrum) and two modulation sidebands, the upper sideband at a frequency equal to the carrier frequency *plus* the information frequency, and the lower sideband at the carrier frequency *minus* the information frequency. This is illustrated in the spectral plot of Figure 11-4(3)(a) showing the discrete spectral lines at ω_c, $(\omega_c + \omega_I)$ and $(\omega_c - \omega_I)$.

The general case is illustrated in Figures 11-4(3)(b) and, 11-4(3)(c). The information spectrum (see Figure 11-4(2)(a)(b) and (c)), a band of frequencies between zero and some maximum information frequency ω_M (much lower than ω_c), is shifted upward to the carrier frequency region by the modulation process. The result of this process is that each frequency component in the modulation waveform appears as a set of upper and lower sidebands around the carrier.

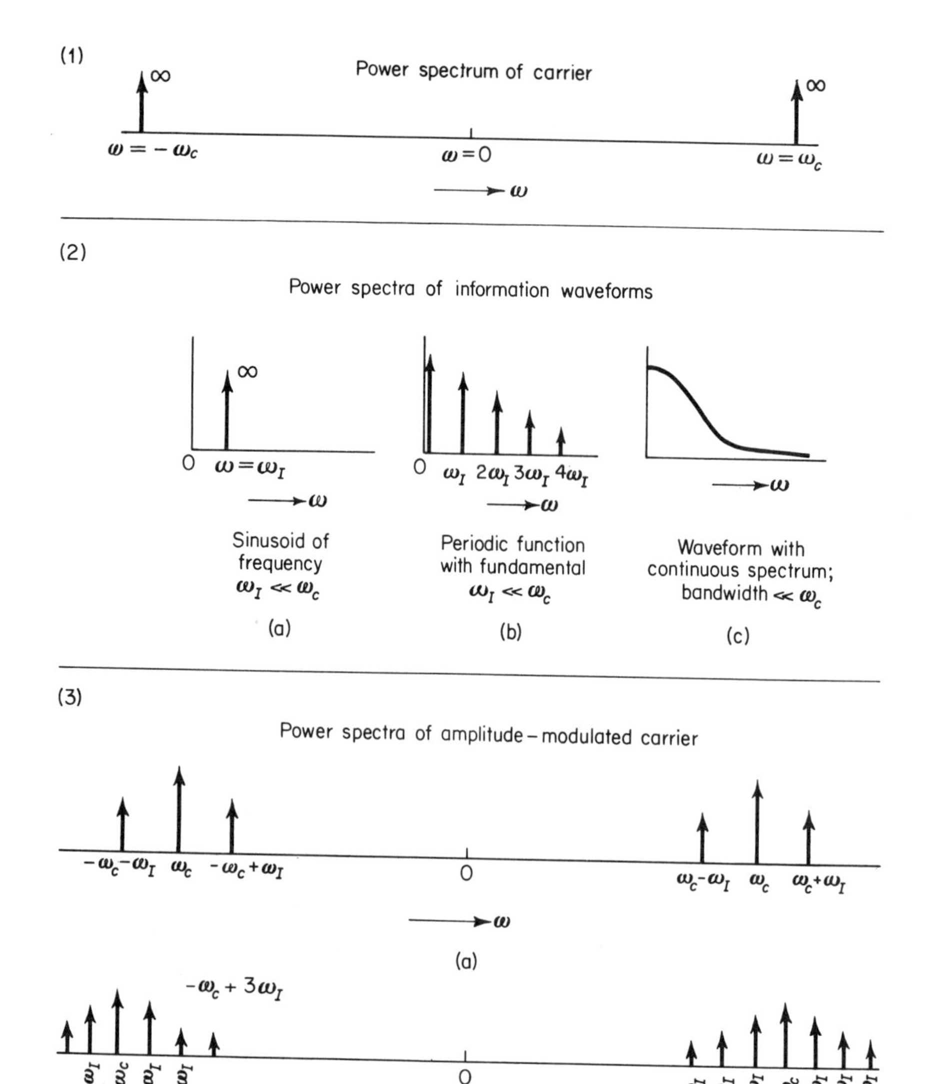

Figure 11-4 Power spectrum of AM signal.

If one imagines the existence of negative frequencies and considers the original information spectrum as a band symmetrical about $\omega = 0$, then the modulation process effectively shifts the center of the band from $\omega = 0$ up to $\omega = \omega_c$, and depresses its power spectral density by a factor 2, leaving its shape completely unaltered.

An important feature of AM is that the spectral shape and the bandwidth of the modulated carrier are the same as those of the original information waveform. As will become apparent later, this is not the case with angle modulation, wherein the bandwidth of the modulated carrier is much greater than the information bandwidth.

In AM, the receiver passband must be opened wide enough to receive the information signal spectrum. It should not be opened wider, however, because the result of this would be the passage of more noise without a corresponding increase in signal power. The tradeoff between rejection of important information components accompanying a decrease in bandwidth and the SNR degradation accompanying an increase in bandwidth is the important consideration in AM.

The demodulation process at the receiver is an attempt to extract the amplitude from the AM carrier. The natural demodulator for this purpose is the linear rectifier or envelope detector (see Sections 3.3.1 and 3.3.3). Suppose this rectifier is of full wave type. The output of such a rectifier with an input consisting of the AM carrier plus narrowband additive noise with zero mean and quadrature components $x_n(t)$ and $y_n(t)$, respectively, is

$$[e_0(t)_{s+n} = c\sqrt{[k(1 + ms_I(t)) + x_n(t)]^2 + [y_n(t)]^2} - ck \qquad (11.13)$$

where c is a constant. In order to simplify analysis, it is usually assumed at this point that the input SNR is high enough to justify neglect of $[y_n(t)]^2$ compared to other terms in the radical of (11.13). This is justified by the fact that SNR must be high for practical modulation systems to perform satisfactorily. With this assumption in (11.13), we obtain

$$[e_0(t)_{s+n} \simeq c\{K[1 + ms_I(t)] + x_n(t)\} - cK = cKms_I(t) + x_n(t) \qquad (11.14)$$

The SNR at the demodulator output is approximately

$$\rho_0 \simeq \frac{[cKms_I(t)]^2}{\langle [x_n(t)]^2 \rangle} \qquad (11.15)$$

Note that the total mean noise power is†

$$\sigma_n^2 = \frac{\langle x_n^2 + y_y^2 \rangle}{2} = \langle x_n^2 \rangle = \langle y_n^2 \rangle \qquad (11.16)$$

Note also that the ratio of the unmodulated carrier power to noise power, designated as the *carrier-to-noise ratio* ρ_c, is

---

† See Section 3.3, in particular the footnote cited after Eq. (3.33).

$$\rho_c = \frac{c^2K^2\,\overline{\cos^2\omega_c t}}{\left\langle\dfrac{x_n^2 + y_n^2}{2}\right\rangle} = \frac{c^2K^2}{2\sigma_n^2} \tag{11.17}$$

Using (11.17) in (11.15), we have

$$\rho_o = 2\rho_c m^2\,\overline{[s_I(t)]^2} \tag{11.18}$$

By Parseval's theorem (Eq. (6.24)),

$$\rho_o = 2m^2\rho_c\,\overline{[s_I(t)]^2} = \frac{m^2\rho_c}{\pi T_s}\int_{-\infty}^{\infty} d\omega\,|S_I(\omega)|^2 \tag{11.19}$$

where $S_I(\omega)$ is the Fourier transform of $s_I(t)$ and T_s is the signal duration.

If $s_I(t)$ is a pure sinusoid of unit amplitude, i.e., if $a_I = 1$ in (11.11.a), then

$$\rho_o = m^2\rho_c \tag{11.20}$$

Equations (11.19) and (11.20) point up the proportionality between the square of the modulation index and the SNR. For a given carrier-to-noise ratio, the SNR is enhanced by increasing m to its maximum allowable value. The latter corresponds to the case of 100 percent modulation (i.e., $m = 1$) since, as previously remarked, higher values of m would run the risk of waveform distortion. If the SNR is high enough, it is desirable to operate as close as possible to 100 percent modulation. For very high SNR, e.g., above 30 or 40 dB, where the risk of distortion due to noise-induced zero axis crossings is small, it should be possible to operate within a few percent of this ideal condition.

The maximum possible SNR in the sinusoidal signal case is obtained by setting $m = 1$ in (11.20), i.e.,

$$(\rho_o)_{\max} = \rho_c \tag{11.21}$$

The fractional rms error in attempting to discern the signal amplitude in noise is a more satisfactory measure of performance than the SNR. The fractional error with large carrier-to-noise ratio is given, with the aid of (11.15), by

$$\mathcal{E} = \sqrt{\frac{\langle([e_o(t)]_{s+n} - cKms_I(t))^2\rangle}{\langle[cKms_I(t)]^2\rangle}} \simeq \frac{\langle[x_n(t)]^2\rangle}{\langle[cKms_I(t)]^2\rangle} = \frac{1}{\sqrt{\rho_o}} \tag{11.22}$$

Thus, according to (11.22), if the carrier-to-noise ratio is very high, the fractional error in amplitude measurement is inversely proportional to the *voltage* or *current* SNR, i.e., the square root of the *power* SNR. This is consistent with the results (8.16) and (8.34), developed for amplitude measurement in general.

Figure 11-5 shows some curves of carrier-to-noise ratio versus "percentage accuracy" (defined as the reciprocal of error as obtained from (11.22), expressed in percentage units) for sinusoidal amplitude modulation with

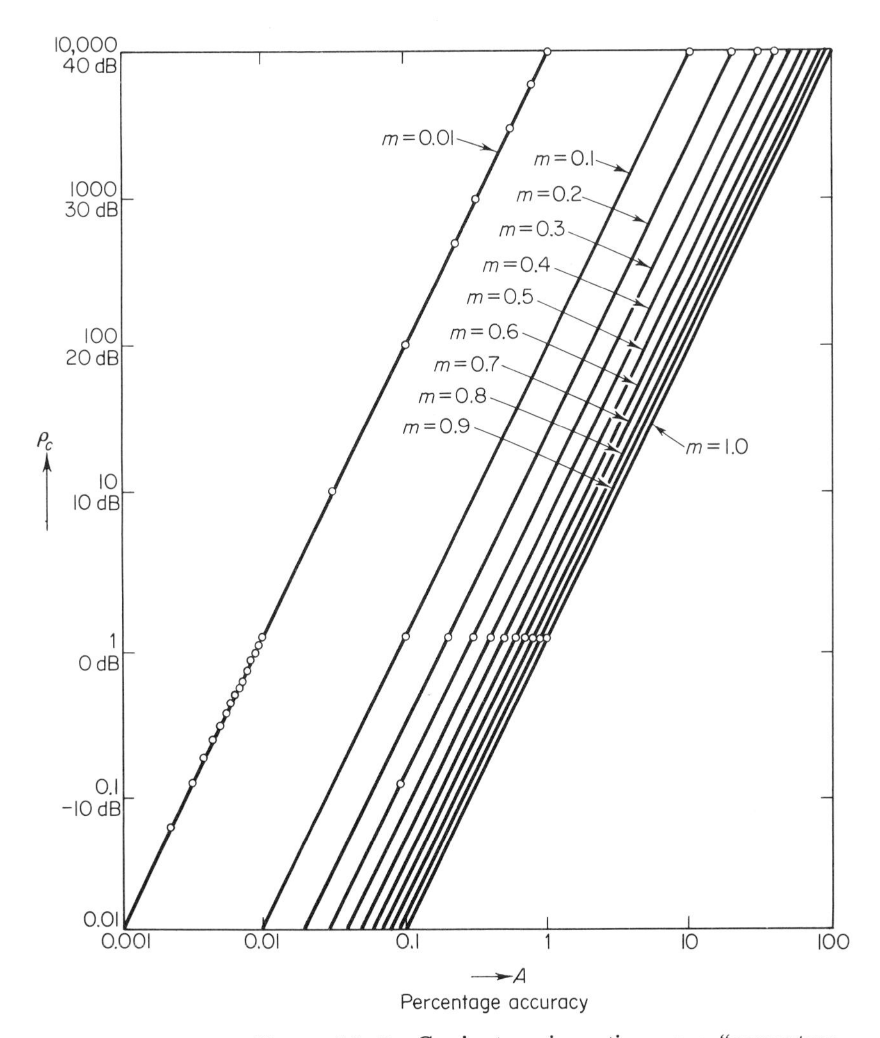

Figure 11-5 Carrier-to-noise ratio versus "percentage accuracy."

various values of the modulation index m. It is apparent from the curves that high accuracies cannot be attained unless carrier-to-noise ratios are substantial and that the carrier-to-noise ratio corresponding to a given accuracy decreases with m. For example, with 100 percent increasing modulation ($m = 1.0$), an accuracy of 10 percent corresponds to a 20 dB carrier-to-noise ratio. With a small percentage modulation, e.g., 10 percent, a 10 percent accuracy corresponds to a carrier-to noise ratio of 40 dB.

It is instructive to relate our discussion of amplitude modulation to Sections 8.2 (especially 8.2.2) and 8.5, in which we treated the optimal measurement of amplitude from the MLE viewpoint, in the static and dynamic cases, respectively. The demodulation process that takes place in the receiver is a practical approximation to the implementation of the ideas set forth in those discussions. The receiver first filters the incoming signal as near optimally as it can, i.e., it passes as much signal energy and rejects as much noise energy as possible. For practical reasons, the bandpass filtering prior to demodulation cannot be idealized matched filtering. First, it must be noncoherent because phase is not a priori known. Secondly, because of dynamic variations in the signal, it must be done with infinite integration time and a finite time constant, which allows it to integrate out noise continuously while losing a minimum of signal information (see Section 6.7). Likewise the low-pass filtering that follows the rectifier is accomplished with a bandwidth large enough to accommodate the slow time variations in the signal. All of this constitutes a very crude way of approximating the optimum filtering of dynamic signals dictated by the theory discussed in Section 8.5. We could not possibly know a priori the signal variations at the receiver; if we did, communication of these variations would be meaningless. Hence the idealized theory, which depends for its efficiency on such a priori knowledge, becomes inapplicable, and we must depend on something less idealized to tell us how the demodulation process should be designed. Referring to Section 8.2.2, if a probabilistic performance measure were desired, we could use (8.34) or (8.36) and Figure 8-3 to evaluate receiver performance. Since (8.36) has a one-to-one correspondence with the input SNR and the latter relates to the output SNR, we know that the probability of the measured amplitude being within some specified region around the true amplitude, as given by (8.36), will be monotonically related to the output SNR ρ_o as given by (11.17) and (11.18).

11.2.4 Phase Modulation (PM)

Phase modulation, which will be abbreviated PM, is a process whereby the signal intelligence waveform is impressed on the carrier as a slow time variation in phase. The modulation-demodulation process is illustrated in Figure 11-6. At the transmitter, the information signal $s_I(t)$ is first multiplied by the *maximum phase deviation* $\Delta\psi$.

The phase modulator, like other modulators, is a nonlinear device followed by a linear filter. Two such modulators are used in the process. The input to the first of these (No. 1) can be viewed as a summation of the form

$$e_{i1}(t) = \cos \omega_c t + \cos [\Delta\psi\, s_I(t)] \tag{11.23}$$

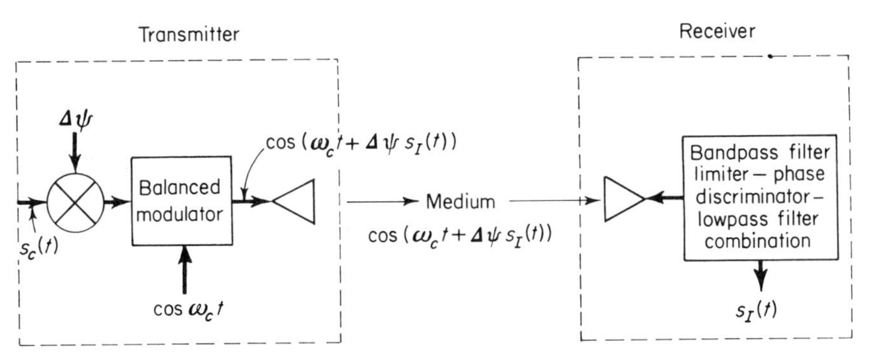

Figure 11-6 Modulation-demodulation process—PM.

A square-law characteristic with a gain $\frac{1}{2}$ acting on $e_{i1}(t)$ would produce an output

$$e_{o1}(t) = \tfrac{1}{2}[e_{i1}(t)]^2 = \tfrac{1}{2}\{1 + \tfrac{1}{2}\cos 2\omega_c t + \tfrac{1}{2}\cos(2\,\Delta\psi\, s_I(t)) + \cos(\omega_c t + \Delta\psi\, s_I(t)) + \cos(\omega_c t - \Delta\psi\, s_I(t))\} \tag{11.24}$$

The nonlinear device output $e_{o1}(t)$ is fed into a bandpass filter centered on the carrier and with a passband sufficiently wide to pass $[\omega_c t \pm \Delta\psi\, s_I(t)]$ but sufficiently narrow to exclude dc terms, terms at frequency $[2\,\Delta\psi s_I(t)]$,† and terms near twice the carrier frequency. The output of No. 1 is

$$e_{o1}(t) = \tfrac{1}{2}\cos(\omega_c t + \Delta\psi\, s_I(t)) + \tfrac{1}{2}\cos(\omega_c t - \Delta\psi\, s_I(t)) \tag{11.25}$$

To obtain the proper PM output, we require another input signal with both carrier and information signal phase shifted by 90°, i.e.,

$$e_{i2}(t) = \sin\omega_c t - \sin(\Delta\psi\, s_I(t)) \tag{11.26}$$

and a duplicate modulator-filter assembly (No. 2). The entire combination constitutes a type of *balanced modulator*, as shown in Figure 11-7. The output of No. 2 with the input $e_{i2}(t)$ is

$$e_{o2}(t) = \tfrac{1}{2}\cos(\omega_c t + \Delta\psi\, s_I(t)) - \tfrac{1}{2}\cos(\omega_c t - \Delta\psi\, s_I(t)) \tag{11.27}$$

The sum or difference $(e_{o2} \pm e_{o1})$ can be used to obtain the PM carrier. We choose the sum, resulting in

$$e_o(t) = \cos(\omega_c t + \Delta\psi\, s_I(t)) \tag{11.28}$$

To discuss PM mathematically, we will specialize to the case of sinusoidal modulation, i.e.,

$$s_I(t) = \cos\omega_I t \tag{11.29}$$

† Note that $[2\,\Delta\psi\, s_I(t)]$ may *not* be substantially less than the carrier frequency. If $\Delta\psi$ were very large it could be near the carrier frequency. This is the case of *wideband PM*, to be discussed later in the present section.

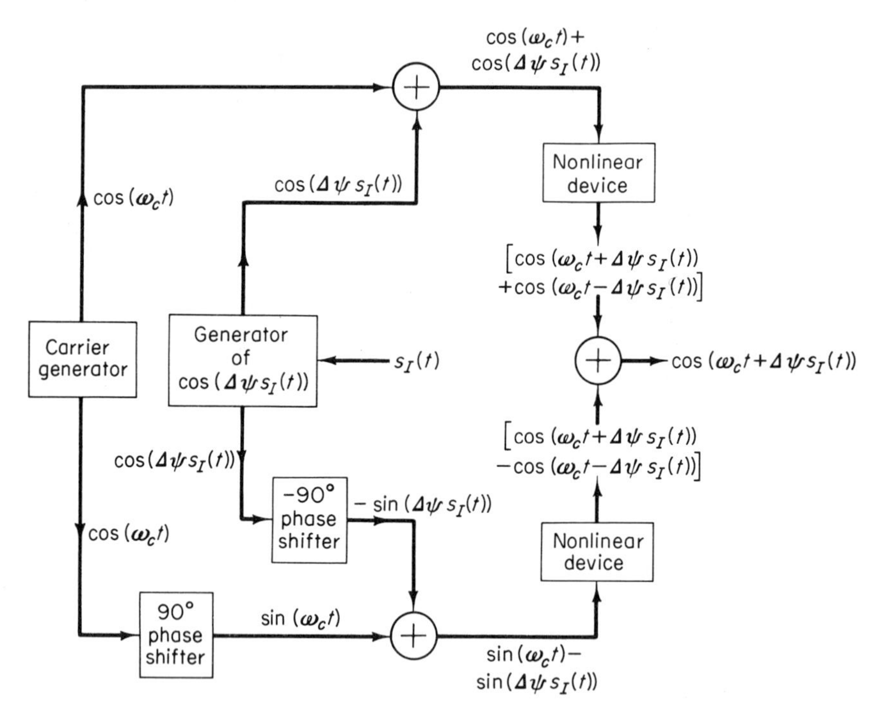

Figure 11-7 Balanced modulator for generation of PM carrier.

The reason for this is the mathematical complexity of the case of an information signal having arbitrary frequency composition. Analysis of the sinusoidal case, although strictly speaking it is not realistic, brings out the important features of PM and is actually used to determine specifications for practical PM systems.†

The PM carrier, with the information waveform (11.29), has the form

$$s_c(t) = \cos(\omega_c t + \Delta\psi \cos \omega_I t) = \cos(\Delta\psi \cos \omega_I t) \cos \omega_c t \\ - \sin(\Delta\psi \cos \omega_c t) \sin \omega_c t = \mathrm{Re}\,\{e^{j\omega_c t}\, e^{j\Delta\psi \cos \omega_I t}\} \tag{11.30}$$

We now adopt a notation that is somewhat standard in discussions of PM, denoting $\Delta\psi$ by β and calling β the *modulation index*.

The functions $\cos(\beta \cos \omega_I t)$ and $\sin(\beta \cos \omega_I t)$ are periodic with fundamental angular frequency ω_I. Their Fourier series expansions are obtained

† See Goldman, 1948, Appendix E, pp. 417–420 or Schwartz, 1959, Sections 3.7 through 3.12, pp. 111–147, where the same sort of information signal model is used in the context of frequency modulation.

from the Fourier series for $e^{j\beta \cos \omega_c t}$, which is†

$$e^{j\beta \cos \omega_I t} = J_0(\beta) + 2\{jJ_1(\beta) \cos \omega_I t - J_2(\beta) \cos 2\omega_I t - jJ_3(\beta) \cos 3\omega_I t + J_4(\beta) \cos 4\omega_I t + jJ_5(\beta) \cos 5\omega_I t - J_6(\beta) \cos 6\omega_I t + \dots\} \tag{11.31}$$

where $J_n(\beta)$ is the nth order Bessel function of the first kind.

From (11.30) and (11.31), it follows that

$$\begin{aligned} s_c(t) = J_0(\beta) \cos \omega_c t &- J_1(\beta)[\sin (\omega_c + \omega_I)t + \sin (\omega_c - \omega_I)t] \\ &- J_2(\beta)[\cos (\omega_c + 2\omega_I)t + \cos (\omega_c - 2\omega_I)t] \\ &+ J_3(\beta)[\sin (\omega_c + 3\omega_I)t + \sin (\omega_c - 3\omega_I)t] \\ &+ J_4(\beta)[\cos (\omega_c + 4\omega_I)t + \cos (\omega_c - 4\omega_I)t] \\ &- J_5(\beta)[\sin (\omega_c + 5\omega_I)t + \sin (\omega_c - 5\omega_I)t] \\ &- J_6(\beta)[\cos (\omega_c + 6\omega_I)t + \cos (\omega_c - 6\omega_I)t] + \dots \end{aligned} \tag{11.32}$$

The PM waveform consists of (1) a carrier term with amplitude $|J_0(\beta)|$ and power proportional to $[J_0(\beta)]^2$, and (2) a set of sum and difference sidebands separated from the carrier frequency by the harmonics of the information frequencies. The sideband corresponding to the nth harmonic of ω_I has amplitude $|J_n(\beta)|$ and power proportional to $|J_n(\beta)]^2$.

Plots of the PM carrier spectrum for various values of the modulation index β are shown in Figure 11-8. The amplitude values are obtained from the Bessel function tables in Abramowitz and Stegun.‡

With sinusoidal modulation, the information bandwidth, in hertz, is defined by

$$B_I = \frac{\omega_I}{2\pi} = f_I \tag{11.33}$$

i.e., B_I is the highest frequency in the information waveform.

The bandwidth of the PM carrier, denoted by $2B_c$, is defined as the separation between the lowest (frequency) and highest (frequency) significant sidebands. We define *significant sideband* to mean a sideband whose amplitude is some minimum number of decibels down from the carrier amplitude. The exact definition of bandwidth is not critical here; the general qualitative arguments are valid for any reasonable definitions of bandwidth.

With these definitions of information bandwidth and modulated carrier bandwidth, the ratio B_c/B_I increases with the modulation index β. For $\beta \ll 1$, B_c and B_I are roughly equal, and a given information bandwidth B_I can be transmitted over a channel with a bandwidth B_I. This situation is equivalent to that prevailing with AM. It is known as *narrowband PM*.

---

† See Schwartz, 1959, pp. 126–127. The left-hand side of Eq. (3.65) in Schwartz can be written in the form $e^{j\beta \cos [\omega_m t - (\pi/2)]}$, from which the equivalent of (11.31) follows.

‡ Abramowitz and Stegun, 1964, pp. 390–408.

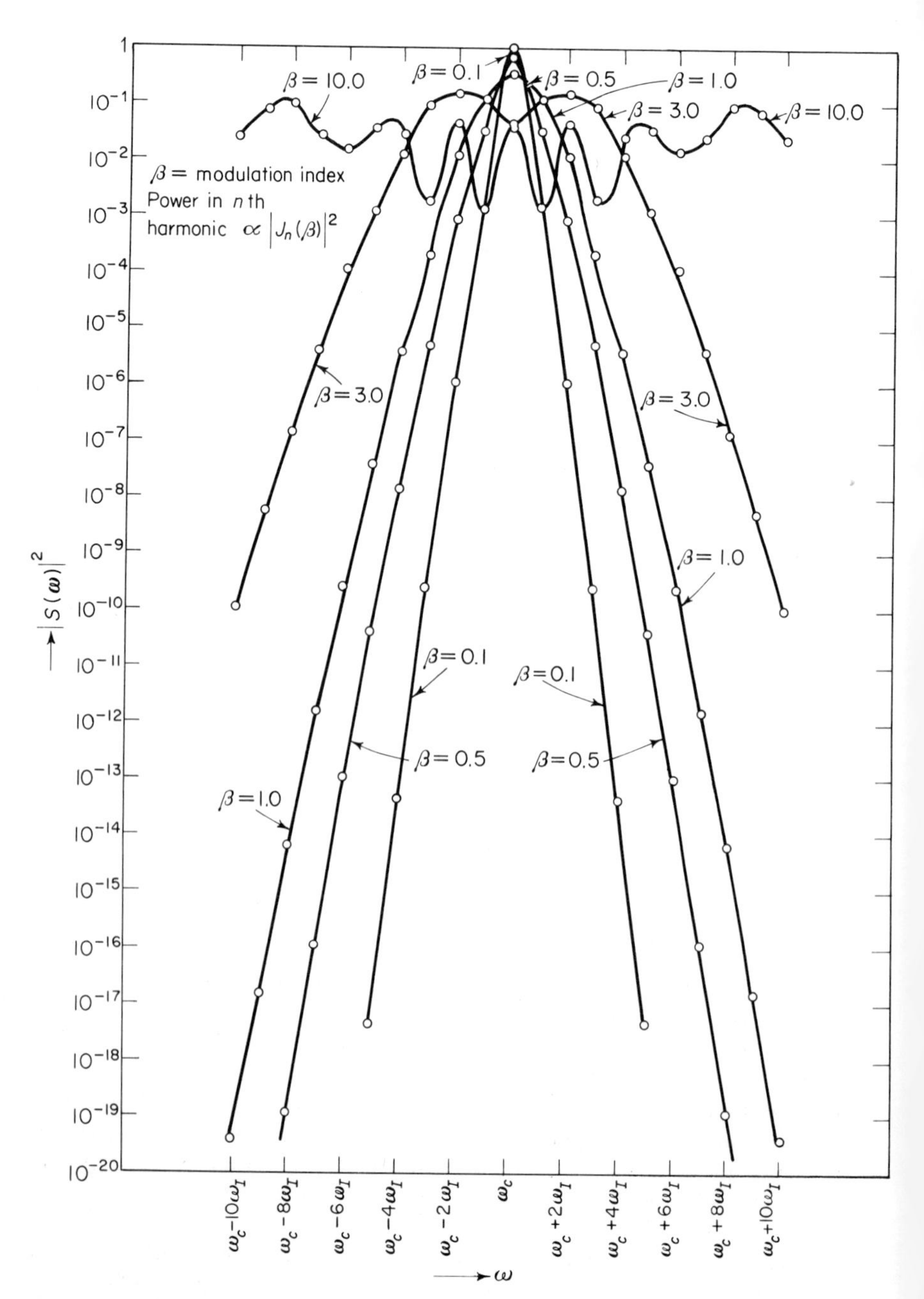

Figure 11-8 Phase modulation spectrum.

To show why narrowband PM is equivalent to AM from a required bandwidth point of view, we return to (11.30) and recognize that the condition

$$\beta \ll 1 \tag{11.34}$$

justifies the approximations $\cos(\beta \cos \omega_I t) \simeq 1$ and $\sin(\beta \cos \omega_c t) \simeq \beta \cos \omega_c t$, from which it follows that

$$\begin{aligned} s_c(t) &= \cos(\beta \cos \omega_I t) \cos \omega_c t - \sin(\beta \cos \omega_I t) \sin \omega_c t \\ &\simeq \cos \omega_c t - \beta \cos \omega_c t \sin \omega_I t \\ &\simeq \cos \omega_c t - \frac{\beta}{2}[\sin(\omega_c + \omega_I)t + \sin(\omega_c - \omega_I)t] \end{aligned} \tag{11.35}$$

Except for phase differences, the narrowband PM waveform (11.35) and the AM waveform with sinusoidal modulation (11.12) are similar. Both waveforms consist of a carrier and two sidebands separated from the carrier by the information frequency. In both cases the (bandpass) bandwidth of the channel required to transmit the information is equal to twice the (lowpass) bandwidth of the information itself.

Such is not the case for *wideband PM*, i.e., the case where β is greater than or at least comparable to unity. The rough sketch curve of (B_c/B_I) versus β

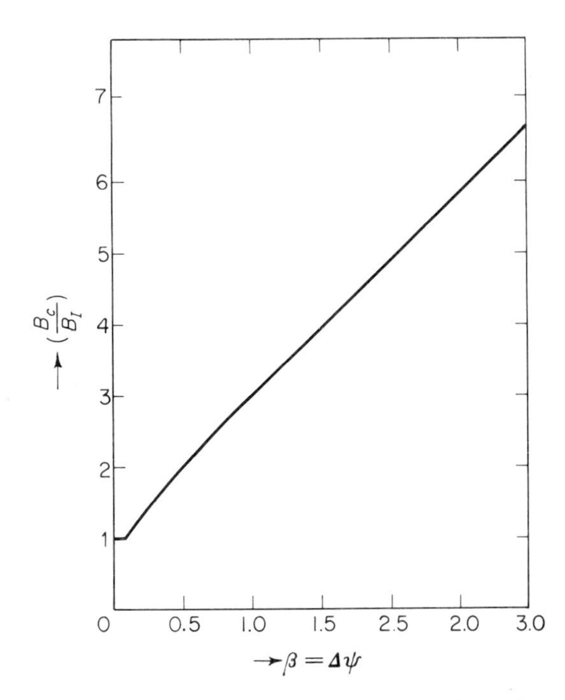

Figure 11-9 Ratio of bandwidth of PM carrier to information bandwidth versus modulation index.

in Figure 11-9† shows that if β is large (e.g., $\simeq 3.0$), then bandwidths well in excess of the information bandwidth are required to transmit a band of information. As β increases toward values much greater than unity, this effect becomes more pronounced. The effect is the basis of the improved discrimination against noise inherent in angle modulation and applies also to FM and to digital modulation techniques.

At the receiving end of a PM system, the PM carrier-plus-noise waveform is fed into a bandpass filter-limiter-phase-discriminator-lowpass filter combination, as shown in Figure 11-10.

The bandpass filter must be wide enough to accept the entire PM carrier, i.e., its bandwidth must be $2B_c$ or greater. In the wideband PM case, this means its bandwidth must be considerably greater than twice the information bandwidth B_I. The optimum bandwidth of the filter is $2B_c$, since a larger bandwidth would allow acceptance of excessive noise without a corresponding increase in information. The limiter smooths out the superfluous amplitude modulation in the PM signal, which does not contribute to its information content but creates dynamic range problems for the phase discriminator. The output of the limiter is a constant amplitude signal whose phase variations are the same as those of the prelimited PM carrier.

The phase discriminator ideally produces at its output a voltage proportional to the phase variations about the unmodulated carrier phase. Since in practice the discriminator output contains fluctuations at frequencies

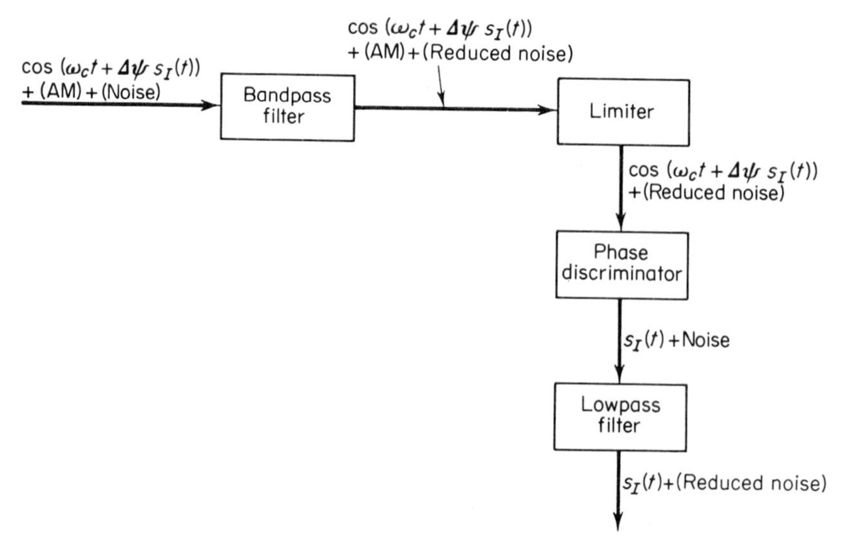

Figure 11-10 Demodulation process in PM receiver.

† To plot Figure 11-9, B_c was interpreted as corresponding to a sideband not more than 30 dB down from the carrier.

above the highest information frequency, the low pass filter is required at the discriminator output to reject these fluctuations. The optimum cutoff frequency of the low pass filter is the highest information frequency, which is $\omega_I/2\pi$ in the case of sinusoidal modulation.

From our point of view, the item of greatest interest is the ideal phase discriminator, which converts an input

$$s_i(t) = a\cos(\psi(t)) \tag{11.36}$$

into an output

$$s_o(t) = \frac{\psi(t) - \omega_c t}{\beta} \tag{11.37}$$

In the noise-free case, where $\psi(t)$ is $(\omega_c t + \beta s_I(t))$ as given in (11.30) and $s_I(t)$ is that given in (11.29) and used in (11.30), we have

$$s_o(t) = \cos\omega_I(t) \tag{11.38}$$

If noise is present, however, the discriminator output is distorted. If the noise can be represented as narrowband noise with inphase and quadrature components $x_n(t)$ and $y_n(t)$, the input to the limiter is

$$\begin{aligned} v_i(t) &= a\cos(\omega_c t + \beta\cos\omega_I t) + x_n(t)\cos\omega_c t + y_n(t)\sin\omega_c t \\ &= [a\cos(\beta\cos\omega_I t) + x_n(t)]^2 + [a\sin(\beta\cos\omega_I t) + y_n(t)]^2 \ldots \\ &\qquad \cos\left[\omega_c + \tan^{-1}\left(-\frac{[a\sin(\beta\cos\omega_I t) + y_n(t)]}{[a\cos(\beta\cos\omega_I t) + x_n(t)]}\right)\right] \end{aligned} \tag{11.39}$$

The limiter smooths out amplitude fluctuations due to noise, producing at the discriminator input a constant amplitude signal with the phase variations of (11.39). The ideal discriminator output is

$$s_o(t) = \frac{1}{\beta}\tan^{-1}\left(-\frac{[a\sin(\beta\cos\omega_I t) + y_n(t)]}{[a\cos(\beta\cos\omega_I t) + x_n(t)]}\right) \tag{11.40}$$

If the argument of the arctangent in (11.35) is small compared to unity, then we can write†

$$s_o(t) \simeq \frac{1}{\beta}\left[U - \frac{U^3}{3} + \frac{U^5}{5} - \frac{U^7}{7} + \ldots\right] \tag{11.41}$$

where

$$U \equiv -\frac{[a\sin(\beta\cos\omega_I t) + y_n(t)]}{[a\cos(\beta\cos\omega_I t) + x_n(t)]}$$

We now assume the carrier-to-noise ratio to be high, implying that

$$|a| \gg |x_n|,\ |y_n| \tag{11.42}$$

To evaluate the SNR at the output of the PM demodulator, we will use the usual simplification in which "noise" is defined as the output in the absence of signal and modulation and "signal" is defined as the output in the absence of noise. These definitions ignore the signal-noise cross terms that actually

† See Pierce, 1929, Nos. 779, 780, p. 92.

exist in the signal-plus-noise output of the demodulator. A rigorous SNR calculation accounting for these terms is mathematically complicated and the simple definition of SNR used here is sufficient to bring out the key feature involved in SNR improvement through angle modulation, the bandwidth-SNR tradeoff, the demonstration of which is the purpose of the calculation to follow.

The discriminator output in the absence of modulation (i.e., $\beta = 0$) and with high carrier-to-noise ratio, from (11.41), is

$$s_o(t) \simeq \frac{1}{\beta}\left\{\left(\frac{y_n}{a + x_n}\right) - \frac{1}{3}\left(\frac{y_n}{a + x_n}\right)^3 + \frac{1}{5}\left(\frac{y_n}{a + x_n}\right)^5 - \frac{1}{7}\left(\frac{y_n}{a + x_n}\right)^2 + \dots\right\}$$
$$\simeq \frac{1}{\beta}\left\{\frac{1}{a}y_n - \frac{1}{a^2}x_n y_n + \frac{1}{a^3}x_n^2 y_n^2 - \frac{y_n^3}{3} + 0\left(\left(\frac{x_n}{a}\right)^4, \left(\frac{y_n}{a}\right)^4, \left(\frac{\sqrt{x_n y_n}}{a}\right)^4\right)\right\} \tag{11.43}$$

where

$$0\left(\left(\frac{x_n}{a}\right)^4, \left(\frac{y_n}{a}\right)^4, \left(\frac{\sqrt{x_n y_n}}{a}\right)^4\right.$$

denotes additional terms of fourth and higher order in the ratios indicated.

Because a low pass filter follows the discriminator, we will calculate the spectrum of the discriminator output and then examine the effect of low pass filtering on a waveform with such a spectrum. The spectrum is best evaluated through the ACF, which is

$$R_{s_o}(\tau) = \langle\overline{s_o(t)s_o(t + \tau)}\rangle = \frac{1}{\beta^2 a^2}\langle\overline{y_n(t)y_n(t + \tau)}\rangle$$
$$- \frac{1}{\beta^2 a^3}\left\{\langle\overline{y_n(t)y_n(t + \tau)x_n(t + \tau)}\rangle + \langle\overline{x_n(t)y_n(t + \tau)y_n(t)}\rangle\right\}$$
$$+ \frac{1}{\beta^2 a^4}\left\{\langle\overline{x_n(t)x_n(t + \tau)y_n(t)y_n(t + \tau)}\rangle\right.$$
$$\left. + \langle y_n(t)[x_n^2(t + \tau)y_n(t + \tau) - \frac{y_n^3}{3}(t + \tau)]\rangle\right\} + \dots \tag{11.44}$$

Assuming that $x_n(t)$ and $y_n(t)$, both Gaussian with zero mean, are uncorrelated and have the same ACF, (see Section 3.3), we have

$$R_{s_o}(\tau) = \frac{\sigma_n^2}{\beta^2 a^2}\rho_{y_n}(\tau) + \frac{\sigma_n^4}{\beta^2 a^4}[\rho_{y_n}(\tau)]^2 + \dots \tag{11.45}$$

where $\rho_{y_n}(\tau)$ is the common normalized ACF of $x_n(t)$ or $y_n(t)$.†

The spectrum of $s_o(t)$ is obtained by taking the Fourier transform of $R_{s_o}(\tau)$. The output noise power of the low pass filter is then obtained by integrating the spectrum over all frequencies. Assuming the filter to be an ideal

† Note from (11.44) that there are other terms in (11.45) of the same (second) order of magnitude as that involving $[\rho_{y_n}(\tau)]^2$. Since we will make use of terms beyond the first only qualitatively these other second order terms are ignored.

filter with cutoff at ω_I, we take the Fourier transform of $R_{s_o}(\tau)$, integrate the result from $\omega = -\omega_I$ to $\omega = \omega_I$, then interchange the order of integration, with the result

$$\begin{aligned} P_n &= \frac{1}{\beta^2}\left\{\frac{1}{2\pi}\left(\frac{\sigma_n^2}{a^2}\right)\int_{-\infty}^{\infty} d\tau\,\rho_{y_n}(\tau)\int_{\omega_I}^{\omega_I} d\omega\, e^{-j\omega\tau}\right. \\ &\quad \left. + \frac{1}{2\pi}\left(\frac{\sigma_n^2}{a^2}\right)^2\int_{-\infty}^{\infty} d\tau[\rho_{y_n}(\tau)]^2\int_{\omega_I}^{\omega_I} d\omega\, e^{-j\omega\tau} + \dots\right\} \\ &= \left\{\frac{2f_I}{\beta^2}\left(\frac{\sigma_n^2}{a^2}\right)\int_{-\infty}^{\infty} d\tau\, p_{y_n}(\tau)\, e^{-j\omega_I\tau}\operatorname{sinc}(\omega_I\tau)\right. \\ &\quad \left. + \frac{2f_I}{\beta^2}\left(\frac{\sigma_n^2}{a^2}\right)^2\int_{-\infty}^{\infty} d\tau[\rho_{y_n}(\tau)]^2\, e^{-j\omega_I\tau}\sin(\omega_I\tau) + \dots\right\} \end{aligned} \tag{11.46}$$

where $f_I = \omega_I/2\pi$, the cutoff frequency in hertz.

The noise entering the limiter-discriminator, whose bandwidth is that of the RF passband of the receiver, has an ACF that is very narrow compared to sinc $(\omega_I\tau)$ (See Section 2.4.4.1), the latter having a width in τ-space of about $1/2f_I$ or π/ω_I while the width of the former is roughly the reciprocal receiver bandwidth. The information frequency ω_I is many times lower than the receiver bandwidth, hence the functions $\rho_{y_n}(\tau)$ and $[\rho_{y_n}(\tau)]^2$ decay to negligible magnitudes while sinc $(\omega_I\tau)$ remains approximately equal to unity; hence recalling that $\beta \equiv \Delta\psi_r$,

$$P_n \simeq \frac{1}{(\Delta\psi_r)^2}\left\{2f_I\left(\frac{\sigma_n^2}{a^2}\right)\int_{-\infty}^{\infty} d\tau\,\rho_{y_n}(\tau)\, e^{-j\omega_I\tau} + 2f_I\left(\frac{\sigma_n^2}{a^2}\right)^2\int_{-\infty}^{\infty} d\tau[\rho_{y_n}(\tau)]^2\, e^{-j\omega_I\tau} + \dots\right\} \tag{11.47}$$

Neglecting the effect of the limiter in altering the spectrum of the noise that enters it, we consider $y_n(t)$ to have a white spectrum from $\omega = 0$ to $\omega = 2\pi B_R$, in which case†

$$\rho_{y_n}(\tau) \simeq \operatorname{sinc}(2\pi B_R\tau) \tag{11.48}$$

By integrating the terms of (11.47) after substitution of (11.48) into these terms, or by simply invoking the Wiener-Khintchine theorem (2.41) and recognizing that these integrals are spectral densities, we obtain

$$P_n \simeq \frac{1}{(\Delta\psi_r)^2}\left(\frac{f_I}{B_R}\right)\left\{\left(\frac{\sigma_n^2}{a^2}\right) + \left(\frac{\sigma_n^2}{a^2}\right)^2 + \dots\right\} \tag{11.49}$$

The signal power output of the limiter-discriminator-filter combination is $s_o(t)$ as given by (11.38) in the absence of noise, i.e.,

---

† Equation (11.48) is easily obtained through (2.48), where

$$\begin{aligned} S_x(\omega) &= \frac{1}{2B_R}, \quad |\omega| \le 2\pi B_R \\ &= 0, \quad |\omega| > 2\pi B_R \end{aligned}$$

$$s_o(t) = \frac{\Delta\psi \cos \omega_I t}{\Delta\psi} = \cos \omega_I t \tag{11.50}$$

Since the lowpass filter passes frequencies up to and including f_I, the waveform (11.50) is not changed by the filter, and hence the signal power at the demodulator output is

$$P_s = \overline{|s_o(t)|^2} = \overline{\cos^2 \omega_I t} = \tfrac{1}{2} \tag{11.51}$$

The demodulator output SNR, from (11.49) and (11.51), is approximately

$$\rho_o = \frac{P_s}{P_n} \simeq \frac{(\Delta\psi)^2 \left(\frac{B_R}{f_I}\right)\left(\frac{a^2}{2\sigma_n^2}\right)}{1 + \frac{1}{2}\left(\frac{2\sigma_n^2}{a^2}\right) + \cdots} \simeq \frac{(\Delta\psi)^2 \left(\frac{B_R}{f_I}\right)\rho_i}{1 + \left(\frac{1}{2\rho_i}\right) + \cdots} \tag{11.52}$$

where $\rho_i \equiv a^2/2\sigma_n^2 =$ input SNR.

A noteworthy point in interpreting (11.52) is to recognize that the input noise power σ_n^2 is equal to the product of a constant noise power density and the receiver's bandpass bandwidth $2B_R$. A somewhat important point is that the receiver bandwidth must be wide enough to accept the enormous frequency deviations implied by a large value of $\Delta\psi$. It is apparent from Figure 11-9 that for large values of $\Delta\psi$ the bandwidth of the incoming PM carrier is very nearly equal to the product of the information bandwidth and twice the modulation index. If the receiver bandwidth is made equal to the bandwidth of the incoming PM carrier, then, it follows that

$$\frac{B_R}{\Delta\psi} \simeq f_I = \frac{\omega_I}{2\pi} \tag{11.53}$$

Using (11.53) in (11.52), we have

$$\rho_o = \frac{\left(\frac{B_R}{f_I}\right)^2 \frac{1}{(2f_I T_s)}\left(\frac{E_s}{N_o}\right)}{1 + \frac{B_R}{f_I}\frac{(f_I T_s)}{(E_s/N_o)} + \cdots} \tag{11.54}$$

where T_s is the signal duration and (E_s/N_o) is the ratio of signal energy

$$E_s = \frac{a^2 T_s}{2}$$

to noise power density

$$N_o = \frac{\sigma_n^2}{2B_R}$$

This ratio is familiar from previous discussions, because it is equivalent to the SNR at the output of an optimum linear filter.

To study the superiority of PM over AM in discrimination against noise, we will compare (11.53) with the demodulator output SNR for an AM system transmitting the same information signal

$$s_I(t) = \cos \omega_I t \tag{11.55}$$

with the same noise power density, the same carrier amplitude, and 100 percent modulation, i.e., the best that can be done with AM. According to (11.21) and (11.17), the maximum attainable output SNR with AM is

$$[(\rho_o)_{\mathrm{AM}}]_{\max} = \rho_c = \left(\frac{E_s}{N_o}\right)\frac{1}{2f_I T_s} \tag{11.56}$$

where it is noted that

$$E_s = \frac{c^2K^2T_s}{2}, \qquad N_o = \frac{\sigma_n^2}{2B_n} = \frac{\sigma_n^2}{2B_R}$$

and the optimum lowpass receiver bandwidth B_R in the AM case is equal to the information bandwidth f_I.

With the aid of (11.56), we can rewrite (11.54) in the form

$$(\rho_o)_{\mathrm{PM}} = \frac{\left(\frac{B_R}{f_I}\right)^2 \rho_c}{1 + \frac{1}{2}\left(\frac{B_R}{f_I}\right)\frac{1}{\rho_c} + \cdots} = \frac{\left(\frac{B_R}{f_I}\right)^2 [(\rho_o)_{\mathrm{AM}}]_{\max}}{1 + \frac{1}{2}\left(\frac{B_R}{f_I}\right)\frac{1}{(\rho_o)_{\mathrm{AM}}} + \cdots}$$

$$\simeq \left(\frac{B_R}{f_I}\right)^2 [(\rho_o)_{\mathrm{AM}}]_{\max} + \frac{1}{[(\rho_o)_{\mathrm{AM}}]_{\max}} \tag{11.57}$$

where $(\rho_o)_{\mathrm{PM}}$ is the SNR for PM and $[(\rho_o)_{\mathrm{AM}}]_{\max}$ is the maximum attainable SNR for AM.

For a given carrier-to-noise ratio (assumed to be very high) PM provides an SNR improvement over AM of roughly $(B_R/f_I)^2$, i.e., the improvement is as the square of the *bandspreading ratio*, defined as the ratio of receiver bandwidth to information bandwidth, or equivalently as the ratio of the bandwidth of the transmitted PM wave to that of the information it carries. Thus it is possible to attain a desired increase in SNR through a sufficient increase in the bandspreading ratio. This cannot be done indefinitely, however. For a given value of ρ_c, as B_R/f_I is increased beyond the point where $B_R/f_I \simeq 4\rho_c$, the denominator terms beyond the first begin to increase in importance and the output SNR $(\rho_o)_{\mathrm{PM}}$ begins to level off and eventually to decrease with further increases in transmission bandwidth.

This effect is known as the *threshold phenomenon* of angle modulation. It also occurs in FM and in other wideband modulation schemes. The higher the carrier-to noise ratio, the larger the bandspreading ratio can be made before the threshold effect occurs. For very small values of ρ_c, the effect will occur for small bandspreading ratios; thus bandspreading cannot be very effective unless ρ_c is quite large.

The lesson to be learned here is that PM is one of a class of wideband modulation schemes wherein an information waveform is coded at the transmitter into a waveform with a much wider bandwidth than that of the information. The increased transmission bandwidth purchases an increased SNR at the receiver demodulator output for a given carrier-to-noise ratio.

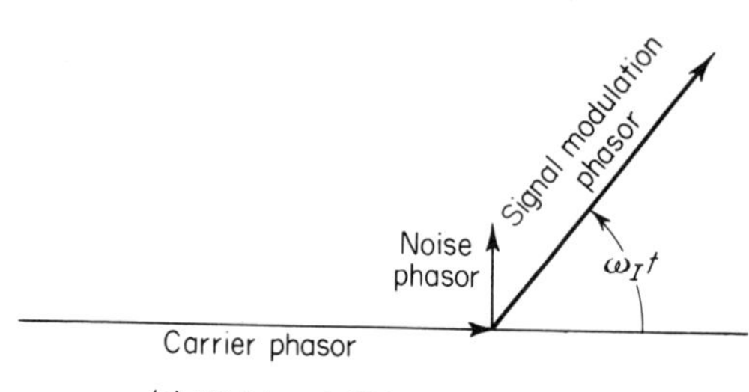

(a) Wideband PM — high SNR

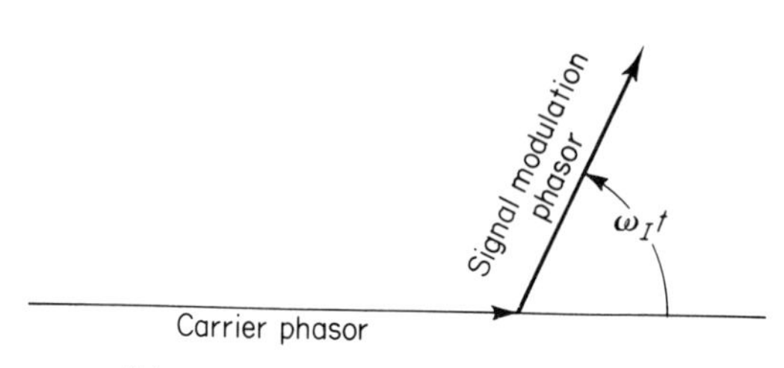

(b) Narrowband PM — noise free

Figure 11-11 Phasor diagrams for PM.

With high SNR, the physical reason for the noise discrimination capabilities of wideband PM with high carrier-to-noise ratio is best illustrated by the phasor diagram of Figure 11-11(a). Only the first term of the series (11.43) is used to describe the PM noise output. The incoming carrier, containing no signal modulation but modulated by *phase* noise, has the approximate form

$$s_c(t) = \cos\left(\omega_c t + \frac{y_n(t)}{a}\right) \tag{11.58}$$

Consider a single frequency component of the noise $y_n(t)$, e.g., the component at the information frequency

$$y_n(t) = y_{n0} \cos(\omega_I t + \psi_n) \tag{11.59}$$

Since

$$\left|\frac{y_{n0}}{a}\right| \ll 1, \qquad \cos\left(\frac{y_n}{a}\right) \simeq 1, \qquad \sin\left(\frac{y_n}{a}\right) \simeq \frac{y_n}{a} \tag{11.60}$$

for high SNR, it follows that, if only this single component were present, the noisy carrier $s_c(t)$ would have the approximate form

$$\begin{aligned} s_c(t) &\simeq \cos \omega_c t - \frac{y_n(t)}{a} \sin \omega_c t \\ &= \cos \omega_c t - \frac{y_{n0}}{2}[\sin((\omega_c + \omega_I)t + \psi_n) + \sin((\omega_c - \omega_I)t - \psi_n)] \end{aligned} \tag{11.61}$$

which is equivalent to a narrowband PM waveform (see Eq. (11.35)), whose phasor is illustrated in Figure 11-11(b). The wideband signal modulation on the carrier produces the phasor M on the figure. The latter will swing widely around the unmodulated carrier phasor c while the noise phasor's swings are small and imperceptible relative to those of the signal modulation, provided the carrier-to-noise ratio is sufficiently high. The more wideband the modulation (i.e., the larger is B_R/f_I), the greater the fluctuations due to the signal. Increases in B_R/f_I, however, do not affect the noise phasor appreciably, implying that such increases should enhance the contrast between signal swings and noise swings, and hence the output SNR.

Since the discriminator in a PM receiver essentially performs a phase measurement, it is legitimate to expect that the discussion of optimal phase measurement in Section 8.3 will be applicable to the PM receiver problem. As in the AM case, the essential conclusion of the theory discussed in Section 8.3 is that one should apply matched filtering to the two quadrature components of the signal before taking their ratio and thereby measuring phase. This effectively maximizes the SNR prior to phase measurement and thereby provides for the most accurate phase measurement possible. The phase discriminator assumed in our theoretical model in the present section attempts to measure the phase by computing the arctangent of the ratio of quadrature components $y(t)$ and $x(t)$ which have been bandlimited by the receiver passband. Again, as in the AM case, the bandpass filtering at the RF or IF stage of the receiver is the practical way to approximate the matched filtering dictated by the theory. True matched filtering is precluded by lack of a priori knowledge of signal variations. In the present case, phase is the quantity to be measured and hence must be preserved as it is in coherent matched filtering. However, the operation deviates from idealized matched filtering in that it must be done with a finite time constant and with large integration time in order to preserve the signal information contained in the dynamic variations of the phase while still obtaining some benefits of noise integration (see Sections 6.7 and 8.5). As in the AM case, one could use either rms error or a probabilistic measure of accuracy in evaluating PM receiver performance in lieu of the demodulator output SNR. The former would be provided by (8.43) and the latter by (8.54) and Figure 8-5. Again, as with AM, both of these criteria vary monotonically with the SNR.

11.2.5 Frequency Modulation (FM)

The only fundamental difference between FM and PM, as pointed out in Section 11.2.4, is the 90° phase difference incurred by *integrating* the phase waveform instead of using it directly as the information bearer. Another effect of this integration process is to make the discriminator output noise

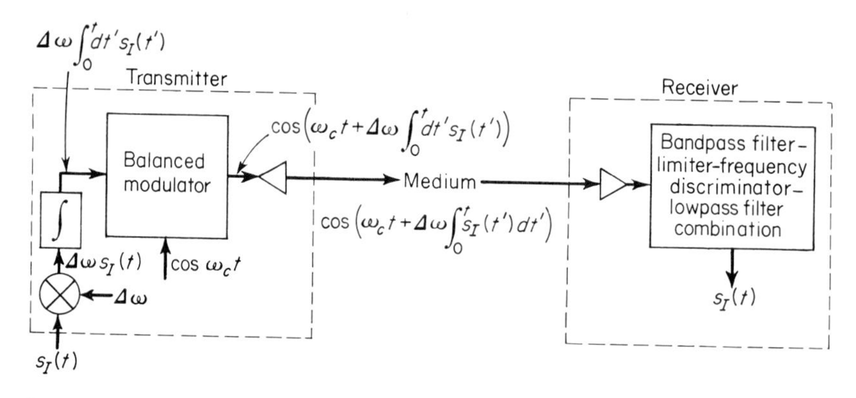

Figure 11-12 Modulation-demodulation process for FM.

spectrum into a ramp function for a flat input noise spectrum. Otherwise the discussions in Section 11.2.4 carry over almost intact into FM.

The block diagram for an FM system, shown in Figure 11-12, differs from that of Figure 11-6 in that the phase discriminator in the receiver becomes a frequency discriminator, which produces a voltage proportional to the time derivative of the phase. This time derivative is known as *instantaneous frequency*, because it can be treated conceptually as if it were a slowly varying carrier frequency. Strictly speaking, of course, it is no frequency at all, and much intellectual energy has been expended in either defending or attacking the use of the term "instantaneous frequency" to describe it.

We will regard the question of the validity of the instantaneous frequency concept as strictly philosophical and concentrate on the purely mathematical definition of $d\psi/dt$ as the information waveform in our discussions of FM.

Confining attention to sinusoidal modulation, we have for the information signal generated at the transmitter

$$s_I(t) = \Delta\omega \cos \omega_I t = \psi(t) - \omega_c t \tag{11.62}$$

where $\Delta\omega/2\pi = \Delta f$ is called the *frequency deviation.*

In FM, this signal undergoes an integration before transmission, i.e., the modulated carrier $s_{mc}(t) = \cos(\psi(t))$, where

$$\begin{aligned}\psi(t) &= \omega_c t + \int_0^t s_I(t')\,dt' = \omega_c t + \Delta\omega \int_0^t \cos \omega_I t'\,dt' \\ &= [\omega_c t + (\Delta\omega/\omega_I)\sin \omega_I t] = \omega_c t + \beta \sin \omega_I t\end{aligned} \tag{11.63}$$

where $\beta = \Delta\omega/\omega_I$ is known as the *modulation index*.

To produce $s_{mc}(t)$ at the transmitter, an integrator in the modulator is required. Otherwise things are much the same as in a PM transmitter.

At the receiver, demodulation is accomplished with a *frequency discriminator* whose ideal output is

$$s_0(t) = \frac{1}{\Delta\omega}\left(\frac{d\psi}{dt} - \omega_c\right) = \cos \omega_I t \qquad (11.64)$$

recognizable as the original information waveform $s_I(t)$.

The FM carrier can be represented in the form

$$\begin{aligned} s_c(t) &= \cos(\omega_c t + \beta \sin \omega_I t) \\ &= \cos(\beta \sin \omega_I t) \cos \omega_c t - \sin(\beta \sin \omega_I t) \sin \omega_c t \end{aligned} \qquad (11.65)$$

The only difference between (11.65) and (11.30) is that the latter involves arguments proportional to $\cos \omega_I t$ while the former involves arguments proportional to $\sin \omega_I t$. The Fourier series representation that replaces (11.31)† is

$$\begin{aligned} e^{j\beta \sin \omega_I t} &= J_0(\beta) + 2[J_2(\beta) \cos 2\omega_I t + J_4(\beta) \cos 4\omega_I t \\ &\quad + J_6(\beta) \cos 6\omega_I t + \ldots] \\ &\quad + 2j[J_1(\beta) \sin \omega_I t + J_3(\beta) \sin 3\omega_I t + \ldots] \end{aligned} \qquad (11.66)$$

and the expression replacing (11.32) is

$$\begin{aligned} s_c(t) = J_0(\beta) \cos \omega_c t &+ J_1(\beta)[\cos((\omega_c + \omega_I)t) - \cos((\omega_c - \omega_I)t)] \\ &+ J_2(\beta)[\cos((\omega_c - 2\omega_I)t) + \cos((\omega_c + 2\omega_I)t)] \\ &+ J_3(\beta)[\cos((\omega_c + 3\omega_I)t) - \cos((\omega_c - 3\omega_I)t)] \\ &+ J_4(\beta)[\cos((\omega_c - 4\omega_I)t) + \cos((\omega_c + 4\omega_I)t)] + \ldots \end{aligned} \qquad (11.67)$$

The amplitudes of the spectral lines of an FM wave are exactly the same as those of a PM wave; thus Figure 11-9 applies to either PM or FM. Comparison of (11.67) and (11.32) shows, however, that the even-order FM sidebands are either in phase or 180° out of phase with their PM counterparts, while the odd order FM sidebands are all in phase quadrature with the corresponding PM sidebands. Otherwise the two forms of angle modulation are spectrally equivalent. In the narrowband FM case, where $\beta \ll 1$, the FM wave has the form

$$s_c(t) = \cos \omega_c t - \frac{\beta}{2}[\cos((\omega_c - \omega_I)t) - \cos((\omega_c + \omega_I)t)] \qquad (11.68)$$

The waveform (11.68) is equivalent in terms of power spectrum to both AM and narrowband PM (see Equations (11.12) and (11.35)), but differs in relative phase of sidebands from both AM and narrowband PM.

In applying the discussion of wideband PM to the case of wideband FM, we first note that in the FM case with high SNR the limiter-discriminator output has the form (the FM counterpart of (11.43))

† See Goldman, 1948, Appendix E, pp. 417–420 or Schwartz, 1959, Eq. (3.65).

$$\begin{aligned} s_o(t) &= \frac{1}{\Delta\omega}\left(\frac{d}{dt} - \omega_c\right) = \frac{1}{\Delta\omega}\frac{d}{dt}\left[\tan^{-1}\left(\frac{y_n(t)}{a + x_n(t)}\right)\right] \\ &= \frac{1}{\Delta\omega}\frac{d}{dt}\left[\left(\frac{y_n}{a + x_n}\right) - \frac{1}{3}\left(\frac{y_n}{a + x_n}\right)^3 + \frac{1}{5}\left(\frac{y_n}{a + x_n}\right)^5 - \cdots\right] \\ &\simeq \frac{1}{\Delta\omega}\left[\frac{1}{a}y'_n(t) - \frac{1}{a^2}(x_n(t)y'_n(t) + x'_n(t)y_n(t)) \right. \\ &\quad \left. + \frac{1}{a^3}(x_n^2(t)\,y'_n(t) + 2x_n(t)y_n(t)x'_n(t) + y_n^2(t)y'_n(t)) + \ldots\right] \end{aligned} \tag{11.69}$$

At this point we would like to determine the SNR at the demodulator output as we did in the PM case. If SNR is again defined as the ratio of signal power in the absence of noise to noise power in the absence of modulation, the analysis proceeds as did that leading from (11.43) through (11.57). Beginning with (11.69), the counterpart of (11.43), we proceed to write the counterpart of (11.44) as follows:

$$\begin{aligned} R_{s_o}(\tau) &= \langle\overline{s_o(t)s_o(t+\tau)}\rangle = \frac{1}{(\Delta\omega)^2a^2}\langle\overline{y'_n(t)y'_n(t+\tau)}\rangle \\ &+ \frac{1}{(\Delta\omega)^4a^4}\{\langle\overline{x_n(t)x_n(t+\tau)}\rangle\langle\overline{y'_n(t)y'_n(t+\tau)}\rangle \\ &+ \langle\overline{x'_n(t)x'_n(t+\tau)}\rangle\langle\overline{y_n(t)y_n(t+\tau)}\rangle \\ &+ \langle\overline{x_n(t)x'_n(t+\tau)}\rangle\langle\overline{y'_n(t)y_n(t+\tau)}\rangle \\ &+ \langle\overline{x'_n(t)x_n(t+\tau)}\rangle\langle\overline{y_n(t)y'_n(t+\tau)}\rangle + \ldots\} + \ldots \end{aligned} \tag{11.70}$$

where the $1/a^3$ terms vanish because of the statistical independence of $x_n(t)$ and $y_n(t)$ and because both have zero mean, and where second order terms beyond those shown have been ignored for the reasons footnoted below Eq. (11.45). The result corresponding to (11.45), obtained from (11.70), is

$$R_{s_o}(\tau) = \frac{1}{(\Delta\omega)^2a^2}R_{y'_n}(\tau) + 0\left(\frac{\sigma_n^4}{a^4}\right) \tag{11.71}$$

where $R_{y'_n}(\tau)$ is the ACF of $y'_n(t)$ and the terms beyond the first are negligible for sufficiently high SNR.

Fourier transforming (11.71) to determine the spectral density of $R_{s_o}(\tau)$, then integrating from $\omega = -\omega_I$ to $\omega = \omega_I$ to account for low pass filtering, again neglecting any effects of the limiter, we obtain the output noise power.

$$P_n \simeq \frac{1}{(\Delta\omega)^2a^2}\frac{1}{2\pi}\int_{-\omega_I}^{\omega_I} d\omega\, G_{y'_n}(\omega) + 0\left(\frac{\sigma_n^4}{a^4}\right) \tag{11.72}$$

where $G_{y'_n}(\omega)$ is the spectral density of $y'_n(t)$.

We will now invoke a well-known relationship between the spectral density of a Gaussian random function $n(t)$ and that of its time derivative $n'(t)$.

This relationship is

$$G'_n(\omega) = \omega^2 G_n(\omega) \tag{11.73}$$

where $G_n(\omega)$ and $G_n'(\omega)$ are spectral densities of $n(t)$ and $n'(t)$, respectively. A simple way to obtain (11.73) is based on the Fourier series representation of a Gaussian noise waveform used in Section 3.1.4. Expressing $n_i(t)$ in the form (3.11) and dropping the subscripts and superscripts (i), we write the time derivative as follows:

$$n'(t) = \frac{2\pi}{T} \sum_{k=1}^{\infty} k\left[-a_k \sin\left(\frac{2\pi kt}{T}\right) + b_k \cos\left(\frac{2\pi kt}{T}\right)\right], \qquad \frac{T}{2} \leq t \leq \frac{T}{2}$$
$$= 0 \qquad , \qquad |t| > \frac{T}{2} \tag{11.74}$$

The ACF of $n'(t)$ is

$$\begin{aligned}\langle\overline{n'(t)n'(t+\tau)}\rangle = \frac{4\pi^2}{T^2} \sum_{\ell,k=1}^{\infty} \ell k \Big\{&\langle a_\ell a_k\rangle \overline{\sin\left(\frac{2\pi k\ell t}{T}\right) \sin\left(\frac{2\pi k(t+\tau)}{T}\right)} \\ &- \langle a_\ell b_k\rangle \overline{\sin\left(\frac{2\pi \ell t}{T}\right) \cos\left(\frac{2\pi k(t+\tau)}{T}\right)} \\ &- \langle b_\ell a_k\rangle \overline{\cos\left(\frac{2\pi \ell t}{T}\right) \sin\left(\frac{2\pi k(t+\tau)}{T}\right)} \\ &+ \langle b_\ell b_k\rangle \overline{\cos\left(\frac{2\pi \ell t}{T}\right) \cos\left(\frac{2\pi k(t+\tau)}{T}\right)}\Big\}\end{aligned} \tag{11.75}$$

If the same assumptions are made concerning the coefficients a_k and b_k as were made in Section 3.1.4 below Eqs. (3.12.a) and (3.12.b),

$$\langle\overline{n'(t)n'(t+\tau)}\rangle = \sum_{k=1}^{\infty} [\sigma_k]^2 \left(\frac{2\pi k}{T}\right)^2 \cos\left(\frac{2\pi k}{T}\tau\right) \tag{11.76}$$

By the same reasoning applied to $n(t)$,

$$\langle n(t)n(t+\tau)\rangle = \sum_{k=1}^{\infty} [\sigma_k]^2 \cos\left(\frac{2\pi k\tau}{T}\right) \tag{11.77}$$

If we now let T approach infinity and Fourier transform both (11.76) and (11.77), then it follows from (2.41) that we will obtain the power spectra of $n'(t)$ and $n(t)$ from (11.76) and (11.77), respectively. The result of these operations is (11.73).

In the analysis of PM in Section 11.2.4, it was assumed that $y_n(t)$ had a flat spectral density. If this assumption is made about $y_n(t)$, we see from (11.73) that the spectrum of $y_n'(t)$ is very different from that of $y_n(t)$, in that it vanishes at zero frequency and increases as the square of the frequency. We note from comparison of (11.43) and (11.69) that the only difference in the lead term of the series for $s_o(t)$ when we go from PM to FM is that $y_n'(t)/\Delta\omega$ replaces $y_n(t)/\beta$. Moreover, we could have written (11.46) in the form

$$P_n = \frac{1}{\beta^2 a^2} \frac{1}{2\pi} \int_{-\omega_I}^{\omega_I} d\omega\, G_{y_n}(\omega) + 0\left(\frac{\sigma_n^4}{a^4}\right) \tag{11.46$'$}$$

Comparing (11.46)′ and (11.72) with the aid of (11.73) and the assumption of a flat spectrum for $y_n(t)$, we arrive at the result

$$\frac{(P_n)_{\text{FM}}}{(P_n)_{\text{PM}}} = \frac{\dfrac{1}{3a^2}\sigma_n^2\left(\dfrac{f_I}{B_R}\right)\dfrac{1}{(\beta)_{\text{FM}}^2} + 0\left(\dfrac{\sigma_n^4}{a^4}\right)}{\dfrac{1}{a^2}\sigma_n^2\left(\dfrac{f_I}{B_R}\right)\dfrac{1}{(\beta)_{\text{FM}}^2} + 0\left(\dfrac{\sigma_n^4}{a^4}\right)} \simeq \frac{1}{3}\frac{(\beta)_{\text{PM}}^2}{(\beta)_{\text{FM}}^2} \tag{11.78}$$

where $\beta_{\text{FM}} \equiv \Delta f/f_I$, the modulation index for FM, and $\beta_{\text{PM}} \equiv \Delta\psi$, the modulation index for PM.

The discriminator output for a pure noise-free modulation input is given by (11.64), and the output signal power is

$$P_s = \overline{(\cos \omega_I t)^2} = \tfrac{1}{2} \tag{11.79}$$

Note from comparison of (11.51) and (11.79) that the output signal power for FM is the same as that for PM. From (11.78) and (11.79), we have (neglecting the effect of terms of order $1/[(\rho_o)_{\text{AM}}]_{\max}$)

$$\begin{aligned}(\rho_o)_{\text{FM}} &= \frac{P_s}{(P_n)_{\text{FM}}} \simeq \frac{3}{(2f_I T_s)}\left(\frac{E_s}{N_o}\right)\left(\frac{\Delta f}{f_I}\right)^2 + 0\left(\frac{1}{[(\rho_o)_{\text{AM}}]_{\max}}\right) \\ &\simeq 3\left(\frac{\Delta f}{f_I}\right)^2[(\rho_o)_{\text{AM}}]_{\max}\end{aligned} \tag{11.80}$$

Comparison of (11.80) with (11.57) yields (with the aid of (11.53))

$$\frac{(\rho_o)_{\text{FM}}}{(\rho_o)_{\text{PM}}} \simeq 3\left(\frac{\Delta f}{B_R}\right)^2 + 0\frac{1}{[(\rho_o)_{\text{AM}}]_{\max}} \simeq 3\left(\frac{\beta_{\text{FM}}}{\beta_{\text{PM}}}\right)^2 \tag{11.81}$$

From (11.80) and (11.81) we see that the output SNR for FM bears approximately the same relationship to the maximum attainable output SNR for AM as does that for PM. The differences between the two cases lie in the factor 3 that appears in the SNR for FM and is absent in that for PM, and in the fact that the modulation index β could be different for FM and PM. The factor 3 is not really fundamental and could be altered by changes in the characteristics of the low pass filter. The modulation index for PM is defined in (11.30) as the maximum phase deviation $\Delta\psi$. In (11.53), it is assumed that the receiver bandwidth is increased over the information bandwidth by an amount equal to the phase deviation. Behind (11.53) lies the tacit supposition that setting B_R equal to $(f_I\,\Delta\psi)$, (or equivalently $(f_I\,\beta)$) will result in the passage into the receiver of the *significant* phase swings that are an important part of the signal information. Guidance for this step is provided by the spectral curve of Figure 11-8, which shows that most of the PM spectrum is contained within $(f_I\beta)$.

In the FM case, the modulation index is defined in (11.63) as the ratio of the maximum frequency deviation to the information frequency. Since FM and PM have the same spectral amplitude lines and differ only in their phases, it would seem (by analogy with PM) that the receiver bandwidth for FM should again be equal to $f_I\beta$, which in this case is Δf, the maximum frequency deviation. Thus $\beta_{\text{FM}} \simeq \beta_{\text{PM}}$, and the difference in output SNR in the PM and FM cases lies in the factor 3, which as stated above, arises from the properties of the low pass filter and should not be considered as fundamental.

The analysis above tells us that FM, like PM, derives its improvements over AM through bandspreading. The terms $0(\sigma_n^4/a^4)$, not carried along in our analysis, have the same effect in FM as in PM, i.e., their inclusion would have resulted in prediction of an eventual decrease in attainable improvement due to further bandspreading and finally a threshold effect, in which further increases in receiver bandwidth would result in a "takeover" of the system by noise and a resulting degradation in SNR.

To complete the discussion of FM, we should note, as we did in the AM and PM cases, that the demodulation procedure, at least in part, crudely approximates the optimum processing dictated by the theory covered in Section 8.7, which simply tells us to maximize SNR through matched filtering before measuring frequency. Again, as in AM and PM, a bandpass filtering operation at RF or IF comes as close to matched filtering as possible under the circumstances. As in the AM case, the filtering in the FM case must be noncoherent because phase is a priori unknown, and it also must be done with a finite time constant in order to preserve the signal information while integrating continuously. In general, the same sorts of compromises with the idealized theory are required as in the AM and PM cases.

REFERENCES

Communication Systems in General

Baghdady, 1961

Balakrishnan, 1965

Jones, Pierce, and Stein, 1964

Schwartz, Bennett, and Stein, 1966

Wozencraft and Jacobs, 1965

Modulation (Amplitude, Frequency, Phase, Pulse)

Baghdady, 1961; many sections, especially Chapter 19

Black, 1953

Downing, 1964

Goldman, 1948; Chapter V

Jones, Pierce, and Stein, 1964; much of Chapter I, especially Sections 1.2, 1.3, 1.4, pp. 26–61

Lathi, 1965; Chapter 11

Lawson and Uhlenbeck, 1950; Chapter 13

Middleton, 1960; Chapters 12, 13, 14, 15

Schwartz, 1959; Chapters 3, 4

Schwartz, Bennett, and Stein, 1966; Chapter 3, 4, 5, 6

Stein and Jones, 1967; Chapters 5, 6

Wozencraft and Jacobs; Sections 8.1, 8.3, 8.4, 8.5

12

APPLICATIONS TO RADIO COMMUNICATION SYSTEMS—II

In this chapter, more material will be presented on the applications of the ideas of noise and statistical communication theory to radio communication systems. The emphasis will be primarily on digital communication systems, wherein the information is carried in the form of sequences of *digits*, or interrupted waveforms modulating the carrier. The pattern of these digits constitutes a codification of the transmitted message. At the receiver, decisions must be made on the type of digit transmitted. If these decisions are correct for each digit, then the entire sequence of digits that constitutes the message will be correctly read. Some of the most interesting problems in statistical communication theory are those arising out of determination of the probabilities of error in reading a digitally-transmitted message at the receiver in the presence of noise.

12.1 BINARY DIGITAL RADIO COMMUNICATION

Digital communication systems are characterized as binary, ternary, quaternary, M'ary, depending on whether decisions must be made between two, three, four, or M possible alternatives to extract information from the received waveform. Attention will be confined in this section to binary systems. These are not only the simplest to analyze but are also in very common use.

In a binary system, the information is coded into a set of *marks* and *spaces*. This terminology is derived from the use of *on-off keying systems*, i.e., those in which one of the two alternative signals is a pulse of RF energy of fixed duration (mark) and the other a blank interval of the same duration (space). However, it has carried over into other types of binary systems and will be used here as a convenient nomenclature for all kinds of binary signalling schemes. The duration of a mark or space signal is called a *baud*.

The material presented in Section 7.2 is directly applicable to the analysis of binary communication systems. All that is required to specialize the discussion of Section 7.2 to binary digital communications is to designate one of the two alternative signals as a mark, the other as a space. With the adoption of this terminology, the entire discussion carries over into the communications context.

12.1.1 Error Probabilities in Binary Signalling Systems

The error probabilities for coherent binary detection in Gaussian noise are given by (7.70.a) and (7.70.b). Those for noncoherent binary detection in Gaussian noise are given by (7.35.a) and (7.35.b) in the case of on-off keying, and by (7.84) in the case of mark and space signals that are of equal energy and mutually orthogonal. Confining attention to Gaussian noise as the interfering agent, we can translate these error probabilities into those in a binary digital communication system perturbed by additive internal receiver noise and external noise arising from sums of independent sources (both of which are Gaussian noise)† and in which fading is negligible. We will consider the most common digital *modems*:‡ on-off keying (OOK), both coherent and noncoherent, frequency shift keying (FSK), both coherent and noncoherent, and phase shift keying (PSK), which by its nature *must* be coherent. Each of the subsections below will treat one of these modems.

In discussion of each modem, it will be noted that the integration time or signal duration time T or T_s that appears in the error probability (see Eqs. (7.35), (7.70), and (7.71)), when measured in seconds, is the reciprocal of the number of bauds per second in the case where ideal matched filtering is applied to the received signals. In a binary signalling system, each baud contains one bit of information. Hence for binary signalling with matched filtering

$$T\,(\text{or } T_s) \equiv \frac{1}{R} \tag{12.1}$$

where R is the information rate in bits per second.

---

† See Sections 3.1.3, 3.5, and 5.5.3.

‡ *Modem* is a term often used by communications engineers to denote modulation technique.

A specialization we can make in the case of binary communication is that the a priori probabilities of mark and space are equal. This is true by design in many practical systems; there is usually no reason to have it otherwise. Thus in Eqs. (7.70. a, b), (7.71. a, b), (7.72), (7.73), (7.74), (7.75), (7.82), and (7.84),

$$P(1) = P(2) = \tfrac{1}{2} \tag{12.2}$$

12.1.2 On-Off Keying

In an on-off keying system, *mark* indicates that a signal was received during a baud and *space* indicates that no signal was received during the baud. The error probabilities for this case are obtained from (7.72) if it is assumed that (12.1) and (12.2) hold, that phase is a priori known and that the dividing point between the mark and space decision regions is at one-half of the mark amplitude (as is optimum if (12.2) holds);

$$P_{SM} = P_{MS} = \frac{1}{2}\left[1 - \operatorname{erf}\left(\frac{1}{2}\sqrt{\frac{\rho_i B_n}{R}}\right)\right] \tag{12.3}$$

where P_{SM} and P_{MS} are, respectively, the error probabilities associated with calling a space a mark and calling a mark a space. These error probabilities are equal in the case under discussion.

If the signal phase is not a priori known, such that detection must be noncoherent, then the error probabilities are obtained from (7.35.a) and (7.35.b), and are in this case (where $K_T = \frac{1}{2}$),

$$P_{MS} = 1 - Q\left(\sqrt{\frac{4\rho_i B_n}{R}}, \frac{1}{2}\sqrt{\frac{\rho_i B_n}{R}}\right) \tag{12.4.a}$$

$$P_{SM} = e^{-(\rho_i B_n/8R)} \tag{12.4.b}$$

The total error probability (from (12.3) and Eqs. (12.4.a, b)) is

$$\begin{aligned} P_\epsilon &= \frac{1}{2}(P_{MS} + P_{SM}) \\ &= \frac{1}{2}\left[1 - \operatorname{erf}\left(\frac{1}{2}\sqrt{\frac{\rho_i B_n}{R}}\right)\right] \qquad \text{for coherent OOK} \\ &= \frac{1}{2}\left[e^{-(\rho_i B_n/8R)} + 1 - Q\left(\sqrt{\frac{4\rho_i B_n}{R}}, \frac{1}{2}\sqrt{\frac{\rho_i B_n}{R}}\right)\right] \qquad \text{for noncoherent OOK} \end{aligned} \tag{12.5}$$

As expected, P_ϵ vanishes for $\rho_i \longrightarrow \infty$ and becomes $\frac{1}{2}$ for $\rho_i \longrightarrow 0$. This reflects the fact that no errors can occur with infinite SNR and the decision is pure guesswork if the SNR is extremely low.

12.1.3 Phase Shift Keying

Phase shift keying (PSK) by its nature must be coherent, because it cannot be implemented without knowledge of phase at the receiver. There is a variation of PSK in which it is not necessary to know the signal phase, known as differentially coherent phase shift keying (DPSK), in which the basic signal property used for modulation is the phase difference between two signals. Only phase *difference*, not absolute phase, must be known at the receiver. This modem will not be discussed here. The reader is referred to the digital communications literature† for information pertaining to DPSK.

The error probabilities for PSK, assuming that mark and space signals have equal energies, hence equal amplitudes (since baud durations are all equal), are obtained from (7.73). With the aid of (12.1) and (12.2), we obtain,

$$P_{SM} = P_{MS} = \frac{1}{2}\left[1 - \operatorname{erf}\left(\sqrt{\frac{\rho_i B_n}{2R}}\sqrt{1-\chi_{12}}\right)\right] \tag{12.6}$$

and

$$P_\epsilon = \tfrac{1}{2}(P_{SM} + P_{MS}) = P_{SM} = P_{MS} \tag{12.7}$$

The mark-space crosscorrelation function χ_{12} for PSK signals, as defined below Eqs. (7.65), is

$$\chi_{12} = \frac{\int_0^T dt' \cos(\omega_c t' + \phi_M)\cos(\omega_c t' + \phi_s)}{\sqrt{\int_0^T dt' \cos^2(\omega_c t' + \phi_M)\int_0^T dt' \cos^2(\omega_c t' + \phi_s)}} \tag{12.8}$$

where it is indicated that the mark and space signals are RF or IF signals of the same frequency ω_c, with phases ϕ_M and ϕ_s, respectively, where $\phi_M \neq \phi_s$. The baud duration T is long compared with $2\pi/\omega_c$; hence

$$\chi_{12} = \frac{\frac{1}{2}\int_0^T dt' \left[\cos(2\omega_c t' + (\phi_M + \phi_s)) + \cos(\phi_M - \phi_s)\right]}{\sqrt{\left[\frac{1}{2}\int_0^T dt' (1 + \cos(2\omega_c t' + 2\phi_M))\right]\left[\frac{1}{2}\int_0^T dt'' (1 + \cos(2\omega_c t'' + 2\phi_s))\right]}}$$

$$\simeq \frac{\frac{T}{2}\cos(\phi_M - \phi_s)}{\sqrt{\left[\frac{T}{2}\right]\left[\frac{T}{2}\right]}} = \cos(\phi_M - \phi_s) \tag{12.9}$$

It is clear from (12.6) and (12.9) that the smallest error probabilities are obtained in the case where

$$\chi_{12} \simeq \cos(\phi_M - \phi_s) = -1 \tag{12.10}$$

or equivalently

$$\phi_M - \phi_s = (2n + 1)\pi \tag{12.10$'$}$$

or‡

$$(\phi_M - \phi_s)_{\text{modulo } 2\pi} = \pi \tag{12.10$''$}$$

Thus it is optimal to design a PSK system with a phase difference of 180° between the mark and space signals. In this case the over-all error probability will be the lowest attainable and will be given by

† See Schwartz, Bennett, and Stein, 1966, pp. 304–310.

‡ The phrase "modulo 2π" restricts the phase to the range from zero to 360°.

$$(P_\epsilon)_{\min} = \frac{1}{2}(P_{SM} + P_{MS}) = P_{SM} = P_{MS} = \frac{1}{2}\left[1 - \operatorname{erf}\left(\sqrt{\frac{\rho_i B_n}{R}}\right)\right] \tag{12.11}$$

Again, as with OOK and as must be the case with our asumptions, $(P_\epsilon)_{\min}$ vanishes for $\rho_i \longrightarrow \infty$ and is equal to $\frac{1}{2}$ for $\rho_i = 0$.

12.1.4 Frequency Shift Keying

Frequency shift keying (FSK) can be either coherent or noncoherent, depending on whether phase is a priori known. In the coherent case, the error probabilities for equal amplitude mark and space signals are obtained from (7.73) and are given by

$$P_\epsilon = P_{SM} = P_{MS} = \frac{1}{2}\left[1 - \operatorname{erf}\left(\sqrt{\frac{\rho_i B_n}{2R}}\sqrt{1 - \chi_{12}}\right)\right] \tag{12.12}$$

where, since $T \gg (2\pi/\omega_M), (2\pi/\omega_s)$,

$$\chi_{12} = \frac{\int_0^T dt' \cos(\omega_M t' + \phi_M) \cos \omega_s t' + \phi_s)}{\sqrt{\int_0^T dt' \cos^2(\omega_M t' + \phi_M) \int_0^T dt' \cos^2(\omega_s t' + \phi_s)}}$$

$$= \frac{\frac{1}{2}\int_0^T dt' [\cos((\omega_M - \omega_s)t' + (\phi_M - \phi_s)) + \cos((\omega_M + \omega_s)t' + (\phi_M + \phi_s))]}{\sqrt{\frac{1}{2}\int_0^T dt' [1 + \cos^2(\omega_M t' + \phi_M)] \frac{1}{2}\int_0^T dt' [1 + \cos^2(\omega_s t' + \phi_s)]}}$$

$$\simeq \frac{1}{T}\int_0^T dt' \cos((\omega_M - \omega_s)t' + (\phi_M - \phi_s))$$

$$= \cos(\phi_M - \phi_s) \operatorname{sinc}((\omega_M - \omega_s)T) + \sin(\phi_M - \phi_s)\left[\frac{\cos(\omega_M - \omega_s)T - 1}{(\omega_M - \omega_s)T}\right]$$

If we set ϕ_M equal to ϕ_s, as can be done in the coherent case under the assumption that the medium introduces no phase shift, then

$$\chi_{12} \simeq \operatorname{sinc}((\omega_M - \omega_s)T) \tag{12.13}$$

The error probability is minimized when χ_{12} is as near to -1 as possible. It is *not* possible to set sinc $((\omega_M - \omega_s)T)$ exactly equal to -1, because the sinc function never attains that value. It *is* possible to equate it to zero, i.e., to make the mark and space signals orthogonal. This will provide a low error probability, given by

$$P_\epsilon = P_{SM} = P_{MS} = \frac{1}{2}\left[1 - \operatorname{erf}\left(\sqrt{\frac{\rho_i B_n}{2R}}\right)\right] \tag{12.14}$$

Orthogonality implies a mark-space separation at least equal to the separation between the peak of χ_{12} and one of its zeros, i.e. (in hertz),

$$\frac{\omega_M - \omega_S}{\pi} = 2(f_M - f_S) = \frac{n}{T} = nR \tag{12.15}$$

where the index n indicates the nth zero of the sinc function.

To pass both mark and space signals, the receiver's bandpass bandwidth must be somewhat larger than $(f_M - f_S)$, or, equivalently, for orthogonal mark and space signals

$$2B_n > |f_M - f_S| = \frac{nR}{2} \tag{12.16}$$

If $2B_n$ were precisely equal to $nR/2$, then (12.14) would take the form

$$P_\epsilon = \frac{1}{2}\left[1 - \operatorname{erf}\left(\frac{\sqrt{\rho_i n}}{2}\right)\right] \tag{12.17}$$

It is evident from (10.98) that increasing n will decrease the error probability. Therefore increased bandwidth improves performance with a fixed input SNR.

To summarize the above, anticorrelation, although it is the optimum condition, cannot be attained with FSK; hence one strives for orthogonality. But orthogonality requires a minimum bandwidth, and the greater the bandwidth, for a fixed value of ρ_i, the smaller the error probability. Thus to achieve a low error probability, a price is paid in bandwidth.

If phase is not a priori known, then noncoherent FSK is used. The error probability in this case is obtained from (7.84), under the assumption that the mark and space signals are mutually orthogonal and have equal energies. From (7.84), we have

$$P_\epsilon = P_{SM} = P_{MS} = \frac{1}{2}e^{-(\rho_0/2)} = \frac{1}{2}e^{-(\rho_1 B_n/2R)} \tag{12.18}$$

As pointed out in connection with coherent FSK, orthogonality requires a minimum frequency spacing between signals as given by (12.15).

12.1.5 Comparison of Binary Modems

In Figure 12-1 a set of curves† is presented in which the error probability P_ϵ is plotted against the parameter $\rho_o = E_s/N_o = 2\rho_i B_n T$ or $2\rho_i B_n/R$ for the following cases:

1. Coherent OOK (from (12.3) or (12.5)).
2. Noncoherent OOK (from Eqs. (12.4.a, b) or (12.5)).
3. PSK—anticorrelated signals (from (12.6) and (12.10)).
4. Coherent FSK—orthogonal signals (from (12.14)).
5. Noncoherent FSK—orthogonal signals (from (12.18)).

† Similar sets of curves showing comparative performance of various binary modems can be found elsewhere in the literature, e.g., Jones, Pierce, and Stein, 1964; Figure 2.3.3, p. 144, and Schwartz, Bennett, and Stein, 1966; Figure 7-5-1, p. 299.

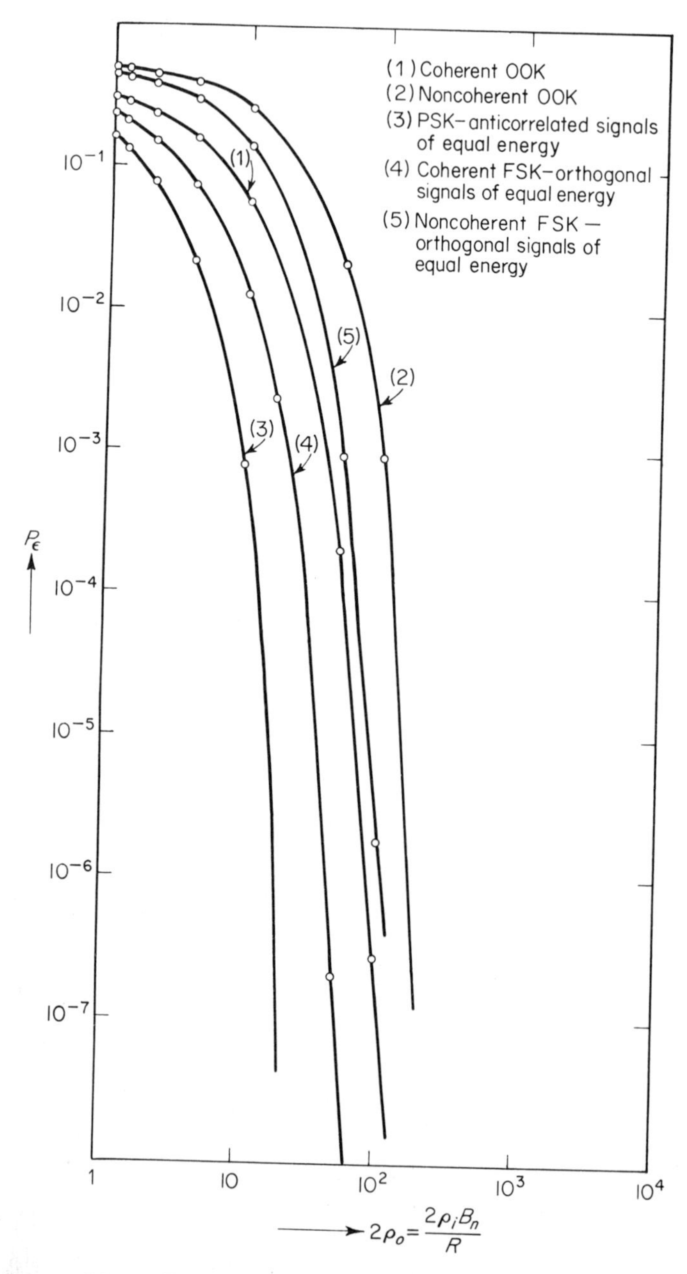

Figure 12-1 Error probabilities for binary communication system.

It is evident from these curves that the comparative performance of these binary digital modems is rather sensitive to the SNR. The region of greatest interest is that of high SNR, wherein error rates are at least as low as 10^{-3}. Systems operating with error probabilities much higher than these are generally regarded as below the threshold of reliability and find little use in the practical communications art.

In the region of interest, we find generally that coherent modems are superior to noncoherent ones (which is to be expected for reasons already discussed; see, for example, Sections 6.3 and 6.4), that PSK and coherent FSK are generally superior to coherent OOK, that coherent PSK is superior to coherent FSK, and that noncoherent FSK is superior to noncoherent OOK.

Some of these conclusions can easily be inferred directly from the equations used to obtain the curves. Comparison of (12.3) with (12.10) shows, for example, that the argument of the error function in the coherent OOK error probability contains a factor $\frac{1}{2}$ that is not present in the PSK error probability, the two probabilities being otherwise identical. Thus for a given SNR, the error function has a larger value for PSK than for OOK; hence the PSK error probability is smaller than the OOK error probability. Comparing(12.14) with (12.11), we note a factor $1/\sqrt{2}$ in the argument of the error function; hence the error rate for coherent FSK is somewhere between that for OOK and that for PSK. The basic reason for this is the fact that the FSK modem treated here employs orthogonal signals, while the PSK case involves anticorrelated signals, the latter being the theoretical optimum.

We note that the error rates in Eqs. (12.5), (12.10), (12.14), and (12.18) are all given in terms of the parameter $\rho_i B_n/R$ and are all monotonically decreasing functions of that parameter. The purpose of this is to point up the fact that a higher SNR is required to maintain a given error probability as one attempts to transmit information at a higher rate. This is the major tradeoff that exists in digital communication systems. If one is content with extremely low information rates, then it is possible to communicate very reliably at low SNR.

To illustrate this point, we first note that $\rho_i B_n/R$ is equivalent to $(a^2/2N_o) \cdot (1/R)$, where a is the signal amplitude and N_o the noise power density. We can regard N_o as a fixed parameter and assume that a is made as large as possible, its highest value being limited by transmission and propagation conditions. The data rate R, then, is the only parameter that can be varied. The curves of Figure 12-2 show the value of $a/\sqrt{2N_o}$ (in seconds) required to attain a given communication reliability (i.e., a given value of P_ϵ) against the information rate R for various values of P_ϵ and for the noncoherent OOK modem which is treated in Section 12.1.2. Figure 12-2 is based on Equations (12.4.a) and (12.4.b), or equivalently in Case (2) of Figure 12-1. The required values of $a/\sqrt{2N_o}$ increase linearly with R for a fixed value of P_ϵ, but the slopes differ quite markedly in different cases, being highest for those modems whose performance is shown by Figure 12-1 to be the poorest.

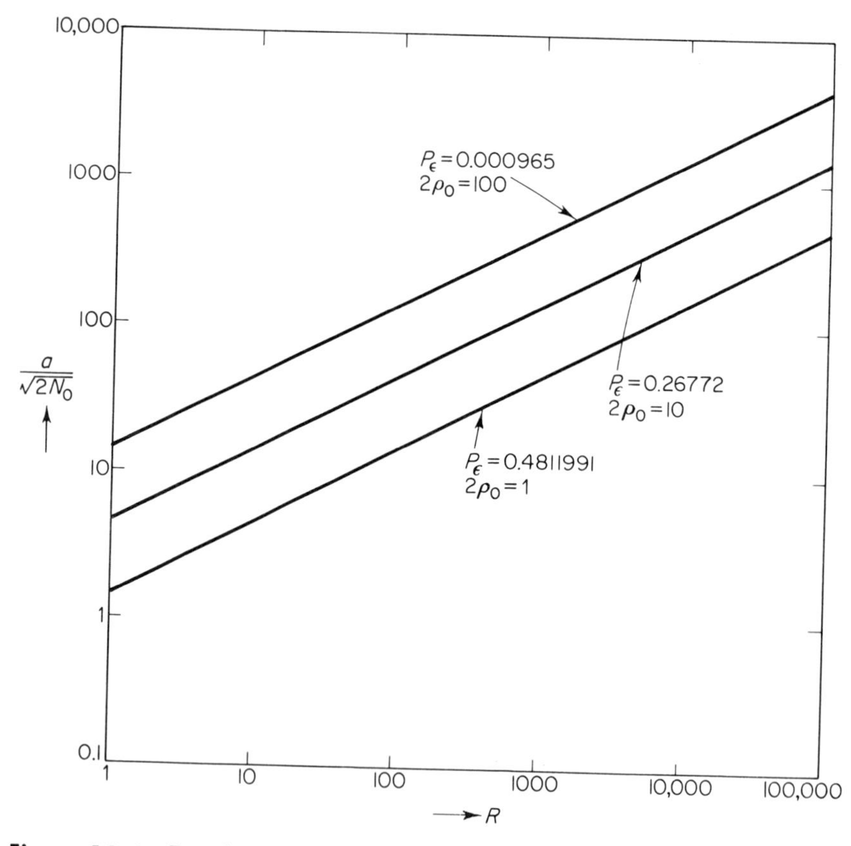

Figure 12-2 Required input SNR versus information rate with fixed error probability—noncoherent OOK.

The results discussed here are based on the assumption that the filtering done on the mark and space signals is ideal matched filtering. If conventional bandpass filters are used, then a performance degradation results, i.e., an increased error rate for a given input SNR, or a larger required input signal power for a given data rate and reliability. In the case of an integrating filter with a finite time constant, the extent of the degradation can be determined from the results given in Section 6.7.

12.2 M'ARY SIGNALLING SYSTEMS

In a M'ary signalling system, information is coded into a set of M possible signals. The problem at the receiver is to determine which of the M signals was transmitted. The signals are distinguished on the basis of a parameter that is different for each of them, e.g., amplitude level, frequency (M'ary FSK), phase (M'ary PSK), possibly a combination of these, or some entirely different parameter.

We can regard M'ary PSK (often called *multiphase*) and M'ary FSK (often called *multitone*) as conceptually simple generalizations of binary PSK and binary FSK, respectively, wherein signals with M (instead of two) different phases or frequencies are transmitted. Determination of the error probabilities for these modems is an application of the analysis of multiple alternative detection presented in Section 7.3. In order to apply that analysis directly, however, we must assume that the M signals all have equal energies and are mutually orthogonal. This is a perfectly reasonable way to design a M'ary signalling system, so such an assumption entails no sacrifice in realism.

The error probabilities for M'ary detection are given by (7.89) in the noncoherent case and (7.92) in the coherent case. In the case of noncoherent multitone, the error probability as given by (7.89) is

$$P_\epsilon^{(M)} = 1 - \sum_{\ell=0}^{M-1} \frac{(M-1)!\,(-1)^\ell e^{-\rho_o/2}\{\ell/(\ell+1)\}}{(\ell+1)!\,(M-1-\ell)!} \tag{12.19}$$

where $\rho_o = \rho_i B_n T$.

It is interesting to investigate the dependence of $P_\epsilon^{(M)}$ on the number of possible signals. We can easily do this by hand for small values of M. The results, for the first few values of M, are as follows:†

$$P_\epsilon^{(2)} = \tfrac{1}{2}e^{-\rho_o/2} \tag{12.20.a}$$

$$P_\epsilon^{(3)} = e^{-\rho_o/2} - \tfrac{1}{3}e^{-2\rho_o/3} \tag{12.20.b}$$

$$P_\epsilon^{(4)} = \tfrac{3}{2}e^{-\rho_o/2} - e^{-2\rho_o/3} + \tfrac{1}{4}e^{-(3/4)\rho_o} \tag{12.20.c}$$

For very large SNR, the lead term of each of the equations (12.20) is the only term of significant magnitude. In this case we can easily extrapolate (12.19) beyond $M = 3$ by calculating the coefficients of the exponentials in the terms where $\ell = 0$ and $\ell = 1$. These coefficients are always equal to -1 and $(M-1)/2$, respectively. Then (12.19) has the approximate form‡

$$P_\epsilon^{(M)} \simeq \frac{(M-1)}{2}\,e^{-\rho_o/2} \quad \text{for very large values of } \rho_o \tag{12.21}$$

The *threshold SNR* ρ_{oT}, defined as the value of ρ_o required to attain a given error probability, is determined by solving (12.21) for ρ_o. The decibel ratio of ρ_{oT} for $M = M_1$ to that for $M = M_2$, with the error probability fixed at the value $P_{\epsilon o}$, is approximately

† Note that (from Eqs. (12.20.a, b, c) when $\rho_o = 0$, $P_\epsilon^{(2)} = \frac{1}{2}$, $P_\epsilon^{(3)} = \frac{2}{3}$, $P_\epsilon^{(4)} = \frac{3}{4}$; that is, for zero SNR, the error probabilities are those attained with a purely random guess as to which of the M signals was transmitted.

‡ For a fixed value of ρ_o, the approximation (12.21) begins to break down as M attains very high values, because the aggregate of additional terms introduced into (12.22), attains a value comparable to the lead term.

$$\mathscr{R}_{M_2/M_1} = \left[\frac{(\rho_{oT})_{M=M_2}}{(\rho_{oT})_{M=M_1}}\right]_{\text{dB}} \simeq 10 \log_{10} \left\{\frac{\ln (M_2 - 1) - \ln (2P_{\epsilon_o})}{\ln (M_1 - 1) - \ln (2P_{\epsilon_o})}\right\} \quad (12.22)$$

The curves of Figure 12-3 are plots of $\mathscr{R}_{M_2/M_1}$ versus M_2 for various values of the error probability P_{ϵ_o}.† In these curves, the case $M_1 = 2$ (i.e., noncoherent binary signalling) is used as the reference case. The curves show a slow increase in $\mathscr{R}_{M_2/M_1}$ with increasing values of M_2. For example, increasing M_2 from 10 to 100 results roughly in a 3 dB increase in the SNR required to attain an error probability of 10^{-6}. This implies that reliable multitone communication becomes increasingly difficult as the number of signals is increased. Our first question might be "Why use multitone in preference to binary FSK if it degrades performance?" The answer is that the increased threshold

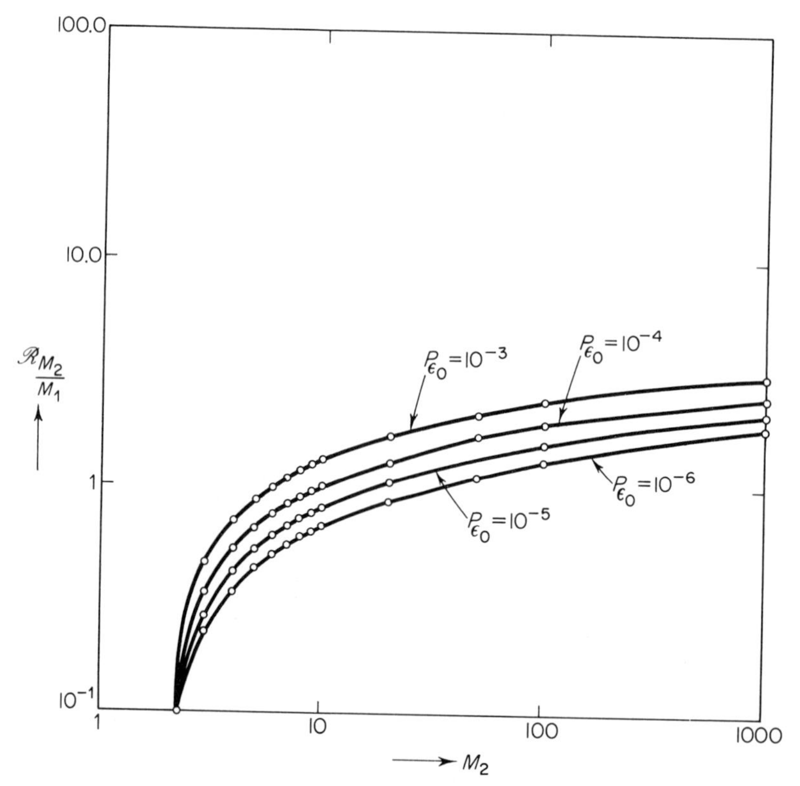

Figure 12-3 Approximate required SNR relative to binary case versus number of signals—M'ary signalling, noncoherent detection.

---

† The curves of Figure 12-3, being based on (12.19), are most accurate for the lower error probabilities (e.g., 10^{-6}), corresponding to high SNR and can be regarded as very rough approximations for the higher error probabilities (e.g., 10^{-3}) and high values of M_2.

SNR is merely a price to be paid for a gain in attainable information rate. This becomes clear when it is recalled that the number of bits of information acquired per baud, according to (3.80) and the assumption that any one of the M possible signals is as likely to have been transmitted as any other, is $\log_2 M$, as compared with $\log_2 2$ in the binary case. Thus, if T is the baud duration, then the information rate is

$$R = \text{number of bits per baud} \times \text{number of bauds per second} = \frac{\log_2 M}{T} \tag{12.23}$$

Equation (12.23) provides us with a one-to-one correspondence between the information rate R and the number of tones M. Comparison of multitone with binary FSK in terms of information rate is achieved by expressing (12.22) in terms of R. From (12.22) and (12.23)

$$\mathscr{R}_{M/2} = \left[\frac{(\rho_{0T})_M}{(\rho_{0T})_2}\right]_{\text{dB}} \simeq 10 \log_{10}\left\{1 - \frac{\ln(2^{RT} - 1)}{\ln(2P_{\epsilon_0})}\right\} \tag{12.24}$$

A set of plots of $\mathscr{R}_{M/2}$ versus RT for various values of the error probability is shown in Figure 12-4.† These curves show the way in which the SNR

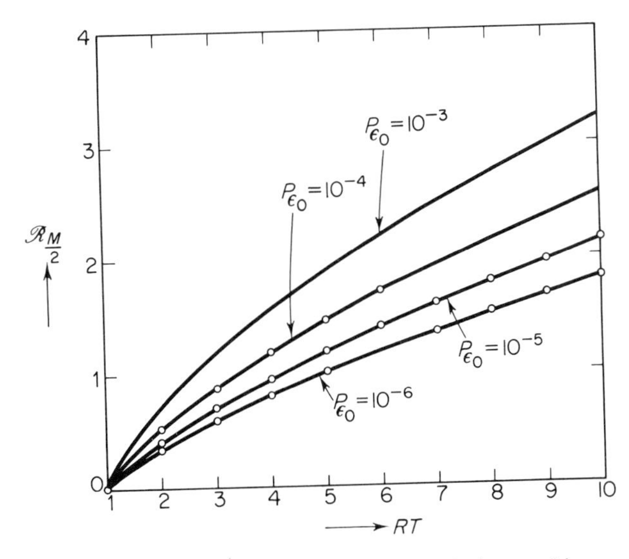

Figure 12-4 Approximate required SNR relative to binary case versus product of information rate and signal duration—M'ary signalling, noncoherent detection.

† Figure 12-4 differs from Figure 12-3 only in that the abscissa is RT rather than M_2. As is the case with Figure 12-3, the accuracy of the curves is greatly reduced for high error probabilities and large RT.

threshold increases as one attempts to increase the information rate by increasing the number of tones.

The same considerations as those discussed above apply to the case of multiphase or coherent multitone. In these cases, the error probabilities are obtained from (7.92)′, i.e.,

$$P_\epsilon^{(M)} = 1 - \frac{1}{2^{M-1}}\left[1 + \operatorname{erf}\left(\sqrt{\frac{\rho_o}{2}}\right)\right]^{M-1} \tag{12.25}$$

where $2\rho_o = 2\rho_i B_n T$, the coherent matched filter output SNR.

Again using (12.23) to relate M to the information rate R, we can plot $P_\epsilon^{(M)}$ against ρ_o for various values of RT. Such a set of curves is shown in Figure 12-5. Some of these curves are redundant with those of Figure 7-9, in which error probability is plotted against SNR for a few small values of M. Figure 12-6 shows error probabilities plotted against ρ_o for M'ary signalling with *noncoherent* detection where RT takes on the same values as in Figure 12-5. The curves of Figure 12-6 are based on (12.21), which only holds for large values of ρ_o. Hence, only the region where ρ_o exceeds 2.0 is shown on the figure.

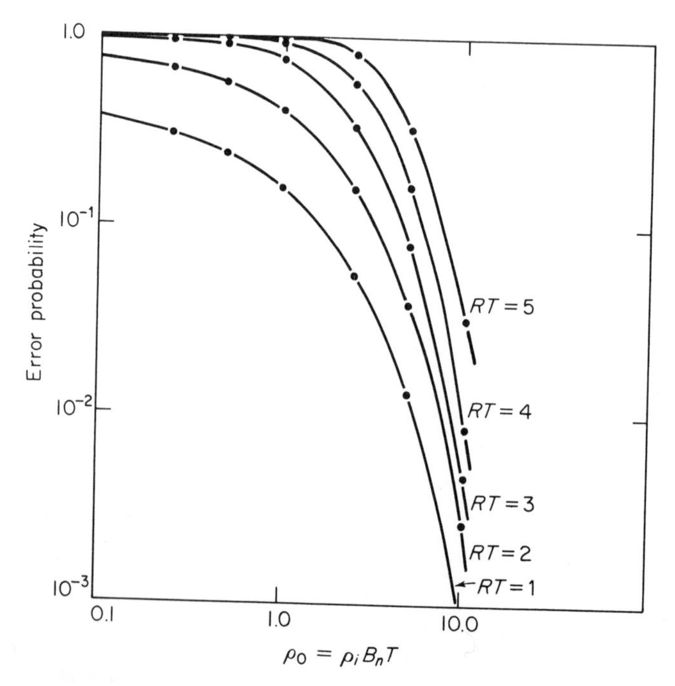

Figure 12-5 Error probabilities for M'ary signalling—coherent detection, equal energy signals.

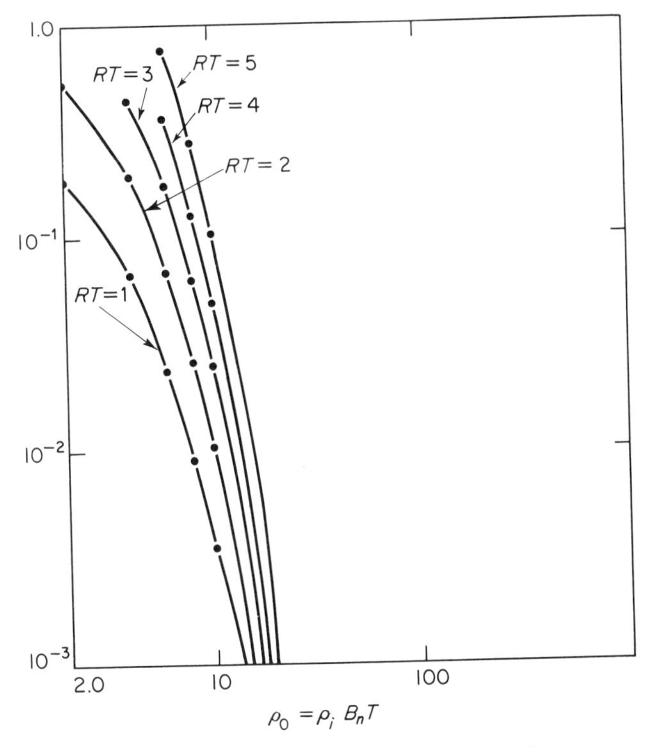

Figure 12-6 Error probabilities for M'ary signalling—noncoherent detection, equal energy signals.

Comparing the curves in Figure 12-5 with those in Figure 12-6, we see that coherent detection is superior to noncoherent detection just as in the binary case. The reason for this superiority is basically the same as in the binary case, the advantage gained through the use of a priori information in the detection process.

12.3 MESSAGE ERROR PROBABILITIES

In the discussions of digital signalling in Sections 12.1 and 12.2, we dealt with *bit error probability*, i.e., the probability of an error in a single bit. In some types of digital signalling, it is essential that an entire message get through correctly, where the message may consist of several mutually statistically independent bits in sequence. It is assumed that an error in only one of the bits disqualifies the entire message. Thus the probability that the message is in error is unity minus the probability that all bits are correct. The probability that a single bit is correct is unity minus the bit error probability. By virtue of the assumption of statistical independence between bits, the probability that all bits are correct is the product of the *correctness probabilities*† for the individual bits. Thus the per message error probability for an N-bit message is given by

$$(P_\epsilon)_{\text{message}} = 1 - [1 - (P_\epsilon)_{\text{bit}}]^N \tag{12.26}$$

It is evident that a message error probability should be greater than a bit error probability. To see *how much* greater, we expand the bracketed expression in (12.26) in a binomial series, with the result

$$(P_\epsilon)_{\text{message}} = \sum_{\ell=1}^{N} (-1)^{\ell+1} \frac{N!}{\ell!(N-\ell)!} [(P_\epsilon)_{\text{bit}}]^\ell \tag{12.27}$$

For low values of N, we have (where $(P_\epsilon)_{\text{bit}} = P$, and $(P_\epsilon)_{\text{message}} = (P_\epsilon)_N$ for a message consisting of N bits in sequence)

$N = 1$:
$$(P_\epsilon)_1 = P \tag{12.28.a}$$
$N = 2$:
$$(P_\epsilon)_2 = 2P - P^2 \tag{12.28.b}$$
$N = 3$:
$$(P_\epsilon)_3 = 3P - 3P^2 + P^3 \tag{12.28.c}$$
$N = 4$:
$$(P_\epsilon)_4 = 4P - 6P^2 + 4P^3 - P^4 \tag{12.28.d}$$

If P is sufficiently small compared to unity, e.g., $\lesssim 10^{-3}$ (common in practical systems), then only the first term of each of the equations (12.28) is significant, and we have, in general, for low values of N

$$(P_\epsilon)_N \simeq NP \tag{12.29}$$

From (12.29) we conclude, for example, that a 10-bit message has an error probability about 10 times larger than the error probability for any of its constituent bits. Another way of looking at this is to observe that a larger SNR is required to transmit an entire N-bit message with a given chance of error than any of its bits with the same chance of error. If the modem is binary noncoherent FSK, for example, then according to (12.18) the value of ρ_o required for a bit error probability P_o is

$$\rho_o = -2 \ln (2\rho_o) \tag{12.30}$$

From (12.29) and (12.30), the value of ρ_o required for an N-bit message error probability P_o is

$$\rho_o = -2 \ln \left(\frac{2P_o}{N}\right) \tag{12.31}$$

A comparison of (12.31) and (12.30) shows that a 10-bit noncoherent FSK message with an error probability of 10^{-3} requires an SNR about 3.9 dB higher than a single bit with the same error probability.

† This terminology is mine and is not in common usage.

12.4 MULTIPATH AND FADING IN COMMUNICATION SYSTEMS

The effects of fading and multipath on radio transmission were discussed in Section 5.5.4.7, later in Section 6.10 in the signal-to-noise ratio context and in Section 7.4 in the context of detection of fading radio signals. The reader is referred to these discussions for background on the effects of fading on communication systems.

The most common methods of dealing with fading in both analog and digital communication systems are the various forms of diversity reception. These were discussed previously in Section 6.10. That discussion applies directly to analog communication systems and will not be repeated here. However, in Section 6.10 the criterion of performance used was the SNR; probabilities were not directly involved in the discussion. Evaluation of communication systems is often based on probabilistic measures of performance; hence it is useful to look at diversity from a probabilistic point of view.

An analog communication system must function at a certain minimum SNR. This might be, for example, the threshold of signal intelligibility for the average listener. In a digital system, the minimum allowable SNR might be that which provides a given error probability, e.g., 10^{-4} for a specified information bandwidth. In either case, it is possible to define the system as "operating satisfactorily" if the SNR is above a certain threshold level (e.g., 10 dB) and as "out of service" when the SNR is below that level.

The effect of slow fading is to transiently depress the signal strength, possibly below the threshold level for satisfactory communication. If the system must operate in a fading environment, it is useful to know the fraction of the time during which the SNR is below threshold. This is known as the *outage time*. If we assume that the fading statistics are stationary, then the fractional outage time is roughly equivalent to the *probability* that the amplitude is below threshold (which can be called *outage probability*). With this point of view, we can evaluate a diversity scheme in terms of its ability to reduce the outage time. To do this, we use the fading statistics (assumed stationary) to calculate the outage probability both with and without the diversity scheme.

As the simplest and most commonly used example of this approach, we will assume Nth order optimal selection diversity combining (see Section 6.10.3.1). We will evaluate the outage probability, first without diversity and then with successively higher orders of optimal selection diversity. If the fading has Rayleigh amplitude statistics, then the outage probability (from 7.94) is

$$(P_\epsilon)^{(1)}_{\text{outage}} = \int_0^{\rho_T} d\rho \, \frac{1}{\langle\rho\rangle} e^{-\rho/\langle\rho\rangle} = 1 - e^{-\rho_T/\langle\rho\rangle} \tag{12.32}$$

where ρ_T is the voltage threshold SNR for satisfactory communication performance.

With Nth order selection diversity, an outage would occur only if all N available diversity branches were below threshold. Thus if the signals in the N branches are uncorrelated, then the outage probability is the probability that *all* branches have a voltage SNR below ρ_T, i.e.,

$$(P_\epsilon)^{(N)}_{\text{outage}} = (1 - e^{-\rho_T/\langle\rho\rangle})^N = [(P_\epsilon)^{(1)}_{\text{outage}}]^N \tag{12.33}$$

If an SNR of 6 dB above $\langle\rho\rangle$ is required for satisfactory performance, then the outage probability without diversity, according to (12.32), is about 98 percent. With a threshold of 3 dB above $\langle\rho\rangle$, this probability is reduced to about 86 percent and with a threshold equal to $\langle\rho\rangle$, it is roughly 64 percent. If two-fold optimal selection diversity is provided, according to (12.33) the outage probabilities in these three cases reduce to (roughly) 97, 75, and 40 percent, respectively. With four-fold selection diversity, they reduce to about 94, 56, and 16 percent respectively. With a 16-fold system, they are about 79, 10, and .07 percent, respectively. This is obviously an enormous reduction from the case where diversity is not employed. The reduction in outage probability can be calculated for other diversity combining schemes, but the example given should be sufficient to illustrate the point.

In digital communication systems, the multipath phenomenon (see Section 5.5.4.7) has important influence on performance in addition to the fading it produces. It causes an effect known as "*intersymbol interference.*

Referring to Figure 12-7, we assume that a mark and a space (e.g., two tones in a binary FSK modem) are transmitted during the periods 0 to τ and τ to 2τ, respectively.

The mark signal travels out over the two paths ABC and $AB'C$. The signal propagating over ABC covers the interval T to $(T + \tau)$ at the receiver while that propagating over $AB'C$ covers the interval $(T + \Delta)$ to $(T + \Delta + \tau)$, where Δ is the relative delay time between paths ABC and $AB'C$. The space signal may travel these same two paths if the atmospheric conditions have not changed over a baud; hence the ABC signal covers the receiver time interval $(T + \tau)$ to $(T + 2\tau)$, while the $AB'C$ signal covers the interval $(T + \tau + \Delta)$ to $(T + 2\tau + \Delta)$.

If the relative time delay Δ is extremely small compared to a baud length, then the mark or space signals propagating over the two paths interfere in such a manner as to produce fading, as described in Section 5.5.4.7.

If Δ is comparable to a baud length or is a substantial fraction of a baud length, however, then the mark signal overlaps the receiver time-slot reserved for the space or the space overlaps that reserved for the mark, depending on whether Δ is positive or negative. In either case, there is interference between these two symbols. The mark-space decision that must be made at the end of each baud is corrupted by the presence of energy rightfully belonging to

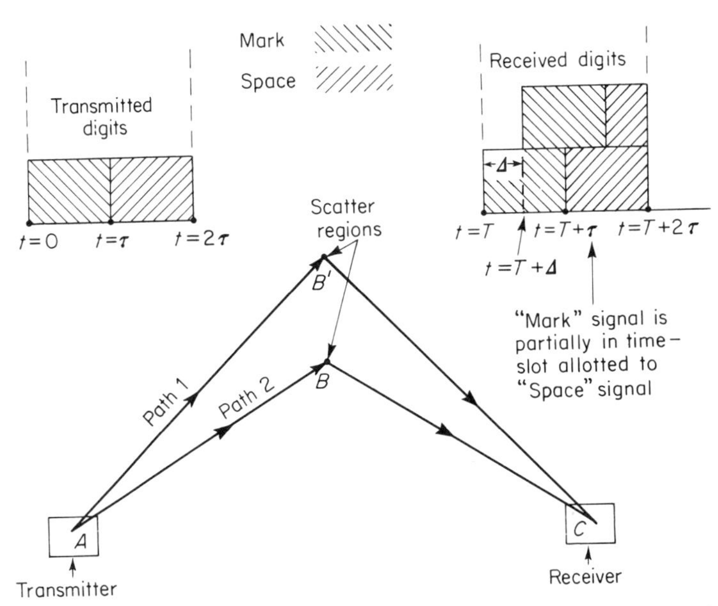

Figure 12-7 Intersymbol interference in digital communication.

other bauds. In the extreme case where there are many paths whose relative delay times overlap several baud intervals, the mark-space decisions would be severely corrupted. If the degree of overlap is not too excessive, i.e., if the total spread in the multipath signals is not excessively large, then a technique known as *Rake*,† can be implemented to combat *multipath*‡ and *multipath fading*.§ This scheme can be regarded philosophically as a sophisticated form of diversity. It can also be regarded (to very rough approximations) as a variation on matched filtering, in which the transmitted signal is accompanied by a wide-band probing signal that determines the impulse response of the communication channel.‖ In the receiver, the probing signal

† The classic reference on Rake is Price and Green, 1958.

‡ The term *multipath* can be used to refer to the propagation phenomenon in which the radio signal travels multiple paths (see Section 5.5.4.7), *or* to its consequence in a receiver, wherein multiple replicas of the transmitted waveform appear with different delays corresponding to different paths.

§ The term *multipath fading* refers to fading that is due to the multipath propagation phenomenon (since there are other possible causes of fading, e.g., the random fluctuations in atmospheric parameters in troposcatter modes (Section 5.5.4.6)).

‖ If the probing signal has an extremely wide band of frequencies (as does an extremely narrow pulse or a wide-band noise) and the channel is the equivalent of a linear filter, then the probing signal appears at the receiver with the characteristics of the channel's impulse response.

is convolved with a replica of the transmitted signal (e.g., mark or space). The convolution of these two signals, which should now resemble the transmitted signal corrupted by the channel, is then crosscorrelated with the incoming information-bearing signal.

When one looks at Rake in this rather oversimplified manner, it is seen to be essentially a matched filter whose reference signal contains the effects of the medium through which the signal has propagated, and therefore resembles the true incoming signal more closely than would a replica of the transmitted signal alone.

There are more mathematically correct ways of viewing Rake systems and there now exist variations on the Rake system described in the original paper by Price and Green. The reader is referred to the literature for further details.†

12.5 ENHANCEMENT OF MESSAGE RELIABILITY

A most important topic in digital communication systems engineering is the improvement of message reliability (that is, minimization of the probability of errors in the message) beyond that provided by straightforward binary or M'ary signalling. A great deal of theoretical and experimental effort in recent years has been expended on development of error-detecting and error-correcting codes. This topic, which could easily occupy a few additional chapters of this book, will not be covered. The brief discussion that follows will concern some very elementary ideas about increasing the reliability of a message without coding techniques.

The crudest form of enhancement of message reliability is repetition of part or all of the message. This method is somewhat akin to that of increasing transmitter power in that it is a "brute-force" method whose efficacy is evident without much theorizing.

Common experience tells us that a repeated message is more likely to be eventually understood than a message delivered only once. If we fail to understand words in an ordinary conversation because of background noise or other distracting sounds, we ask the speaker to repeat himself. Some people with an especially strong desire to be understood will repeat themselves without being asked. In either case, the listeners are less likely to miss words or phrases than would be the case without repetition.

To place our discussion on a more quantitative level, consider a single bit in a message transmitted over a noisy radio channel. Having made a mark-space decision at the receiver, we would like assurance that it was a

† See Price and Green, 1958; Kailath, Chapter 6 in Baghdady, 1961; Bello and Raemer, 1962.

correct decision. Absolute assurance is unattainable in the presence of noise. But we can check the decision by arranging to have the bit transmitted twice, with the understanding that we will accept a pair of mark decisions (or a pair of space decisions) as a final indication that a mark (or space) was transmitted. If one mark decision and one space decision are made, we are left with an ambiguity. We then have a number of choices. For example, we may provide a feedback channel, that is, a receiver-to-transmitter link through which we can communicate in the reverse of the usual direction. If such a channel exists, we can ask for a retransmission whenever a split decision occurs and accept the decision made on the retransmitted message.

It is here that interesting tradeoff considerations begin to appear. For example, a feedback channel in general contains noise, and hence the repetition request may not be detected. Another repetition request may be required. A natural alternative is to ignore or discard the feedback channel and simply prearrange that each bit be transmitted *three times* and that "majority decision logic" be used, wherein two- or three-mark decisions will be accepted as a mark and two- or three-space decisions will be accepted as a space.

The tradeoff here involves that most valuable commodity, time. Triple transmission of each bit requires a threefold increase in message transmission time. This must be compared with the time required for double transmission of each bit with a feedback repeat request only when a split decision occurs. The required message transmission time with the feedback technique is at least twice the original message duration, but may be much greater if split decisions occur often.

A highly oversimplified model will illustrate the method a communication systems engineer might use to do a comparative analysis of these two reliability enhancement techniques.

Let the per-bit error probability be denoted by $P_{\epsilon 1}$. Suppose a "mark" bit is transmitted twice. The probability that it will be incorrectly interpreted as a space on *both* transmissions is $P_{\epsilon 1}^2$, assuming statistical independence between the noise waveforms on the two successive transmissions. The probability that it will be correctly read as a mark on both transmissions is $(1 - P_{\epsilon 1})^2$. The probability that the first decision will be "mark," the second "space" is $(1 - P_{\epsilon 1})P_{\epsilon 1}$. Since a split decision can occur in two ways, mark space or space mark, and each of these two ways has the same probability of occurrence the probability of a split decision is $2P_{\epsilon 1}(1 - P_{\epsilon 1})$.

Suppose the mark bit is transmitted three times. The possible sets of decisions are (where M and S denote mark and space respectively) MMM, MMS, MSM, SMM, SSM, SMS, MSS, and SSS. The first of these sets of decisions occurs with probability $(1 - P_{\epsilon 1})^3$ and the last with probability $P_{\epsilon 1}^3$. The second through the fourth each has probability $(1 - P_{\epsilon 1})^2 P_{\epsilon 1}$, and the fifth through the seventh each has probability $P_{\epsilon 1}^2(1 - P_{\epsilon 1})$. An error will occur if a majority of the three decisions are in error, that is, if one of the events *SSS*, *SSM*, *SMS*, or *MSS* occurs. The error probability for the case of three prearranged transmissions of a"mark" bit, being the sum of probabilities of the various types of error, is

$$P_{\epsilon 3}^{(M)} = P_{\epsilon 1}^3 + 3P_{\epsilon 1}^2(1 - P_{\epsilon 1}) = 3P_{\epsilon 1}^2 - 2P_{\epsilon 1}^3 \tag{12.34}$$

It is obvious that the same analysis, and hence the result (12.34), applies to the case of three transmissions of a "space" bit. Acceptable per-bit error probabilities in actual systems are typically of the order of 10^{-4} to 10^{-3} or lower; hence a perfectly adequate approximation to (12.34) for the error probability with either a mark or a space transmitted three times is

$$P_{\epsilon 3}^{(M)} = P_{\epsilon 3}^{(S)} \simeq 3P_{\epsilon 1}^2 \tag{12.35}$$

Assuming (as is usually the case in practice) that marks and spaces are equally likely, the over-all error probability is

$$P_{\epsilon 3} = \tfrac{1}{2}P_{\epsilon}^{(M)} + \tfrac{1}{2}P_{\epsilon}^{(S)} = 3P_{\epsilon 1}^2 - 2P_{\epsilon 1}^3 \simeq 3P_{\epsilon 1}^2 \tag{12.36}$$

Returning to the case of two transmissions of a mark with a feedback channel to request a third transmission, we note that an error occurs if the pair of decisions *SS* is made (probability $P_{\epsilon 1}^2$) or if the decision is split (probability $2P_{\epsilon 1}(1 - P_{\epsilon 1})$) *and* the decision on the requested retransmission is in error (probability $P_{\epsilon 1}$). The same set of probabilities applies to the case in which a space is transmitted. The feedback channel is assumed to be noise-free,† such that no errors can occur on the retransmission request. Then the over-all per-bit error probability is

$$P_{\epsilon 2R} = P_{\epsilon 1}^2 + 2P_{\epsilon 1}(1 - P_{\epsilon 1})P_{\epsilon 1} = 3P_{\epsilon 1}^2 - 2P_{\epsilon 1}^3 \simeq 3P_{\epsilon 1}^2 \tag{12.37}$$

Comparision of (12.37) with (12.36) tells us that the over-all error probability with two transmissions and a repeat request in case of a split decision is identical with that obtained with three prearranged transmissions of the bit.

To complete our story, we consider the transmitted bit as part of an N-bit message (with N not very large) whose over-all error probability is well approximated through (12.29), which results in a message error probability roughly equal to

$$P_{\epsilon} \simeq NP_{\epsilon 3} \simeq NP_{\epsilon 2R} = N(3P_{\epsilon 1}^2 - 2P_{\epsilon 1}^3) \simeq 3NP_{\epsilon 1}^2 \tag{12.38}$$

From (12.38), it follows that the per-bit error probability can be expressed in terms of the message error probability by the approximate expression

$$P_{\epsilon 1} \simeq \sqrt{\frac{P_{\epsilon}}{3N}} \tag{12.39}$$

† Since much less information is transmitted over the feedback channel than the direct channel, it is possible to arrange matters so that the error-producing effects of the feedback channel noise will be much smaller than those of the direct channel noise. Consequently, the "noise-free" assumption is not necessarily a drastic oversimplification.

and that this holds in both the feedback channel case and the case of three prearranged transmissions of each bit.

It is instructive to compare the time required to transmit the entire message with and without the feedback channel. Without the feedback channel, each message bit is transmitted 3 times; hence the total N-bit message transmission time is

$$T_{M3} = 3NT_b \tag{12.40}$$

where T_b is the baud duration.

In the feedback channel case, each double transmision of the bit involving two-mark or two-space decisions requires a time interval $2T_b$, and each such transmission that leads to a split decision requires $2T_b + (T_b + \tau_1)$, where $(T_b + \tau_1)$ represents the sum of the time expended in a repeat request, denoted by τ_1, and that expended on retransmission of the requested bit, T_b. Thus the average time required to transmit the message is

$$\begin{aligned} T_{M2R} = N[(2T_b)\ \text{Prob}\{\text{2-mark or 2-space decisions}\} \\ + (3T_b + \tau_1)\ \text{Prob}\{\text{split decision}\}] \end{aligned} \tag{12.41}$$

The probabilities indicated in (12.41), in order of their appearance, are $[P_{\epsilon 1}^2 + (1 - P_{\epsilon 1})^2]$ and $[2P_{\epsilon 1}(1 - P_{\epsilon 1})]$. Substituting these probabilities into (12.41), we have

$$T_{M2R} = 2NT_b\left\{1 + (P_{\epsilon 1} - P_{\epsilon 1}^2)\left(1 + \frac{\tau_1}{T_b}\right)\right\} \simeq 2NT_b \tag{12.42}$$

where $P_{\epsilon 1} \lesssim 10^{-3}$.

We have tacitly assumed that the repeat-request time τ_1 has an order of magnitude no greater than that of T_b, which would certainly be the case in most applications. Eq. (12.42) tells us that the additional time for repeat requests and retransmissions is virtually negligible under the assumption of a sufficiently small per-bit error probability. This consideration leads to the conclusion that the feedback technique would reduce the total transmission time by 33 percent relative to that required without the feedback channel, assuming the baud duration to be the same in both cases.

Before we make further comparisons between the feedback and non-feedback techniques, we should consider a third alternative. We know that an increase in baud length with fixed signal power and fixed noise power density will decrease error probability. Hence a straightfoward increase in baud length might be as effective in reducing errors as message repetitions.

To make this comparison, suppose we consider total transmission time to be the basic constraint. We have an N-bit message to transmit and have exactly T s in which to transmit it. We can make direct use of all the available time and transmit the N bits without repetitions, thus establishing a baud length T/N or an information rate $R = N/T$.

The per-bit error probability for a case in which N is not too large is (from 12.29)

$$P_{\epsilon 1} \simeq \frac{P_\epsilon}{N} \tag{12.43}$$

where P_ϵ is the message error probability.

Suppose we place a specification on the *message* error probability P_ϵ rather than the per-bit error probability $P_{\epsilon 1}$. Hence a fixed parameter in our analysis is P_ϵ/N. We will compare the per-bit error probabilities in the 3 cases considered by equating P_ϵ/N in (12.39) and (12.43), resulting in

$$3(P_{\epsilon 1})_A^2 = 3(P_{\epsilon 1})_B^2 = (P_{\epsilon 1})_C \tag{12.44}$$

where the subscripts have the following meanings:
A = 3 repetitions, B = 2 repetitions with feedback repeat requests, and C = no repetitions.

For purposes of discussion, we go to the special case of noncoherent FSK. From (12.18), we obtain

$$P_{\epsilon 1} = \frac{1}{2} e^{-\rho_i B_n T_b/2} \tag{12.45}$$

where the parameter R that appears in (12.18) is recognized as the reciprocal baud length $1/T_b$, which is *not* equivalent to the actual information rate when bits are repeated.

We now account for the difference in transmission time between the three techniques under investigation, noting that with fixed time T for transmission of the entire message, the baud lengths in the three different cases are as follows:

$$(T_b)_A = \frac{T}{3N} \tag{12.46.a}$$

$$(T_b)_B \cong \frac{T}{2N} \tag{12.46.b}$$

$$(T_b)_C = \frac{T}{N} \tag{12.46.c}$$

We will now fix B_n and allow ρ_i, the input SNR, to be different in the three cases under comparison. Substituting (12.45) and Eqs. (12.46) into (12.44), we have

$$\frac{3}{4} e^{-(B_nT/3N)(\rho_i)_A} = \frac{3}{4} e^{-(B_nT/2N)(\rho_i)_B} = \frac{1}{2} e^{-(B_nT/2N)(\rho_i)_C} \tag{12.47}$$

We will now make a comparison of the minimum values of ρ_i required to attain the specified message error probability with fixed values of T, N, and B_n. From (12.47) we have

$$\frac{2}{3}(\rho_i)_A = (\rho_i)_B = (\rho_i)_C + \left(\frac{2N}{B_nT}\right) \ln\left(\frac{3}{2}\right) \tag{12.48}$$

Comparison of cases A and B requires no specification of the values of the parameters N, B_n, and T. Regardless of the values of these fixed parameters, the first equation in (12.48) tells us that the use of the feedback channel to reduce the required number of repetitions decreases the minimum required input SNR by a factor 1.5, or about 2 dB. Thus the feedback channel allows us to transmit a message with a given reliability and within a given time but with a transmitter power about 2 dB lower than that required with 3 message repetitions and no feedback channel.

A numerical comparison between the case of 3 repeated bits (case A) and that of nonrepeated bits (case C) requires a knowledge of the parameters N, B_n, and T. Consider, for example, a 500-bit message, transmitted within a period of 1000 s with a receiver noise bandwidth of 1 KHz. The parameter, $2N/B_nT$, then is 10^{-3}. In this case the second term on the right-hand side of (12.48) is negligible, and we have, to very good approximation,

$$(\rho_i)_A \simeq \tfrac{3}{2}\,(\rho_i)_C \tag{12.49}$$

Thus for this particular combination of parameter values, three repetitions cost us about 2 dB in increased threshold SNR. From the magnitudes of the parameters involved, it would appear that 2 dB is a good estimate for a wide range of values of $2N/B_nT$. If $2N/B_nT$ were to be very large, the difference in threshold SNR between the repetition and nonrepetition cases would be somewhat larger than 2 dB. Since 2 dB is not a spectacular number except near the very limit of transmitter power capability, it would appear that in many cases it does not make an enormous difference in a theoretical sense whether the energy available for transmission of messages is expended in increased bit length or in repetition of bits. In a given case, of course, one or the other of these techniques may be advantageous from a practical design point of view.

12.6 REMARKS ON PULSE MODULATION

Throughout our discussions of analog modulation schemes in Chapter 11 and digital modulation schemes in the present chapter no mention has been made of pulse modulation, a discussion of which contains elements of both analog and digital concepts. Although pulse modulation is too extensive a topic to cover in depth without adding substantially to the length of this book, it is worthwhile to mention a few highlights of this interesting and important area of communications engineering.

Basically, pulse modulation is a procedure wherein the information waveform is sampled, usually at uniform intervals conforming to the Nyquist rate. The modulation waveform then becomes a train of pulses rather than a continuous signal.

A number of alternatives exist for handling the pulses, arising from the sampling procedure. For example, one can modulate a carrier directly with the train of sample pulses. At the receiver, a straightforward demodulation process theoretically equivalent to that used in ordinary AM will result in the recovery of the train of sample pulses. The original information waveform is then recovered from the pulse train.

This technique, known as *pulse amplitude modulation*, or PAM† is the most direct method of pulse modulation. An alternative technique, known as *pulse duration modulation* (PDM) or *pulse width modulation* (PWM) begins with the generation within the transmitter of a secondary pulse train from the train of sample pulses. The amplitudes of the secondary pulses are all equal, but their durations are proportional to the amplitudes of the primary (sample) pulses from which they were derived. The secondary pulses are used to modulate the carrier to be transmitted. At the receiver, the demodulation process involves the transformation of pulse durations back into sample amplitudes, which in turn are used to reconstruct the original information waveform.

Another technique, known as *pulse position modulation* (PPM) also involves a secondary pulse train generated within the transmitter and used to modulate the transmitted carrier. In this case, amplitudes and durations of the secondary pulses are all equal, but the position within the interpulse interval of a designated reference point on a secondary pulse (for example, center, leading edge, or trailing edge) relative to a given zero-reference point is proportional to the amplitude of the primary pulse from it was derived.

Each of the pulse modulation techniques above corresponds to standard analog modulation techniques. PAM corresponds to conventional AM and, like AM, involves no bandwidth-SNR tradeoff. PDM and PPM, however, both correspond to angle modulation, that is, FM or PM. Both involve an expenditure of bandwidth in exchange for a gain in SNR (see Sections 11.2.4 and 11.2.5). The reader is referred to the literature for extensive discussions of this and other important aspects of pulse modulation.‡

A most important pulse modulation technique is known as *pulse code modulation* or PCM, which consists of sampling and quantization of the sample values (see Section 3.6.2) followed by digital transmission of the quantization levels. At the receiver, the digital indications of the quantization levels of the samples can be used to reassemble the original signal.

---

† To be precise, we should use the designation PAM/AM, because the PAM train modulates the carrier directly without the use of FM or other techniques. This is not always the case. An example is PAM/FM, wherein a PAM waveform is frequency modulated for transmission.

‡ See, for example, Black, 1953, Chapter 3, pp. 30–34, and Chapters 15 through 19; Schwartz, 1959, Chapter 4; Schwartz, Bennett, and Stein, 1966, Chapter 6; Rowe, 1965, Chapter 5; Panter, 1965, Chapter 20.

The obvious advantage of PCM over purely analog modulation schemes is the increased transmission accuracy provided by the digitization of the signal. An equally obvious disadvantage is the "quantization error," or the inherent loss in accuracy that accompanies any quantization process. We alluded to these ideas in a general way in Section 3.6.2.

In a PCM system, the signal amplitude is quantized into M regions. Each sample falls into a given quantization region. Let us assume that the entire message is to be transmitted in real time. A digital message is transmitted after each sampling instant, informing the receiver that the most recent sample was within a given quantization region. For real-time transmission, we must require that the digital message for each sample have duration no greater than the sampling interval $1/2B_I$, where B_I is the information bandwidth, and it is assumed that sampling is done at the Nyquist rate. Since (with a fixed signal amplitude) the minimization of error probability is achieved by maximizing the signal duration, the total available time should be used to transmit the message. The duration of the digital message specifying the quantization level of each sample, then, is $1/2B_I$.

Suppose we begin the discussion by assuming two-level quantization, a very crude indication of signal amplitude. If the amplitude falls below a certain fixed value A_0, the signal has quantization level 1. If it is above A_0, it has quantization level 2. Our digital message must indicate whether the signal is at level 1 or level 2. This is a one-bit message. From Section 12.1 we can see that all the standard techniques for binary digital transmission (for example, OOK, FSK, PSK) yield a per-bit error probability that is a single-valued function of the parameter $\rho_i B_n/R$. We denote this per-bit error probability by $P_{\epsilon I}$ and the duration of the original information waveform by T_I. We observe that real-time transmission of the quantization levels of all the $(2B_I T_I)$ samples that make up the information waveform requires a message containing $(2B_I T_I)$ bits. For the very low per-bit error probabilities characterizing digital data transmission, (12.29) tells us that the message error probability is approximately $2B_I T_I P_{\epsilon 1}$.

To facilitate the discussion, we specialize immediately to binary noncoherent FSK as the modem to be used. From (12.18) and (12.29) the message error probability is

$$P_\varepsilon \simeq B_I T_I\, e^{-\rho_i B_n/2R} = B_I T_I\, e^{-\rho_i B_n/4M B_I} \tag{12.50}$$

where it is noted that the information rate R in this case is equivalent to the Nyquist sampling rate $2B_I$.

Now suppose we wish to improve the accuracy of the signal amplitude measurement by increasing the number of quantization levels to 4. In this case, the message informing the receiver in which quantization level the amplitude of a sample lies requires $\log_2 4 = 2$ bits. For real time transmission, this implies that we must transmit a total of $4B_I T_I$ bits within the time interval T_I, just twice the number required with two-level quantization.

Further refinement of the amplitude measurement through 8-level, 16-level, or 32-level quantization increases the number of bits-per-sample to $\log_2 8 = 3$, $\log_2 16 = 4$ or $\log_2 32 = 5$, respectively. In general, if the number of quantization levels is 2^M, then the number of bits-per-sample is $\log_2 2^M = M$. Thus with 2^M quantization levels the transmission rate R is $2B_IM$, and the error probability for the message indicating the quantization level of a sample is roughly $MP_{\epsilon 1}$. If binary noncoherent FSK is used to transmit each bit, then from (12.18) the error probability for the entire message reporting quantization levels of all samples is given by

$$P_\epsilon \simeq B_IT_IM\, e^{-\rho_i B_n/4MB_I} \tag{12.51}$$

More generally, with K quantization levels, the number of bits per sample is $\log_2 K$ and the message error probability is approximately

$$P_\epsilon \simeq B_IT_I \log_2 K\, e^{-\rho_i B_n/4B_I \log_2 K} \tag{12.52}$$

We conclude from (12.52) that the message error probability increases with the number of quantization levels in two ways. First there is the increase with $\log_2 K$ due to the increased number of bits-per-sample and secondly the increase due to the increased per-bit error probability. Also, the message error probability increases with the information bandwidth B_I in two ways, through the increase in the required number of samples per second and again through the increase in per-bit error probability.

Tradeoff studies can be carried out through (12.52), for example, determination of the increase in minimum SNR required to attain a specified message error probability (other parameters remaining fixed) when the number of quantization levels is increased. Since the dependence on number of levels is a logarithmic one, it is rather slow, and beyond a certain value of K one can change the quantization considerably before an appreciable effect on the error probability is observed.

12.7 COMMUNICATION SYSTEM DESIGN STUDIES

Throughout this book, we have often alluded directly or indirectly to the idea of a system design compromise, often called a *tradeoff* by engineers. The major themes of this book, in fact, have been the simple theoretical ideas characteristically used by radar and communication systems engineers to analyze the parameter tradeoffs that arise in the design of an actual system. If there is a single central theme that permeates most system design studies, it is the fact that "optimization," as defined through highly idealized theory, must always be compromised by the constraints of the environment. Thus a design study often focusses on compromises, guidelines being provided by some sort of optimization theory but tempered by qualitative knowledge acquired through experience.

This final section will be devoted to an example of the way in which systems engineers might use some of the concepts developed in the book to perform a design study of a large-scale complex communication system. Suppose that the system to be dealt with is one involving an orbiting satellite gathering data on the upper atmosphere, for example, continuous readings of temperature, pressure, electron density, or other physical parameters that can be measured directly with instruments on the satellite. This data must be transmitted to one of N ground stations $G_1, G_2, \ldots, G_N$, at which the data undergoes a certain amount of processing followed by transmission of the processed data to a central ground station G_c. The entire system is managed and coordinated from G_c. There are personnel at G_c and at each of the stations $G_1, \ldots, G_N$. Two-way communication links must be maintained between each of these stations and the central station, in order that information, instructions, and replies can flow between stations. Some control can be exercised over the satellite from the ground, for example, signals activating and deactivating some of the instruments. Thus, both space-to-ground and ground-to-space communication links exist between the satellite and each of the ground stations.

This is a highly complex system. In the absence of a well-developed set of rules set up by national and international communications agencies, the number of design options open to the systems engineer would be staggering. Fortunately or unfortunately, depending on the viewpoint, these rules limit the number of options by specifying certain frequency regions, modes of transmission, etc. However, even within this framework, there are many choices to be made. The purpose of the design analysis is to determine as quantitatively as possible, accounting for whatever factors can be specified, the relative performance quality of the alternative methods of accomplishing the task. To perform the analysis, the engineer must view the system as an integrated whole, and hence he must integrate those aspects of the analysis based on different disciplines, for example, communication theory, propagation, antennas, and circuit analysis.

It is evident that the first step in a design analysis is to define criteria of merit. To do the analysis meaningfully, we must be able to determine not only that one technique is better than another, but also decide *how much* better. This requires that we have a merit criterion that can be characterized in terms of numerical values.

Let us begin by assuming that operations analysts have already performed a study of the over-all system, whose purpose is to deliver to the central station G_c a certain amount of scientific data with specified accuracies within a specified period of time and within a specified cost. From diverse considerations, scientific, technical, economic, and possibly political, these analysts have arrived at certain specifications that apply to the "communication subsystem." This phrase describes the portion of the system that involves communication. It is important to recognize that communication as a function is auxiliary to the main purpose of the system but that it is vital to the system's operation. Our operations analysts may have told us, for example, that a satellite will be gathering continuous readings of temperature, pressure, and electron density throughout certain portions of its trajectory but that its program of measurements will be changed from time to time by means of instructions from the ground stations. In order to effectuate this sort of flexibility in the pattern of measurement activity, the ground-to-satellite links must operate with good reliability. The over-all system considerations may necessitate an assurance that the ground-to-satellite instruction messages contain errors no more than 1 percent of the time. This means that the message error probabilities should be no greater than 10^{-2}.

The operations analysts may also tell us that the space-to-ground links must be 99 percent reliable; that is, only one message in a hundred may contain an error. This again corresponds to a message error probability of 10^{-2}.

When these specifications are known, it is possible for communication systems engineers to determine the design requirements on various parameters that have a bearing on the error probability. The first step, for example, might be to write an equation for the received power in terms of transmitted power. Before this can be done, it may be necessary to know what frequency region is to be used on each of the links, since this determines the kind of theoretical laws relating received power to transmitted power (see Section 5.5). A study of ionospheric transmission properties (see Section 5.5.4.5) might tell the communications engineer that a frequency in the SHF region will be required on the ground-to-satellite† and satellite-to-ground† links in order to adequately penetrate all of the ionospheric layers. On the other hand, the use of exceptionally high frequencies may be impractical because of attenuation in the part of the atmosphere below ionospheric altitudes. These frequencies may also be precluded by power generation limitations or for other reasons relating to the state of the equipment art.

In a large and important system development project, it is likely that a group of radio propagation specialists would conduct an initial study to determine the relative advantages of different frequencies. Others knowledgeable on the subject of equipment limitations might contribute to this study, whose end result might be a decision on an "optimum" frequency to use on

† The generic term to be used in what follows to denote both space-to-ground and ground-to-space links is *space-ground links.*

the links between space and ground. The "optimization" would not be based entirely on idealized theory but would also include the economic and political constraints of the real world (for example, the radio spectrum allocations of regulatory agencies).

Based on the results of this study, the communications specialist may be told that the frequency to be used on the links between ground and space is (say) 10 GHz. He also knows that at that frequency, propagation between ground and space in either direction may be viewed theoretically as propagation in infinite free space. An analysis based on the beacon equation (5.14) should be sufficiently accurate for all practical purposes.

From (5.14) we have, for the power P_R received at the peak of the antenna beam in terms of the transmitted power P_T at transmitter-to-receiver distance r (in m)

$$P_R = \frac{P_T A_{eT} A_{eR}}{c^2 r^2} f^2 \tag{12.53}$$

where A_{eT} and A_{eR} represent effective aperture areas (in m^2) of transmitting and receiving antennas respectively, c is the free space velocity of light (in m/s) and f the radio frequency (in Hz).

Suppose it is a system design specification that all space-ground links must be capable of maintaining communication over the maximum separation distance between a satellite and a ground station and that the latter is 1000 km. With this value of r and the value $f = 10$ GHz, for example, (12.53) becomes

$$P_R = (1.11)(10^{-9}) P_T A_{eT} A_{eR} \tag{12.54}$$

The effective aperture area of the satellite antenna is always severely limited by size and weight considerations, whereas the ground station antenna can have a much larger aperture. Although (12.54) tells us that the product $A_{eT} A_{eR}$ should be as large as possible, there is another consideration that often limits the desirability of a very large aperture. If the ground antenna aperture is too large, then the directivity may be excessively high (see Eq. (5.9)). This may necessitate an elaborate tracking system to keep the moving space vehicle within the beam. A particular ground station may be required to communicate with a number of widely separated vehicles. If the beam is directional, it should not be required to swing too rapidly from one direction to another within a short period of time. These operational requirements may dictate a solid angle of coverage that is not too small, for example, $(\pi/10)$ sr, or 1/20 of a hemisphere.† From (5.9) the angle of coverage is estimated through

$$\Omega = \frac{\lambda^2}{A_e} = \frac{c^2}{f^2 A_e} \tag{12.55}$$

† Hemispherical coverage would be desirable, but is impracticable at this frequency because the corresponding aperture size would be ridiculously small.

At a frequency of 10 GHz, (12.55) and the requirement of a solid angle of $\pi/10$ dictates that the ground station antenna have an effective aperture area of about 29 sq cm. A circular dish antenna with this effective aperture has a diameter in the general neighborhood of 10 cm. A similar requirement of wide-angle coverage should be imposed on the space vehicle. A tendency of the vehicle to roll or wobble could render a highly directive antenna totally impracticable in maintaining communication between the vehicle and a single station. Gyro stabilization may be required to keep the beam pointed always in the same direction regardless of the gyrations of the vehicle. In many cases, it is far simpler to provide wide angular coverage. This is usually no problem, because size and weight limits and other aerodynamic considerations are such that the maximum allowable vehicle antenna is too small to be directive.

Let us assume that both ground and space vehicle coverage angles are $\pi/10$ sr. Then from (12.55) and (12.54) we conclude that

$$P_R \simeq (10^{-14})P_T \tag{12.56}$$

that is, the received power is roughly 140 dB below the transmitted power.

To relate (12.56) to communications reliability (that is, the ability of the system to sustain communication with a low frequency of errors), we must use it to find the familiar parameter (E_S/N_0), the ratio of signal energy to noise power density for a single bit. Note that this parameter is equivalent to $(2\rho_i B_n T)$ or $(2\rho_i B_n/R)$, in which form we have expressed it earlier in this chapter. As has been observed in previous discussions in this book, it is always the parameter that determines error probabilities in detection and digital communication in the presence of Gaussian noise. From (3.62), (5.19), and Section 5.5.3.1, the noise power density N_0 is

$$N_0 = \left(kT_{abs} + \frac{\sigma_{ne}^2}{2B_n}\right) F_n \tag{12.57}$$

where k = Boltzmann's constant $= 1.38 \times 10^{-23}$ J/deg, T_{abs} = absolute temperature, F_n is the noise figure, and σ_{ne}^2 the noise power entering the receiver from external sources, that is, atmospheric and extraterrestrial noise.

To determine the relative contributions of internal and external noise, we must consult experimentally derived curves like those of Figures 5.3 and 5.4, showing median atmospheric and extraterrestrial noise versus frequency with a half-wave dipole receiving antenna. These curves do not include the frequency region of interest, but by a simple extrapolation of Figure 5.4 we can deduce that extraterrestrial noise is somewhere between 35 and 85 dB below ideal receiver noise. The latter is defined in terms of power density as the value of kT_{abs} corresponding to $T_{abs} = 290$°K, i.e., 4×10^{-21} W/Hz or -240 dBw/Hz. At 10 GHz, extraterrestrial noise is far more significant than atmospheric noise. We can consider the external noise at the frequency of interest as arising entirely from extraterrestrial sources. On the space vehicle receiver, where size and weight limitations preclude the use of low noise amplification techniques, we can assume that the internal receiver noise is not substantially less than ideal noise. Hence, if the antenna gain in the direction of the noise source is not enormously greater than that of a half-wave dipole, which is the case in our example,† the internal receiver noise still overwhelms the maximum external noise that can be expected. Thus it is the internal noise ($kT_{abs}F_n$) that limits the receiver's sensitivity. If we assume a temperature of 290°K and a noise figure of 4 dB in the space vehicle receiver, N_0 has a value of about 10^{-20} Ws.

The ground receiver installation may employ low noise amplification techniques. The ground antenna, pointing upward, is likely to encounter a much higher extraterrestrial noise level than the space vehicle antenna. If external noise is at the high end of its range of probable values, it may equal or exceed internal noise if the latter has been reduced by 20 to 30 dB through low noise techniques.

To stay on the conservative, or "worst-case" side of the situation, let us assume that N_0 for the ground receiver, whether internal or external noise predominates, is only 10 dB below that for the space vehicle receiver. The ground receiver noise power density, then, is 10^{-21}.

Combining (12.56), (12.57), and the parameter values assumed above, we arrive at the following expressions:

$$\left(\frac{E_S}{N_0}\right)_{\substack{\text{space}\\ \text{vehicle}\\ \text{receiver}}} \simeq (10^6)\frac{P_T}{R} \tag{12.58.a}$$

$$\left(\frac{E_S}{N_0}\right)_{\substack{\text{ground}\\ \text{receiver}}} \simeq (10^7)\frac{P_T}{R} \tag{12.58.b}$$

where R = information rate in bits per second.

Equations (12.58.a) and (12.58.b) provide us with a means of relating transmitted power and information rate to the error probability under the assumption that fading is not an important consideration.‡ The choice of digital modem for the ground-to-space and space-to-ground links can be guided by these relationships, combined with curves like those of Figure 12.1. If considerations of engineering convenience dictate the use of a binary modem but allow a free choice between modems, we might look at Figure 12.1 and decide to use PSK since it provides the lowest per-bit error probability for a given SNR in the error probability region of interest (that is, below 10^{-3}). However, the use of PSK may be precluded by the requirement

† The peak gain in our example, from (5.13) and the assumed parameter values, is about 40, or 16 dB. That of a half-wave dipole is 1.64 or slightly greater than 2 dB. Thus, ideal noise exceeds the highest external noise level to be expected by at least 20 dB.

‡ We will assume here that previous studies have indicated that fading is likely to be negligible in our space-ground links.

of a priori knowledge of signal phase. Unpredictable and significantly large phase shifts may occur over the long transmission paths characteristic of the space-ground links. A modem based on the direct use of absolute phase information† may be impractical.

The next best binary modem, according to Figure 12.1, is coherent FSK. This modem requires very tight control of signal phase, because the reference signal for matched filtering must be phase-synchronized with the incoming signal (see Chapter 6 or 7). Hence, we may be forced to consider the use of noncoherent FSK. Although theoretically inferior to both PSK and coherent FSK, it may be the best compromise with the engineering "facts of life," in that it does not require tight control of phase‡ and provides a reasonably low error probability for a given SNR (lower, for example, than that provided by noncoherent OOK).

We recall that a message error probability no greater than 10^{-2} was the specification imposed on the space-ground links. From (12.29), we know that the message error probability is roughly the product of the per-bit error probability and the number of message bits. To relate Eqs. (12.58) to a message error probability of 10^{-2}, we must first determine the information content of the message in bits and divide 10^{-2} by that number. This will yield the approximate per-bit error probability. We then examine Figure 12.1 and find the value of (E_S/N_0) corresponding to that error probability for noncoherent FSK. This value of (E_S/N_0) can then be inserted into (12.58.a) and (12.58.b) to determine the minimum acceptable ratio of transmitted power to information rate required for ground-to-space and space-to-ground communication, respectively (with the specified reliability).

We assume that the ground-to-space messages are largely sets of instructions with no more than 100 bits per message. The space-to-ground messages originate from analog waveforms obtained from instrument outputs. Suppose these waveforms are of 100-s duration at most and have bandwidths no greater than 100 Hz. To enhance the reliability of the messages conveying the information contained in these instrument outputs, it is decided that they will be sampled at the Nyquist rate, quantized, and transmitted to ground by 5-bit PCM (see Section 12.6). This implies 32 quantization levels. The number of bits in the most informative of these messages is (200 samples/s $\times$ 5 bits/ sample $\times$ 10 s) $=$ 10 kilobits.

From (12.29) and the message information content as given above, we conclude that the per-bit error probabilities in the ground-to-space and space-to-ground links must be no greater than 10^{-4} and 10^{-6} respectively. The requirement on the former corresponds to $(E_S/N_0) \simeq 34.2$, or about 15.3

† A modem based on phase *difference*, for example, differentially coherent PSK (DPSK) may be practical in this case. See Schwartz, Bennett, and Stein, 1966, pp. 304–310.

‡ Such control may require phase-locked loops, which introduce an additional order of complexity into the system design.

dB with noncoherent FSK. That on the latter correponds to $(E_S/N_0) \simeq 52.6$, or about 17.2 dB, again with noncoherent FSK.

From Eqs. (12.58) and the above values of (E_S/N_0), we conclude that P_T/R must be about $3.4(10^{-5})$ for the ground-to-space link and about $5.3(10^{-6})$ for the space-to-ground link.

We can summarize these results by writing the following expressions:

$$(P_T)_{\mathrm{GdBw}} \simeq -45 + 10 \log_{10} R_{\mathrm{GS}} \tag{12.59.a}$$

$$(P_T)_{\mathrm{SdBw}} \simeq -53 + 10 \log_{10} R_{\mathrm{SG}} \tag{12.59.b}$$

where $(P_T)_{\mathrm{GdBw}}$, $(P_T)_{\mathrm{SdBw}}$, R_{GS}, and R_{SG} are, respectively, ground transmitter power in dB over 1 watt, space transmitter power in dB over 1 watt, ground-to-space information rate in bits-per-second and space-to-ground information rate in bits-per-second.

Equations (12.59) tell us how much transmitted power is required to communicate information at a given rate over the space-ground links with the specified reliability and system parameter values arrived at through design studies. These equations tell us, for example, that rates of 100 kilobit/s would require a transmitted power of 5 dBw, or about 3 W on the ground-to-space link, and −3 dBw, or about 500 mW on the space-to-ground link. These figures are not very large and might be attainable in a practical system. But suppose we wanted a much higher information rate, say, 100 megabit/s. In this case, the transmitted power for the space-to-ground link must be at least 500 W, which may be impossible to attain due to size and weight limitations of the space vehicle antenna. In such a case, we must either reduce the information rate or loosen the reliability requirements. If we are absolutely constrained to the specified reliabilities and are also constrained to a maximum of (say) 5 mW of power, the maximum attainable space-to-ground information rate will be 1 kilobit/s.

To show how difficult of attainment some communication system design goals may be, suppose we retained the same values of all parameters in (12.59.b) except the separation distance between ground station and space vehicle. In the discussion above, we assumed this distance to be 1000 km, somewhat characteristic of the order of magnitude of maximum separation distances in satellite communication links (other than synchronous satellites, where the distances are much greater).

But suppose we were to replace our satellite by a deep-space vehicle at a distance of (say) 100 million km. This would increase the required transmitter power by a factor 10^{10}, or 100 dB. Suppose we allowed a reduction of 20 dB through employment of an extremely high-gain antenna on the ground receiver. In this case, if the maximum possible space-to-earth transmitter power were (say) 1 W, the maximum possible information rate would be $2(10^{-3})$ bits/s, or equivalently about 1 bit every 8 min. If we insisted on an information rate as high as (say) 10 kilobits/s, we would need a transmitted power of 5 MW. This illustration points up the difficulties that can be encountered in deep-space communication, where hours, days, months, or even years may be required to communicate significant amounts of information reliably between earth and far-ranging space vehicles unless an enormous transmitter power can be attained.

At this point, we will leave the ground-space links and turn our attention to the links between the ground stations. These may be required to support several kinds of communication, for example, voice channels to allow station personnel to talk to each other and data channels for automatic transmission of satellite data to the central station for processing.

Confining our attention to the set of data channels between closely spaced (< 50 km) stations, suppose we begin with the requirement obtained from our operations analysts that each interstation channel be capable of transmitting data at 50-km distances with a per-bit error probability no greater than 10^{-4}.

The decision concerning the frequency band to use, if it is an open one, should be based on considerations like those noted in Table 5.1. In the case at hand, VLF and LF might be ruled out immediately because the cost and complexity of the large antenna structures that would be required are not justified in this case by a need for long-distance coverage. EHF and SHF may also be ruled out, this time because their ability to penetrate the ionosphere is not needed and their other advantages may not be worth the price in equipment complexity. Thus the choice is between MF, HF, VHF, and UHF. The first of these involves (primarily) ground-wave transmission, and its reliability does not depend on the state of the ionospnere. Atmospheric noise is the most important source of error to be overcome. The reliabilities of HF and VHF are heavily dependent on ionospheric state. By deciding on HF or VHF in lieu of MF, we substitute ionospheric fading for atmospheric noise as a source of degradation in reliability. We can also use UHF line-of-sight, which incurs a requirement for tall antennas but overcomes the dependence on ionospheric state.

We could go on indefinitely, but the above remarks should be sufficient to convey the message that the choice of an "optimum" frequency region, or an "optimum" specific frequency within that region, often depends on a great many variables that cannot be specified precisely and deterministically, such as ionospheric state.

Let us assume for purposes of discussion that our team of analysts has made a study of the frequency "optimization" problem and has arrived at UHF line-of-sight as the mode of transmission to be used and 2 GHz as the specific frequency.

Again, as with the ground-space links, we can use the beacon equation (5.14) or (12.53) to do a rough calculation of the power requirements. In this case, however, we should note that the wave interference phenomenon

(Section 5.5.4.2) may be present and that its presence will modify the beacon equation, which is strictly applicable only to an unbounded medium.

Let us refer back to Section 5.5.4.2 to consider the importance of wave interference. We first note that the raido wavelength corresponding to 2 GHz is 15 cm. If the heights of transmitting and receiving antennas are of the same order of magnitude as the wavelength (in this case anything less than a few meters) and if the ground reflection coefficient is high enough to produce a reflected wave, then interference *is* important. Let us assume that the environmental conditions and antenna design are such that it is *not* important (for example, directive antenna elements many times higher than the wavelength and poorly reflecting terrain) and that an analysis has satisfied us that it need not be accounted for. In this case, we can use (12.53) directly and go through the same line of reasoning as was traversed in the analysis of the space-ground links. In this case, the transmitting and receiving antennas are allowed to be very directive. We can point the transmitting and receiving antennas directly at each other, adjusting their directions to the specific pair of stations that are in contact. At a ground station, we could even have an aggregate of 4 to 8 stationary antennas each covering a small range of azimuthal angles. This feature allows us to take advantage of the high gain attainable at UHF with relatively small antenna elements.

Suppose we decide on a coverage angle of $\pi/100$ sr. Then we can deduce from (12.55) that the effective aperture area for both antennas is 0.715 m^2. From (12.53), using the parameter values $r = 50$ km, $f = 2$ GHz, and A_{eT} and A_{eR} both equal to 0.715 m^2, we obtain (very roughly)

$$P_R \simeq 10^{-8} P_T \tag{12.60}$$

From (12.60) and (12.57), using the values $T_{abs} = 290°\text{K}$, $F_n = 4$ dB, and the assumption that external noise σ_{ne}^2 is negligible compared with internal noise, we have (again very roughly)

$$\left(\frac{E_S}{N_0}\right) \simeq 10^{12} \frac{P_T}{R} \tag{12.61}$$

In this case, we should account for fading in our calculations of error probability. Let us assume that various system considerations again (as in the space-ground links) lead us to noncoherent FSK as the modem to be used. In this case, with the additional assumption of Rayleigh fading, (7.98) provides us with the error probability as a function of the parameter $\langle\rho\rangle$, defined as the average SNR (that is, averaged with respect to the fading statistics). Specifying a per-bit error probability of 10^{-4} (as indicated earlier) and applying (7.98), we arrive at the conclusion that $\langle\rho\rangle$ must be at least 10^4 or 40 dB in order to attain such an error probability. This is in constrast to a 15.3-dB requirement on the SNR with noncoherent FSK and no fading. Thus, if we were to calculate error probability *without* fading and thereby determine the required transmitter power, we would obtain

$$(P_T)_{\text{dBw}} \simeq -105 + 10 \log_{10} R \tag{12.62}$$

Guided by our error probability calculation *with* fading and not allowing for the use of diversity or some other antifading technique (see Section 12.4) to improve performance, we would conclude that a "fade margin" of at least 25 dB should be added to the required transmitted power. If diversity were to improve the signal strength by 10 to 15 dB, we could restore most of this margin and settle for 10 to 15 dB of required power added to the right-hand side of (12.62). Since it appears from a glance at (12.62) that we can attain enormous information rates with small power, it is wiser to provide a "safety factor" by retaining all 25 dB of margin, resulting in

$$(P_T)_{\text{dBw}} \simeq -80 + 10 \log_{10} R \tag{12.63}$$

From (12.63), we conclude that a data rate of 100 megabits/s requires a transmitted power of only a single watt. Even if only 1 mW were available, we could still attain data rates as high as 500 kilobits/s. The task of communicating at high data rates with small amounts of power is seen to be many orders of magnitude simpler over short-range ground-ground links than is the same task over the enormous distances involved in ground-space links.

It would be possible to continue this discussion at great length, treating such matters as tradeoffs between binary and M'ary modems for transmission of digital data, the choice between various continuous and pulse modulation techniques for transmission of analog information, further effects of fading and intersymbol interference due to propagation irregularities, the use of error-detecting and error-correcting codes, effects of radio interference from other transmitters within the environment, and an infinity of other problems that enter into the design of complex communication systems. Many of the questions considered in design studies such as that of our illustrative example involve propagation and environment matters. However, statistical communication theory plays the major part in the analysis. Its use provides the system designer with a catalyst that brings together diverse considerations and attempts to answer the question of how well the communication function can be performed subject to practical constraints, both those imposed by nature and those imposed by man.

REFERENCES

Digital Communication Systems in General

See the reference list in Chapter 3, (matched filters, etc.) and Chapters 4 and 7 (detection and estimation theory). For references to certain topics that are directly pertinent to digital communication systems, see especially Helstrom, PGCS, 1960. See also:

Baghdady, 1961; Chapter 6 by W. M. Siebert, Chapter 9 by W. L. Root

Jones, Pierce, and Stein, 1964; Chapter II

Schwartz, Bennett, and Stein, 1966; Part III by S. Stein, especially Chapters 7, 8

Stein and Jones, 1967; Chapters 9 through 17

Wozencraft and Jacobs, 1965; many sections in this book relate to digital communication systems, see especially Chapter 4

M'ary Signalling Systems

See the reference list in Chapter 7 on multiple alternative detection. Some of the references cited are repeated here.

Helstrom, PGCS, 1960

Jones, Pierce, and Stein, 1964; Section 2.6, pp. 181–196

Kotelnikov, 1960;

Middleton, 1960; Section 23.1, pp. 1024–1045

Nuttall, 1962

Pierce, 1966

Schwartz, Bennett, and Stein, 1966; Section 2.7, pp. 85–99

Stein and Jones, 1967; Chapter 14

Wozencraft and Jacobs, 1965; many sections, see especially Section 4.2, pp. 212–222

Multipath and Fading in Communication Systems; also Random Channels

Bello, 1963

Jones, Pierce, and Stein, 1964; Sections 2.9.2, 2.9.3, 2.9.4, pp. 215–254

Kailath, 1960

T. Kailath; Chapter 6 of Baghdady, 1961, "Channel Characterization: Time-Variant Dispersive Channels," pp. 95–124

Schwartz, Bennett, and Stein, 1966; Part III, by S. Stein, especially Chapters, 9, 10, 11

Stein and Jones, Chapters 16, 17

Wozencraft and Jacobs, 1965; Section 7.4, pp. 527–563

Pierce, 1966

PROBLEMS

CHAPTER 1

1.1 Consider a system whose input and output waveforms, denoted by $x_i(t)$ and $x_o(t)$, respectively, are as shown below:

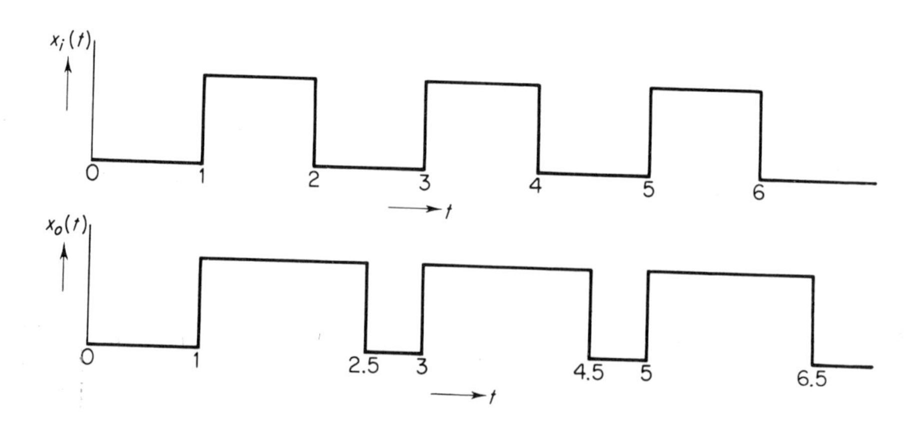

Assuming that the system is linear, determine its frequency response function.

1.2 The behavior of a linear system is characterized by the following differential equation involving input $x_i(t)$ and output $x_o(t)$.

$$\frac{d^2x_o}{dt^2} + 3\frac{dx_o}{dt} + x_o = x_i + \frac{dx_o}{dt}; \qquad x_o(0) = 0; \qquad \left(\frac{dx_o}{dt}\right)_{t=0} = 0$$

Find the output in response to a unit step function input applied at $t = 0$.

1.3 Find the output of the system of Problem 1.2 in response to an input of the form

$$x_i(t) = u(t) \sin (2t)$$

where $u(t)$ is the unit step function applied at $t = 0$.

1.4 Given the system whose differential equation and initial conditions on $x_o(t)$ are those of Problem 1.2, with input

$$x_i(t) = 5u(t) \sin \left(3t + \frac{\pi}{6}\right)$$

Find the output using Fourier or Laplace transform methods.

1.5 The behavior of a system is characterized by the following differential equation and initial conditions:

$$\frac{d^2x_o}{dt^2} - \frac{dx_o}{dt} + 4x_o = 2x_i; \qquad x_o(0) = 0; \qquad \left(\frac{dx_o}{dt}\right)_{t=0} = 0$$

Find the output in response to a unit step function input applied at $t = 1$.

1.6 Find the frequency response function of the system of Problem 1.5.

1.7 A linear system's differential equation is

$$\frac{d^2x_o}{dt^2} + 4\frac{dx_o}{dt^2} + 4x_o = u(t) \cos\left(\frac{5\pi t}{4}\right)$$

Find the transient and steady state outputs of this system by Fourier or Laplace transform methods if all of the derivatives of $x_o(t)$ are zero at $t = 0$ and $x_o(0) = 1$.

1.8 Consider a linear filter with the frequency response function

$$F(\omega) = |F(\omega)| \, e^{j\psi(\omega)}$$

where

$$|F(\omega)| = 1; \qquad |\omega| \leq W$$
$$0; \qquad |\omega| > W$$

$\psi(\omega) = -a\omega$, where a is a positive constant (this is called an "ideal lowpass filter"). Find the impulse response of this filter and determine whether it is causal. Explain your answer.

1.9 The transfer function of a linear system is

$$T(-js) = \frac{3s + 4}{s^2 + 7s - 8}$$

Find the impulse response of the system.

1.10 Consider a radio antenna with a rectangular aperture, illuminated as follows:

$$\hat{\psi}_j(x, y, 0) = e^{-a|x|-b|y|}; \qquad \frac{-A}{2} \leq x \leq \frac{A}{2}$$
$$\frac{-B}{2} \leq y \leq \frac{B}{2}$$

$$0; \qquad |x| > \frac{A}{2}$$

$$|y| > \frac{B}{2}$$

where a and b are positive constants. Find the radiation pattern $\psi(\theta, \phi)$

1.11 Determine (approximately) the time and frequency ambiguity functions $\rho(\tau)$ and $\psi(f)$, respectively, for the waveform

$$v(t) = A\cos(\omega_1 t + \phi_1)\cos(\omega_2 t + \phi_2)[u(t) - u(t - T)]$$

where $\omega_2 \gg \omega_1$, $T = 20\pi/\omega_1$. Use your results to discuss the time and frequency resolution of $v(t)$.

1.12 Determine (approximately) the time-frequency ambiguity function $\Lambda(\tau, f)$ for the waveform of Problem 1.11. Discuss the problem of compromise or "tradeoff" between time and frequency resolution for this waveform.

1.13 Consider the waveform

$$v_1(t) = [u(t) - u(t - 12)](e^{5t} - e^{-7t})$$

Represent $v_1(t)$ in terms of its time-domain samples, after estimating the highest significant frequency in the waveform.

1.14 Consider the waveform

$$v_2(t) = u(t)e^{-3t}\cos\left(2t + \frac{\pi}{4}\right)$$

Represent $v_2(t)$ (approximately) in terms of its frequency-domain samples.

1.15 Apply the duration bandwidth uncertainty principle to the following waveforms:

(a) A rectangular pulse of duration τ: $\left[u\left(t + \frac{\tau}{2}\right) - u\left(t - \frac{\tau}{2}\right)\right]$

(b) A Gaussian pulse: $e^{-(t/\tau)^2}$

(c) A pulse of the form: $\left[u\left(t + \frac{\tau}{2}\right) - u\left(t - \frac{\tau}{2}\right)\right]\cos\left(\frac{\pi t}{\tau}\right)$

In each case, find $\overline{(\Delta t)^2}$ and $\overline{(\Delta\omega)^2}$ and discuss the variation of these quantities with the parameter τ.

CHAPTER 2

2.1 Twenty identical balls are placed within a container and marked with numbers 1 through 20. Another 20 balls are placed in a different container, and also marked with numbers 1 through 20. A blindfolded man simultaneously picks one ball out of each container. What is the probability that he will pick 2 balls with the same number?

2.2 In Problem 2.1, suppose that the man picks two *pairs* of balls at random out of each container.

(a) What is the probability that he will pick two identically numbered pairs?
(b) What is the conditional probability that he will pick two consecutively

numbered balls from container #2 given that he has picked balls numbered 1 and 2 from container #1?
(c) If he has picked #3 and #5 from container #1, what is the probability that a third ball picked from container #1 will have a number in excess of 5?

2.3 In the configuration of Problems 2.1 and 2.2, what is the probability that a single ball picked out of container #1 will be *either* ball #13 or ball #14? How does this compare with the probability that a *pair* of balls picked from container #1 will be #13 and #14?

2.4 A batch of 1000 electronic devices is turned out on an assembly line. In each device there are three components labelled C_1, C_2, and C_3. The following are known for each of the devices: (1) the probability that C_1 will fail, (2) the conditional probability that C_2 will fail *if* C_1 fails, and (3) the conditional probability that C_3 will fail *if* C_1 and C_2 *both* fail. Determine the probability that in any given one of the 1000 devices, at least one of the three components will operate successfully.

2.5 In Problem 2.4, assume that the 1000 devices are statistically identical and mutually statistically independent. Determine the probability that four or more of the devices will fail given that the failure of any one of the three components C_1, C_2, or C_3 will cause failure of the device.

2.6 In a four-engine aircraft, given that the probability of failure of a single engine is .0000001 and that failure of any one engine is statistically independent of that of any other, what is the probability of failure of exactly two engines? Three engines? Four engines?

2.7 In a digital communication system (to be encountered in Chapter 12), it is extremely important that a certain message reach its intended receiver *error-free*. Consequently, a number of repetitions of the message are made, to insure that it will eventually be received correctly. The messages are repeated under identical conditions and are statistically independent of each other. Because of noise, each repetition has a probability of error equal to 0.01. How many repetitions are required to attain a value of 10^{-8} for (a) the probability that *exactly one* of the repetitions be error-free and (b) the probability that *at least* one of the repetitions be error-free?

2.8 In Problem 2.7, if there are $2N$ repetitions, what is the probability that *at least* N of them are correct? Do the problem for general N, then for $N = 5$, $N = 10$, and $N = 20$.

2.9 The probability *density* function of the output of an electronic device at a given instant of time is given by

$$p(x) = \begin{cases} 0; & x < x_1 \\ \dfrac{A(x - x_1)}{(x_2 - x_1)}; & x_1 \leq x < x_2 \\ A; & x_2 < x < x_3 \\ \dfrac{A(x - x_4)}{(x_3 - x_4)}; & x_3 \leq x \leq x_4 \\ 0; & x > x_4 \end{cases}$$

Find the probability *distribution* function and plot it against x.

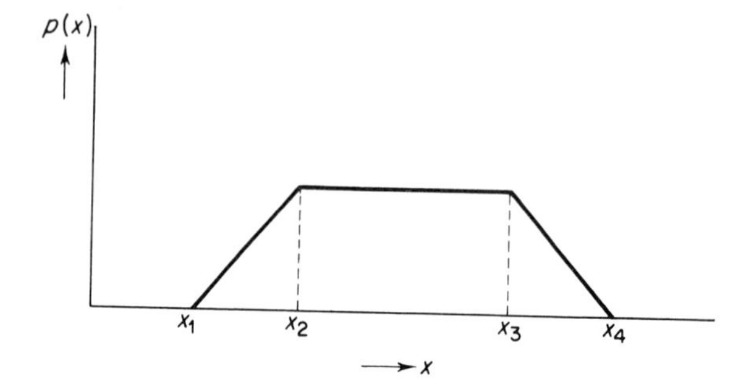

2.10 In Problem 2.9, calculate the first five quantities defined in Table 2.1

2.11 Consider the probabilities for the outcomes of a throw of two dice. Could you characterize them in terms of probability density functions? What form would the PDF take? Depict the PDF graphically.

2.12 The time function $f(t)$ is a single cycle of a sine wave with fixed amplitude A and fixed period T, but unknown starting point t_1. Given that t_1 has a PDF of the form

$$p(t_1) = \frac{1}{3T}; \qquad 0 \leq t_1 \leq 3T$$

$$0; \qquad t_1 < 0; \qquad t_1 > 3T$$

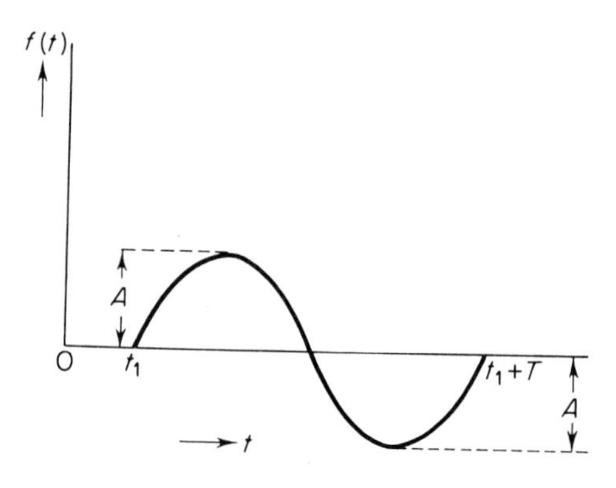

Find the PDF for the value of $f(t)$ at an arbitrary instant of time between 0 and $3T$.

2.13 Given the function $f(t)$ in Problem 2.12, find (for a fixed value of t_1 within its allowed range) the time-averages $\overline{f(t)}$, $\overline{f^2(t)}$, and $\overline{[f(t) - \overline{f(t)}]^2}$, where the time-averaging is done over the interval 0 to $3T$. Then, find the ensemble averages $\langle f \rangle$, $\langle f^2 \rangle$ and $\langle [f - \langle f \rangle]^2 \rangle$ at an arbitrary instant of time between 0 and $3T$. What do your answers tell you about the ergodicity and stationarity of the random process of which $f(t)$ is considered a sample function?

2.14 Find the autocorrelation function (as defined in Section 2.4.2.3) of the function $f(t)$ of Problems 2.12 and 2.13.

2.15 Given a random time function $x(t)$ whose approximate normalized autocorrelation function is as shown below:

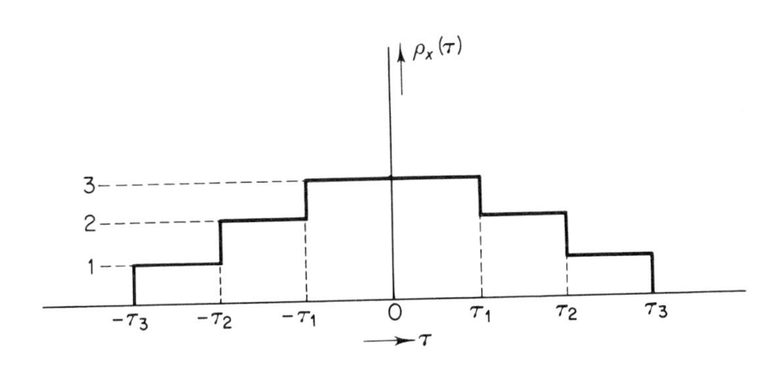

Find the normalized spectrum $S_x(\omega)$ and compare it with that of Eq. (2.55) in the cases where (a) $T_c = \tau_1$, (b) $T_c = \tau_2$, and (c) $T_c = \tau_3$. What are the important differences in these two types of spectra?

2.16 Find the autocorrelation function and power spectrum of the function $f(t) = s(t) + n(t)$ where

$$\begin{aligned} s(t) = {} & 3 \cos(4\pi t) + 4 \cos(6\pi t) \\ & -16 \cos(8\pi t) + 5 \sin(4\pi t) \\ & + 12 \sin(12\pi t) \end{aligned}$$

and where $n(t)$ is a random noise, uncorrelated with $s(t)$, whose power spectrum is

$$G_n(\omega) = \begin{cases} 10; & -5 \le \omega \le 5 \\ 0; & |\omega| > 5 \end{cases}$$

2.17 Consider the signal-plus-noise waveform $f(t) = s(t) + n(t)$, where

$$s(t) = A \cos(\omega_0 t) + B \cos(2\omega_0 t + \phi_0)$$

and where A, B, and ϕ_0 are constants and $n(t)$ is a noise with a normalized ACF of the form

$$\rho(\tau) = e^{-|\tau|/T}$$

uncorrelated with the signal $s(t)$. The waveform $f(t)$ is the input to a linear filter with impulse response

$$h(t) = u(t)(e^{-at} - e^{-b(t-c)})$$

where a, b, and c are positive constants and $a < b$. Find the normalized ACF and power spectrum of the signal-plus-noise output.

CHAPTER 3

3.1 An electronic device is subject to internal noise (#1) and noise from two mutually statistically independent external noise sources (#2 and #3), both statistically independent of #1. Each of these separate noise contributions has Gaussian statistics. Noise #2 has a finite mean, equal to 3, while noises #1 and #3 both have zero mean. The rms values of #1, #2 and #3 are 1, 2 and $\frac{1}{2}$ respectively. Find the first order PDF of the sum of the three noise contributions.

3.2 In Problem 3.1, suppose #2 and #3 are *not* mutually statistically independent. The normalized cross-correlation function between them (after the mean is subtracted from #2) is

$$\rho_{\substack{1\\2\\3}}(\tau) = e^{\frac{-|\tau|}{T}}, \qquad \text{where } T = 3T_o$$

Find the normalized autocorrelation function and normalized power spectrum of the sum of the three noise contributions. The conditions stated in Problem 3.1 all remain the same except for the deviations explicitly indicated here.

3.3 Find the second order PDF of the sum of the three noise contributions in Problem 3.2.

3.4 The square wave signal shown below is superposed on a stationary random noise whose autocorrelation function is that given by Eq. (2.56). Find the frequency response function of the optimum linear filter (i.e., optimum in the sense of Section 3.2.1) for this signal-plus-noise waveform.

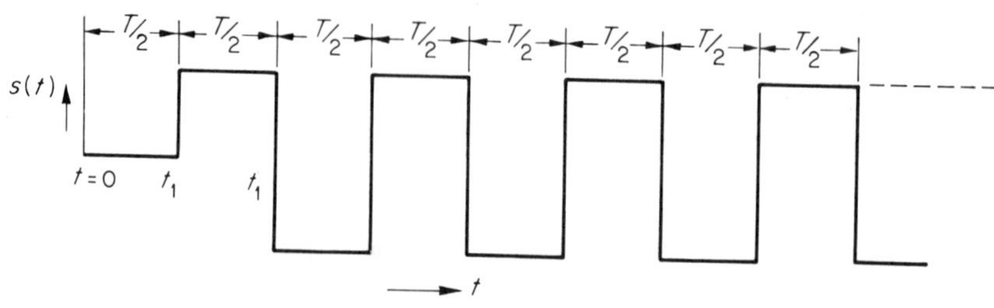

3.5 The signal-pulse signal shown below is superposed on a stationary random noise with autocorrelation function as given by Eq. (2.62). Find the frequency response of the optimum linear filter for this signal-plus-noise waveform.

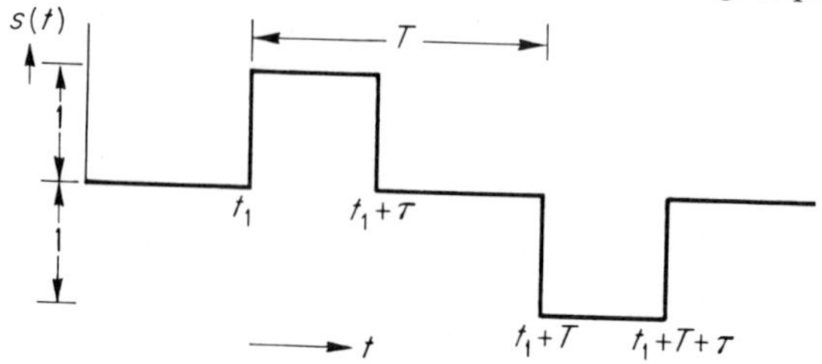

3.6 In Problem 3.4, what is the SNR degradation incurred by designing the filter as if the noise were white, i.e., optimizing the filter for signal-plus-white noise instead of signal-plus-the noise actually present?

3.7 Repeat Problem 3.6 substituting the words "Problem 3.5" for "Problem 3.4".

3.8 Repeat the derivations of the PDF's at the output of various rectifiers in Section 3.3 in the case where the input signal-plus-noise is first passed through a linear filter with impulse response $h(t)$ before entering the rectifiers; i.e., obtain your results in terms of the input to the linear filter, *not* the direct input to the rectifier.

3.9 Find the first order PDF of the sum of N statistically independent identical random functions, each with the first order PDF:

$$p(x) = \frac{1}{x_0}; \qquad \frac{-x_0}{2} \leq x \leq \frac{x_0}{2}$$

$$0; \qquad |x| > \frac{x_0}{2}$$

Do this problem for $N = 5$. From your result, does it appear that you can approximate the PDF for $N = 6$, 7, or 8 without calculating it?

3.10 A man begins with the knowledge that exactly two persons out of a specified group of 100 have a particular characteristic with no external signs (e.g., a particular occupation or disease). He eventually determines which two of the 100 people have this characteristic. How much information (in the sense of Section 3.6) has he acquired in the process?

3.11 A sequence of messages each of which specifies one of 10 mutually statistically independent and equally probable alternatives, one of which *must* occur, is transmitted over a communication link. Each message requires .03 seconds to transmit. Find the rate of transmission of information in bits-per-second. How would your answer change if 5 of the 10 alternatives had a probability of occurrence of $\frac{1}{20}$ and the other 5 had a probability of $\frac{3}{20}$?

CHAPTER 4

4.1 A sonar operator attempting to detect signals indicating the presence of submarines in a background of zero mean additive Gaussian noise sets up the following decision rule: if the observed signal at the receiver is above X_1 volts, he will conclude "submarine present"; if it is below X_2 volts (where $X_2 < X_1$), he will conclude "no submarine present." If it is between X_1 and X_2, he will reserve judgment. Derive expressions for the error probabilities for this type of detection system. Compare your results with those for the standard signal-presence detection system as discussed in Section 4.1.

4.2 In a certain type of binary "sequential" detection system, a sequence of observations is made, and a tentative decision on signal presence or absence is made after each observation. Each decision is followed by assignment of a new set of a priori probabilities of signal presence or absence, from which a new optimum detection threshold is set. Let the threshold assignment be based on the ideal observer criterion. Discuss the basic design of such a system for detection of a train of pulses in additive Gaussian noise.

4.3 In Problem 4.2, suppose that each observation is that of a single pulse. The pulses are rectangular, each of amplitude 1 V and duration of 0.1 ms. The noise has zero mean, rms value equal to 1 V and a correlation time of 10^{-5}s. The pulse repetition frequency is 200 V. Determine the error probability as a function of the number of pulses observed. Assume that initially the presence and absence of the pulse train are equally likely and that the pulse positions, if they are present, are a priori known at the receiver.

4.4, 4.5, 4.6, 4.7, 4.8 Compare quantitatively the performance of the three optimal detection schemes discussed in this chapter (ideal observer, Neyman-Pearson and minimum average loss or risk) in the detection situations indicated in the following five problems. In the ideal observer case, consider .001 as the maximum acceptable overall error probability. In the Neyman-Pearson case, require a false rest probability no greater than .001 with a false alarm probability of .01. In the minimum average loss case, assign a loss of 1000 to false rests and 100 to false alarms and require that L not exceed unity.

Then, in each case, determine the minimum signal power required to attain the specified performance level.

4.4 Detection of a single rectangular pulse of duration 6 in additive zero-mean Gaussian noise with rms equal to 5. The a priori probabilities $P^{(s)}$ and $P^{(n)}$ are $\frac{1}{3}$ and $\frac{2}{3}$, respectively.

4.5 Detection of a train of 10 pulses with the same signal and noise parameters as in Problem 4.4 and with a pulse repetition frequency of 100 per second. In this case $P^{(s)} = P^{(n)} = \frac{1}{2}$.

4.6 Detection of a sine wave in an additive zero-mean Gaussian noise with rms 2, where the sine wave frequency is not a priori known, but has an a priori PDF that is uniform between 55 and 58 Hz and zero outside that range. In this case, $P^{(s)} = \frac{3}{4}$, $P^{(n)} = \frac{1}{4}$.

4.7 Detection of a signal of the form

$$s(t) = A[u(t) - u(t - T)]\, e^{-at} \cos(\omega_0 t + \phi)$$

in additive zero-mean Gaussian noise with rms σ_n, where T, a, and ω_0 are a priori known, but ϕ is not. The a priori PDF for ϕ is uniform from 0 to 2π; $P^{(s)} = P^{(n)} = \frac{1}{2}$.

4.8 Assume the same conditions as in Problem 4.7 but phase PDF is changed to the following:

$$P(\phi) = \begin{cases} \dfrac{1}{\pi}; & 0 \leq \phi \leq \pi \\ 0; & \pi < \phi \leq 2\pi \end{cases}$$

The SNR is assumed to be very small. Formulate this problem and obtain an approximate solution. Compare your results with those of Problem 4.7

4.9 An observer is attempting to detect a rectangular pulse of known amplitude V and unknown duration τ in additive zero-mean Gaussian noise with rms σ_n; $P^{(s)} = P^{(n)} = \frac{1}{2}$. The a priori PDF for τ is

$$P(\tau) = \begin{cases} \dfrac{1}{(\tau_2 - \tau_1)}; & \tau_1 \leq \tau \leq \tau_2 \\ 0; & \tau < \tau_1; \quad \tau > \tau_2 \end{cases}$$

Determine the likelihood ratio for this detection process and describe the basic conceptual design of the detection system.

4.10 Repeat Problem 4.9 where all conditions are the same except that τ is now a priori known, but V is unknown. The a priori PDF for V is Gaussian, with a nonzero mean, i.e.,

$$P(V) = \frac{1}{\sqrt{2\pi\sigma_v^2}} e^{-(V-\langle V\rangle)^2/2\sigma_v^2}$$

4.11 In the detection situation described in Problem 4.9, suppose it is decided to search for the unknown parameter as discussed in Section 4.1.3.2. If only a single processing channel were available, how much expenditure of time would be required to attain an error probability as low as $P_{\varepsilon 1}$, where the decision threshold for signal-plus-noise is set at twice the noise power level. Do this problem without specifying numerical values of parameters. Then, choose parameter values, set $P_{\varepsilon 1}$ equal to .001, and obtain a numerical answer.

4.12 In Problem 4.11, suppose that three processing channels are available. Determine the minimum attainable error probability with the expenditure of time obtained as the answer to Problem 4.11. All other parameters are the same as in Problem 4.11.

4.13 A unit step function $u(t)$ is the signal input to an RC lowpass filter, whose impulse response is given by (1.17). Additive Gaussian noise also enters the filter. The noise has a spectrum that is flat from $-W$ to $+W$ where $W \gg 1/RC$, and has a rms value σ_n. How could you use the output in response to the unit step input to measure the filter time constant RC, using the measurement philosophy discussed in Section 4.2.1?

4.14 Discuss the measurement of the phase ϕ of the sine wave

$$\sin(\omega_0 T + \phi)$$

from the viewpoint of Section 4.2.2, where the noise is additive zero-mean Gaussian with a spectral shape given by Eq. (2.54)

4.15 Consider the signal

$$f(t) = A\left[u(t) - u\left(t - \frac{100\pi}{\omega_0}\right)\right] \cos(\omega_0 t + \phi)$$

in additive zero-mean Gaussian noise with rms σ_n and spectral shape given by Eq. (2.61). What is the minimum attainable rms error in the measurement of the amplitude A from the viewpoint of Section 4.2.3?

CHAPTER 6

6.1 Explain the significance of the assumption (6.1) in analysis of radio signals. If this assumption were dropped, how would each of the equations (6.9) through (6.18) be changed? Would the analysis in Section 6.2 still be useful?

6.2 How would the abandonment of the assumption (6.21) influence the equations

(6.22) through (6.28) and the overall utility of the analysis leading from (6.22) through (6.28)?

6.3 Using the definition of bandwidth given by (6.25), calculate approximately the bandwidths of the following types of bandpass filters.

(a) "Butterworth":

$$|\hat{F}(\omega)| = \frac{1}{1 + \left(\frac{\omega}{W}\right)^{2n}}; \qquad n = 1 \quad \text{and} \quad n = 2$$

(b) "Chebyshev":

$$|\hat{F}(\omega)|^2 = \frac{1}{1 + \epsilon^2 C_n^2\left(\frac{\omega}{W}\right)}; \qquad n = 0, 1$$

where

$$C_n\left(\frac{\omega}{W}\right) = \cos\left[n \cos^{-1}\left(\frac{\omega}{W}\right)\right]$$

(c) "Double tuned Gaussian":

$$|F(\omega)| = |e^{-(\omega-\omega_1)^2/2\sigma^2} + e^{-(\omega+\omega_1)^2/2\sigma^2}|$$

How do your calculated bandwidths compare with "half-power" point bandwidths?

6.4 From the results obtained in Problem 6.3, discuss the relative noise-reduction effectiveness of these three types of filters in terms of output SNR as a function of input SNR.

6.5 Compare the output SNR of an ideal bandpass filter (uniform response over the passband, zero outside the passband; see Problem 1.8) with that of a coherent matched filter, where the input signal is a rectangular RF pulse whose radio frequency is at the center of the filter's passband. Assume the bandwidth of the ideal filter to be equal to the reciprocal of the input pulse duration.

6.6 Consider a signal entering a radio receiver, consisting of two rectangular RF pulses each of 10 ms duration, each with a peak power of 1 μW and with centers separated by 10 ms. The disturbing noise is additive bandlimited white Gaussian with a bandwidth of 10 MHz and with a power density of 1 μW per Hz. The signal is fed into a linear RF (or IF, assuming that the signal is heterodyned first) filter of the *RLC* type, designed to provide the highest output SNR possible with that type of filter, consistent with the requirement that the two pulses be clearly resolved at the filter output. Roughly, how high an output SNR can be attained?

6.7 Do Problem 6.6 where all parameters and conditions are unchanged except that the incoming pulses are now of Gaussian rather than rectangular shape. The pulse duration is now defined as twice the "standard deviation" σ of the pulse, determined as if the pulse shape were a probability density function (see Table 2.1). By this definition, the duration of each of the two pulses, again, is 10 ms.

6.8 Consider a train of rectangular RF pulses in additive zero-mean bandlimited white Gaussian noise. At the receiver, the pulses, if they are present, are *believed* to be uniformly spaced and to have a pulse repetition frequency (PRF) of 500 per second. Due to certain propagation effects, however, there is a sinusoidal oscillation in the PRF, with the characteristics shown in the diagram.

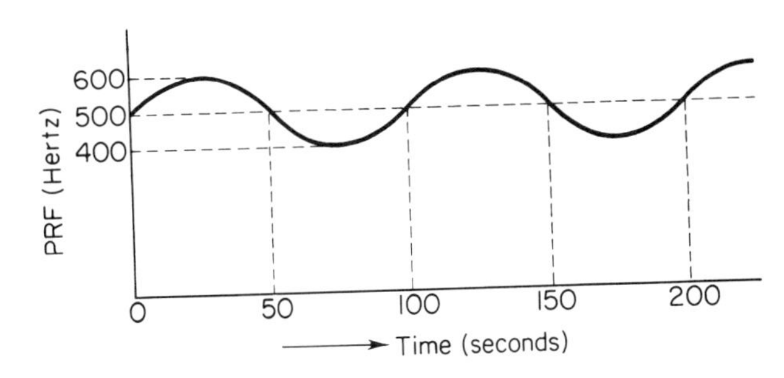

The pulse amplitude and duration are 10 volts and 10 microseconds, respectively, and the rms noise level is 15 volts. The noise bandwidth greatly exceeds the signal bandwidth. The pulse duration, the starting time of the first pulse, and the radio frequency are a priori known at the receiver, but the RF phase is completely unknown. Find the attainable SNR at the output of a matched filter or crosscorrelator operating for 10,000 seconds, and compare it with that attainable at the output of an autocorrelator operating for the same period, given that the oscillation in the PRF is *not* a priori known at the receiver.

6.9 In Problem 6.8, leave all conditions and parameter values unchanged except for a priori knowledge of radio frequency. The true radio frequency is 100 MHz, but it is a priori known at the receiver only that the radio frequency is somewhere between 98 and 102 MHz. Superpose this condition on the others and repeat Problem 6.8.

6.10 In a radio communication system, it is required that the SNR at the input to the second detector in the receiver be at least 10 dB. The system is designed to this specification in the absence of fading. It is then found that it must operate in a fading environment where the basic system parameters (e.g., transmitter power, receiver noise, etc.) cannot be changed. The depth of fading is such that a 10 dB margin or "safety factor" in the SNR is required in the system design. Discuss the ways of accomplishing this with the methods discussed in Section 6.10.3. Assume that RF phase is not a priori known at the receiver.

6.11 Generalize the analysis in Section 6.10.2 to allow for the possibility that the signals in the different diversity branches are mutually correlated, and show how your result modifies Eqs. (6.99), (6.100), (6.101), (6.102) and (6.103).

CHAPTER 7

7.1 A receiving system is designed to detect radio transmissions whose existence is a priori considered as 50% probable. Its antenna scans horizontally at 10 revolutions per minute. Its antenna aperture, 2m wide, is rectangular and uniformly illuminated, resulting in a sinc function pattern (see Section 1.3.2.1), which you can approximate as flat over a certain range.

Roughly how much transmitter power is required to detect this transmission *within a single antenna rotation?* Successful detection is defined as corresponding to an error probability of 10^{-3} with a detection threshold based on ideal observer theory. Assume a priori knowledge of all signal parameters except RF phase (which is uniformly distributed from 0 to 2π) at the receiver.

7.2 In Problem 7.1, suppose that the receiving antenna is omnidirectional. The transmission to be detected is an RF pulse train. The pulse duration is 100 μs, peak pulse power is 1 kW and the pulse repetition frequency is 4000 per second. Other parameter values, conditions, and criteria are identical with those of Problem 7.1. What is the probability of detection of the transmissions?

7.3 Do Problem 7.1 in the case where phase is a priori known at the receiver, other conditions being unchanged.

7.4 Do Problem 7.2 in the case where the phase is a priori known at the receiver, other conditions being unchanged.

7.5 Discuss the basic conceptual design of a system for detecting the presence of radio signals in Gaussian noise, where the optimization is based on decision—theoretic arguments like those in Section 4.1, but where the false *alarm* probability is minimized with fixed false *rest* probability.

7.6 Discuss as quantitatively as you can the validity of the statement in the last sentence in Section 7.1.3, "Whether one uses U itself or its square makes little difference in the detection process." If you can't demonstrate this in a general way, try convincing yourself with a few examples.

7.7 Repeat the development in Section 7.1.2 using an integrator with a finite time constant T_c, i.e., a factor e^{-t'/T_c} appears in the integrals beginning with that of Eq. (7.11). Discuss the effect of this factor on the results.

7.8 Do Problem 7.7 but substitute the words "Section 7.1.3" for "Section 7.1.2."

7.9 Consider a "sequential" binary detection system (see Problem 7.2). Minimization of the overall error probability is used as a basis for optimization of the decision rule, where the two possible signals are initially assumed to be equally probable. After each reading the detection threshold is changed in accordance with a new computation of a priori probabilities based on the readings that have been taken thus far. Discuss the basic design of such a system where the two possible signals are RF tones whose radio frequencies differ by an amount $\Delta\omega$, and whose phase is a priori known at the receiver. Additive bandlimited white Gaussian noise is the disturbing agent. How would you determine the decrease in error probability as the number of readings is increased?

7.10 Consider the problem of determining which of two possible signals $s_1(t)$ or $s_2(t)$ is present in a background of additive white Gaussian noise with zero mean and rms σ_n. The two signals are:

$$s_1(t) = [u(t) - u(t - T)] \cos^2(\omega_0 t)$$
$$s_2(t) = [u(t) - u(t - T)] \sin^2(\omega_0 t)$$

where ω_o and T are a priori known and $T \gg 2\pi/\omega_o$. The two signals are equally likely. Determine the error probabilities.

7.11 In Problem 7.10, suppose that T is a priori known but ω_o is *a priori* unknown and has the PDF

$$P(\omega_0) = \frac{1}{\sqrt{2\pi\sigma_\omega^2}} e^{-(\omega_0 - \bar{\omega}_0)^2/2\sigma_\omega^2}$$

Derive the optimum processing operations (from the viewpoint of minimization of error probability).

7.12 Consider three signals of the form

$$s_{\substack{1\\2\\3}}(t) = A \cos(\omega_0 t + \phi_{\substack{1\\2\\3}}) \cos(\omega_{\substack{1\\2\\3}} t)$$

where $\omega_0 \gg \omega_{\substack{1\\2\\3}}$ and where ω_1, ω_2, and ω_3 are all different.

The interfering noise is additive bandlimited white Gaussian with bandwidth much larger than signal bandwidths; zero-mean and rms value are equal to σ_n. The phases ϕ_1, ϕ_2, and ϕ_3 are all equal and the parameters A, ϕ_1, ϕ_2, ϕ_3, ω_1, ω_2, and ω_3 are all a priori known. Using the detection philosophy indicated in Section 7.3, determine the minimum value of A required for a $X\%$ probability of successfully determining which of the three signals is present, if they are (a priori) equally likely.

7.13 Do Problem 7.12 in the case where the phases ϕ_1, ϕ_2, and ϕ_3 are a priori unknown but all have PDF's that are uniform from 0 to 2π.

7.14 Suppose you were to attack the detection problem indicated in Problem 7.12 with a single matched filter whose reference frequency is switched among ω_1, ω_2 and ω_3. If you want an error probability $P_{\epsilon 0}$ and if A is equal to A_0, determine the time required to accomplish the detection process.

CHAPTER 8

8.1 Show that the linear filtering process that optimizes amplitude measurement (with phase a priori known; Section 8.2.1) of a sinusoid in additive Gaussian noise from the MLE viewpoint would also turn out to be optimum from the "minimum variance" viewpoint (i.e., the linear filter that minimizes the variance of the output). Note: For a rigorous proof you may be forced to consult other literature. Try to do the problem heuristically, even if the best solution you can produce is one based on comparison between the output variance with the optimum filter indicated in Section 8.2.1 and the output variance with other linear filters.

8.2 A signal entering a radio receiver is as follows:

$$s(t) = [u(t) - u(t - 10^{-2})]\ Re\ (e^{j2\pi(10^7)t}[20 + (5 + j4)e^{j2\pi(10^2)t}])$$

It is desired to measure the (real) amplitude of the envelope of this signal as a function of time in a background of zero-mean additive bandlimited white Gaussian noise with rms = 1, and bandwidth large compared to that of $s(t)$. Two ways of doing this are suggested as follows:

(a) Drive the signal through an *RLC* bandpass filter of appropriate bandwidth, followed by straightforward amplitude measurement.

(b) Divide the signal into time slots of appropriate duration and perform optimal amplitude measurement (as discussed in Section 8.2) on each time slot.

Compare (quantitatively) the performance of these two methods.

8.3 Do Problem 8.2 replacing amplitude measurement with phase measurement.

8.4 Repeat Problem 8.2 replacing $s(t)$ with

$$s(t) = [u(t) - u(t - 3(10^{-2})]$$
$$\times Re\left[\left(6 \cos\left(2\pi(10^2)t + \frac{\pi}{4}\right) - 3 \sin\left(8\pi(10^2)t - \frac{\pi}{4}\right)\right)e^{j2\pi(10^8)t}\right]$$

and specifying that both amplitude and phase are to be measured.

8.5 In Problem 8.2, the probability that the measured amplitude will be within 3% of the true amplitude must be at least 99% in order to meet a set of accuracy specifications. Find the minimum input SNR required to meet that specification with techniques (a) and (b) in Problem 8.2. What is the difference in this threshold SNR (in dB) with these two methods of measurement?

8.6 Consider the RF signal $\cos(2\pi(10^7)t + \frac{\pi}{4})$ modulated by a dc signal voltage $\frac{1}{2}$ plus a square wave of amplitude $\frac{1}{2}$ and fundamental period 6s, as shown in the diagram.

You are required to measure the amplitude of this waveform as a function of time in additive zero-mean Gaussian noise with rms value $\frac{1}{10}$ volt, using the measurement philosophy of Section 8.2. How many cycles of the square wave must elapse before you can be 99% certain that the measured amplitude is within 2% of its true value? Assume an a priori knowledge of the radio frequency and the signal wave shape.

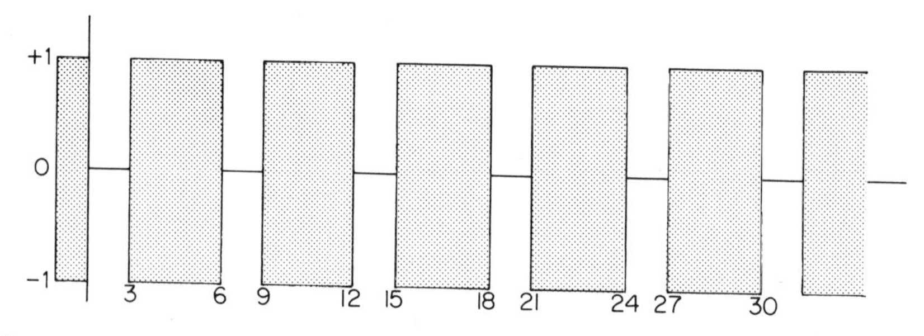

8.7 Do Problem 8.6 where phase replaces amplitude as the parameter to be measured.

8.8 Consider the signal-plus-noise waveform of Problem 8.6. Determine the expected accuracy in measuring the first three Fourier coefficients of the signal envelope.

8.9 Do Problem 8.8 substituting the words "x and y values at the sample points corresponding to the Nyquist sampling rate" for the words "first three Fourier coefficients."

8.10 Consider a signal 10 cos $(\omega_0 t + \phi)$ whose radio frequency is increasing linearly as follows:

$$(\omega_0 t) = 10^7[1 + .001t]$$

Use the ideas of Section 8.7 to envision a system to measure the radio frequency. Determine (roughly) as a function of time how accurately this measurement can be made in the presence of additive zero-mean Gaussian noise with rms value 2 and a bandwidth far in excess of the signal bandwidth.

CHAPTER 9

9.1 Consider a pulsed search radar operating at 10 GHz whose antenna scans horizontally in a circular pattern at a rate of 10 rotations per minute. Transmitted peak power is 1 kW and the transmitted pulses are rectangular and of duration 1 ms. The antenna aperture is rectangular and uniformly illuminated in the horizontal direction, with a horizontal aperture dimension of 5 m. Peak antenna gain is 30 dB over an isotropic radar for either transmission or reception. Miscellaneous additional system losses total to 6 dB. Receiver noise power density (additive Gaussian) is 1 micromicrowatt per Hz. Find the maximum range for detection of a target with a 0.3 m^2 radar cross-section within a period of 3 minutes. Consider the target as roughly stationary during the 3-minute period and consider its return signal phase as a priori known at the receiver. Successful detection is defined as 98% detection probability. The target is definitely known a priori to be stationary, to be somewhere within the region over which the beam scans, and to be at a range somewhere between those ranges for which a 1 m^2 target would produce received SNR's of 0 dB and 30 dB.

9.2 In Problem 9.1, suppose that, at a range of 30 km, the target begins receding from the radar at a speed of 400 km per hour. All parameter values are the same as in Problem 9.1 except that the transmitter power is now unspecified. Also, in the present problem, assume that a priori knowledge at the receiver is limited to the fact that the target is receding from the radar at a speed somewhere between 200 and 600 km per hour from an initial range of between 30 and 50 km from the radar. Also assume that return signal phase is unknown. What peak transmitter power is required to provide a 98% detection probability?

9.3 In Problem 9.1, suppose the requirement were to measure the range to within 2%. All other parameters and conditions remain the same (except that pulse duration may be varied subject to an invariant signal energy condition). Find the maximum range at which such a range measurement can be made, based on the measurement philosophy of Section 9.3.1.

9.4 In Problem 9.2, all parameters and conditions remain the same but an additional requirement for measurement of the range-rate (known to be constant) to within 3% is imposed. Determine the transmitter power required to do this.

9.5 Do Problem 9.2, changing the pattern of target motion to that shown in the diagram below while leaving the other parameters and conditions unchanged.

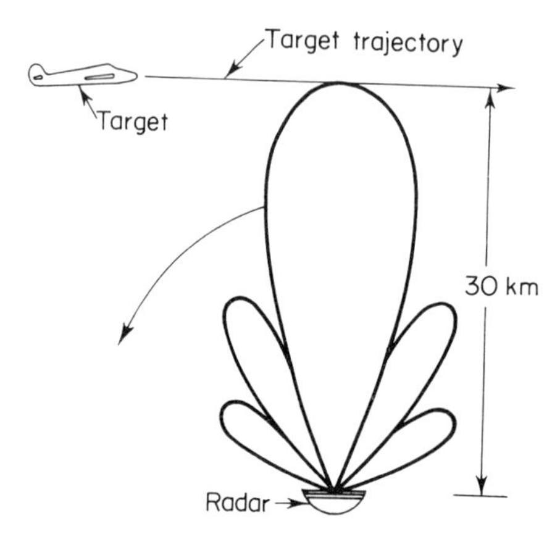

Assume a target velocity and assign values for the size of range, range-rate, and angle regions in which the target is a priori known to be situated when it begins this motion. Determine the required transmitter power as a function of these parameters.

9.6 Combine the situations depicted in Problems 9.1, and 9.2, i.e., consider that two targets are present, one as in Problem 9.1, the other as in Problem 9.2, and that the task is to detect *both* of them. How (quantitatively) would this affect the answer to Problem 9.1?

9.7 How would the situation of Problem 9.6 affect the answer to Problem 9.2?

9.8 Perform the calculation indicated at the end of Section 9.3.4.1 as an "exercise for the reader."

9.9 Carry out the analysis in the early part of Section 9.3.4.2 on angle measurement through amplitude with a scanning antenna (Eqs. (9.42) through (9.51)), using a Gaussian pattern shape in lieu of the sinc function used in the text. What fundamental differences exist between results obtained with these two different pattern shapes?

9.10 Carry out the analysis in the latter part of Section 9.3.4.2 on the differential amplitude method of angle measurement (Eqs. (9.52) through (9.54)) using a sinc (θ/θ_o) pattern shape in lieu of a Gaussian shape. What fundamental differences exist between the results obtained with those two different pattern shapes?

CHAPTER 10

10.1 Discuss quantitatively the effect on a radar's range resolution capabilities (from the viewpoint of ambiguity functions) of a low frequency sinusoidal modulation on the transmitted radar pulse of duration τ_p. The signal representation is as follows:

$$s(t) = [u(t + \tau_{p/2}) - u(t - \tau_{p/2})] \sin(\omega_o t + \phi_o) \sin(\omega_1 t + \phi_1)$$

where ω_o, the radio frequency, is much greater than ω_1, the modulating frequency (see Problem 1.11).

10.2 Discuss quantitatively (again, as in Problem 10.1, in terms of ambiguity functions) the effect of a slow linear variation of the radar frequency of the transmitted pulse on the range-rate resolution capabilities of a CW Doppler radar.

10.3 Consider a pulsed radar system. Discuss quantitatively the effect of a modulation like that of Problem 10.1 on the possibilities of attaining both high range resolution and high range-rate resolution.

10.4 Suppose that both the modulation effect introduced in Problem 10.1 and that of linear frequency variation as in Problem 10.2 are present in a pulsed Doppler radar. How does this influence the radar's ability to attain both high range resolution and high range-rate resolution?

10.5 Do Problem 1.15 in the context of monostatic radar.

10.6 Generalize the development in Section 10.1.4 (text and Eqs. (10.25a) through (10.27b)) to include the possibility of a specular component superposed on the random component of the fading. Do this in general at first, then specialize to the limiting cases where the ratio of specular to random component is (a) small and (b) large.

10.7 Do Problem 9.1 for a bistatic radar, where the geometry is as indicated in the diagram.

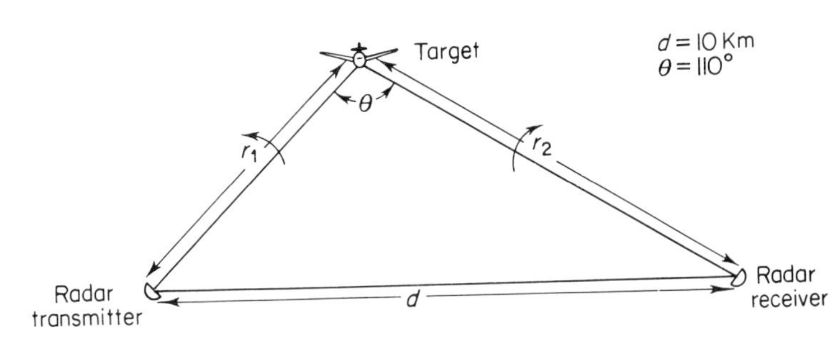

Applicable parameter values, such as transmitter power, frequency, noise parameters, peak gains of transmitting, receiving antennas (both beams pointed directly at the target), etc., are the same as those of the monostatic radar of Problem 9.1. Other conditions and detection criteria are also identical with those of Problem 9.1. The oblique cross-sections of the target for transmission and reception are 0.5 m^2 and 0.3 m^2, respectively. Define "range" as $\sqrt{r_1 r_2}$.

10.8 Do Problem 9.2 for the bistatic radar configuration of Problem 10.7 where all parameters and conditions are the same as in Problem 10.7 or Problem 9.2 where applicable, except that r_1 is now specified as 3 km within the time interval during which the detection process is taking place. The target is moving horizontally at 500 km per hour during this period, and neither the antenna gains at target position nor the oblique cross-sections change significantly during this time interval.

10.9 Extend the discussion of radar range and range-rate resolution in Section 10.1. 2.1 to bistatic radar.

CHAPTER 11

11.1 Discuss in a general way the application of Eqs. (11.1) and (11.2) to the following classes of radio communication systems:

(a) Voice transmission over a HF channel.
(b) TV transmission over a UHF channel.
(c) Transmission of analog waveforms with bandwidth 1 kHz over a VLF or LF channel.

11.2 There is a variation of an AM communication system in which the carrier is removed from the signal and only the modulation sidebands are transmitted ("double-sideband suppressed carrier"—DSB-SC). How (in general "block-diagram" terms) is this accomplished and what are the major advantages of this scheme over straightforward AM?

11.3 Another variation of AM is "single-sideband" (SSB), in which only the sidebands above or below the carrier frequency are transmitted. How (again in "block-diagram" terms) is this accomplished and what are its advantages over straightforward AM?

11.4 In a certain radio communication system constrained to use conventional AM (i.e., double-sideband with carrier) it is necessary to accommodate 32 voice messages each of 5 m duration and each of bandwidth 4 kHz. A SNR of at least 10 dB at the receiver input is specified for minimally reliable voice transmission. The conditions and parameters (e.g., transmitted power, receiver noise level, etc.) are such that 7 dB is the maximum achievable SNR if each voice message has its bandwidth allocation of 4 kHz and is transmitted in real time. Suggest 3 possible arrangements (i.e., sets of allocations of time and bandwidth) to accommodate this communication traffic without increasing the power in the transmitted signal or reducing the noise power density.

11.5 Do Problem 11.4 given that the system employs single-sideband transmission, all other conditions remaining the same.

11.6 Consider a PM communication system. Suppose that Eq. (11.29) was modified to read as follows:

$$s_I(t) = \cos(\omega_I t) + \alpha \sin(\omega_I t)$$

where α is a real constant. How would this influence the analysis that follows (11.29)?

11.7 Again consider a PM communication system. Eq. (11.29) is modified as follows:

$$s_I(t) = \cos(\omega_I t) + \alpha \cos(2\omega_I t)$$

where α is a real positive constant far below unity. Carry through an approximate analysis to replace that following (11.29).

11.8 Extend the analysis in Section 11.24, proceeding from Eq. (11.44) to Eq. (11.52) and replacing the ideal filter by an RC lowpass filter.

11.9 In the analysis of PM in Section 11.2.4, the effect of the limiter in altering the spectrum of the noise traversing it has been neglected. Consider the signal-noise passage properties of a limiter and describe how the analysis should be modified to include this effect. [Note: This is a difficult problem if it is done quantitatively and will require consultation with other literature. Before going to the literature, make an attempt to reason qualitatively about the action of the limiter on signal and noise spectra.]

11.10 Do Problem 11.6 for FM.

11.11 Do Problem 11.7 for FM.

CHAPTER 12

12.1 Consider a digital radio communication system constrained to use on-off keying. It is desired to transmit a 10-megabit message within a 10-minute period, with a per-bit error probability $\lesssim 10^{-4}$. The receiver noise is additive zero-mean Gaussian, with a total noise power of 100 $\mu\mu$W. The loss between transmitter and receiver is 40 dB. How much transmitter power (in watts) is required in each of the "on" bits to achieve the specified requirements if, at the receiver, the phase is a priori (a) known, and (b) unknown?

12.2 Differentially coherent phase shift keying (DPSK) was mentioned in Section 12.1.3. Derive the expressions for the error probabilities consulting the cited references for aid or confirmation after you have tried to solve the problem yourself.

12.3 Determine the SNR required for an error probability of 10^{-4} in a binary PSK system where the crosscorrelation function χ_{12} can have an arbitrary value. Plot the SNR against χ_{12}.

12.4 Do Problem 12.3 for a binary coherent FSK system and compare your results with those of Problem 12.3.

12.5 In a digital communication system involving an airborne transmitter and a ground-based receiver, the line-of-sight velocity of the transmitting station is a priori known to be within $\pm$ 20% of 300 miles per hour approaching. The radio frequency is 100 MHz. How (quantitatively) does this influence the performance of the communication system, which is assumed to be constrained to use noncoherent binary FSK?

12.6 Given a binary PSK system, frequency is known to great precision but phase is a priori known only to within $\pm$ 10°. How (quantitatively) does this compromise the system's performance?

12.7 Do Problem 12.5 for M'ary noncoherent FSK, where M = 4.

12.8 Do Problem 12.5 for M'ary PSK, where M = 4.

12.9 Do Problem 12.6 for M'ary PSK, where M = 3.

12.10 A 1-megabit message is transmitted on a digital communication link (additive zero-mean Gaussian noise) at a rate of 1 kilobit per second. The modem used is binary coherent FSK and the probability that the message is error-free is .98. Suppose that the message is repeated three times. Find the probability that at least one of the repetitions is error-free. Do this again for four repetitions.

12.11 In Problem 12.10, suppose that system limitations are such that message repetition is *not* possible. If it *is* possible to increase the transmitted power, what increase is required to achieve the same error probability reduction that was achieved by repeating the message three times? Four times?

12.12 Describe briefly how you might extend the discussion in Section 11.2.3 to encompass PAM systems.

12.13 Discuss briefly how the ideas of bandwidth-SNR tradeoff as discussed in Section 11.2.4 apply to pulse position modulation and pulse duration modulation systems.

APPENDIXES

I

IMPEDANCE MATCHING†

Consider a voltage source feeding a load impedance $Z_L = R_L + jX_L$. The source voltage is V_s and the source impedance is $Z_s = R_s + jX_x$. The load impedance can be controlled by the designer. He wants to set it at a value such that the maximum possible power will be transferred from the source to the load.

The average power dissipated by the load is given by

$$P_L = \text{Re}\,(V_L I_L^*) \tag{I.1}$$

where V_L is the voltage drop across the load and I_L is the current through the load. The complex conjugate of I_L is equal to $V_s^*/(Z_s^* + Z_L^*)$ and the voltage V_L is equal to $[V_s/(Z_s + Z_L)]Z_L$. Therefore, from (I.1)

$$P_L = \frac{|V_s|^2}{|Z_s + Z_L|^2}\,\text{Re}\,(Z_L) = \frac{|V_s|^2 R_L}{(R_s + R_L)^2 + (X_s + X_L)^2} \tag{I.2}$$

P_L is maximized by setting to zero its partial derivatives with respect to R_L and X_L. The result is that power transfer is maximized when

$$R_s = R_L, \qquad X_s = -X_L \tag{I.3}$$

or equivalently

$$Z_s = Z_s^* \tag{I.4}$$

---

† This appendix on an extremely basic and well-known topic is included for the reader's convenience in covering certain parts of the book where impedance matching ideas and specifically Eqs. (I.4) and (I.5) are used in discussions.

Under the conditions (I.4) the power dissipated across the load, or equivalently the maximum power that can be delivered to the load, is

$$P_L = \frac{|V_s|^2}{4R_s} \tag{I.5}$$

II

MATCHED FILTERING OF PULSES WITH UNKNOWN POSITION AND DURATION

Consider the two pulses #1 and #2, one of which is the signal pulse and the other the reference pulse. Pulse #1 is of duration T_1 and centered at time t_1. Pulse #2, shorter than #1, has duration T_2, and is centered at t_2. We will fix the position of #1 and allow #2 to vary its position relative to #1; i.e., fix t_1 and vary t_2. We are interested in the total overlap interval between the two pulses, which will be designated as L.

Five different classes of relative pulse positioning are possible, as follows:

$$
\begin{aligned}
&(1)\quad t_2 + \frac{T_2}{2} < t_1 - \frac{T_1}{2}; \qquad \text{no overlap,}\quad L = 0 \\
&(2)\quad t_1 - \frac{T_1}{2} \le t_2 + \frac{T_2}{2} \le t_1 + \frac{T_1}{2}, \qquad t_2 - \frac{T_2}{2} < t_1 - \frac{T_1}{2}; \\
&\qquad\qquad \text{partial overlap,}\quad L = (t_2 - t_1) + \frac{(T_2 + T_1)}{2} \\
&(3)\quad t_1 - \frac{T_1}{2} \le t_2 + \frac{T_2}{2} \le t_1 + \frac{T_1}{2}, \qquad t_2 - \frac{T_2}{2} \ge t_1 - \frac{T_1}{2}; \\
&\qquad\qquad \text{complete overlap,}\quad L = T_2 \\
&(4)\quad t_1 - \frac{T_1}{2} \le t_2 - \frac{T_2}{2} \le t_1 + \frac{T_1}{2}, \qquad t_2 + \frac{T_2}{2} > t_1 + \frac{T_1}{2}; \\
&\qquad\qquad \text{partial overlap,}\quad L = -(t_2 - t_1) + \frac{(T_2 + T_1)}{2} \\
&(5)\quad t_2 - \frac{T_2}{2} > t_1 + \frac{T_1}{2}; \qquad \text{no overlap,}\quad L = 0
\end{aligned}
\tag{II.1}
$$

or, in terms of the parameter $(t_2 - t_1)$,

$$
\begin{aligned}
&L = 0, && (t_2 - t_1) < -\frac{(T_1 + T_2)}{2} \\
&L = \frac{(T_1 + T_2)}{2} + (t_2 - t_1), && -\frac{(T_1 + T_2)}{2} \leq (t_2 - t_1) < -\frac{(T_1 - T_2)}{2} \\
&L = T_2, && -\frac{(T_1 - T_2)}{2} \leq (t_2 - t_1) \leq \frac{(T_1 - T_2)}{2} \\
&L = \frac{(T_1 + T_2)}{2} - (t_2 - t_1), && \frac{(T_1 - T_2)}{2} < (t_2 - t_1) \leq \frac{(T_1 + T_2)}{2} \\
&L = 0, && (t_2 - t_1) > \frac{(T_1 + T_2)}{2}
\end{aligned}
\tag{II.2}
$$

The overlap time L is a trapezoidal function of $(t_2 - t_1)$, the separation between pulse centers.

From (II.2) we can draw the following conclusions: (Let $\bar{t}_r$ and $\bar{t}_p$ be reference and input signal pulse centers, respectively, and let T_r and T_p be reference and input signal pulse lengths, respectively. Let the noise integration time be T_r in all cases).

Case (a): Reference and signal pulses are of equal length, $T_r = T_p$ (illustrated in Figure 6-3).

The noise integration time is T_p in this case. Then the filter output SNR is a *tent function* (see Figure 6-3). The SNR is proportional to L_0^2/T_p. Then

$$
\begin{aligned}
&\rho_o = 0, && (\bar{t}_r - \bar{t}_p) < -T_p \\
&\rho_o = \kappa\rho_i B_n[T_p + (\bar{t}_r - \bar{t}_p)], && -T_s \leq (\bar{t}_r - \bar{t}_p) \leq 0 \\
&\rho_o = \kappa\rho_i B_n[T_p], && \bar{t}_r - \bar{t}_p \\
&\rho_o = \kappa\rho_i B_n[T_p - (\bar{t}_r - \bar{t}_p)], && 0 < (\bar{t}_r - \bar{t}_p) \leq T_p \\
&\rho_o = 0, && (\bar{t}_r - \bar{t}_p) > T_p
\end{aligned}
\tag{II.3}
$$

where $\kappa = 2$ in the coherent filter case, and $\kappa = 1$ for a noncoherent filter.

Case (b): Reference pulse is shorter than signal pulse, $T_r = cT_p$, where $c < 1$ (illustrated on lower part of Figure 6-4).

Then the noise integration time is cT_p. The SNR is proportional to

$$\rho_o = 0, \qquad (t_r - t_p) < -\left(\frac{1 + c}{2}\right)T_p \frac{L^2}{cT_p}$$

Then

$$
\begin{aligned}
&\rho_o = \kappa\rho_i B_n T_p \left\{\frac{[(1 + c)/2]^2}{c}\right\}\left\{1 + \frac{(\bar{t}_r - \bar{t}_p)}{[(1 + c)/2]T_p}\right\}, \\
&\qquad\qquad -\left(\frac{1 + c}{2}\right)T_p \leq (\bar{t}_r - \bar{t}_p) \leq -\left(\frac{1 - c}{2}\right)T_p \\
&\rho_o = \kappa\rho_i B_n T_p[c], \qquad -\left(\frac{1 - c}{2}\right)T_p < (\bar{t}_r - \bar{t}_p) \leq \left(\frac{1 - c}{2}\right)T_p
\end{aligned}
\tag{II.4}
$$

$$\rho_o = \kappa\rho_i B_n T_p \left\{\frac{[(1+c)/2]^2}{c}\right\}\left\{1 - \frac{(\bar{t}_r - \bar{t}_p)}{[(1+c)/2]T_p}\right\},$$
$$\left(\frac{1-c}{2}\right)T_p < (\bar{t}_r - \bar{t}_p) \le \left(\frac{1+c}{2}\right)T_p$$

$$\rho_o = 0, \qquad (\bar{t}_r - \bar{t}_p) > \left(\frac{1+c}{2}\right)T_p$$

Case (*c*): Reference pulse is longer than signal pulse, $T_p = cT_r$, where $c < 1$ (illustrated on upper part of Figure 6-4).

Then noise integration time is (T_p/c). The SNR is proportional to cL^2/T_p. Then

$$\rho_o = 0, \qquad (\bar{t}_r - \bar{t}_p) < -\left(\frac{1+c}{2c}\right)T_p$$

$$\rho_o = \kappa\rho_i B_n T_p\left[\left(\frac{1+c}{2}\right)^2 c\right]\left\{1 - \frac{(\bar{t}_r - \bar{t}_p)}{[(1+c)/2c]T_p}\right\},$$
$$-\left(\frac{1+c}{2c}\right)T_p < (\bar{t}_r - \bar{t}_p) \le -\left(\frac{1-c}{2c}\right)T_p$$

$$\rho_o = \kappa\rho_i B_n T_p[c], \qquad -\left(\frac{1-c}{2c}\right)T_p < (\bar{t}_r - \bar{t}_p) \le \left(\frac{1-c}{2c}\right)T_p \tag{II.5}$$

$$\rho_o = \kappa\rho_i B_n T_p\left[\left(\frac{1+c}{2}\right)^2 c\right]\left\{1 + \frac{(\bar{t}_r - \bar{t}_p)}{[(1+c)/2c]T_p}\right\},$$
$$\left(\frac{1-c}{2c}\right)T_p < (\bar{t}_r - \bar{t}_p) \le \left(\frac{1+c}{2c}\right)T_p$$

$$\rho_o = 0, \qquad (\bar{t}_r - \bar{t}_p) > \left(\frac{1+c}{2c}\right)T_p$$

CALCULATION OF OUTPUT SNR OF AN AUTOCORRELATOR

If the input is a radio signal-plus-noise, we can express the output in terms of complex envelopes, as follows:

$$\begin{aligned} v_o(\tau) &= \int_0^T dt' \operatorname{Re}\{[\hat{s}_i(t') + \hat{n}_i(t')]\, e^{j\omega_c t'}\} \operatorname{Re}\{[\hat{s}_i^*(t' + \tau) + \hat{n}_i^*(t' + \tau)]\, e^{j\omega_c(t'+\tau)}\} \\ &= \tfrac{1}{2} \operatorname{Re}\{\hat{v}_o(\tau)\, e^{j\omega_c \tau}\} \end{aligned}$$

where

$$\hat{v}_o(\tau) = \{[\hat{v}_{ss}(\tau) + \hat{v}_{sn}(\tau) + \hat{v}_{nn}(\tau)]\, e^{j\omega_c \tau}\} \tag{III.1}$$

and where

$$\hat{v}_{sn}(\tau) = \int_0^T dt'\, \hat{s}_i(t')\hat{s}_i^*(t' + \tau)$$

$$\hat{v}_{sn}(\tau) = \int_0^T dt'\, [\hat{s}_i(t')\hat{n}_i^*(t' + \tau) + \hat{n}_i(t')\hat{s}_i^*(t' + \tau)]$$

$$\hat{v}_{nn}(\tau) = \int_0^T dt'\, \hat{n}_i(t')\hat{n}_i^*(t - \tau)$$

The autocorrelator signal output, in terms of the variable τ, is thus a narrow-band signal centered at frequency ω_c.

If the radio signal is periodic, e.g., a sinusoid of frequency ω_c, then the autocorrelator output is also periodic with the same period; in fact, each harmonic in the signal would appear as an output of the autocorrelator. For this reason, autocorrelators can be used to find hidden periodicities in noisy signals. This can be done by varying the delay time τ until a value of τ is found for which the output is large.

To determine the output SNR of an autocorrelator, we first calculate the ensemble mean of the output, from (III.I),

$$\begin{aligned}\langle \hat{v}_o(\tau)\rangle &= \hat{v}_{ss}(\tau) + \int_0^T dt'\, \hat{s}_i(t')\langle \hat{n}_i^*(t'+\tau)\rangle + \int_0^T dt'\, \langle \hat{n}_i(t')\rangle \hat{s}_i^*(t'+\tau) \\ &\quad + \int_0^T dt'\, \langle \hat{n}_i(t')\hat{n}_i^*(t'+\tau)\rangle \\ &= \hat{v}_{ss}(\tau) + \int_0^T dt'\, \langle \hat{n}_i(t')\hat{n}_i^*(t'+\tau)\rangle \\ &= \hat{v}_{ss}(\tau) + \langle \hat{v}_{nn}(\tau)\rangle \end{aligned} \tag{III.2}$$

In the case where the noise output voltage or current has a finite mean, the noise output power can be defined as the average power minus the signal power.†

Using this definition of noise, the output SNR is

$$\begin{aligned}\rho_o &= \frac{|\hat{v}_{ss}(\tau)|^2}{\langle |\hat{v}_o(\tau)|^2 - |\hat{v}_{ss}(\tau)|^2\rangle} \\ &= \frac{|\hat{v}_{ss}(\tau)|^2}{\{\langle |\hat{v}_{sn}|^2\rangle + 2\mathrm{Re}\,[\hat{v}_{ss}\langle \hat{v}_{sn}\rangle + \hat{v}_{ss}\langle \hat{v}_{nn}^*\rangle + \langle \hat{v}_{sn}\hat{v}_{nn}^*\rangle] + \langle |\hat{v}_{nn}|^2\rangle} \end{aligned} \tag{III.3}$$

The noise terms involve the following (possibly) nonzero averages. (Note that $\langle \hat{v}_{sn}\rangle = 0$ if signal and noise are uncorrelated.)

$$\langle \hat{v}_{nn}\rangle = \int_0^T dt'\, \langle \hat{n}_i(t')\hat{n}_i^*(t'+\tau)\rangle = \hat{\rho}_n(\tau)T\sigma_n^2 \tag{III.4}$$

$$\begin{aligned}\langle |\hat{v}_{sn}|^2\rangle = \int_0^T\int_0^T dt'\, dt''\, \{&\hat{s}_i(t')\hat{s}_i^*(t'')\langle \hat{n}_i^*(t'+\tau)\hat{n}_i(t''+\tau)\rangle \\ &+ \hat{s}_i(t''+\tau)\hat{s}_i^*(t'+\tau)\langle \hat{n}_i^*(t'')\hat{n}_i(t')\rangle\}\end{aligned} \tag{III.5}$$

$$\langle \hat{v}_{sn}\hat{v}_{nn}^*\rangle = \int_0^T\int_0^T dt'\, dt''\, \hat{s}_i(t')\langle \hat{n}_i^*(t'+\tau)\hat{n}_i^*(t'')\hat{n}_i(t''+\tau)\rangle \tag{III.6}$$

$$\langle |\hat{v}_{nn}|^2\rangle = \int_0^T\int_0^T dt'\, dt''\langle \hat{n}_i(t')\hat{n}_i^*(t'+\tau)\hat{n}_i^*(t'')\hat{n}_i(t''+\tau)\rangle \tag{III.7}$$

According to a set of relationships that apply to product averages of complex envelopes of zero mean narrowband Gaussian random processes,‡

$$\langle \hat{n}_i(t_1)\hat{n}_i(t_2)\rangle = \langle \hat{n}_i^*(t_1)\hat{n}_i^*(t_2)\rangle = 0 \tag{III.8}$$

$$\begin{aligned}\langle \hat{n}_i(t_1)\hat{n}_i^*(t_2)\hat{n}_i^*(t_3)\hat{n}_i(t_4)\rangle &= \langle \hat{n}_i(t_1)\hat{n}_i^*(t_3)\rangle\langle \hat{n}_i(t_4)\hat{n}_i^*(t_2)\rangle \\ &\quad + \langle \hat{n}_i(t_1)\hat{n}_i^*(t_2)\rangle\langle \hat{n}_i(t_4)\hat{n}_i^*(t_3)\rangle\end{aligned} \tag{III.9}$$

$$\langle \hat{n}_i(t_1)\hat{n}_i^*(t_2)\hat{n}_i(t_3)\rangle = \langle \hat{n}_i^*(t_1)\hat{n}_i^*(t_2)\hat{n}_i(t_3)\rangle = 0 \tag{III.10}$$

---

† It can be defined in other ways as well, e.g., noise power in the zero signal case.

‡ Freeman, 1958, Section 8.5, pp. 245–248. The last result appearing on p. 248 of Freeman, when applied to complex envelopes of zero-mean Gaussian processes, leads to III.9.

Also we define the complex envelope ACF $\hat{R}_n(\tau)$ and the normalized ACF $\hat{\rho}_n(\tau)$ as follows:

$$\underset{s}{\hat{R}_n}(\tau) = \sigma^2_{\underset{s}{n}}\hat{\rho}_{\underset{s}{n}}(\tau) = \langle \underset{\hat{s}}{\hat{n}_i}(t')\underset{\hat{s}}{\hat{n}_i^*}(t' + \tau)\rangle \tag{III.11}$$

Applying (III.8), (III.9), (III.10), (III.11) in Eqs. (III.5) through (III.7), we have

$$\langle|\hat{v}_{sn}|^2\rangle = \sigma_n^2 \iint dt'\, dt''\, \hat{s}_i(t')\hat{s}_i^*(t'')\hat{\rho}_n(t' - t'') \tag{III.5)$'$}$$

$$\langle\hat{v}_{sn}\hat{v}_{nn}^*\rangle = 0 \tag{III.6)$'$}$$

$$\langle|\hat{v}_{nn}|^2\rangle = \sigma_n^4\{T^2\,|\hat{\rho}_n(\tau)|^2 + \int_0^T\int_0^T dt'\, dt''\,|\hat{\rho}_n(t' - t'')|^2\} \tag{III.7)$'$}$$

If the noise bandwidth is assumed to be very large compared to the signal bandwidth, then *only* within the integrands of (III.5)′ and (III.7)′ can we use the approximation†

$$\hat{\rho}_n(\tau) \approx \frac{1}{B_n}\delta(\tau) \tag{III.12}$$

from which we have, with the aid of (6.32)

$$\langle|\hat{v}_{sn}|^2\rangle \simeq \frac{2\sigma_n^2}{B_n}E_s \tag{III.5)$''$}$$

$$\langle|\hat{v}_{nn}|^2\rangle = \sigma_n^4\,|\hat{\rho}_n(\tau)|^2 T^2 + \frac{\sigma_n^2}{B_n}T \tag{III.7)$'''$}$$

and hence

$$\rho_o \simeq \frac{E_s^2\,|\hat{\rho}_s(\tau)|^2}{(\sigma_n^2 E_s/2B_n) + (\sigma_n^4 T^2/4)\,|\hat{\rho}_n(\tau)|^2 + (\sigma_n^4 T/4B_n) + \sigma_n^2 E_s T\,\mathrm{Re}\,(\hat{\rho}_s(\tau)\hat{\rho}_n^*(\tau))} \tag{III.13}$$

or, equivalently,

$$\rho_o \simeq (\rho_o)_{\mathrm{coh}}\left\{\frac{|\hat{\rho}_s(\tau)|^2}{1 + (1/2\rho_i)[1 + B_n T\,|\hat{\rho}_n(\tau)|^2 + (\rho_o)_{\mathrm{coh}}\,\mathrm{Re}\,(\hat{\rho}_s(\tau)\hat{\rho}_n^*(\tau))]}\right\} \tag{III.13)$'$}$$

where $(\rho_o)_{\mathrm{coh}}$ is the output SNR of a coherent matched filter, as given by (6.53),‡ where ρ_i is the input SNR, and where

$$\hat{\rho}_s(\tau) = \frac{\int_0^T dt'\, \hat{s}_i(t')\hat{s}_i^*(t' + \tau)}{\int_0^T dt'\,|\hat{s}_i(t')|^2}$$

---

† Equation (III.12) is derived from (6.21) through the definition of a complex envelope given by (6.3). The factor $\frac{1}{2}$ that appears in (6.21) is not present in (III.12), because (III.12) applies to complex envelopes and (6.21) to real waveforms (see Eq. 10.35).

‡ That is, $(\rho_o)_{\mathrm{coh}} = 2\rho_i B_n T$.

IV

CALCULATION OF RMS ERROR IN OPTIMAL PHASE MEASUREMENT

If a noise $n_i(t)$ is added to the signal, the cosine and sine filter noise outputs are [see Eq. (8.39)]

$$n_{os}(T) = \int_0^T dt'\, n_i(t') \sin \omega_c t' \tag{IV.1.a}$$

$$n_{oc}(T) = \int_0^T dt'\, n_i(t') \cos \omega_c t' \tag{IV.1.b}$$

From (8.40)

$$\psi_{\text{MLE}} = \tan^{-1}\left[-\left(\frac{-(aT/2)\sin\psi + n_{os}(T)}{(aT/2)\cos\psi + n_{oc}(T)}\right)\right] \tag{IV.2}$$

The rms error in MLE measurement of phase is given by

$$\epsilon_\psi = \sqrt{\langle(\psi_{\text{MLE}} - \psi)^2\rangle} = \sqrt{\left\langle\left[\tan^{-1}\left(\frac{\sin\psi - (2n_{os}(T)/aT)}{\cos\psi + (2n_{oc}(T)/aT)}\right) - \psi\right]^2\right\rangle} \tag{IV.3}$$

To calculate the phase error, we first invoke the geometric series within the argument of the arctangent functions in (IV.3), as follows:

$$\tan^{-1}\left(\frac{\sin\psi + N_s}{\cos\psi + N_c}\right) = \tan^{-1}\left[\tan\psi - \frac{N\sin(\psi - \psi_n)}{\cos^2\psi}\left(1 - \frac{N\cos\psi_n}{\cos\psi} + \frac{N^2\cos^2\psi_n}{\cos^2\psi} - \cdots\right)\right], \quad \text{if } |N_c| < \cos\psi \tag{IV. 4.a}$$

where

$$N_s \equiv \frac{-2n_{os}(T)}{aT} = N \sin \psi_n$$

$$N_c \equiv \frac{2n_{oc}(T)}{aT} = N \cos \psi_n$$

or, alternatively,

$$\tan^{-1}\left(\frac{\sin \psi + N_s}{\cos \psi + N_c}\right) = \text{ctn}^{-1}\left(\frac{\cos \psi + N_c}{\sin \psi + N_s}\right)$$
$$= \text{ctn}^{-1}\left[\text{ctn}\, \psi - \frac{N \sin (\psi_n - \psi)}{\sin^2 \psi}\left(1 - \frac{N \sin \psi_n}{\sin \psi} + \frac{N^2 \sin^2 \psi_n}{\sin^2 \psi} - \dots\right)\right],$$
$$\text{if } |N_s| < \sin \psi \tag{IV.4.b}$$

The rms phase error is calculated by expanding the arctangent and arctangent functions in their respective power series, then using the binomial expansion, as follows:

$$\begin{aligned}
\psi_{\text{MLE}} - \psi &= \tan^{-1}[\tan \psi + N \sin (\psi_n - \psi) f(\psi)] - \tan^{-1}[\tan \psi] \\
&= \{[\tan \psi + N \sin (\psi_n - \psi) f(\psi)] - \tfrac{1}{3}[\tan \psi + N \sin (\psi_n \\
&\quad - \psi) f(\psi)]^3 + \tfrac{1}{5}[\tan \psi + N \sin (\psi_n - \psi) f(\psi)]^5 - \dots\} \\
&\quad - \{[\tan \psi] - \tfrac{1}{3}[\tan \psi]^3 + \tfrac{1}{5}[\tan \psi]^5 \dots\} \\
&\simeq N f(\psi) \sin (\psi_n - \psi) - \frac{1}{3}\left[\frac{3!}{2!1!} \tan^2 \psi\, N \sin (\psi_n - \psi) f(\psi)\right] \\
&\quad + \frac{1}{5}\left[\frac{5!}{4!1!} \tan^4 \psi\, N \sin (\psi_n - \psi) f(\psi)\right] \\
&\quad - \frac{1}{7}\left[\frac{7!}{6!1!} \tan^6 \psi\, N \sin (\psi_n - \psi) f(\psi)\right] + \dots + \dots + 0(N^2)
\end{aligned} \tag{IV.5.a}$$

where

$$f(\psi) = \frac{1}{\cos^2 \psi}\left(1 - N\frac{\cos \psi_n}{\cos \psi} + N^2\frac{\cos^2 \psi_n}{\cos^2 \psi} - \dots\right) \simeq \frac{1}{\cos^2 \psi} + 0(N)$$

Neglecting all terms of order N^2 or higher, we obtain

$$\begin{aligned}
\psi_{\text{MLE}} - \psi &\simeq \frac{N}{\cos^2 \psi} \sin (\psi_n - \psi)[1 - \tan^2 \psi + \tan^4 \psi \\
&\quad - \tan^6 \psi + \dots] + 0(N^2) \simeq N \sin (\psi_n - \psi)
\end{aligned} \tag{IV.5.b}$$

valid if $|N| \ll |\cos \psi|$ and $\tan^2 \psi < 1$ [e.g., $-(\pi/4) \leq \psi \leq (\pi/4)$; $(3\pi/4) \leq \psi \leq (5\pi/4)$].

Applying the same reasoning with the arccotangent representation, we obtain

$$\begin{aligned}\psi_{\text{MLE}} - \psi &= \text{ctn}^{-1}[\text{ctn}\,\psi - N\sin(\psi_n - \psi)g(\psi)] - \tan^{-1}[\text{ctn}\,\psi] \\ &\simeq -\Big\{- Ng(\psi)\sin(\psi_n - \psi) - \frac{1}{3}\Big[-\frac{3!}{2!1!}\text{ctn}^2\,\psi\, N\sin(\psi_n \\ &\quad - \psi)g(\psi)\Big] + \frac{1}{5}\Big[\frac{5!}{4!1!}\text{ctn}^4\,\psi\, N\sin(\psi_n - \psi)g(\psi)\Big] \\ &\quad - \ldots + 0(N^2)\Big\}\end{aligned} \tag{IV.6.a}$$

where

$$\begin{aligned}g(\psi) &= \frac{1}{\sin^2\psi}\left(1 - N\frac{\sin\psi_n}{\sin\psi} + N^2\frac{\sin^2\psi_n}{\sin^2\psi} - \ldots\right) \\ &\simeq \frac{1}{\sin^2\psi} + 0(N)\end{aligned}$$

Again neglecting terms $0(N^2)$, we have

$$\begin{aligned}\psi_{\text{MLE}} - \psi &\simeq N\frac{\sin(\psi_n - \psi)}{\sin^2\psi}[1 - \text{ctn}^2\,\psi + \text{ctn}^4\,\psi - \text{ctn}^6\,\psi + \ldots] \\ &\qquad + 0(N^2) \simeq N\sin(\psi_n - \psi)\end{aligned} \tag{IV.6.b}$$

valid if $|N| \ll |\sin\psi|$ and $\text{ctn}^2\,\psi < 1$ [e.g., $(\pi/4) \leq \psi \leq (3\pi/4)$; $(5\pi/4) \leq \psi \leq (7\pi/4)$].

From Eqs. (IV.5.b) and (IV.6.b), we see that the absolute value of the difference between the maximum likelihood estimator of ψ and the true value of ψ is approximately $|N\sin(\psi_n - \psi)|$ for all values of ψ, provided the SNR is sufficiently high to warrant neglect of terms $0(N^2)$ in the power series expansions in Eqs. (IV.5) and (IV.6). Using the definition of N_s and N_c given below (IV.4.a), and the definition of ϵ_ψ in (IV.3), we have

$$\begin{aligned}\epsilon_\psi &\simeq \sqrt{\langle N^2\sin^2(\psi_n - \psi)\rangle} \\ &= \frac{1}{aT}\sqrt{[2(\langle n_{os}^2\rangle + \langle n_{oc}^2\rangle) + 2(\langle n_{os}^2\rangle)\cos 2\psi - 4\langle n_{oc}n_{os}\rangle\sin 2\psi]}\end{aligned} \tag{IV.7}$$

From (IV.1.a), (IV.1.b) and (6.21), we have

$$\langle n_{os}^2\rangle = \langle n_{oc}^2\rangle \simeq \frac{\sigma_n^2 T}{4B_n} \tag{IV.8.a}$$

$$\langle n_{oc}n_{os}\rangle = 0 \tag{IV.8.b}$$

from which it follows that

$$\epsilon_\psi \simeq \frac{1}{\sqrt{2\rho_i B_n T}} \tag{IV.9}$$

BIBLIOGRAPHY

J. Aarons (ed.), *Radio-Astronomical and Satellite Studies of the Atmosphere* (NATO Conference, June 17–June 29, 1962), North Holland Publishing Company, Amsterdam, 1963.

J. Aarons, W.R. Barron and J.P. Castelli, "Radio Astronomy Measurements at VHF and Microwaves," *Proc. IRE*, Vol. 46, No. 1, January, 1958, pp. 325–35.

M. Abramowitz and I.A. Stegun (ed.), *Handbook of Mathematical Functions*, Dover, N.Y., 1965.

N. Abramson, *Information Theory and Coding*, McGraw-Hill, N.Y., 1963.

Y.L. Alpert, *Radio Wave Propagation and the Ionosphere*, U.S.S.R., English Translation published in U.S. by Consultants' Bureau, N.Y., 1963.

J.A. Aseltine, *Transform Method in Linear System Analysis*, McGraw-Hill, N.Y., 1958.

E. Baghdady (ed.), *Lectures on Communication System Theory*, McGraw-Hill, N.Y., 1961.

A.V. Balakrishnan (ed.), *Space Communication*, McGraw-Hill, N.Y., 1963.

———, *Advances in Communication Systems*, Vol. 1, Academic Press, N.Y., 1965.

B. Barrow, "Error Probabilities for Data Transmission over Fading Radio Paths," Doctoral Dissertation, Delft, Netherlands, 1962.

———, "Diversity Combination of Fading Signals with Unequal Mean Strengths," *IEEE Trans. Commun. Sys.*, CS-11, No. 1, March, 1963, pp. 73–78.

D.K. Barton, *Radar System Analysis*, Prentice-Hall, Englewood Cliffs, N.J., 1964.

Y. Beers, *Introduction to the Theory of Errors*, Addison-Wesley, Reading, Mass., 1953.

P. Bello, "Joint Estimation of Delay, Doppler and Doppler Rate," *IRE Trans. Info. Thry.*, IT-6, No. 3, June, 1960, pp. 330–41.

———, "On the Approach of a Filtered Pulse Train to a Stationary Gaussian Process," *IRE Trans. Info. Thry.*, IT-7, No. 3, July, 1961, pp. 144–49.

———, "Characterization of Randomly Time-Variant Linear Channels," *IEEE Trans. Commun. Sys.*, CS-11, No. 4, December, 1963, pp. 360–90.

———, J. Ekstrom and D. Chesler, "Study of Adaptable Communication Systems," RADC-TDR 62-314, Applied Research Laboratory, Sylvania Electronic Systems, Waltham, Mass., June, 1962.

——— and B. Nelin, "Pre-detection Diversity Combining with Selectively Fading Channels," *IRE Trans. Commun. Sys.*, CS-10, No. 1, March, 1962, pp. 32–42.

——— and ———, "The Influence of Fading Spectrum on the Binary Error Probabilities of Incoherent and Differentially Coherent Matched Filter Receivers," *IRE Trans. Commun. Sys.*, CS-10, No. 2, June, 1962, pp. 160–68.

P. Bello and H. Raemer, "Performance of a Rake System over the Orbital Dipole Channel," *Proceedings of Eighth National Communications Symposium*, Utica, N.Y., October, 1962.

J.S. Bendat, *Principles and Applications of Random Noise Theory*, Wiley, N.Y., 1958.

W.R. Bennett, "Methods of Solving Noise Problems," *Proc. IRE*, Vol. 44, No. 5, May, 1956, pp. 609–38.

———, *Electrical Noise*, McGraw-Hill, N.Y., 1960.

L.L. Beranek, *Acoustics*, McGraw-Hill, N.Y., 1954.

R.S. Berkowitz (ed.), *Modern Radar*, Wiley, N.Y., 1965.

N. Blachman, *Noise and its Effect on Communication*, McGraw-Hill, N.Y., 1966.

H.S. Black, *Modulation Theory*, Van Nostrand, Princeton, N.J., 1953.

R.B. Blackman and J.W. Tukey, *The Measurement of Power Spectra*, Dover, N.Y., 1958.

E.V. Bohn, *The Transform Analysis of Linear Systems*, Addison-Wesley, Reading, Mass., 1963.

H.G. Booker and W.E. Gordon, "A Theory of Radio Scattering in the Troposphere," *Proc. IRE*, Vol. 38, No. 4, April, 1950, pp. 401–12.

D.G. Brennan, "Linear Diversity Combining Techniques," *Proc. IRE*, Vol. 47, No. 6, June, 1959, pp. 1075–1102.

L. Brillouin, "The Negentropy Principle of Information," *J. Applied Phys.*, Vol. 24, No. 9, September, 1953, pp. 1152–63.

———, *Science and Information Theory*, Academic Press, N.Y., 1956.

G.M. Brown (ed.), "Space Radio Communication" (*URSI Symposium*, Paris, September, 1961); Elsevier, Amsterdam, 1962.

C.A. Burrows (chairman) and S.S. Attwood (ed.), "Radio Wave Propagation," *Report of Progagation Committee of National Defense Research Committee*, Academic Press, N.Y., 1949.

J.J. Bussgang and M. Leiter, "Error Rate Approximations for Differential Phase-Shift Keying," *IEEE Trans. Commun. Sys.*, CS-12, No. 1, March, 1964, pp. 18–27.

——— and D. Middleton, "Optimal Sequential Detection of Signals in Noise," *IRE Trans. Info. Thry.*, IT-1, No. 3, December, 1955, pp. 5–18.

C. Cherry (ed.), "Information Theory" (*Proceedings of 4th London Symposium on Information Theory*, September, 1960), Butterworth, Washington, 1961.

R.V. Churchill, *Modern Operational Mathematics in Engineering*, McGraw-Hill, N.Y., 1944.

D. Corson and P. Lorrain, *Introduction to Electromagnetic Fields and Waves*, Freeman, San Francisco, 1962.

C.A. Coulson, *Waves*, Oliver and Boyd, London, 1941.

E.J. Craig, *Laplace and Fourier Transforms for Electrical Engineers*, Holt, Rinehart and Winston, Inc., N.Y., 1964.

H. Cramer, *Mathematical Methods of Statistics*, Princeton University Press, Princeton, 1951.

———, *The Elements of Probability Theory*, Wiley, N.Y., 1955.

W.B. Davenport and W.L. Root, *Theory of Random Signals and Noise*, McGraw-Hill, N.Y., 1958.

K. Davies, "Ionospheric Radio Propagation," *National Bureau of Standards Monograph 80*, April 1, 1965.

R. Deutsch, *Estimation Theory*, Prentice-Hall, Englewood Cliffs, N.J., 1965.

J.L. Doob, *Stochastic Processes*, Wiley, N.Y., 1953.

J.J. Downing, *Modulation Systems and Noise*, Prentice-Hall, Englewood Cliffs, N.J., 1964.

B. Dwork, "Detection of a Pulse Superimposed on Fluctuation Noise," *Proc. IRE*, Vol. 38, No. 7, July, 1950, pp. 771–76.

R.C. Emerson, "First Probability Densities for Receivers with Square-Law Detectors," *J. Applied Phys.*, Vol. 24, No. 9, September, 1953, pp. 1168–76.

R.M. Fano, *Transmission of Information*, MIT Press, Cambridge, Mass., 1961.

A. Feinstein, *Foundations of Information Theory*, McGraw-Hill, N.Y., 1958.

W. Feller, *An Introduction to Probability Theory and its Applications*, 2nd edition, Vol. 1, Wiley, N.Y., 1957.

J.J. Freeman, *Principles of Noise*, Wiley, N.Y., 1958.

T. Fry, *Probability and its Engineering Uses*, Van Nostrand, Princeton, N.J., 1928.

D. Gabor, "Theory of Communication," *Journal of* (*British*) *Institution of Electrical Engineers*, Vol. 93, Part III, 1946, pp. 429–41.

J. Galejs, H. Raemer and A. Reich, "Detection of Weak Wideband Signals," *Cook Technological Review*, Vol. 4, No. 3, December, 1957, Cook Technological Center, Morton Grove, Illinois.

S. Goldman, *Frequency Analysis, Modulation and Noise*, McGraw-Hill, N.Y., 1948.

———, *Information Theory*, Prentice-Hall, Englewood Cliffs, N.J., 1956.

S.J. Goldstein, "A Comparison of Two Radiometer Circuits," *Proc. IRE*, Vol. 43, No. 11, November, 1955, pp. 1663–66.

S.W. Golomb (ed.), *Digital Communications with Space Applications*, Prentice-Hall, Englewood Cliffs, N.J., 1964.

J. Granlund and W. Sichak, "Diversity Combining for Signals of Different Medians," *IRE Trans. Commun. Sys.*, CS-9, No. 2, June, 1961, pp. 138–44.

E. Guillemin, *Introductory Circuit Theory*, Wiley, N.Y., 1953.

———, *Theory of Linear Physical Systems*, Wiley, N.Y., 1963.

J.C. Hancock, *Introduction to the Principles of Communication Theory*, McGraw-Hill, N.Y., 1963.

J. Hancock and P. Wintz, *Signal Detection Theory*, McGraw-Hill, N.Y., 1966.

W.W. Harman, *Principles of the Statistical Theory of Communication*, McGraw-Hill, N.Y., 1963.

L. Harris and A.L. Loeb, *Introduction to Wave Mechanics*, McGraw-Hill, N.Y., 1963.

R.V.L. Hartley, "Transmission of Information," *Bell System Technical Journal*, I, 1928, pp. 535–63.

C.W. Helstrom, "Resolution of Signals in White Gaussian Noise," *Proc. IRE*, Vol. 43, No. 9, September, 1955, pp. 1111–18.

———, *Statistical Theory of Signal Detection*, Pergamon Press, N.Y., 1960 (designated as Helstrom, 1960).

———, "Comparison of Digital Communication Systems," *IRE Trans. Commun. Sys.*, C-8, No. 3, September, 1960, pp. 141–50 (designated as Helstrom, *PGCS*, 1960).

K. Henney (ed.), *Radio Engineering Handbook*, McGraw-Hill, N.Y., 1959.

E.H. Holt and R.E. Haskell, *Foundations of Plasma Dynamics*, Macmillan, N.Y., 1965.

W.H. Huggins and D. Middleton, "A Comparison of the Phase and Amplitude Principles in Signal Detection," *Proc. Nat'l Electronics Conference*, Vol. II, 1955.

Institute of Radio Engineers (Proceedings), "Special Issue on Matched Filters," IT-6, No. 3, June, 1960 (designated as *IRE, PGIT*, June, 1960).

Institute of Radio Engineers (Proceedings), "Special Issue on Scatter Propagation," October, 1955.

International Telephone and Telegraph, "Reference Data for Radio Engineers," 4th edition, IT & T, N.Y., 1956 (designated as *ITT*, 1956).

W. Jackson, (ed.), *Communication Theory* (Papers from Symposium on Applications of Communication Theory, (British) Institute of Elec. Engrs., London, September 22–26, 1952), Butterworth, Washington, 1953.

E. Jahnke and F. Emde, *Tables of Functions*, Dover, N.Y., 1945.

H. M. James, N.B. Nichols and R.S. Phillips, *Theory of Servomechanisms* (MIT Rad. Lab. Series, Vol. 25), McGraw-Hill, N.Y., 1947.

F. Jenkins and H. White, *Fundamentals of Physical Optics*, McGraw-Hill, N.Y., 1937.

J.J. Jones, A. Pierce and S. Stein, "Communication Theory and Application to Digital Communication Systems," Research Note 509, Project 132, Applied Research Laboratory, Sylvania Electronic Systems, Waltham, Mass., October 13, 1964.

G. Joos, *Theoretical Physics*, Hafner, N.Y., 1934 (translation by I.M. Freeman).

E.C. Jordan, *Electromagnetic Waves and Radiating Systems*, Prentice-Hall, Englewood Cliffs, N.J., 1950.

M. Kac and A.J.F. Siegert, "On the Theory of Noise in Radio Receivers with Square-Law Detectors," *J. Applied Phys.*, Vol. 18, No. 4, April, 1947, pp. 383–97.

T. Kailath, "Correlation Detection of Signals Perturbed by a Random Channel," *IRE Trans. Info. Thry.*, IT-6, No. 3, June, 1960, pp. 361–66.

E.I. Kelly, I.S. Reed and W.L. Root, "The Detection of Radar Echoes in Noise," Part I, *J. Soc. Indus. and Applied Math.* (*SIAM*), Vol. 8, June, 1960, pp. 309–41.

D.E. Kerr (ed.), *Propagation of Short Radio Waves*, McGraw-Hill, N.Y., 1951.

G.A. Korn and T.M. Korn, *Mathematical Handbook for Scientists and Engineers*, McGraw-Hill, N.Y., 1961.

V.A. Kotelnikov, *Theory of Optimum Noise Immunity*, McGraw-Hill, N.Y., 1960.

J. Kraus, *Antennas*, McGraw-Hill, N.Y., 1950.

J.H. Laning and R.H. Battin, *Random Processes in Automatic Control*, McGraw-Hill, N.Y., 1956.

B.P. Lathi, *Signals, Systems and Communication*, Wiley, N.Y., 1965.

J.L. Lawson and G.E. Uhlenbeck, *Threshold Signals* (MIT Rad. Lab. Series, Vol. 24), McGraw-Hill, N.Y., 1950.

W. Lee, *Statistical Theory of Communication*, Wiley, N.Y., 1960.

Y.W. Lee, T.P. Cheatham and J.B. Wiesner, "Application of Correlation Analysis to the Detection of Periodic Signals in Noise," *Proc. IRE*, Vol. 38, No. 10, October, 1950, pp. 1165–71.

R.M. Lerner, "Signals with Uniform Ambiguity Functions," *IRE Trans. Nat'l Conv. Record 6*, Part 4, March, 1958, pp. 27–36.

———, "A Matched Filter Detection System for Complicated Doppler-Shifted Signals," *IRE Trans. Info. Thry.*, IT-6, No. 3, June, 1960, pp. 373–85.

R.B. Lindsay, *Introduction to Physical Statistics*, Wiley, N.Y., 1941.

M. Loeve, *Probability Theory*, Van Nostrand, Princeton, N.J., 1955.

R.W. Lucky and J.C. Hancock, "On the Optimum Performance of N'ary Systems Having Two Degrees of Freedom," *IRE Trans. Commun. Sys.*, CS-10, No. 2, June, 1962, pp. 185–92.

W.A. Lynch and J.G. Truxal, *Introductory System Analysis*, McGraw-Hill, N.Y., 1961.

V.I. Maeda and S. Silver (eds.), "Space Radio Science" (*Progress in Radio Science*, Vol. III, 1960–1963), Elsevier, Amsterdam, 1965.

J.I. Marcum, "Tables of the *Q*-Function," *Rand Corporation Memorandum*, RM-399, January, 1950.

———, "A Statistical Theory of Target Detection by Pulsed Radar," *IRE Trans. Info. Thry.*, IT-6, No. 2, April, 1960, pp. 59–144 (sometimes designated together with Swerling paper in same issue as "Marcum and Swerling, 1960").

R.B. Marsten (ed.), "Communication Satellite Systems Technology" (*AIAA Communication Satellite Systems Conference*, May 2–4, 1966), Academic Press, N.Y., 1966.

J.A. McFadden, "The Axis-Crossing Intervals of Random Functions," *IRE Trans. Info. Thry.*, IT-2, No. 4, December, 1956, pp. 146–50.

———, "The Axis-Crossing Intervals of Random Functions," *IRE Trans. Info. Thry.*, IT-4, No. 1, March, 1958, pp. 14–24.

D. Menzel (ed.), *The Radio Noise Spectrum*, Harvard University Press, Cambridge, Mass., 1960.

D. Middleton, *Introduction to Statistical Communication Theory*, McGraw-Hill, N.Y., 1960.

——— and D. Van Meter, "Detection and Extraction of Signals in Noise from the Point of View of Statistical Decision Theory-I," *J. Soc. Indus. and Applied Math.* (*SIAM*), Vol. 3, No. 4, December, 1955, pp. 192–253.

——— and ———, "Detection and Extraction of Signals in Noise from the Point of View of Statistical Decision Theory-II," *J. Soc. Indus. and Applied Math.* (*SIAM*), Vol. 4, No. 2, June, 1956, pp. 86–119.

——— and J.H. Van Vleck, "Theoretical Comparison of Visual, Aural and Meter Reception of Pulsed Signals in the Presence of Noise," *J. Applied Phys.*, Vol. 17, No. 3, November, 1946, pp. 940–72.

A. Mood, *Introduction to the Theory of Statistics*, McGraw-Hill, N.Y., 1964.

R.K. Moore, *Wave and Diffusion Analogies*, McGraw-Hill, N.Y., 1964.

G.H. Munro, "Scintillation of Radio Signals from Satellites," *J. Geophys. Res.*, Vol. 68, No. 1, April, 1963, pp. 1851–60.

J. Neyman, *First Course in Probability and Statistics*, Holt, Rinehart and Winston, Inc., N.Y., 1950.

——— and E.S. Pearson, "On the Use and Interpretation of Certain Test Criteria for Purposes of Statistical Inference," *Biometrica*, Vol. 20A, No. 175, 1928, Part I, pp. 175–240; Part II, pp. 263–94.

N.J. Nilsson, "On the Optimum Range Resolution of Radar Signals in Noise," *IRE Trans. Info. Thry.*, IT-7, No. 4, October, 1961, pp. 245–53.

K.A. Norton, "The Propagation of Radio Waves Over the Surface of the Earth and in the Upper Atmosphere," *Proc. IRE*, Vol. 24, No. 10, October, 1936, pp. 1367–87.

———, "The Calculation of Ground-Wave Field Intensity over a Finitely Conducting Spherical Earth," *Proc. IRE*, Vol. 29, No. 12, December, 1941, pp. 623–39.

A.H. Nuttall, "Error Probabilities for Equicorrelated N'ary Signals Under Phase-Coherent and Phase-Incoherent Reception," *IRE Trans. Info. Thry.*, IT-8, No. 4, July, 1962, pp. 305–14.

H. Nyquist, "Certain Topics on Telegraph Transmission Theory," *Trans. AIEE*, Vol. 47, No. 2, April, 1928, pp. 617–44.

P. Panter, *Modulation, Noise and Spectral Analysis*, McGraw-Hill, N.Y., 1965.

A. Papoulis, *Probability, Random Variables and Stochastic Processes*, McGraw-Hill, N.Y., 1965.

E. Parzen, *Modern Probability and Its Applications*, Wiley, N.Y., 1960.

M.E. Pedinoff and A.A. Ksienski, "Multiple Target Resolution of Data Processing Antennas," *IRE Trans. Antennas and Propagation*, AP-10, No. 2, March, 1962, pp. 112–26.

W.W. Peterson, T.C. Birdsall and W.C. Fox, "The Theory of Signal Detectability," *IRE Trans. Info. Thry.*, IT-4, 1954 Symposium on Info. Theory, September 15–17, 1954, pp. 171–212.

B.O. Pierce, *A Short Table of Integrals*, Ginn and Co., Boston, 1929.

J.N. Pierce, "Theoretical Diversity Improvement in Frequency-Shift Keying," *Proc. IRE*, Vol. 46, No. 5, May, 1958, pp. 903–10.

———, "Ultimate Performance of M'ary Transmissions on Fading Channels," *IEEE Trans. Info. Thry.*, IT-12, No. 1, January, 1966, pp. 2–4.

——— and S. Stein, "Multiple Diversity with Non-Independent Fading," *Proc. IRE*, Vol. 48, No. 1, January, 1960, pp. 89–104.

D.J. Povejsil, R.S. Raven, and P. Waterman, *Airborne Radar*, Van Nostrand, Princeton, N.J., 1961.

R. Price and P. Green, "A Communication Technique for Multipath Channels," *Proc. IRE*, Vol. 46, No. 3, March, 1958, pp. 555–70.

H.R. Raemer and A. Reich, "Correlators and Radiometers Detect Weak Signals," *Electronics*, May 22, 1959, pp. 58–60.

G. Raisbeck, *Information Theory*, MIT Press, Cambridge, Mass., 1964.

S. Ramo and J. Whinnery, *Fields and Waves in Modern Radio*, Wiley, N.Y., 1953.

E. Reich and P. Swerling, "Detection of a Sine Wave in Gaussian Noise," *J. Applied Physics*, Vol. 24, No. 3, March, 1953, pp. 289–96.

S. Reiger, "Error Probabilities of Binary Data Transmission Systems in the Presence of Random Noise," *IRE Nat'l. Conv. Rec.*, Part 8, March, 1953, pp. 72–79.

J. Reinties and G.T. Coate (editors) and MIT Radar School Staff, *Principles of Radar*, McGraw-Hill, N.Y., 1952.

F.M. Reza, *An Introduction to Information Theory*, McGraw-Hill, N.Y., 1961.

S.O. Rice, "Mathematical Analysis of Random Noise," *Bell System Technical Journal*, Vol. 23, No. 3, July 1944, pp. 282–332 and Vol. 24, No. 1., Jan. 1945, pp. 46–156; in Wax, 1954, pp. 133–294.

L. Ridenour (ed.), *Radar Systems Engineering* (MIT Rad. Lab. Series, Vol. 1), McGraw-Hill, N.Y., 1947.

R. Rowe, *Signals and Noise in Communication Systems*, Van Nostrand, Princeton, 1965.

L.I. Schiff, *Quantum Mechanics*, McGraw-Hill, N.Y., 1949.

L.S. Schwartz, *Principles of Coding, Filtering and Information Theory*, Cleaver-Hume, London, 1963.

M. Schwartz, *Information Transmission, Modulation and Noise*, McGraw-Hill, N.Y., 1959.

———, W. Bennett and S. Stein, *Communication Systems and Techniques*, McGraw-Hill, N.Y., 1966.

R.J. Schwartz and B. Friedland, *Linear Systems*, McGraw-Hill, N.Y., 1965.

R. Scott, *Linear Circuits*, Addison-Wesley, Reading, Mass., 1960.

F.W. Sears, "Principles of Physics," Vol. III, *Optics*, Addison-Wesley, Reading, Mass., 1948.

S. Seely, *Radio Electronics*, McGraw-Hill, N.Y., 1950.

I. Selin, *Detection Theory*, Princeton University Press, Princeton, N.J., 1965 (designated as Selin 1965).

———, "Detection of Coherent Radar Returns of Unknown Doppler Shift," *IEEE Trans. Info. Thry.*, IT-11, No. 3, July, 1965, pp. 396–400 (designated as Selin PGIT, 1965).

C. Shannon, "A Mathematical Theory of Communication," *Bell System Technical Journal*, Vol. 27, No. 3, July, 1948, pp. 379–423.

——— and W. Weaver, *The Mathematical Theory of Communication*, University of Illinois, Urbana, 1949.

W. Siebert, "A Radar Detection Philosophy," *IRE Trans. Info. Thry.*, IT-2, No. 3, September, 1956, pp. 204–21.

A.J.F. Siegert, "Passage of Stationary Processes through Linear and Nonlinear Devices," *IRE Trans. Info. Thry.*, IT-3, No. 4, March, 1954, pp. 4–25.

S. Silver, *Microwave Antenna Theory and Design*, McGraw-Hill, N.Y., 1949.

H.H. Skilling, *Fundamentals of Electric Waves*, Wiley, N.Y., 1942.

M. Skolnick, *Introduction to Radar Systems*, McGraw-Hill, N.Y., 1962.

D. Slepian, "Estimation of Signal Parameters in the Presence of Noise," *IRE Trans. Info. Thry.*, PGIT-3, March, 1954, pp. 68–89.

R.A. Smith, "The Relative Advantages of Coherent and Incoherent Detectors: A Study of Their Output Noise Spectra Under Various Conditions," *Proc. IEE* (British), Part III, Vol. 98, September 1951, pp. 401–406 and Part IV, Vol. 98, October, 1951, pp. 43–54.

L.D. Smullin and H.A. Haus (editors), *Noise in Electron Devices*, Technology Press, Cambridge, Mass. and Wiley, N.Y., 1959.

I.N. Sneddon, *Fourier Transforms*, McGraw-Hill, N.Y., 1958.

F.G. Splitt, "Comparative Performance of Digital Data Transmission Systems in the Presence of CW Interference," *IRE Trans. Commun. Sys.*, CS-10, No. 2, June, 1962, pp. 169–76.

H. Staras, "Diversity Reception with Correlated Signals," *J. Applied Phys.*, Vol. 27, No. 1, January, 1956, pp. 93–94.

S. Stein, "Unified Analysis of Certain Coherent and Non-coherent Binary Communication Channels," *IEEE Trans. Info. Thry.*, IT-10, No. 1, January, 1964, pp. 43–51.

——— and J. Jones, *Modern Communication Principles*, McGraw-Hill, N.Y., 1967.

J.L. Stewart and E.C. Westerfield, "A Theory of Active Sonar Detection," *Proc. IRE*, Vol. 47, No. 5, May, 1959, pp. 872–88.

H.L. Stiltz (ed.), *Aerospace Telemetry*, Prentice-Hall, Englewood Cliffs, N.J., 1961 (Vol. II, 1966).

R.L. Stratonovich (ed.), *Topics in the Theory of Random Noise* (USSR), Gordon and Breach, N.Y., 1963 (Vol. I, Translation by R.A. Silverman).

J.A. Stratton, *Electromagnetic Theory*, McGraw-Hill, N.Y., 1941.

S. Sussman, "Least Square Synthesis of Radar Ambiguity Functions," *IRE Trans. Info. Thry.*, IT-8, No. 3, April, 1962, pp. 246–54.

P. Swerling, "Maximum Angular Accuracy of a Pulsed Search Radar," *Proc. IRE*, Vol. 44, No. 9, September, 1956, pp. 1146–55.

———, "Detection of Fluctuating Pulsed Signals in the Presence of Noise," *IRE Trans. Info. Thry.*, IT-3, No. 3, September, 1957, pp. 175–78.

———, "Parameter Estimation for Waveforms in Additive Gaussian Noise," *J. Soc. Indus. and Applied Math.* (*SIAM*), Vol. 7, No. 2, June, 1959, pp. 152–66.

———, "Probability of Detection of Fluctuating Targets," *IRE Trans. Info. Thry.*, IT-6, No. 2, April, 1960, pp. 269–308 (sometimes designated together with Marcum paper in same issue as "Marcum and Swerling," 1960).

———, "Detection of Radar Echoes in Noise Revisited," *IEEE Trans. Info. Thry.*, IT-12, No. 3, July, 1966, pp. 348–61.

F.E. Terman, *Radio Engineering*, McGraw-Hill, N.Y., 1947.

———, *Electronic and Radio Engineering*, McGraw-Hill, N.Y., 1955.

F.J. Tischer, *Basic Theory of Space Communications*, Van Nostrand, Princeton, N.J., 1965.

J. Topping, *Errors of Observation and Their Treatment* (monograph), The Institute of Physics, London, 1955.

D.G. Tucker, N.M. Graham and S.J. Goldstein, "A Comparison of Two Radiometer Circuits," *Proc. IRE*, Vol. 45, No. 3, March, 1957, pp. 365–66.

G.L. Turin, "An Introduction to Matched Filters," *IRE Trans. Info. Thry.*, IT-6, No. 3, June, 1960, pp. 311–29.

———, "On Optimal Diversity Reception I," *IRE Trans. Info. Thry.*, IT-7, No. 3, July, 1960, pp. 154–66.

———, "On Optimal Diversity Reception II," *IRE Trans. Commun. Sys.*, CS-10, No. 1, March, 1962, pp. 22–31.

H. Urkowitz, "Filters for Detection of Small Radar Signals in Clutter," *J. Applied Phys.*, Vol. 24, No. 8, August, 1953, pp. 1024–31.

A. Van der Ziel, *Noise*, Prentice-Hall, Englewood Cliffs, N.J., 1954.

M.E. Van Valkenberg, *Network Analysis*, Prentice-Hall, Englewood Cliffs, N.J., 1955.

L.A. Wainstein and V.D. Zubakov, *Extraction of Signals from Noise*, Prentice-Hall, Englewood Cliffs, N.J., 1962.

Wald, A., Statistical Decision Functions, Wiley, N. Y., 1950 .

N. Wax (ed.), *Selected Papers on Noise and Stochastic Processes*, Dover, N.Y., 1954.

E.C. Westerfield, R.H. Prager and J.L. Stewart, "Processing Gains Against Reverberation (Clutter) Using Matched Filters," *IRE Trans. Info. Thry.*, IT-6, No. 3, June, 1960, pp. 342–48.

D.V. Widder, *The Laplace Transform*, Princeton University Press, Princeton, N.J., 1946.

P.M. Woodward, *Probability and Information Theory*, McGraw-Hill, N.Y., 1953 (2nd Edition, Pergamon Press, N.Y., 1964).

J.M. Wozencraft and I.M. Jacobs, *Principles of Communication Engineering*, Wiley, N.Y., 1965.

A.M. Yaglom, *An Introduction to the Theory of Stationary Random Functions*, Prentice-Hall, Englewood Cliffs, N.J., 1962.

AUTHOR INDEX

SUBJECT INDEX